VACUUM ULTRAVIOLET RADIATION PHYSICS

VACUUM ULTRAVIOLET RADIATION PHYSICS

Proceedings of the 10th VUV conference

Paris, July 27 – 31, 1992

Editors

Francois J. Wuilleumier

Yves Petroff

Irène Nenner

World Scientific
Singapore • New Jersey • London • Hong Kong

Published by

World Scientific Publishing Co. Pte. Ltd.

P O Box 128, Farrer Road, Singapore 9128

USA office: Suite 1B, 1060 Main Street, River Edge, NJ 07661

UK office: 73 Lynton Mead, Totteridge, London N20 8DH

VACUUM ULTRAVIOLET RADIATION PHYSICS:
Proceedings of the 10th VUV Conference

ISBN 981-02-1252-6

Printed in Singapore by Utopia Press.

INTRODUCTION

The 1992 International Conference on Vacuum Ultraviolet Radiation Physics (VUV-10) was held in Paris, on July 27–31. The five-day Conference was attended by about 600 participants from 31 countries. It was held in Maison de la Chimie, near the Invalides. The synopsis of the Conference Program is given at the beginning of this book. There were 7 Plenary Lectures presented in general sessions, about 60 Progress Reports presented in three parallel sessions pertaining to the major areas covered by the Conference and contributed papers. Most of the contributed papers (a total of about 550 papers were submitted) were presented in three poster sessions; a few of them were given an oral presentation in the parallel sessions.

The program had a wide scope, ranging from pure spectroscopy on atoms and molecules to the most advanced X-ray and high-T_c superconductor studies, and from free atoms and molecules to ions in plasmas. The place given to atomic and molecular physics was larger than usual. This is due to the building of the new storage rings making available synchrotron radiation beams with higher photon flux and brightness over a larger photon energy range. It reflects the strong developments in these fields, especially in Europe, over the past few years. A significant part of the Conference was devoted to recent progress in techniques and instrumentations, covering the use of synchrotron radiation as well as laser sources. This book presents first the Plenary Lectures, then the Progress Reports that have been ordered according to the various areas.

The Conference organizers gratefully acknowledge the assistance of the International Advisory Board for their helpful suggestions, the efficient work of the Program Committee in contributing to the selection of the Invited Papers, and the members of the local Organizing Committee for refereeing and ordering the Contributed Papers. The members of the Organizing Committee deserve also special recognition for their assiduous labor in planning and implementing the meeting. Finally, the Editors would like to thank all the participants in VUV-10 for providing such a strong scientific menu.

The Editors:

F.J.Wuilleumier
Laboratoire de Spectroscopie Atomique et Ionique
Université Paris-Sud, Orsay

Y. Petroff
Laboratoire pour l'Utilisation du Rayonnement
Electromagnétique, Orsay

I. Nenner
DSM/DRECAM, Service de Photons, Atomes et
Molécules, CEN Saclay and LURE

ACKNOWLEDGEMENTS

The organization of this conference was made possible thanks to the support of :

- Centre National de la Recherche Scientifique
  (Sciences Physiques et Mathématiques, Sciences Chimiques)
- Commissariat à l'Energie Atomique
  (Direction des Sciences de la Matière, Direction du Cycle du Combustible)
- European Synchrotron Radiation Society
- Laboratoire pour l'Utilisation du Rayonnement Electromagnétique
- VUV International Advisory Board,

and of the following companies :

- Caburn UHV
- Canon Inc.
- Gammadata Mätteknik
- Grenoble Modular Instruments
- Hamamatsu
- Instruments S.A. Division Jobin Yvon
- Kohzu Seiki Corporation
- Mipsys
- Omicron
- Oxford Instruments SARL
- Scienta Instrument AB
- X-Ray Instrumentation Associates
- Lancôme
- Les Galeries Lafayettes
- Meca 2000

The help of the French Ministère de la Recherche et de l'Espace for the printing of these Proceedings is gratefully acknowledged.

INTERNATIONAL ADVISORY BOARD

Chairman: L. Hedin, Lund, Sweden

M. Y. Amusia, Russia
E. T. Arakawa, USA
Z. Bao, People's Republic of China
N. G. Basov, Russia
A. Bradshaw, Germany
R.C. Elton, USA
C. S. Fadley, USA
M. Finkenthal, Israel
V. A. Fomichev, Russia
S. Hagström, USA
H. Inokuchi, Japan
Janos Kirz, USA
G. V. Marr, Scotland
Y. Petroff, France
M. Pouey, France
R. Rosei, Italy
S. Sato, Japan
G. A.Sawatzky, The Netherlands
J. B. West, England
G. Zimmerer, Germany

LOCAL ORGANIZING COMMITTEE

Y. Petroff, Orsay, France, Chairman

I. Nenner, Saclay, France, Vice-Chairwoman

F. J. Wuilleumier, Orsay, France, Vice-Chairman

H. Bergeron, Orsay, France
M. Berland, Orsay, France
R. Boschiero, Saclay, France
D. Chandesris, Orsay, France
M.E. Couprie, Saclay, France
G. Dujardin, Orsay, France
O. Dutuit, Orsay, France
V. Fériès, Orsay, France
P. Lagarde, Orsay, France
P. Lablanquie, Orsay, France
M. Le Monze, Orsay, France
L.A. Lompré, Saclay, France
F. Mérola, Orsay, France
F. Rousseau, Bagneux, France
M. Sommer, Orsay, France
P. Thiry, Orsay, France

PROGRAM COMMITTEE

SCIENTIFIC PROGRAM

	MONDAY 27	TUESDAY 28	WEDNESDAY 29	THURSDAY 30	FRIDAY 31
8h30	**Registration** **Opening** *Chairman* M.L. COHEN **SCHRIEFFER**	*Chairman* S. LEACH **KROTO** **ZEWAIL**	*Chairman* L. HEDIN **FRIEDEL** **BRADSHAW**	*Chairman* B. SONNTAG **AMUSIA** **SCHMIEDESKAMP**	*Room A* Biology — *Chairman* **Hedman** Photoemission — *Chairman* I. LINDAU **Shen**, Gutierrez, Björneholm, Shikin *Room B* Molecules IV — *Chairman* D. HANSON **Möller**, **White**, Merkt, Lambert, Tuckett *Room C* Laser — *Chairman* **Miyazaki**, **L'Huillier**, **Jaeglé**, **Benerofe**
10h30	COFFEE BREAK	COFFEE BREAK	COFFEE BREAK	COFFEE BREAK	COFFEE BREAK
11h00	*Room A* Super conductors I — *Chairman* A.J. ARKO **Onellion**, **Varma**, Schroeder, Claessen *Room B* Molecules I — *Chairman* D. THOMAS **Morin**, **Ruhl**, Müller-Dethlefs, Poliakoff *Room C* Optics Development — *Chairman* C. KUNZ **Aristov**, Tommasi, Hussain, Couprie	*Room A* Fullerenes — *Chairman* G. KAINDL **Martins**, **Fink**, Krummacher, Debever *Room B* Source Development — *Chairman* J. SAMSON **Pestman**, **Kincaid**, **Marin** *Room C* Interfaces I — *Chairman* P. PERFETTI **Kanski**, **Andersen**, **Reihl**	*Room A* Adsorbates — *Chairman* G. TOURILLON **Stöhr**, **Dujardin**, **Arvanitis**, **Seki** *Room B* Atoms II — *Chairman* G. WENDIN **Richter**, **Ederer**, Krause, Combet Farnoux, Tulkki *Room C* Correlation effects — *Chairman* Y. BAER **Arko**, **Malterre**, Laubschat, Janowitz	*Room A* Dichroism I — *Chairman* G. DA SILVA **Van Der Laan**, **Schneider**, Brouder, Koide *Room B* Molecules III — *Chairman* K. CODLING **Bancroft**, **Yagishita**, **Holub-Krappe** *Room C* Diffraction and Holography — *Chairwoman* M.C. ASENSIO **Osterwalder**, **Terminello**, Fadley	*Chairman* **MENZEL**
12h30	LUNCH	LUNCH	LUNCH	LUNCH	
14h00	POSTER Session I	POSTERS Session II		POSTERS Session III	
16h00	COFFEE BREAK	COFFEE BREAK		COFFEE BREAK	
16h30	*Room A* Super conductors II — *Chairman* P. LAGARDE **Tjeng**, **Pompa**, Tjeng, Katayama *Room B* Molecules II — *Chairman* C.A. de LANGE **McKoy**, **Pratt**, **Hepburn**, Schwentner, Ukai *Room C* Atoms I — *Chairman* F. SCHLACHTER **Greene**, **King**, **Schmidt**, Becker	*Room A* Magnetism I — *Chairman* W. GUDAT **Johnson**, Clemens, Kachel, Himpsel *Room B* Microscopy — *Chairman* G. MARGARITONDO **Ade**, **Rudolph**, **De Stasio**, Perera *Room C* Interfaces II — *Chairman* T. ISHII **Louie**, **Dumas**, **Modesti**, Long, Palomares		*Room A* Dichroism II — *Chairman* C. FADLEY **Secchi**, **Tobin**, Bansmann, Hague, Ma *Room B* Atoms III — *Chairman* J. WEST **Bizau**, Pahler, **Ivanov**, Kennedy *Room C* Lithography & Fluorescence — *Chairman* M. TISSIER **Kitayama**, **Forsyth**, **Kubota**, **Shi**, Böske, Mikhailin	
18h30					

CONTENTS

Introduction v

I. PLENARY LECTURES

1. Photoionization of Atoms, Ions and Multiatomic Formations 3
M. Ya Amusia

2. Chemistry in Femtotime 20
A. H. Zewail

3. High Resolution Soft X-Ray Spectroscopy of Molecules, Surfaces and Solids 29
A. M. Bradshaw

4. The Physics of Elementary Clusters 43
J. Friedel

5. Experiments with Circularly Polarized Radiation 49
B. Schmiedeskamp

II. SOURCES AND OPTICS

6. Very Low Emittance Storage Rings: Goals and Problems 65
P. Marin

7. High Resolution Monochromators for Undulator Radiation 73
W. B. Peatman & F. Senf

8. 116 nm H_2 Laser Pumped by a Traveling-Wave Photoionization Electron Source 85
S. J. Benerofe, G.-Y. Yin & S. E. Harris

9. Coherent XUV–Soft X-Ray Generation by Subpicosecond Visible and Ultraviolet Lasers 96
K. Miyazaki, H. Sakai, H. Takada & G. U. Kim

10. Germanium Laser at 236 Å 106
P. Jaeglé, A. Carillon, H. Z. Chen, P. Dhez, L. Dwivedi, J. Jacoby, G. Jamelot, J. Zhang, M. H. Key, A. Kidd, A. Klisnick, R. Kodama, J. Krishnan, C. L. S. Lewis, D. Neely, P. Norreys, D. O. Neill, G. J. Pert, S. A. Ramsden, J. P. Raucourt, G. J. Tallents & J. Uhomoibhi

11. VUV Radiation Produced by Harmonic Generation in Rare Gases 114
A. L'Huillier, Ph. Balcou & L. A. Lompré

III. ATOMS

12. Present Trends in the Theory of Atomic Photoionization 125
C. H. Greene & F. Robicheaux

13. Many-Electron Interactions in Single and Multiple Photoionization of Metallic Vapors 135
M. Richter

14. Threshold Behaviour of Photodouble Ionization in Atoms 145
G. C. King

15. Complete Fragmentation Patterns in Two-Step Double Photoionization of Atoms 154
V. Schmidt

16. Recent Developments in Photoionization of Excited Atoms and Ions 165
J. M. Bizau, D. Cubaynes & F. J. Wuilleumier

17. Photoionization of Positive and Negative Ions 178
V. K. Ivanov

IV. MOLECULES AND CLUSTERS

18. High Resolution Core Level Photoemission of Molecules 191
G. M. Bancroft, J. D. Bozek, J. N. Cutler, Y. F. Hu, Z. F. Liu, D. G. J. Sutherland, K. H. Tan & J. S. Tse

19. Angle Resolved Photoion Spectroscopy of Core-Excited Free-Molecules 201
A. Yagishita, E. Shigemasa, J. Adachi & N. Kosugi

20. Core Ionization in the Gas Phase, Probed by Multicoincidence Techniques 211
P. Morin, M. Lavollée & M. Simon

21. Core Excitation in Atomic and Molecular Clusters 226
E. Rühl, C. Heinzel, A. P. Hitchcock & H. Baumgärtel

22. Electronic Excitations in Rare Gas Clusters Studied by VUV Fluorescence Spectroscopy 236
M. Joppien, R. Müller & T. Möller

23. Ion Rotational Distributions at Near-Threshold Photoelectron Energies 247
K. Wang, J. Stephens & V. McKoy

24. Threshold Photoelectron Spectroscopy with Coherent Vacuum Ultraviolet: Studies of Ionic Excited States 257
W. Kong, D. Rodgers & J. W. Hepburn

25. Rotationally Resolved Photoionization with Coherent VUV Radiation 267
R. T. Wiedmann, R. G. Tonkyn & M. G. White

26. Multiphoton Ionization and Ion-Pair Formation in Molecular Hydrogen 277
S. T. Pratt, E. F. McCormack, J. L. Dehmer & P. M. Dehmer

V. ELECTRONIC STRUCTURES IN SOLIDS

27. Photoelectron Spectroscopy in Heavy Fermions: Inconsistencies with the Kondo Model 291
A. J. Arko, J. J. Joyce, R. R. Blyth, P. C. Canfield, J. D. Thompson, R. J. Bartlett, Z. Fisk, J. Lawrence, J. Tang & P. Riseborough

28. Angle-Resolved Photoemission of Single Crystal C_{60} 301
Z. X. Shen

29. Full Multiple Scattering Interpretation of the Cu L_3 Edge XANES in High T_c Superconductors 310
M. Pompa, A. Bianconi, S. Della Longa & P. Lagarde

30. High-Resolution Photoemission Studies of Correlations and Phase Transitions in One and Two Dimensional Systems 320
D. Malterre, B. Dardel, M. Grioni, P. Weibel & Y. Baer

31. Collapse of Mott-Hubbard Framework by Hole Doping in $Bi_2Sr_2CaCu_2O_8$ 330
T. Takahashi, S. Suzuki, T. Kusunoki & H. Katayama-Yoshida

32. High Resolution Photoemission Study of $Bi_2Sr_2CaCu_2O_{8+x}$ 337
M. Onellion

VI. SURFACES

33. High Resolution Absorption Spectroscopy of Molecular Absorbates after Core Level Excitation 347
D. Arvanitis, H. Rabus, T. Lederer & K. Baberschke

34. Final-State Effects in Photoemission from Metal-Semiconductor Interfaces 357
J. Kanski, K. Karlsson, O. Nyqvist, P. O. Nilsson, U. O. Karlsson, M. C. Håkansson & S. P. Svensson

35. Inverse Photoemission of Alkali-Metal Overlayers on Semiconductors 367
B. Reihl, R. Dudde, L. S. O. Johansson & K. O. Magnusson

36. Photon Stimulated Reactions in Adsorbed and Condensed Atoms and Molecules Studied by Ion Desorption 378
G. Dujardin, L. Hellner, L. Philippe & M. J. Besnard-Ramage

37. Photoemission and NEXAFS Studies of Organic Molecular and Polymeric Materials 385
K. Seki

38. High Energy Electron Loss and UV Photoemission Spectroscopies H/Si(111)-(1×1) Surfaces 395
P. Dumas

39. Excited States in Electronic Structure Calculations 408
S. G. Louie

VII. MAGNETIC DICHROISM

40. Magnetic Circular Dichroism in Angle-Resolved Photoemission from Ferromagnetic Surfaces in the X-Ray and VUV Regime 421
C. M. Schneider, J. Kirschner & D. Venus

41. Soft X-ray Linear Dichroism 431
M. Sacchi

42. Magnetic Dichroism Theory of Core Level Spectroscopy 441
G. Van Der Laan

43. Magnetic Circular Dichroism in X-Ray Absorption and Core-Level Photoemission of Fe/Cu(001) 449
J. G. Tobin, G. D. Waddill & D. P. Pappas

44. Spin Polarized Photoemission Studies of Interfacial and Thin Film Magnetism 453
P. D. Johnson, Y. Chang & N. B. Brookes

VIII. ELECTRON DIFFRACTION AND HOLOGRAPHY

45. Surface and Near-surface Atom Imaging with Photoelectron Holography 467
L. J. Terminello, J. J. Barton & D. A. Lapiano-Smith

46. Valence Photoelectron Diffraction: From Partial Densities of States to Fermi Surfaces 475
J. Osterwalder, A. Stuck, T. Greber, P. Aebi, L. Schlapbach & S. Hüfner

IX. FLUORESCENCE

47. Soft X-Ray Fluorescence Spectroscopy Excited by Synchrotron Radiation: Threshold Effects 489
D. L. Ederer, K. E. Miyano, W. L. O'Brien, T. A. Callcott, Q.-Y. Dong, J. J. Jia, D. R. Mueller & E. T. Arakawa

48. Excitation Dynamics in Ionic Crystals 501
C.-S. Shi, Y. Dong & X.-P. Lu

49. Auger-Free Luminescence from Large Gap Insulators 511
S. Kubota

X. MICROSCOPY

50. Recent Developments in Scanning Soft X-Ray Microscopy 523
H. Ade, C. Jacobsen, J. Kirz, C.-H. Ko, S. Lindaas, S. Williams, X. Zhang, C. Buckley, E. Anderson, K. Goncz & S. Rothman

51. Photoemission Spectromicroscopies of Neurons 534
G. De Stasio, P. Perfetti, D. Mercanti, M. T. Ciotti, S. Hardcastle, S. F. Koranda, B. P. Tonner, C. Capasso, W. Ng, A. K. Ray-Chaudhuri, S. Liang, S. Singh, F. Cerrina & G. Margaritondo

52. X-Ray Microscopy: Current Status 544
D. Rudolph, G. Schmahl, B. Niemann, P. Guttmann, J. Thieme, G. Schneider, C. David, M. Diehl & T. Wilhein

I. PLENARY LECTURES

PHOTOIONIZATION OF ATOMS, IONS AND MULTIATOMIC FORMATIONS

Amusia M. Ya*
Institute for Theoretical Physics, Frankfurt University
6000 Frankfurt am Main 11, Germany

ABSTRACT

The ideas which determine the understanding of the photoionization of atoms, ions and multiatomic formations are considered. The manifestations of the multielectron nature of this phenomenon are displayed and discussed at length. Theoretical approaches, their achievements and shortcomings are presented. Attention is given to specific features of new objects of photoionization studies, such as excited states of atoms and negative ions. Many-body diagrams which forms the basis of theoretical investigations are presented for a multitude of different processes which accompany the absorbtion of a photon. The perspectives of the domain are outlined.

1. Introduction

Photoionization is a process in which one or more electrons are removed from the target under photon action. The investigation of it gives information on the electronic structure of the target. Since for atoms, ions and multiatomic formations the interconstituent interaction (i.e. that between electrons and atomic nuclei) is well known, the deviations form the simple hydrogen-like picture of the photoionization cross section (jumps at thresholds, fast decrease above threshold) are due to multiparticle structure of the target. For different systems, which are the objects of consideration in this paper, the photoionization process has a number of common features, which are essential in its investigation.

1.1. Main physical conceptions, important for understanding of the photoionization process:

a) <u>Self-consistent field</u> - a common statical field acting upon each electron and created by the nuclei and all electrons of the system under consideration. The mostly well known and precise is the Hartree-Fock (HF) approximation for this field.

b) <u>Polarization of atomic electron cloud</u> - the deformation of electron system due to action of the incoming

---

*Permanent address: A. F. Ioffe Physical-Technical Institute, St. Petersburg K-21, Russia 194021

photon. Big maxima in the response function, it is in target polarizabilities, present so-called collective oscillations. Damping of these oscillations is due to either inhomogeneity of electron distribution or to interaction with "two-electron-two vacancy" excitation. These osicllations are described in so-called Random-Phase Approximation (RPA) or Random Phase Approximation with Exchange - RPAE [1].

c) Polarizational potential - additional to the self-consistent HF potential acting upon the photoelectron and the vacancy and appearing due to deformation of electron distribution either by photoelectron or the vacancy.

d) Rearrangement is the modification of the field acting upon the outgoing electron due to creation of an extra vacancy. Described by the generalized vesions of RPAE: GRPAE I and GRPAE II.

e) Interchannel interaction - interelectron interaction leading to autoionization, decay of vacancies, creation of satellites.

1.2. Theoretical approaches and the mathematical apparatus.

In the discussion of photoionization in this paper the many-body diagrammatical technique is used, which employes the terms vacancy, electron, effective external field, effective interaction interaction. The word "effective" emphasizes the role of virtual electron excitation in modification of external field and interelectron inter-action. All objects - atoms, ions, and multiatomic formations - are described theoretically in the frameworks of HF, RPAE and GRPAE. A very essential specific feature of systems with comparatively low electron number is the difficulty to draw the border between one electron and multelectron effects. In fact, in Term-Dependent HF version a considerable fraction of electron correlations is included.

1.3. Main Scientific Results

In investigation of photoionization performed during the last decade essential results were obtained:

- The understanding was achieved of the multielectron nature of giant resonacnes in photoionization cross sections of all systems discussed in this paper.
- It was demonstrated that multielectron shells qualitatively modify the photoionization cross section of few electron shells, leading to new continuum spectrum resonances.
- It was shown that together with main lines in photoelectron spectrum satellite or shadow lines of multielectron nature exist.
- Strong interaction of two or even more continuous spectra grounded on different levels, leading to new resonances was observed.
- Autoionizational resonances coming from the interaction of discrete "two electron-two vacancy" excitation with one

electron-one vacancy" ontinuum were observed.

- It was demontrated that the Auger-line is shifted to the higher energy when inner vacancy is created near its threshold due to Post Collision Interaction.
- It was shown that multielectron (at least two-electron) excitation or ionization plays an important role, determining to large extent the total cross section of the photoionization process.
- The strong similarity was observed of photoionization cross sections in different phase states of the same composition of atoms, e.g. gas, mixtures and solid bodies for intermediate and inner shells.

2. RPAE (RPA) Correlations

Most essential and nontrivail feastures of the photoionization process are connected with correlational or collective effects in electron motion. RPAE forms the most convenient basis to study these effects in multiparticle systems: electron gas, atoms, solid bodies. This approach takes into account the direct elimination or excitation of an electron after absorbing a photon and an indirect action - elimination of the considered electron by the alternative part of the atomic self-consistent field coming due to virtual or real excitation of any other electrons after absorbing a photon [1]. The total photo-ionization amplitude $D(\omega)$ is determined by the integral equation:

$$(\varepsilon/D(\omega)/i) = (\varepsilon/d/i) + \sum_{\varepsilon'>F,\ i'\leq F} [(\varepsilon'/D(\omega)/i')(\omega-\varepsilon'-I_{i}')^{-1}(\varepsilon'i/u/i'\varepsilon) + (i'/D(\omega)/\varepsilon')(\omega+\varepsilon'+I_{i}')^{-1}(i'i/u/\varepsilon'\varepsilon)]. \qquad (1)$$

Here u stands for a combination of the direct and exchange Coulomb interelectron interaction, $i,i' \leq F$ denotes initially occupied atomic levels, while $\varepsilon,\varepsilon' > F$ denotes vacant discrete and continuous spectrum states. Diagrammatically, this equation is presented in the folllowing form:

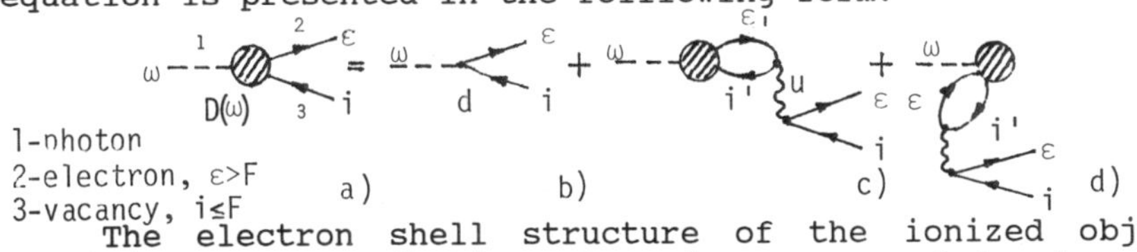

(2)

The electron shell structure of the ionized object amplifies the contribution of a given group of level, e.g. those belonging to the same shell.

Term-dependent Hartree-Fock approximation includes, in the language of diagrams the following equations:

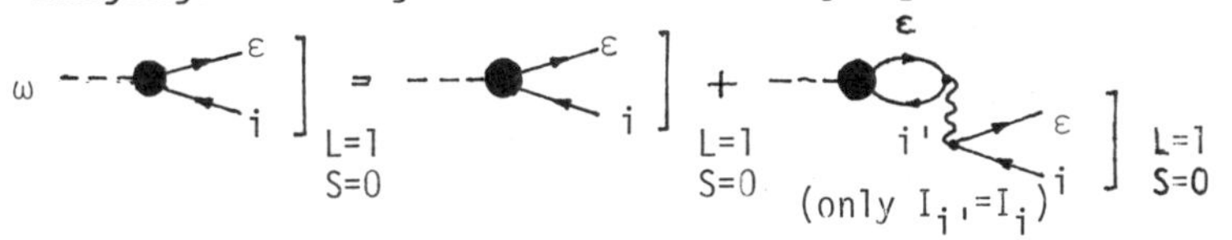

(3)

where I_i denotes the ionization potential of the electron on the level i. Semiphenomenological choice of the self-consistent field can effectively include much more terms which in the language of diagrams belongs to multielectron correlations. The degree of collectivization in electron motion depends upon the number of electrons in a given shell (or neighboring shells) and on the role of nuclear charge. Intrashell correlation, apart from that accounted by (3), includes time reverse diagrams (e.g. (2a)) and terms with i'= i. It is plenty of examples of successful application of RPAE to different atoms, mainly with closed and semiclosed shells, which demonstrated the high degree of collectivization [2].

Recent impressive example is the giant resonance in the fullerene C_{60}, where an RPA calculation [3] leads to a resonance at ≈ 23 eV with a tremendous oscillator strength of about 180, while the total number of molecular electrons is 240 (4 from each of the atoms). The results of calculations are qualitively confirmed by experiment [4]. The theory of collective effects in photabsorbtion of metallic clusters and fullerene can be improved considerably by consistent incluaion of the exchange effects. A reasonable model would be with a positive charge homogeneously (except the surface region) smeared over the cluster volume. Then HF and RPAE should be constructed for electrons moving in such a smeared "nuclear" field.

From RPAE point of view not only intra-, but also intershell correlations are essential, their inclusion completely modifying the photionization cross section of a given shell due to the action of neighbors. Intershell interaction leads also to dissolving of discrete excitations with even big oscillator strength into a comparatively broad continuum [5]. RPAE corrections affects not only the cross section but all other characteristics of the photoization process, such as photoelectron angular distribution anisotropy parameters and spin orientation.

At first glance, as it is seen from (2) and (3), RPAE describes only one-electron photoionization. However, double vacancy state creation after Auger-decay or due to inelastic scattering of the photoelectron with the outer shell is included in RPAE.

This method was applied recently with success not only to closed shell atoms, but also to a number of new objects, such as atoms with semifilled shells, excited atoms and negative ions. Straightforward application of RPAE to objects with semiflled shells require generalization of RPAE for systems with two kinds of electrons - "up" and "down" electrons, according to their spin projections. Then each shell splits into two levels with different energies. Thus, a semifilled shell object may be considered as having only completely filled levels, with "up" or "down" electrons on it. Applying RPAE, Mn and Eu atoms cross sections and other photoionization characteristics were calculated and it was demonstrated the dominant role of the transition to the discrete vacant level in the semifilled shell, namely 3p → 3d in Mn. However not only this transition but correlations in the semifilled $3d^5$ shell are very important [6].

Excited atoms give examples of systems, the photoionization cross section of which is determined by both the self consistent field and the direct effect of atomic electrons, including virtual excitations, upon the ionizing one [7]. Unexpectedly strong proved to be the correlation effects in so called simple negative ions: those which have either filled or semifilled electron shells, like, e.g. I^- or Si^-. The photoionization cross section of outer and intermediate electron shells is strongly affected by the neighboring electrons.

3. Beyond RPAE: Inclusion of Rearrangement Effects (GRPAE I and II)

From the pure physical point of view it is evident, that the ionized or virtually excited electron is moving not in the frozen core field of an atom with a vacancy. In fact as soon as being created the vacancy affects also the other electron states, leading to so-called statical rearrangement. Two major effects determine the picture of this process:

a) the screening of the electron-vacancy interaction due to virtual excitations of other atomic electrons,

b) the shift of the ionization potential due to polarization of electron shells by inner (or intermediate) vacancy. Diagrammatically, statical rearrangement is illustrated by the following pictures:

a)

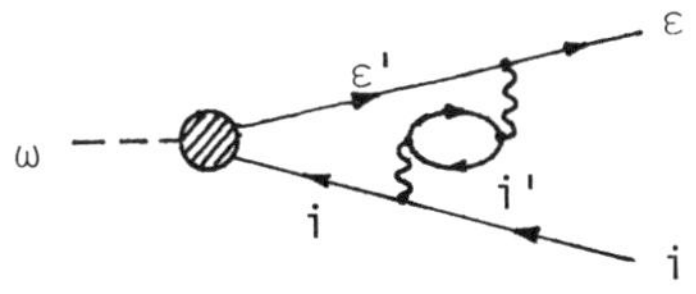

(4)

which describes screening of the electron-vacancy interaction, and

b)

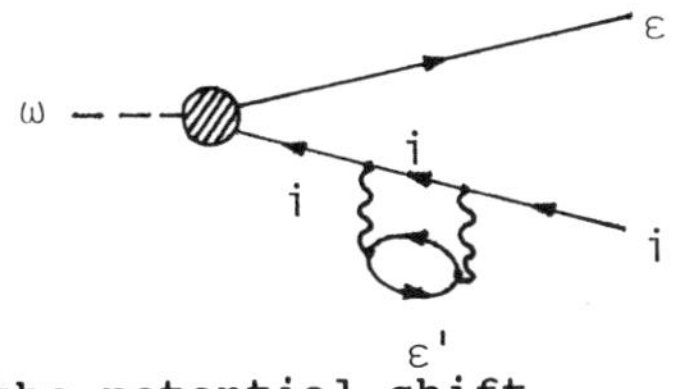

(5)

describing the potential shift.

It is essential to have in mind that statical rearrangement corresponds not to these diagrams (even in the considered order of the perturbation theory) but only to parts of them, where instead of summation over all intermediate vacancy states only one term with vacancy i is taken into account.

By including corrections of this type ot the RPAE scheme,

the generalized RPAE (or GRPAE) is formed, in which the photoionization amplitude D(w) is given by the equation:

$$(\underset{\sim}{\varepsilon}|\tilde{D}(\omega|\underset{\sim}{i}) = (\underset{\sim}{\varepsilon}|d|\underset{\sim}{i}) + \sum_{\varepsilon>F,\, i'\leq F} [(\underset{\sim}{\varepsilon}|\tilde{D}(\omega)|\underset{\sim}{i}')(\omega-\varepsilon'-I_{\underset{\sim}{i}'})^{1}(\underset{\sim}{\varepsilon}'\underset{\sim}{i}|u|\underset{\sim}{i}'\underset{\sim}{\varepsilon})+$$

$$+(\underset{\sim}{i}'|\tilde{D}(\omega|\underset{\sim}{\varepsilon}')(\omega+\varepsilon'+I_{\underset{\sim}{i}'})^{-1}\ (\underset{\sim}{i}'\underset{\sim}{i}|u|\underset{\sim}{\varepsilon}'\underset{\sim}{\varepsilon})]. \qquad (6)$$

The sign ~ under the electron or vacancy emphasize that atomic states are calculated in the HF field of an atom with a vacancy, namely $\underset{\sim}{\varepsilon}$, $\underset{\sim}{i}$ in the presence of the vacancy i and $\underset{\sim}{\varepsilon}'$,$\underset{\sim}{i}'$ in the presence of the vacancy i'.

The inclusion of statical rearrangement lead to considerable effects for near threshold regions of intermediate and inner shells: the vacancy field became screened thus decreasing its strength and the cross section at threshold. This rearrangement is mostly essential for slow photoelectrons. It must be of importance for multiatomic formations also, where many outer electrons are moving in a common field and are ready to change their states due to inner or intermediate shell vacancy formation.

Another version of rearrangement effect takes place when a deep vacancy is created which can decay by Auger effect. In this case the outgoing electron, if sufficiently slow, can find itself instantly not in a single, but in a double (or multiple) vacancy field. This is stronger than that of a single vacancy, so this dynamical rearrangement leads to increase of the cross section at threshold, as compared to the RPAE case. Diagrammatically, the process can be represented by the following picture

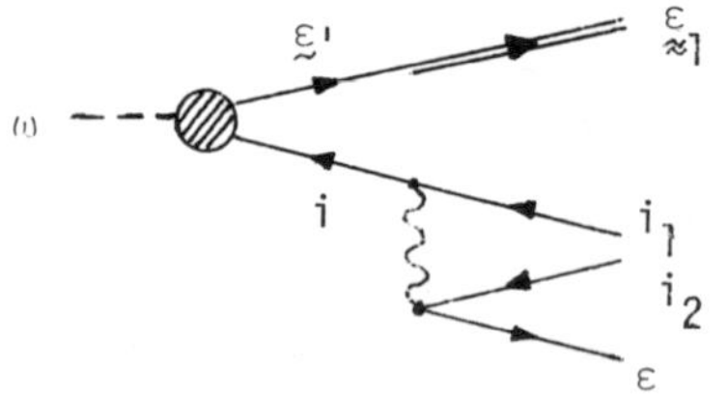

═══ - overlap integral $(\underset{\sim}{\varepsilon}'|\underset{\approx}{\varepsilon}_1")$ (7)

Here the doubled line starting from the inner vacancy i decay moment denotes that the photoelectron moves in a double (i_1 and i_2) instead of a single i- vacancy field. This leads to an extra factor in the diagram (7): an overlap integral $(\underset{\sim}{\varepsilon}'/\underset{\approx}{\varepsilon}_1)$ of wave-functions of electrons, moving in the single vacancy i (denoted by ~) or in the double vacancy i_1 and i_2 (denoted by ≈) fields, respectively. If the wave functions of the final state electron, and of the intermediate RPAE states are calculated taking into account the possibilty of Auger vacancy decay, the GRPAE II is formed and the corresponding amplitude is determined by (6), with electron (vacancy) function calculated in HF field of an atom having $i_1 i_2$ vacancies instead of i.

4. Further Beyond RPAE Frame

The corrections taken into account by GRPAE I,II are due to inclusion, as it is seen from (4), (5) and (7), of "two-electron-two vacancy" excitations. The precise inclusion of these is now impossible because it requires to take into account the interaction of two electrons moving in the field of an atom with vacancies, i.e. to find a solution of at least the three-body Schrodinger equation, which must then be incorporated into a rather complicated scheme, including multielectron correlations. The three body problem itself is not yet solved for such systems as He, or e + H scattering. On the other hand, it seems, that the main features of photoionization cross section are reasonably described in RPAE frame and even GRPAE corrections are essential only close to ionization threshold. Of course, there are exceptions: the photoionization cross section of La atom above 4d threshold is entirely determined by statical rearrangement. Without it instead of a broad maximum above threshold a small monotonically decreasing cross section would be obtained. So, even the desription of photoionization cross section require the development of a theory which would include GRPAE effects not semiphenomenologically, but by direct summation of corresponding perturbation theory diagrams.

When considering more delicate than the total cross section characteristics of the photoionization process, the role of virtual excitations, other than that included in RPAE, become evident. The simplest way to demonstrate it is to investigate small partial cross sections and their characteristics, such as angular anisotropy parameter β. A good example here is the $5s^2$ subshell of Xe, where the cross section can be understood only if "two electron-two vacancy" excitations are taken into account. In describing this small (about 1% ÷ 4% of the total value) cross section several effects proved to be important: the alteration due to virtual "two electrons-two vacancy" excitations of the direct 5s-electrons photoionization amaplitude, the intershell interaction between 5p(4d) and 5s atomic electrons, and the pure 5s-vacancy in the final state which become mixed with two electron excitations of the outer 5p-shell.

Another case of obvious importance of "two electron-two vacancy" virtual excitations is the formation and photoionization of complex (Ca^--like) negative ions, where the following corrections must be taken into account:

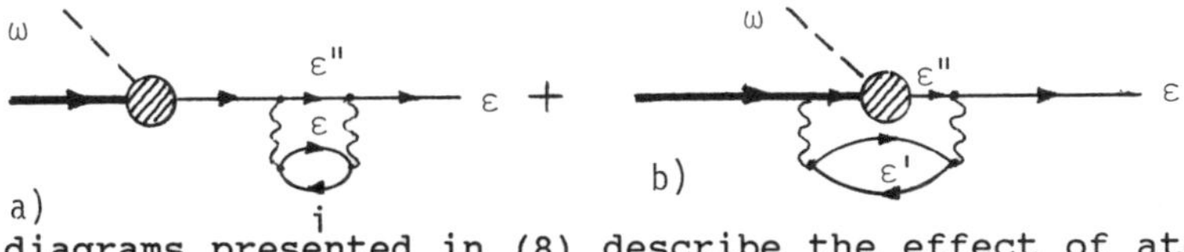

(8)

The diagrams presented in (8) describe the effect of atomic polarization upon the extra electron photoionization process. The heavy line in (8) denotes, that this is an outer electron in a negative ion, formed only when the polarization of the atomic core by the extra electron is taken into account. It is essential, that corrections presented in (8) for Ca^- like ions must be taken into account nonperturbatively. In the existing up to now calculations [7] term 8b is neglected. It is of interest to investigate not only the photodetachment of the outer electron, modified by the virtual RPAE-type excitation of intermediate and inner shells, but also that for the inner shell electrons.

An extremely interesting object of photoionization study is Xe^- [9]. It could be formed by adding an extra electron to the long living metastable state of Xe atom: $5p^{-1}6s\ ^3P$. The time of life of this state is big. If an electron is placed to the 6p state with the same spin projection as 6s, its transition to $5p^{-1}$ is forbidden by spin conservation.

5. Direct Knock Out.

It appeared that on its way out of the atom the photoelectron can ionize the outer shell with rather high probability. Good examples are atoms Xe and Ba [10], in which the photoelectron moving from the 4d-shell has a 30-40% probability to excite or ionize outer atomic electrons [11]. This process may be even more probable for negative ions and multiatomic formations, with many loosely bound electrons.

In the lowest order in the interaction between the photoelectron and outer electrons the knock out process is presented by a diagram (9)

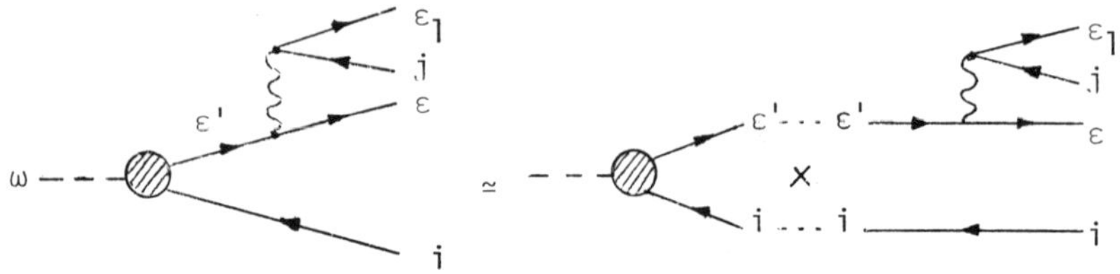

(9)

For good estimations the knock out amplitude may be presented as a product of the inner electron photoionization amplitude and that for photelectron inelastic scattering.

The respective corrections to the one-electron elimination amplitude come from next order in the perturbation theory. This is exemplified by the following picture:

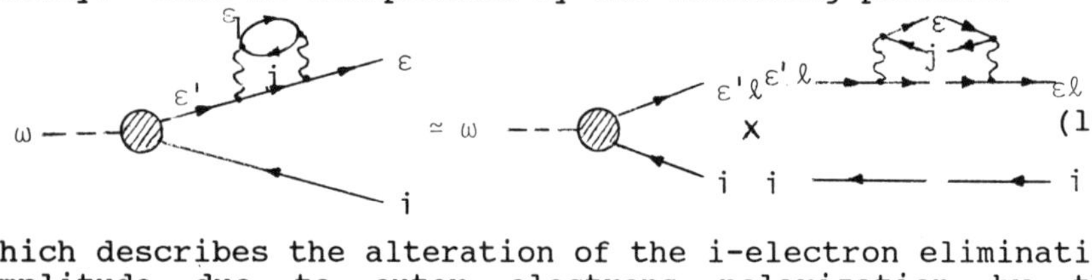

(10)

which describes the alteration of the i-electron elimination amplitude due to outer electrons polarization by the photoelectron. Iterating the polarizational part of (10), an expression for the i-electron ionization cross section is derived:

$$\sigma_i^+ \approx \sigma_i^{RPAE} \times \exp[-2\text{Im}\delta_\ell] \approx \sigma_i^{RPAE} [1-2\ \text{Im}\delta\_\ell], \quad (11)$$

the last equality being valid for small $\text{Im}\delta_\ell$, where δ_ℓ is the photoelectron's ℓ-partial wave elastic scattering phaseshift. The cross section for double electron ionization and ionization with excitation due to the photoelectron inelastic scattering is given by

$$\sigma^{++} + \sigma^{+*} \approx \sigma^{RPAE} [1 - \exp(-2\text{Im}\delta_\ell)] \approx 2\ \text{Im}\delta_\ell\ \sigma^{RPAE}. (12)$$

The direct knock-out is accompanied by some other interesting processes:

a) If the photoelectron energy is close to the electron ionization or excitation thresholds in the cross section, a Wigner-Baz singularity appears modified due to the fact that the outgoing electron is moving in a Coulomb instead of a shrot range field.

b) The secondary photon emission - both direct and indirect, i.e. via other electrons virtual excitation can take place, which is described by daigrams (13a) and (13b), respectively:

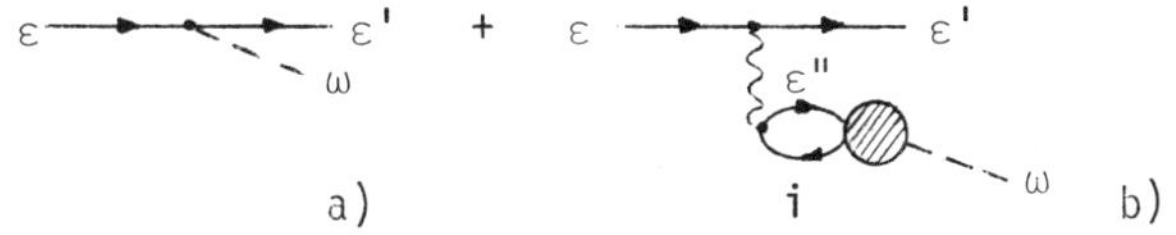

(13)

c) Post-Collision Interaction due to inelastic photoelectron scattering on other atomic electrons can take place, when its energy and the autoionization resonance energies are close. This will modify the energy distribution of the photoelectrons. Such a process is represented by the diagram

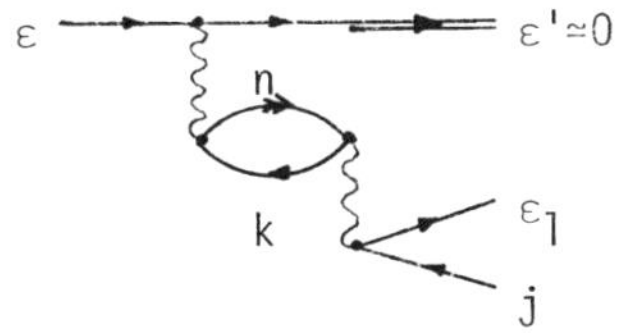

(14)

where kn is the resonance state with the energy $\omega_{ex(kn)}$ and j denotes the outer vacancy, which is created in the process of autoionizational decay. The doubled arrow denotes the excited discrete level, while doubled line emphasize the instant change of the field acting upon the outgoing photoelectron, which after inelastic scattering loses almost all of its energy.

d) Multi-step processes can become important, which is exemplified by the Mn atom case, where the photoelectron from an inner shell can excite the most intensive discrete level

3p → 3d decaying subsequently into semifilled shell excitation 3d → εf, εp and only after that to other channels:

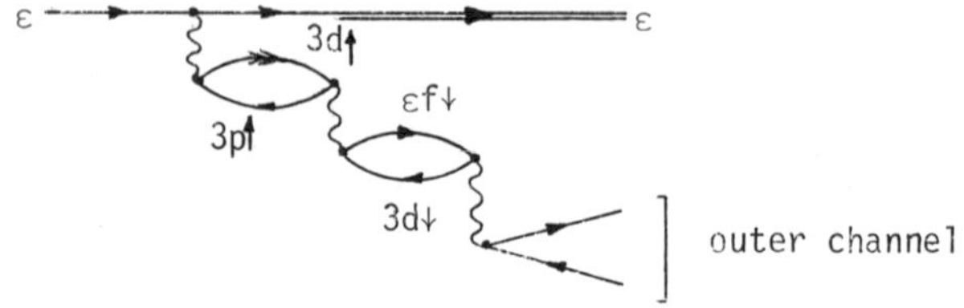

(15)

6. Satellites and Shadows - Well above and Near Threshold.

In photoionization not only a simple vacancy may be created in atoms, but more complex states, e.g. a vacancy accompanied by a discrete electron-vacancy excitations. These are called satellites (or shadows). Well above ionization threshold, when the interaction of photoelectron and vacancy is neglected, satellites appear due to "shake up" while shadows - via quasi-Auger decay. The simplest diagram for the "shake up" is (16a), and for quasi-Auger is (16b):

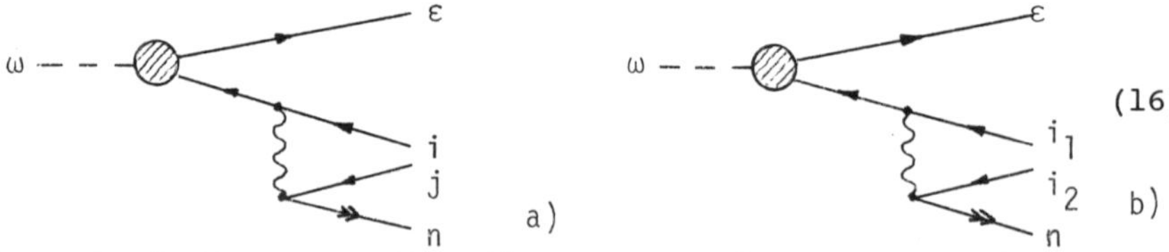

(16)

Here in both cases the final state is reached via the i-vacancy formation stage. While (16) presents the lowest order perturbative description of these processes (formation of i is described in RPAE), the probability to excite satellite (shadow) states, with n being a discrete level, often is big. So, they must be accounted for nonperturbatively.

Rather interesting is the investigation of photoionization process, in which the atom is left in one of the satellite (shadow) states. Not rare are cases when the direct formation of the satellite (shadow) states is impossible. An example is the photoionization of Ar, which leaves it in $3p^{-2}$ 3d(4d,5d...)^{2}S states being shadows of the $3s^{-1}$ "main" state. The probability to have the ion in a shadow state, just as in the main state is in the near threshold frequency ω region completely determiend by the resonating action of outer $3p^6$ electrons [12]. For high ω the ratio $\sigma_{sat}(\omega)/\sigma_{main}(\omega)$ became almost independent upon ω.

For low ω other states and other kinds of satellits can be excited. It is because the probability for the outgoing photoelectron to excite a discrete state is not low, if the photoelectron is not too fast. Corresponding simplest diagram is presented by (17):

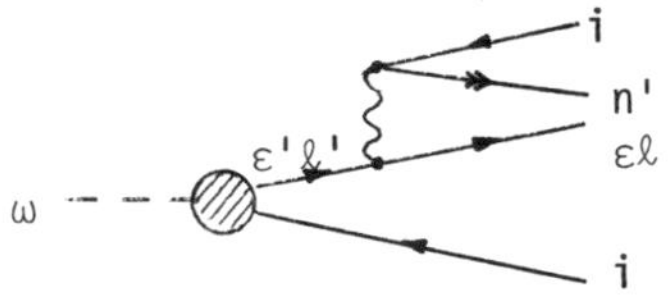

(17)

Note that in the collision the energy and angular momentum of the photoelectron is altered, thus leading to alteration of its angular distribution. In such a process, contrary to (16a) not only monopole excitations (satellites) are created. While proceeding not via the i-state, the process presented by (17) leads to satellites, with total momentum and spin different from that of vacancy i[13]. In (17) the excitation probability is strongly enhanced if the process goes via a "door way" state-discrete excitation in", instead that of the continuous spectrum iε'. In (17) only those "door way" states are included, which are described in the RPAE (GRPAE) framework.

7. Formation of Multiply-Charged Ions

The simplest way to form a multiple-charged ion is to proceed step by step via Auger processes. Of course, each state reached by an allowed Auger decay is dominative. But it is almost self evident that not all possible degrees of ionization can be reached in such a way. On contrary, there are photon frequency regions and ionization degrees which can be understood theoretically only if other mechanisms of multiple ionization, namely shake off, quasi Auger decay and direct knock out by the photoelectron are taken into account.

In investigation of multiply charged ion formation a similarity was observed between σ^{+n} and partial photoionization $\sigma_{n\ell}$ cross sections [14]. Its reason is not completely understood. But in noble gases the ns-electron RPAE partial photoionization cross section σ_{ns} calculated with HF thresholds above the double electron threshold almost coincide with the double electron photoionization cross section σ^{++} of the outer np^6 electrons. It is unclear whether such a coincidence exists for systems.

Some qualitative features of the A^{+n} formation process are well understood. An example is the transition from dominating A^{++} formation via Auger effect to A^{+} via autoionization on the way from Xe to Eu in the 4d-shell frequency region. While in Xe the photoionization of 4d electron dominates, in Eu the discrete transition 4d → 4f with subsequent autoionization of one of the seven 4f-electrons is most important.

Of great interest for understanding the electronic structure of atoms and multiatomic formations is the non Auger double charged ion formation. The few existing calculations describes this process in the lowest order of perturbation theory in interelectron interaction. This is definitely insufficient: in near threshold region, where both outgoing electrons are slow their interaction must be taken into account nonperturbatively, by including the sequence, may be infinite, of the following diagrams:

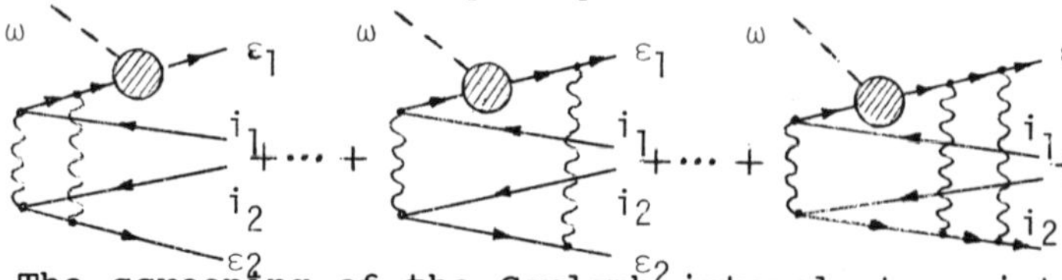

(18)

The screening of the Coulomb interelectron interaction by virtual excitations of target electronsis is also of importance.

The role is not yet clarified of "doorway states" - resonances in "oneelectron-one vacancy" channel, leading to enhancement of the double electron photoionization probability. The double ionization deserves further theoretical efforts and new experimental data are required, mainly for complex atoms, where the "door way" can have resonances and virtual excitations of the core are important.

The angular and energy distributions of both electrons are particularly interesting, being strongly affected by the interelectron interaction. Near inner shell ionization threshold the double charge ion production is determined to large extent by the Post-Collision Interaction effects. Note that the direct production of A^{+n} ($n \geq 3$) is not yet investigated theoretically.

Recently it was demonstrated experimentally [15] that a rather big contribution to A^{+2} is coming from the decay of a discrete state in a complicated process, represented by a second instead of a first order diagram:

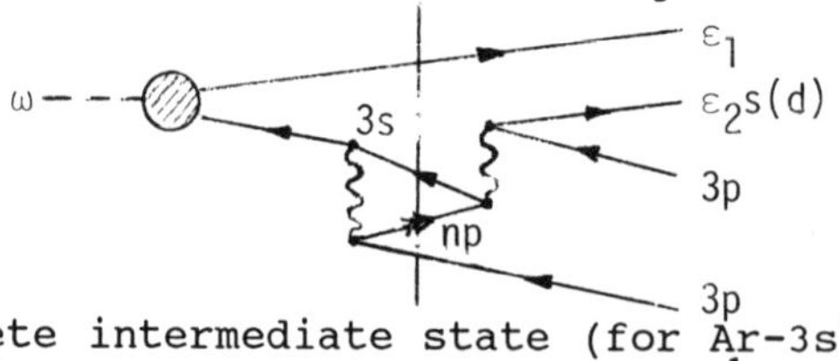

(19)

The discrete intermediate state (for Ar-$3s^{-1}3p^{-1}$np) which is a satellite to the main (for Ar-$3s^{-1}$) line, serves as a resonance, modifying the energy distribution of the ionized electrons in the edge (ε_2-slow) part of the spectrum,. For the Ar case 30-40% of A^{++} are produced via the resonance decay of a satellite [15].

Of special interest is the A^{++} formation in negative ions, excited atoms or equivalent states in multiatomic formations. It is because the outer electrons in these cases are less bound to the system as compared to atoms. For example, it was demonstrated by calculations, that at least for some excited states double electron photoionization, including the elimination of the excited electron is by orders of magnitude more probable, than the ionization of an inner electron itself [16].

8. Near- and Below Threshold Processes

If the formation of A^{++} takes place near the inner (or intermediate) shell vacancy threshold, the spectrum of emitted electrons is considerably modified due to Auger vacancy decay, which leads to increse of the force, acting upon the slow photoelectron. This effect is called Post Collision Interaction (PCI) and is in the lowest order diagrammatically represented by the picture:

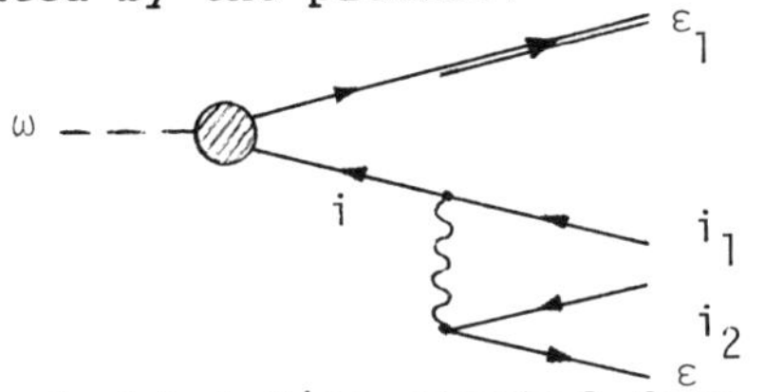

(20)

The doubled line started from the moment of vacancy i decay into i_1, $i_2\varepsilon_2$ state demonstrates that the photoelectron initially moving in the field of i, instantly find itself in the field of $i_1 i_2 \varepsilon_2$ state. The theoretical investigation of this process is essentially simplified, if the Auger-electron ε_2 is fast enough and much faster than the photoelectron ε_1, which is slow. The alteration of the field due to decay leads to extra attraction acting upon electron ε_1 and as a result to it dexceleration or even capture to one of the discrete levels in the field of two vacancies, thus leading to A^{+*} instead of A^{++}. The PCI effects are particularly strong near thresholds of rapidly decaying vacancies, such as e.g. 4p subsehll in Xe. However in this case the Auger-electron is slow, so the interaction of both outgoing electrons must be included.

For frequencies below inner shell threshold as a first step of the process an electron is excited to one of the discrete levels, which is followed by the vacancy decay or something more complicated. The excited electron can be either a passive spectator or an active participator of the subsequent process. Very often it is an active participator, what is illustrated by the decay of the $2p^{-1}4d$ state of Ar, where the presence of the excited electron 4d almost completely suppresses the 2p-vacancy Auger decay [17]. This process is represented by the diagram:

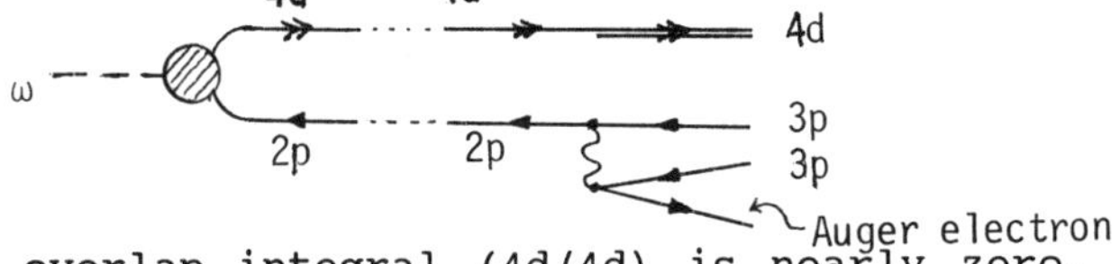

(21)

The overlap integral (4d/4d) is nearly zero, and most probable is the decay accompanied by the electron 4d transition to higher excited levels - 5d, 6d... Note that due to PCI the slow electrons lose its energy becoming slower, or even intercepted to high discrete levels, while the electron at a low level tries to acquire energy and to go higher[17]

The Auger electron angular distribution can be essentially altered due to participation of the excited electron even if initially an s-vacancy is formed: the excited electron and one of the final state vacancies can easily exchange angular momentum.

In the subthreshold processes the radiative decay can be also strongly affected. The exchange of angular momentum opens forbidden channels, for example, the „is →js + γ" transition:

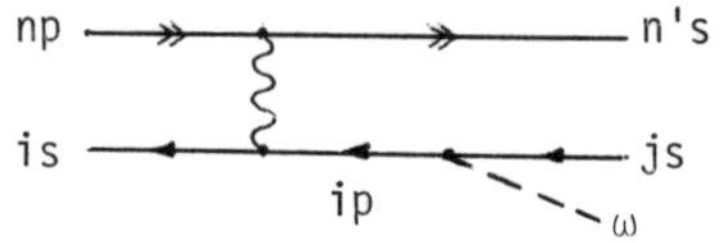

(22)

The presence of an excited electron strongly affects the probability of the decay if it is of correlative nature. An example is presented of Xe, described by the following diagram:

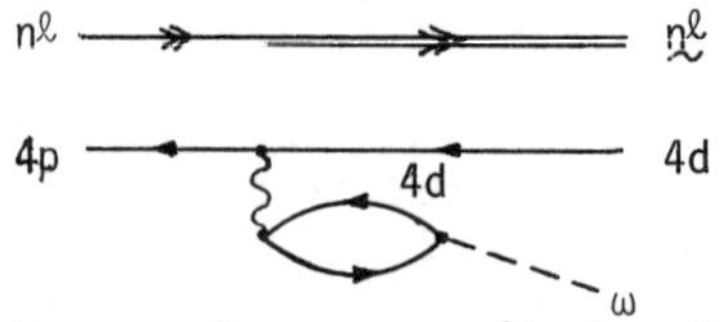

(23)

The presence of an excited electron modifies the transition energy, as well as the virtual excitation enegy, as compared to its values in an ion, which considerably alters the correltional contribution to the decay process.

The below-threshold excitation leads to formation of states with several vacancies and excited electrons, their decay enriching the Auger spectrum with new lines, which give information on the photoionization mechanism.

The presence of excited electrons opens new decay channels for states which are more complex than (22). As an example the radiative decay of "two-vacancy-one electron" state $4s^{-2}5p$ into $4p^{-2}4d$ in Kr can serve [18]. The simplest respective diagram looks as follows:

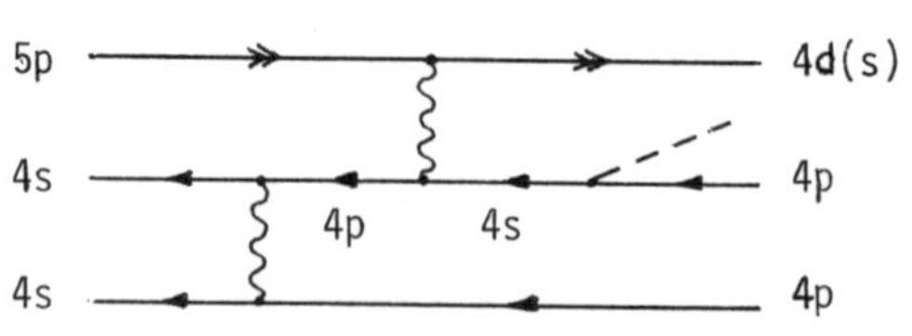

(24)

The initial state $4s^{-2}$ 5p appeared after the $3d^{-1}$ vacancy Auger-decay of the $3d^{-1}$ 5p electron-vacancy excitation of Kr [18]. While for inner shell vacancies the excited electron is a spectator, for intermediate shells the competition is very strong between independent vacancy Auger decay, i.e. the spectator picture and that of participator, with mainly the autoionizational decay of the initially created state.

Interesting is the autoionization of not only the initially created state, but of that formed after the Auger decay, i.e. the autoionization of two vacancy-one electron states, which is described in the lower order by diagrams:

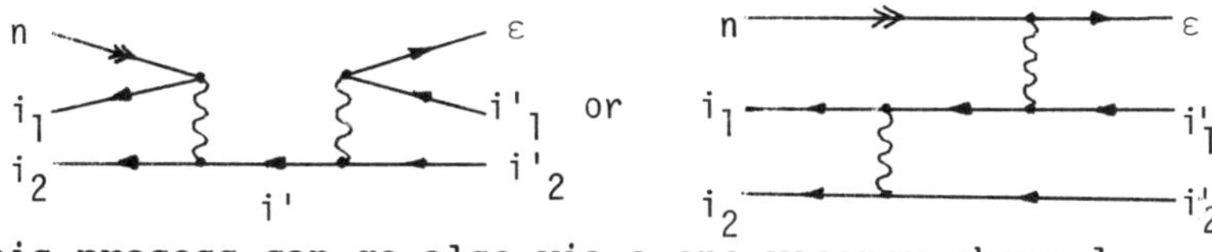

(25)

This process can go also via a one-vacancy channel.

The theory of all these correlational processes is at the very beginning: only shake-type and lowest order perturbative approaches are developed. However, it is evident that higher order corrections are important: screening of the electron-vacancy and vacancy-vacancy interactions and the polarization of the core by vacancies and/or the excited electron.

9. New Objects and Processes for Photoionization Studies

In order to summarize this paper, which was devoted mainly to discussion of interesting mechanisms and processes instead of concrete results we would like to mention those new objects of investigation which seem to be perspective. It does not mean that the field of atoms is completely developed. But for them more important are rather complicated processes or detailed study of the photoionization, while for these new objects even the absolute cross sections are of interest.

The new objects are a) negative ions, b) atoms in their excited states, c) positive ions, d) spin-oriented atoms, e) empty inner shell or hollow atoms. The a) and b) objects were discussed above. For c) two features considerably modifies the photoionization cross section - the possibility for a transition with big oscillator strength to the level occupied in atom and the modification of the first excited states wave functions due to the increase of the field acting upon the excited or outgoing electron in the ion. This leads to considerable redistribution of the continuous photoionization spectrum, making it rather sensitive to the theoretical models used in calculations. The enhancement of discrete transiton icnreases the intensity of the resonant processes. This leads in some cases to very big values of the photoionization cross section, up to thousands Mb in resonances.

It was shown recently that spin oriented states of atoms, i.e. those in which all or many electrons have the same spin projections, are comparatively long living - microseconds and even more. It makes them rather interesting objects of photoionization investigation, their electronic structure being quite different from that of corresponding atoms.

If long living atomic states could be formed, in which a number of inner levels are vacant, while the electrons occupy outer levels (object e)), it would be interesting to study its photoionization. The role of electron correlations would be much bigger than in the ground state of the corresponding atom.

New process only recently attracting attention are the satellites photoionization and decay, double and multiple electron photoionization, PCI, near threshold phenomena. Of great interest for double electron ionization are the energy and angular distributions, as well as the alteration of these due to different resonances.

10. New Goals of the Theory of Photoionization

It is necessary to develop an approach which would permit to include consistently "two electron-two vacancy" excitations into the existing RPAE frame in order to derive the generalized schemes of it. It is important to include both the discrete and continuum "two electron-two vacancy" excitations describing resonance and nonresonance effects. This step will inevitably require solution of at least two-dimensional equations, while until now after separating the angular parts only one-dimensional equations are solved.

The other problem in making the next step beyond RPAE is to include in HF and the exchange part of RPAE the effective interelectron interaction derived in RPAE. To perform such a step consistently is rather difficult. However, there are serious grounds to believe, that after performing this step the accuracy of ab initio calculations of photoionization cross section and other characteristics of this process will be as high as 1%.

11. General Goals.

Entirely, the goal of investigation of photoionization of different objects, such as atoms, ions and multiatomic formation is to improve our understanding of atomic electron structure and of the photoionization process itself.

Along with permanent improvement of the accuracy, achieved in description of photoionization of atoms and ions, it is very important in describing photoionization of multiatomic formations to reach the level already achieved for isolated atoms.

Acknowledgement

The author is grateful to the Alexander von Humboldt Foundation, which make his stay and research in Germany possible. He is also grateful to the Institute of Theoretical Physics, Frankfurt University, for hospitality. Many thanks due to Mary Kliwer (University of Nevada, Reno) for preparation of the manuscript.

References

1. D. Pines and P. Nozieres, The Theory of Quantum Liquids

(Benjamin, N.Y., 1966).
2. M. Ya. Amusia, Atomic Photoeffect (Plenum Press, N.Y. - London, 1990).
3. G. F. Bertsch, A. Bulgac, D. Tomanek and Yang Wang, Phys. Rev. Lett. 67 (1992) 2090.
4. I. V. Hertel, H. Steger, J. deVries, B. Weisser, C. Menzel, B. Kamske and W. Kamke, Phys. Rev. Lett. 68 (1992) 784.
5. T. E. Hansen, T. Brilly, E. T. Kennedy and G. O'Sullivan, Phys. Rev. Lett. 63 (1989) 1934.
6. M. Ya. Amusia, V. K. Dolmatov and M. M. Mansurov, J. Phys. B: At. Mol. Opt. Phys. 23 (1990) L491.
7. M. Ya. Amusia and N. B. Avdonina, Z. Phys. D - Atoms, Molecules and Clusters 14 (1989) 191.
8. S. Froese-Fisher and J. E. Hansen, Phys. Rev. A44 (1991) 1559.
9. H. Haberland, T. Kolar and T. Reiners, Phys. Rev. Lett. 63 (1989) 1219.
10. B. Kammerling, H. Kossman and V. Schmidt, J. Phys. B: At. Mol. Opt. Phys. 22 (1989) 841.
11. M. Ya. Amusia, G. F. Gribakin, V. K. Ivanov and L. V. Chernysheva, J. Phys. B: At. Mol. Opt. Phys. 23 (1990) 385.
12. a. W. Wijesundera and H. P. Kelly, Phys. Rev. A39 (1989) 634.
b. V. L. Sukhorukov, B. M. Lagutin, H. Schmoranzer, I. D. Petrov and K.-H. Schartner, J. Phys. B: At. Mol. Opt. Phys. (1992), in press.
13. U. Becker and D. A. Shirley, Physica Scripta, T31 (1990) 56.
14. P. Zimmerman, Comments on Atomic and Molecular Physics, XXIII (1989) 45.
15. U. Becker, R. Wehlitz, Physica Scripta T41 (1992) 127.
16. Z. Felfli and S. T. Manson, Phys. Rev. Lett. 68 (1992) 1687.
17. M. Meyer, E. V. Raven, B. Sonntag and J. E. Hansen, Phys. Rev. A43 (1991) 177.
18. M. Ya. Amusia, V. A. Kilin, A. Ehresmann, H. Schmoranzer and K.-H. Schartner, J. Phys. B: At. Mol. Opt. Phys. (1992), in press.

CHEMISTRY IN FEMTOTIME

AHMED H. ZEWAIL
Arthur Amos Noyes Laboratory of Chemical Physics
California Institute of Technology
Pasadena, California 91125 U.S.A.

This summary of the conference overview lecture* is concerned with chemistry on the femtosecond (10^{-15} s) time scale, the time window that allows one to observe and study the fundamental motions of atomic and molecular reactions. The time resolution in chemistry has continued to improve throughout this century. In the 1950's, the techniques of flash photolysis and the relaxation method made it possible to resolve the kinetics of intermediates on a millisecond to microsecond time scale. With the invention of lasers in 1960, the time resolution improved to nanoseconds and reached femtoseconds in the 1980's - an improvement of nine orders of magnitude! On the femtosecond time scale, the *transition states*, which are the most *elementary structures* in a chemical change, are frozen out. These states describe the passage of the nuclei through ephemeral configurations which are far from the stable equilibrium configurations of reactants or products. In these states the entire fate of the reaction is decided.

There are two general classes of elementary reactions, those described as a full collision (bond breakage and bond formation) and those as a half collision (bond breakage only). In both cases, the lifetimes of the transition states are ultrashort ($<10^{-12}$ s), and to freeze out the atomic motion the experimental time resolution has to be better than 10^{-12} s. An estimate of the time scale may be made from knowledge of the typical velocity of nuclear motion in a reaction, $\sim 10^5$ cm/s. Since the dynamics of the chemical bond must be probed on its scale of distance (~ 1 Å), the time scale must be of the order of 100 fs. The motion, if resolved, reflects the nature of the chemical forces (potential) and may be related to quantum or classical theories of the dynamics.

To resolve the motion, we have to consider the following three requirements:

a) The zero-of-time of the reaction must be established with the required time resolution – synchronous triggering or clocking of the reacting molecule(s);

b) The probing window must also have the required time resolution - - synchronous probing of the system in the transient configuration;

c) The motion must be followed with an ability to observe changes as the reaction proceeds from reactants to products – fingerprinting of the motion (spectroscopy) or photographic imaging of the motion (diffraction).

In femtochemistry, these requirements are met by combining ultrafast laser and electron techniques with molecular beam/gas phase methods. In the course of this development at Caltech, there were several technological achievements that have aided the fast progress in the field. First, the development of dye lasers (and more recently Ti:sapphire lasers) was particularly important as these lasers provided pulses of less than 100 fs in duration; the state of the art is now 6 fs, using pulse compression and other methods. Second, the development of continuum generation by focusing a ps (fs) pulse into a liquid solution has opened the door to reliable wavelength tunability. Third, molecular beam techniques combined with multiphoton ionization, laser-induced fluorescence and photoelectron analysis, make it feasible to detect a few thousand atoms or molecules, and in some cases single atom detection is possible. Finally, the use of CCD cameras in diffraction experiments has made it possible to record diffraction patterns with single electron sensitivity.

We will focus here on a few examples from elementary dynamical studies of reactions in gases, clusters, and solutions. We conclude with recent advances. Some of the concepts and methodology are discussed in several review articles and these articles (see below) form the basis for the discussion given for some of the reactions studied:

A. Reactions involving direct bond breakage. For example, ICN $\rightarrow$ I+CN; $I_2 \rightarrow 2I$; $Bi_2 \rightarrow 2Bi$; $CH_3I \rightarrow CH_3 + I$; and $ICF_2CF_2I \rightarrow I + CF_2CF_2I$.

B. Reactions involving two potential energy curves - covalent/ionic resonance. For example, the observed resonance behavior for NaI $\rightarrow$ Na+I. The bond, in this case, changes from covalent to ionic character with time.

C. Reactions with saddle-point potentials. This is best illustrated by the potential for IHgI $\rightarrow$ I+HgI or I+Hg+I. The coherent trajectories of the reaction are observed as the system passes from the transition state (saddle point) to final products.

D. Reactions involving simultaneous bond breakage and bond formation; bimolecular processes. For example, the reactions of $Br + I_2 \rightarrow BrI + I$ and $H + CO_2 \rightarrow OH + CO$.

Of fundamental importance to all of these experiments is the initial localization of atoms by coherent preparation of the nuclear configuration. This is a consequence of the ultrashort time duration of the initiating pulse and the accurate establishment of the zero-of-time. But what about the uncertainty in energy and momentum dictated by the uncertainty principle? Since $\Delta E \Delta t \geq \hbar/2$, one may think that the energy resolution becomes poorer as Δt becomes shorter. While true that $\Delta E \Delta t \geq \hbar/2$, this is not a correct picture for two reasons. First, to observe the motion, one must localize the nuclei in the spatial domain and the shorter the pulse the *better* the localization; long pulses will prepare a whole set of extended nuclear configurations and there is essentially no motion left to observe! The second point of relevance here is *coherence*. In the usual way of thinking about excitation of energy levels, one neglects the relative phases of the levels and considers only the energy E_n of the nth eigenstate. However, to probe nuclear motion in a reaction, a wave packet must be formed. This wave packet is prepared with all phases ($e^{-iE_n t/\hbar}$) well-defined, and the packet is a coherent superposition of the n eigenstates:

$$\Psi_{packet}(t) = \sum_n C_n \psi_n(R) e^{-iE_n t/\hbar} , \tag{1}$$

where the C_n's are coefficients which give the contribution $| C_n |^2$ of the nth eigenstate in the wave packet. Thus, the shorter the pulse the greater the number of eigenstates involved and in the time domain the wave packet has all the Fourier components of the energy spectrum. In fact, from fs experiments one can obtain the E_n's and the potential governing the motion for systems with bound energy states. Clear examples of this were shown with I_2 and ICl in their bound B and A potentials, respectively. It is also possible to observe perturbations of the nuclear motion by neighboring atoms or molecules if the same systems are studied in clusters or in solutions. Such studies have been made for I_2 in Ar solutions and in Ar clusters.

The simple correspondence made above between time and energy for bound states does not hold when studying systems with transient intermediates. The initial state of the system contains no information on the intermediates and these intermediates must be isolated in *time* for any dynamics to be meaningful. When transient species were studied on the millisecond or microsecond time scale with flash photolysis the unusual discrepancies in Arrhenius plots of ln (rate constant) versus inverse temperature were resolved simply because the different intermediates of the system had to be isolated before assigning a unique kinetic parameter.

A final point regarding the time and energy descriptions is particularly relevant to theoretical studies. When we discuss reactions, we think of motions in time, e.g. atom A approaching molecule BC, etc. Our thinking in time does not need an "inversion" from energy space to the time domain, and if the experiments are performed on the fs time scale (wave packet preparation), the system is localized (particle-like; short de Broglie wavelength) and one can, indeed, speak of motion. In principle, one can solve for all the energy states E_n at infinite time and then construct a time-dependent picture for the motion on the fs time scale, an opaque process unnecessary for thinking of experiments dealing with the dynamics of femtosecond atomic and molecular motions.

Recent Developments

A. Dynamics of Highly-Excited Systems: Rydberg Electrons

Highly-excited molecules have large Rydberg orbitals and the interaction of the electron with the nuclear motion leads to fascinating dynamics. For example, the electron may "crash" into the core valence state and bond breakage may take place. These processes are model systems for transition state dynamics and for dissociation on a potential of two or more coordinates. We have recently initiated studies on a number of systems where chemical reactions proceed from Rydberg states. Examples include:

Methyl iodide $\rightarrow CH_3 + I$

Cyclobutanones and cyclopentanones $\rightarrow CO +$ fragments

The high excitation energy required ~ 8-10 eV is achieved by multiphoton excitation. Probing in these experiments is typically achieved by ionizing the molecule from the Rydberg state with a femtosecond pulse. Single photon excitation is possible by doubling and tripling the frequencies before interaction with the molecules (see e.g., J. Hepburn paper in this book).

B. Femtosecond Control of Reactions

The ability to probe nuclear motions on the fs time scale raises another interesting question: can the timing (and phase) of these localized excitations be used to control the outcome of a chemical reaction? In a recent publication from this group this idea was first demonstrated on the reaction,

$$Xe + I_2 \rightarrow [Xe^{+} \cdot\cdot I^{-} \cdot\cdot I]^{\ddagger} \rightarrow XeI + I. \qquad (2)$$

The yield of the product XeI could be controlled by varying the timing between a pair of femtosecond pulses. The key here is the ability to catch the wave packet in the course of the motion at a given configuration and induce the reaction by controlling the timing.

C. *From Elementary to Complex Systems*

One novel feature of fs resolution is the ability to resolve the primary event(s) of the motion even in complex systems. Therefore, one should be able to observe the dynamics as the wave packet's initial coherence decays due to the large (almost infinite) number of molecular modes. Localization of energy in certain modes and resonance phenomena may be observed, and one can directly examine the nature of the reaction coordinate. Studies of this type have recently been made for systems undergoing proton transfer and for isomerization reactions.

D. *Reactions in Micro- and Macro- Clusters, and in Solutions*

En route to the condensed phase, clusters offer an interesting opportunity for studying solvation dynamics under conditions of controlled size and density. In contrast to confined systems, reactions in solutions exhibit dynamics in an "open" system under the influence of a large number of incoherent collisions. We have addressed the issue of solvation on the fs and ps time scale by examining micro- and macro- clusters, and solutions of the same composition at different pressures. One such example is the breakage of the iodine bond in the presence of one (or a few) argon atoms or when I_2 is caged in a cluster of 100 or more Ar atoms. Comparison with fs studies of I_2 in dense solutions of Ar was also made. The nuclear dynamics have therefore been studied from the isolated system limit out towards the condensed phase limit.

E. *Ultrafast Diffraction and Molecular Structure*

Spectroscopic probing of atomic motion is based on changes of the spectra with time. In the reaction $A+BC \rightarrow AB+C$, the spectrum of BC is different from that of the $[A{\cdot\cdot}B{\cdot\cdot}C]^{\ddagger}$ transition species or the AB product. It is therefore possible to observe the motion by tuning the probe to transitions of the AB product or to those of the transition states. It is impossible to image directly atomic motions because the wavelength of the photon (typically ~ 6000 Å) is much longer than internuclear distances. With electrons, however, the wavelength is given by

$$\lambda_{de\ Broglie} = \frac{h}{m_e v}, \tag{3}$$

or with the relativistic correction

$$\lambda_{de\ Broglie} = h\,[2m_e E + E^2/c^2]^{-1/2}, \tag{4}$$

where m_e is the mass, v_e its velocity, and c is the speed of light. When accelerated to kiloVolt energies, λ may be made much smaller than the molecular size, reaching ~ 0.05 Å.

Our idea was to generate ps and fs electron pulses and image the nuclear motion directly with electron diffraction. This technique is quite general and now makes it possible to determine structures on an ultrashort time scale. The electron diffraction images published recently from this laboratory utilize a newly designed two-dimensional CCD camera capable of single electron detection. From the images, one can obtain the bond lengths of the system under study directly.

The field of femtochemistry continues to provide exciting and new developments, and research groups (in theory and experiment) around the world are making important contributions. As Porter has recently pointed out, we have now reached the "limit of chemistry". With fs resolution, the passage of fundamental reactions through a transition state may be observed and studied. Is there a point in going to the attosecond? The answer may be yes, as we discussed in one of the review articles below, but there is plenty to do in this age of femtosecond chemistry.

Acknowledgement

This work was supported by a grant from the U.S. Air Force Office of Scientific Research and the National Science Foundation; Contribution No. 8733.

Some Review Articles

1. A. H. Zewail, *Science* **242** (1988) 1645.
2. A. H. Zewail and R. B. Bernstein, *Chem. & Eng. News* **66** (November 7, 1988) 24; see updated version in Ref. 9.
3. L. R. Khundkar and A. H. Zewail, *Ann. Rev. Phys. Chem.* **41** (1990) 15.
4. M. Gruebele and A. H. Zewail, *Phys. Today* **43**(5) (1990) 24.
5. A. H. Zewail, *Scientific American* **263**(6) (1990) 76.
6. I. W. M. Smith, *Nature* **343** (1990) 691.
7. A. H. Zewail, *Faraday Discuss. Chem. Soc.* **91** (1991) 207.
8. J. M. Thomas, *Nature* **351** (1991) 694.
9. G. Porter, in *The Chemical Bond: Structure and Dynamics*, ed. A. H. Zewail (Academic Press, Inc., Boston, 1992) p. 113.
10. J. C. Polanyi, *ibid*, Ref. 9.
11. A. H. Zewail, M. Dantus, R. M. Bowman, and A. Mokhtari, *J. Photochem. Photobiol. A: Chem.* **62/3** (1992) 301.

References to Recent Work from This Group, Relevant to This Talk and Not Covered in The Above Review Articles

Highly-Excited Systems

R. M. Bowman, M. Dantus, and A. H. Zewail, *Chem. Phys. Lett.* **174** (1990) 546; M. Dantus, M. H. M. Janssen, and A. H. Zewail, *Chem. Phys. Lett.* **181** (1991) 281; M. H. M. Janssen, M. Dantus, and A. H. Zewail, *J. Chem. Phys.*, to be published; L. Bañares, S. Pedersen, and A. H. Zewail, to be published.

Femtosecond Laser Control of Chemical Reactions

E. D. Potter, J. L. Herek, S. Pedersen, Q. Liu, and A. H. Zewail, *Nature* **355** (1992) 66; see also J. J. Gerdy, M. Dantus, R. M. Bowman, and A. H. Zewail, *Chem. Phys. Lett.* **172** (1990) 99.

From Elementary to Complex Systems

J. L. Herek, S. Pedersen, L. Bañares, and A. H. Zewail, *J. Chem. Phys.*, in press; see also L. Bañares, A. A. Heikal, and A. H. Zewail, *J. Phys. Chem.* **96** (1992) 4127; S. Pedersen, L. Bañares, and A. H. Zewail, *J. Chem. Phys.*, to be published.

Reactions in Micro- and Macro- Clusters, and in Solutions

D. M. Willberg, M. Gutmann, J. J. Breen, and A. H. Zewail, *J. Chem. Phys.* **96** (1992) 198; M. Gutmann, D. M. Willberg, and A. H. Zewail, *J. Chem. Phys.*, in press; E. Potter, Q. Liu, and A. H. Zewail, Chem. Phys. Lett., to be published; Y. Yan, R. M. Whitnell, K. R. Wilson, and A. H. Zewail, *Chem. Phys. Lett.* **193** (1992) 402; A. H. Zewail, M. Dantus, R. M. Bowman, and A. Mokhtari, *J. Photochem. Photobiol. A: Chem.* **62/3** (1992) 301.

Ultrafast Diffraction and Molecular Structure

J. C. Williamson, M. Dantus, S. B. Kim, and A. H. Zewail, *Chem. Phys. Lett.* **196** (1992) 529; J. C. Williamson and A. H. Zewail, *Proc. Natl. Acad. Sci.* **88** (1991) 5021.

HIGH RESOLUTION SOFT X-RAY SPECTROSCOPY OF MOLECULES, SURFACES AND SOLIDS

A. M. BRADSHAW
Fritz-Haber-Institut der Max-Planck-Gesellschaft,
Faradayweg 4-6, D-W-1000 Berlin 33, Germany

ABSTRACT

Recent advances in monochromator design and storage ring performance have led to significant improvements in spectral resolution in the soft x-ray region. First results also show that the use of such monochromators on undulators can give even higher spectral resolution combined with higher photon flux. Some of the new possibilities opened up by this development, particularly in spectroscopy with soft x-rays, form the subject of this article.

The soft x-ray region of the electromagnetic spectrum (conventionally 100 - 1000 eV photon energy) covers the K shell excitation thresholds of the first row elements as well as the L shell thresholds of the elements Si to Ca and the first transition metal series. Radiation in this wavelength range is thus of particular importance for molecular spectroscopy, surface chemistry, microscopy and surface analysis, polymer science, solid state physics and materials research[1]. Despite the fact that dedicated synchrotron radiation sources have been available for some years, many experiments in these areas could not be performed optimally - and in some cases not at all - due to low spectral resolution and/or insufficient photon flux. However, developments in grazing-incidence grating monochromators as well as concomitant advances in storage ring performance have recently led to substantial improvements in **resolution** in the soft x-ray region. Taking a simple figure of merit, the $E/\Delta E$ resolution attained at the N K-edge ($h\nu = 410$ eV) has increased from a value of $\sim 5 \times 10^2$ to values exceeding 5×10^3.

A simple example illustrates this point rather vividly. In 1938 Magnusson measured the near-edge N1s absorption spectrum of the molecule N_2O using an x-ray tube and spectrograph[2]. He resolved the two chemically shifted $N1s \rightarrow \pi^*$ resonances and observed excitations into Rydberg states and the continuum (without actually being able to assign any of the features). The spectrum taken with synchrotron radiation in the 1970's showed, of course, a substantial improvement in the signal-to-noise ratio but no significant increase in spectral resolution[3]. Only in the last two years has the N_2O absorption spectrum been measured at an instrumental resolution sufficient to resolve all the details of the excitations below the absorption edges[4]. In fact, high resolution monochromators are now used at several synchrotron radiation facilities for photoabsorption and photoionisation studies of atoms, molecules, surfaces and solids in which the instrumental resolution is less - or at least comparable to - the natural linewidth. This high performance has been employed not only to resolve the complex spectra due to excitations into valence and Rydberg states (including vibrational fine structure) below the K absorption edges of simple molecules, but also to investigate substrate core level shifts in surface chemistry. In the latter case, which is only one of many examples in the surface/solid state area, diagnostic studies of such chemical shifts as a function of adsorbate coverage and temperature can give valuable information on surface structure.

Substantially increased photon **flux** at high resolution becomes possible when such monochromators are used in conjunction with insertion devices (wigglers or undulators). These linear, periodic magnetic structures are placed in the straight sections of a storage ring causing the electron or positron beam to oscillate about its mean position and to produce an intense radiation beam in the forward direction. In the undulator limit (maximum angle of electron beam deflection $<$divergence of emitted radiation) a very substantial increase in spectral brilliance is achieved compared to the normal synchrotron radiation emitted by bending magnets. As a result of interference the radiation is concentrated in narrow bands (harmonics) the energy of which can be varied by changing the pole gap of the magnetic structure. On storage rings built to cover the soft x-ray range with bending magnet radiation (electron energy 0.5 - 0.8 GeV), the first harmonic of a typical undulator does not extend beyond a photon energy of $\sim$ 100 eV. To cover the soft x-ray range with undulator radiation, storage rings with energies of 1.5 - 2.5 GeV are required. The increase in spectral brilliance produced by such an undulator can - by proper matching of the monochromator - be used to increase photon flux or to optimise resolution, or to do both. Alternatively, for experiments requiring spatial resolution it is possible to achieve smaller focussed spot sizes than has been hitherto possible.

So far two high resolution grazing-incidence monochromator/undulator combinations for the photon energy range above 100 eV are in routine operation on electron storage rings[5,6]. A major advance in this area will occur when the so-called third generation storage rings for the soft x-ray region come into operation. These will have typically twelve straight sections so that the various undulators provide the primary sources of radiation. The storage ring optics will have a very low electron beam emittance (product of cross-section and divergence) in order to achieve maximum spectral brilliance of the emitted radiation. At the time of writing the ALS in Berkeley is due to begin operation within the next few month; ELETTRA in Trieste, MAX II in Lund and the recently funded BESSY II in Berlin will follow.

Without going into great detail it is appropriate to mention the general problems associated with soft x-ray grating monochromator design and to describe briefly the recent breakthroughs which have led to the significant increase in spectral resolution referred to above. (See also the paper of Peatman in this volume[7]). Firstly, we note that grating monochromators are still necessary in this spectral region. The lowest photon energy at which crystal monochromators with a naturally occurring diffracting material (beryl) can be used is 820 eV. Synthetic multilayer structures or organic crystals do not give the necessary resolution; the latter are also strongly susceptible to radiation damage and often contain unwanted absorption edges. Secondly, in contrast to lower photon energies, near-to-grazing geometries are required for both gratings and mirrors in soft x-ray monochromators in order to maintain high reflectivity. Thus, optical aberrations (even for perfect optics), the effect of finite slit dimensions and figure errors of mirrors and gratings become more serious than in the normal-incidence region. Moreover, the problem of matching the monochromator to the source also becomes critical. Liouville's theorem tells us, for instance, that we cannot radically reduce the image size at the entrance slit in the dispersion plane without concomitantly increasing the angular acceptance.This in turn would require a large grating in a grazing geometry if over-illumination is to be avoided.

Recently, two monochromator types have been successful in producing a particularly high spectral resolution ($E/\Delta E \sim 10^3$-10^4). The plane grating monochromator (PGM) in the SX-700 version described by Petersen[8] is based on the concept of a fixed virtual source focussed onto a fixed exit slit by an ellipsoidal mirror (Fig. 1). The plane mirror and the plane grating produce out of the real source a virtual source in one of the two foci of the ellipsoid. The photon energy is scanned by rotating the grating and the plane mirror such that the monochromatic image of the source remains fixed in the other focus of the ellipsoid, S1, where

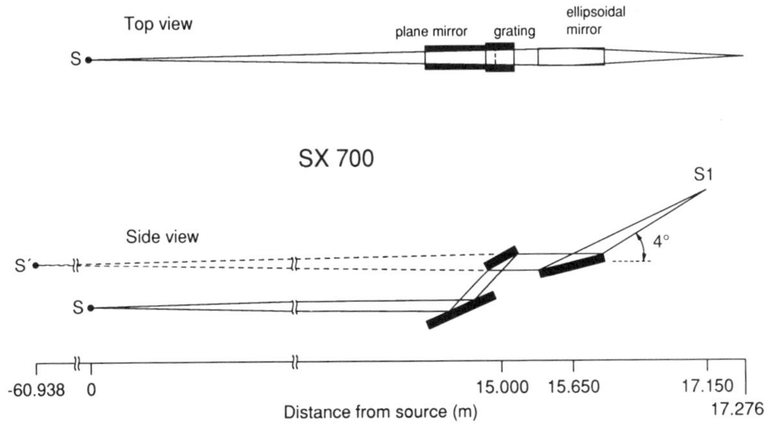

Fig. 1: Layout of the SX-700 monochromator at BESSY. After Petersen[8].

the exit slit is located. Resolution is normally limited by the vertical source dimensions (since there are no entrance slits) as well as by the figure error of the ellipsoidal mirror. The recent drastic improvement in resolution with this type of monochromator at BESSY has been achieved by (i) placing it at a point on the storage ring where the vertical source size is smallest, (ii) by reducing further the vertical source size with special machine optics and (iii) by employing an ellipsoidal mirror with a figure error of ± 0.65 arcsec rms. Clearly, the high accuracy production of aspherical surfaces is extremely difficult and has limited the resolution on the first SX-700 installed at BESSY[9]. The calculated resolution of the second instrument ($E/\Delta E \sim E^{-1/2}$) together with some data points is shown in Fig. 2[10]. Note that further improvements in resolution can be made by shadowing off parts of the ellipsoidal mirror, thereby effectively reducing the figure error at the cost of flux. An alternative version of the SX-700, in which the ellipsoidal mirror has been replaced by two plane-elliptical mirrors for separate horizontal and vertical focussing, is in operation at Lund[11]. The main advantages of this arrangement are a rectangular image at the exit slit and a larger horizontal acceptance.

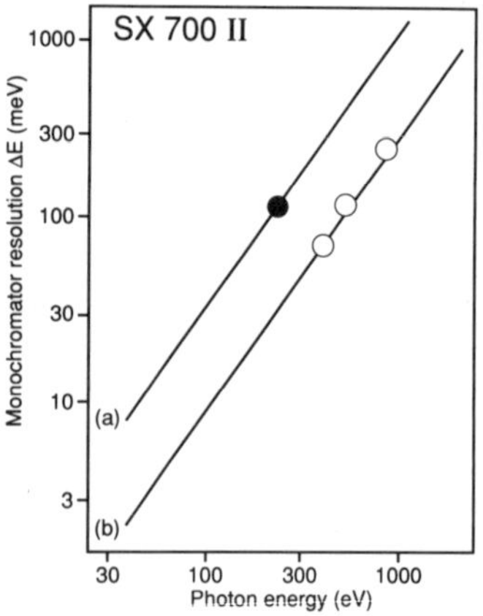

Fig. 2: Calculated resolution of the SX-700-II monochromator at BESSY (Freie Universität Berlin) together with experimental data points; 1221 lines mm^{-1} grating; a) normal operation of storage ring and monochromator, b) second order, "small source" mode of storage ring and 90 % of the ellipsoidal mirror shadowed in order to increase resolution. After Domke *et al.*[10].

Although the spherical grating monochromator (SGM) has also been around for many years, the recent breakthrough achieved by Chen and Sette[12] largely results from simplicity of design. Here, the monochromator itself consists of entrance slit, spherical grating and translating exit slit, as shown in Fig. 3; there are no mirrors which might degrade the resolution. This is actually the version of the Chen SGM on the X1B undulator beam line at Brookhaven. The original version - the DRAGON - is on the UV ring at this facility. Theoretical resolution curves are shown in Fig. 4. With 5μm entrance and exit slits the resolution is al-

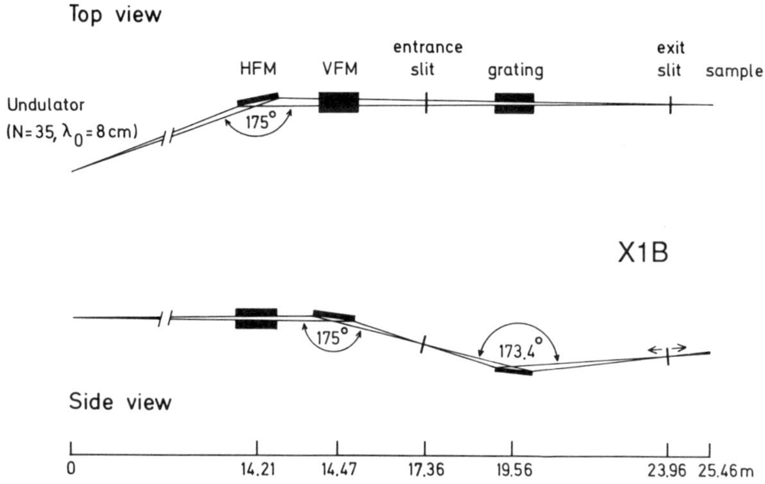

Fig. 3: Layout of the X1B SGM at Brookhaven. After Chen and Sette[12] and Randall *et al.*[6].

most entirely determined by the figure error of the grating which is of the order of 0.5arcsec rms. Note that there are also stringent requirements on the figure error of the vertically focussing pre-mirror so that flux losses are minimised when the entrance slits are narrowed down (5 - 10 μm) for the highest possible resolution. Because of the problem of measuring resolution generally in the soft x-ray region, only some very conservative estimates from experiments on X1B are included in Fig. 4.The high performance of the SGM's at Stanford[13] and Photon Factory[5] should also be noted at this point.

Because of the simple optical design and the use of an entrance slit the ultimate resolution of an SGM of the Chen-type is probably higher than that of the SX-700. On the other hand, its disadvantages include (for practical reasons) the translation of the exit slit as well as the changing divergence as a function of photon energy and the relatively strong contribution of higher orders. The latter can be suppressed on the SX-700 by a suitable combination of mirror and grating angles, albeit at the cost of resolution and/or flux. The SX-700 also enables a very wide spectral range to be covered with one grating. Despite their relative advantages and disadvantages it is, however, true to say that the recent realisation of the SGM and PGM designs under almost optimal conditions has given a very strong boost to synchrotron radiation research, in particular to spectroscopy with soft x-rays. Of the many examples that could be used in this article, the emphasis in the following will be on those where high resolution has been the decisive factor in obtaining new, important information on the particular physical phenomenon or chemical system under investigation.

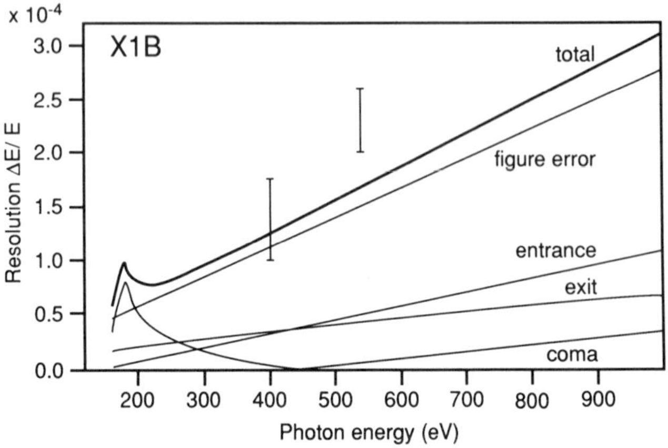

Fig. 4: Calculated resolution of the X1B SGM for 5μm entrance and exit slits; 0.5 arcsec figure error on the 800 lines mm^{-1} grating. Vertical bars represent resolution measurements using the K-edge absorption spectra of N_2 and O_2; the uncertainty is due to the unknown natural linewidth. After Randall *et al.*[6].

As a first example, Fig. 5 shows the C1s absorption spectrum of the CH_4 molecule[14] which reveals considerably more detail than previous EELS data. Due to lack of space the corresponding CD_4 spectrum, which is useful for assignment, has been omitted. For the same reason only a brief description of some of the more interesting features will be given here. The first small feature is the C1s-3s(a_1) Rydberg transition which is formally forbidden, but obtains intensity via vibronic coupling to the C1s-3p(t_2) transition at higher energy. Vibrational fine structure due to the ν_4 mode of t_2 symmetry which couples the two electronic states is discerneable. The complex spectral profile between 287.6 eV and ca. 289.0 eV is associated with the C1s-3p(t_2) transition. Because of the threefold degeneracy and the two t_2 modes (ν_3 and ν_4) the excited state constitutes a complicated Jahn-Teller system which leads to a large manifold of vibronic final states. This is confirmed by comparison with CD_4. The transitions above 289 eV correspond to higher members of the ns and np Rydberg series. Calculations by Schirmer *et al.*[14] indicate that the higher np Rydberg excitations contain considerable C1s-2t_2 valence character.

Another example of vibronic coupling is provided by the high resolution C K-edge absorption spectrum of C_2H_4. Figure 6 shows the complete spectrum below threshold from Ma *et al.*[4] as well as a close-up of the C1s-π* band[15]. In C_2H_4 the two equivalent C1s hole sites give rise to two nearly degenerate excited states,

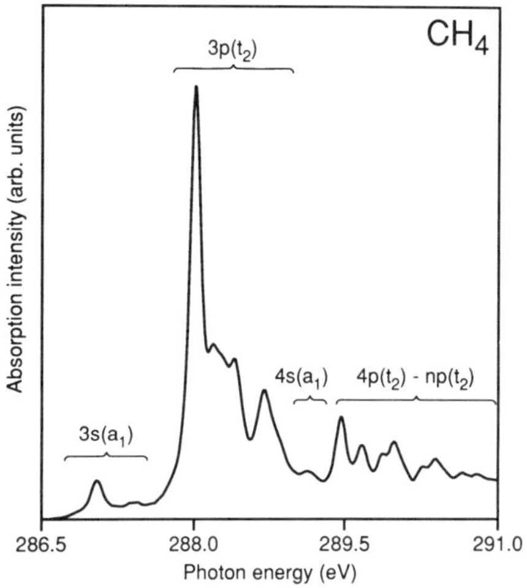

Fig. 5: The carbon 1s photoabsorption spectrum of CH_4 in the region below the ionisation threshold taken on the Dragon monochromator at Brookhaven. After Schirmer *et al.*[14].

B_{2g} and B_{1u}, the former of which is dipole-forbidden. Because of their near-degeneracy the two states can, however, couple vibronically via the anti-symmetric C-H stretching and bending modes ν_{11} and ν_{12} which belong to B_{3u}. Thus, the vibrational fine structure is expected to contain a contribution from these modes as well as from the totally symmetric modes. This form of symmetry breaking via an asymmetric in-plane distortion is a characteristic effect in the excitation of equivalent core systems and provides a mechanism for dynamic localisation of the core hole. Using the approach described in Ref. 15 *ab initio* calculations have been performed to determine the vertical electronic excitation energies, the dipole-transition moments and the coupling constants. According to these calculations there is substantial excitation of the symmetric C-H stretch ν_1, the C-C stretch ν_2, the anti-symmetric stretch ν_{11} and the torsional mode ν_4. The calculated band envelope is in good agreement with experiment not only for C_2H_4[15] (Fig. 6) but also for C_2D_4 and $C_2H_2D_2$[16]. The $C_2H_2D_2$ case is particularly interesting because it allows in principle the C1s hole on $-CH_2$ and $-CD_2$ to be distinguished in the experimental spectrum. The question of assignment in the C_2H_4 spectrum of Fig. 6 (top) has recently been discussed by Arvanitis *et al.*[17] who measured the absorption spectrum of the physisorbed species. Due to the fact that the molecule is oriented and that many of the Rydberg/valence features in the energy range 287 - 290 eV survive, they can be assigned by comparing their polarisation dependences with that of the C1s→π* transition.

Vibrational fine structure in the core level photoelectron spectra of small molecules containing C, N and O has also been a subject of some interest for many years[18]. Unfortunately, the spectral resolution with conventional labora-

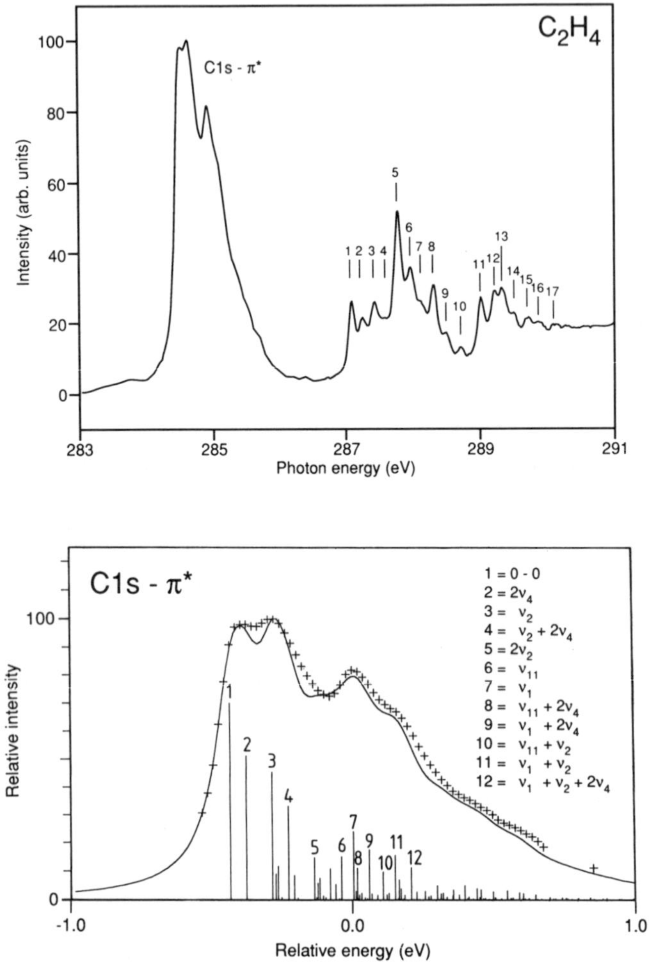

Fig. 6: The carbon 1s photoabsorption spectrum of C_2H_4 in the region below the ionisation threshold (above) and an expanded view of the C1s-π* transition with calculated vibrational fine structure (below). After Ma *et al.*[4] and Gadea *et al.*[15].

tory sources, even monochromatised AlK_α, has rarely been sufficient to resolve the individual lines. The exception is CH_4 where the vibrational spacing (ν_1 mode) is 390 meV[18]. Recent work shows that the new generation of soft x-ray monochromators in conjunction with undulator radiation produces sufficient flux at high resolution ($\Delta E <$ 100 meV) to make such measurements possible (see e.g. the CH_4 photoemission spectrum in Ref. 19). The reader is also referred to the Si2p spectra of SiH_4 and other Si compounds by Bancroft and co-workers[20] at lower energy. A very recent example is provided by the C1s spectrum of CO (Fig. 7)[21] which was taken at 304.2 eV, corresponding to the σ^* shape resonance. The latter perturbs strongly the branching ratios of the individual lines in the vibrational fine structure. Note that the fit requires asymmetric line shapes to account for post-collision interaction, an effect which severely distorts the corresponding zero kinetic energy (ZEKE) photoelectron spectrum[22]. The vibrational energy from Fig. 7 is determined to be 300 $\pm$ 10 meV agreeing with the (Z+1) equivalent core NO ($2\pi^{-1}$) value of 290 meV.

Surface core level shifts produced by the chemisorption of atoms and molecules on clean metal and semiconductor surfaces provide another application of high resolution in the grazing incidence region. Figure 8 shows, for example, Pd $3d_{5/2}$ photoelectron spectra from Andersen and co-workers[23] of the clean Pd{100} surface as well as at three distinct CO coverages, corresponding to well-characterised ordered overlayers. The clean surface spectrum shows a chemical shift to *lower* binding energy associated with the outermost Pd layer. Adsorption of CO induces large changes in the core level emission: intensity is transferred from the low binding energy side to the high binding energy side. Furthermore, increased CO coverage leads to increased intensity at higher binding energy and a splitting into two components. A decomposition of the spectra into bulk and surface peaks is also shown in Fig. 8. In the $(2\sqrt{2}x\sqrt{2})R\,45°$ structure (coverage, θ = 0.5) all the molecules are in equivalent bridge sites giving a single, uniform surface core level shift to *higher* binding energy. The $(3\sqrt{2}x\sqrt{2})R45°$ structure is characterised by heavy domain walls, giving rise to two different types of surface Pd atoms which are coordinated either to one or to two CO molecules. A second chemically shifted peak results at still higher binding energy. The super-heavy domain walls of the $(4\sqrt{2}x\sqrt{2})R\,45°$ structure contain more Pd atoms coordinated to two CO molecules so that the highest binding energy component increases further in intensity. The technique can thus be used - in conjunction with LEED - for surface structural studies. A more recent example from the same group is provided by the adsorption system Al{111}-Na[24]. It is not clear at present to what extent very high resolution ($\Delta E <$ 200 meV) is required for studies on low Z **adsorbate** core levels (C, N and O). For many studies an overall resolution in the range 200 - 500 meV may be adequate, for example to resolve the N1s photoelectron lines in the system Ni{100}-N_2[11,25]. As in the case of near-edge photoabsorption spectra of adsorbates[26], more systematic studies are required.

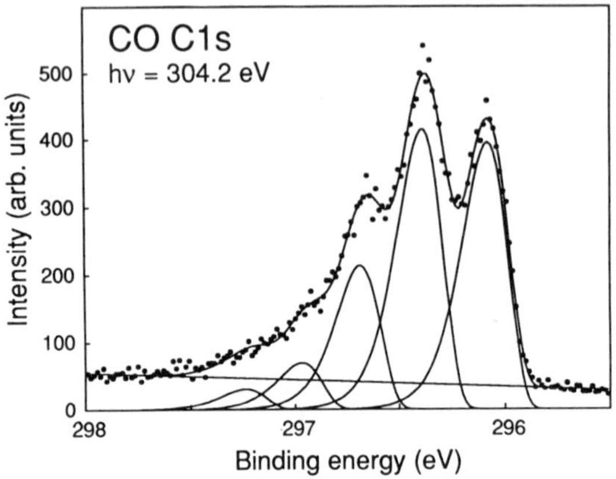

Fig. 7: The C1s photoelectron spectrum of CO (single hole state) at hν = 304.2 eV taken on the X1B undulator beam line. After Randall *et al.*[21].

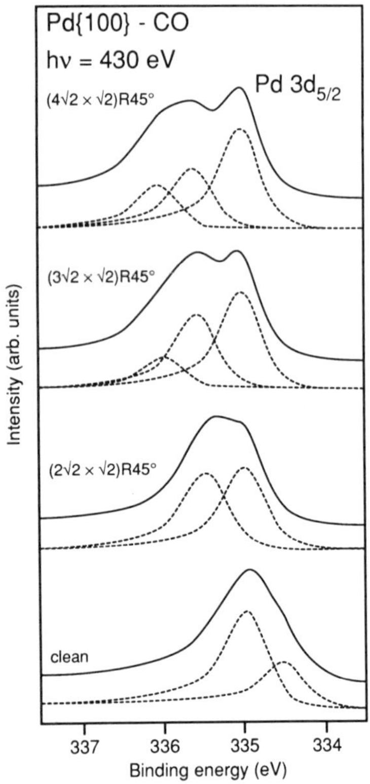

Fig. 8: Pd $3d_{5/2}$ photoelectron spectra from the clean Pd{100} surface and from the same surface covered with three ordered CO overlayers as designated. After Andersen *et al.*[23].

High resolution photoabsorption spectroscopy has provided valuable information on the electronic structure near the Fermi edge in the superconducting copper oxide compounds (e.g. Ref. 27 and references therein). The technique has also recently been applied[28] by Chen *et al.* to the compounds K_xC_{60} found to be superconducting below 19.3 K for x = 3[29]. As shown in Fig. 9, the lowest energy peak

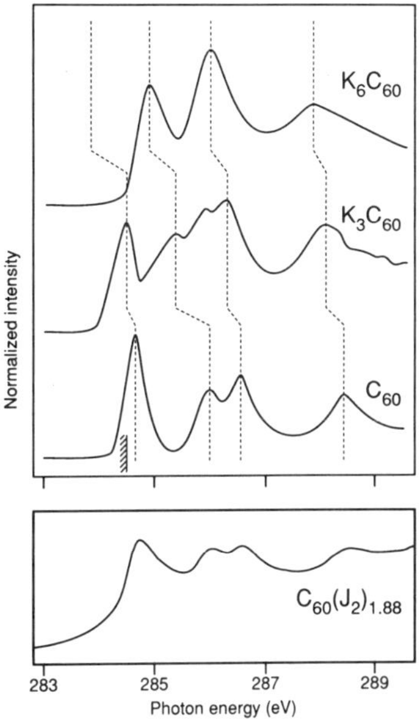

Fig. 9: C1s photoabsorption spectra for C_{60}, K_3C_{60}, K_6C_{60} and $C_{60}(I_2)_{1.88}$. After Chen *et al.*[28] and Werner *et al.*[30].

in the spectrum of C_{60} corresponds to the C1s-LUMO transition. Formation of the K compounds produces pronounced changes in the spectrum. In particular, there are large non-rigid-band shifts towards threshold between the three phases, giving rise to half and complete filling of the LUMO in the K_3C_{60} and K_6C_{60} phases, respectively. The data indicate that there is strong hybridisation of C2p and K states with charge transfer from K to the fullerene. This is not the case in the recently discovered iodine compound where charge transfer might be expected in the reverse direction. The C1s absorption spectrum for $C_{60}(I_2)_{1.88}$[30] (Fig. 9, bottom) shows only a very small rigid band shift and minor changes in relative amplitude compared to C_{60}.

A further application in the area of solid state is the investigation of lanthanides with 4f resonant photoemission. Elemental Ce metal, for example, is characterised by an isostructural γ-α phase transition intimately connected with the strong interaction between the localised 4f electrons and the conduction band states. In normal photoemission at low energies it is very difficult to completely eliminate valence band contributions and obtain the pure f spectrum as well as to separate bulk and surface effects. These difficulties may be avoided by using 4d→4f and 3d→4f resonant photoemission which enhances the 4f signal at photon energies where the cross-sections of the valence band states are low. These experiments have, however, only just become possible because of the availability of soft x-ray monochromators with high resolution. Figure 10 shows resonant pho-

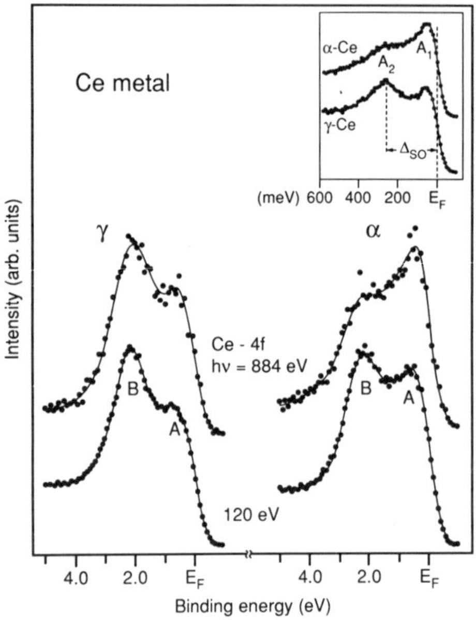

Fig. 10: Resonant photoemission spectra of γ- and α-Ce each taken at the 4d→4f (hν = 120 eV) and 3d→4f (hν = 884 eV) resonances. Inset: high resolution spectra of the A peak at hν = 122 eV. After Weschke *et al.*[31].

toemission spectra of γ- and α-Ce metal each taken at the 4d→4f (hν = 120 eV) and 3d→4f (hν = 884 eV) resonances by Weschke *et al.*[31]. The features A and B are assigned to $4f^1$ and $4f^0$ final states. The A feature actually consists of a spin-orbit split doublet which can be resolved at hν = 120 eV (inset) with an overall instrumental resolution of 40 meV. The resolution has, however, been deliberately worsened in the main spectrum in order to facilitate comparison with the hν = 884 eV spectra. We note that the latter are expected to be much more bulk-sensitive than the 120 eV spectra. For α-Ce a drastic difference is observed in the relative intensity of peak B in the bulk-sensitive and surface-sensitive spectra indicating the presence of a γ-like surface layer. Moreover the "bulk" spectrum of α-Ce now agrees better with theory, showing the expected degree of hybridisation between the 4f and conduction band states.

These various examples of photoabsorption and photoemission experiments demonstrate the power of the new generation of high resolution soft x-ray monochromators. As already demonstrated in one specific case (vibrational fine structure on the C1s photoelectron line of CO), the combination with undulator radiation opens up whole new perspectives that will only be fully realised on third generation sources.

Acknowledgements

The author's own work cited in this short review article has been performed jointly with the group of J. Feldhaus at the Fritz-Haber-Institut, partly in cooperation with L. Cederbaum, C. T. Chen, W. Eberhardt and J. Schirmer. He is also grateful to H. Ågren, J. Andersen, K. Baberschke, G. Kaindl, D. Menzel, A. Nilsson, J. Nordgren and R. Schlögl for useful discussions as well as to W. B. Peatman for a critical reading of the manuscript. Financial support from the German Federal Ministry of Research and Technology (BMFT) through contract numbers 05 5EBFXB 2 and 05 EBFXB 6 is gratefully acknowledged.

References

1. For reviews of recent work performed with synchrotron radiation in the soft x-ray region the reader is referred to the two series *Handbook on Synchrotron Radiation* (North Holland, Amsterdam) and *Synchrotron Radiation Research* (Plenum Press, New York) as well as to the proceedings of recent VUV conferences, e.g. *Physica Scripta T* 31 (1990).
2. T. Magnusson, *Arkiv f. Mat. Astron och Fysik* 26B(1938) 1.
3. A. Bianconi, H. Petersen, F. C. Brown and R. Z. Bachrach, *Phys. Rev. A* 17 (1978) 1907.
4. Y. Ma, C. T. Chen, G. Meigs, K. Randall and F. Sette, *Phys. Rev. A* 44 (1992) 1848.
5. A. Yagishita, S. Masui, T. Toyoshima, H. Maezawa and E. Shigewasa, *Rev. Sci. Instrum.* 63 (1992) 1351.
6. K. J. Randall, J. Feldhaus, W. Erlebach, A. M. Bradshaw, W. Eberhardt, Z. Xu, Y. Ma and P. D. Johnson, *Rev. Sci. Instrum.* 63 (1992) 1367.
7. W. B. Peatman, this Proceedings volume.
8. H. Petersen, *Opt. Commun.* 40 (1982) 402; *Nucl. Instrum. Methods A* 246 (1986) 260.
9. W. Braun, H. Petersen, J. Feldhaus, A. M. Bradshaw, E. Dietz, J. Haase, I. T. McGovern, A.Puschmann, A. Reimer, H. H. Rotermund and R. Unwin, *SPIE Proc.* 447 (1984) 117.
10. M. Domke, T. Mandel, A. Puschmann, C. Xue, D. A. Shirley, G. Kaindl, H. Petersen and P. Kuske, *Rev. Sci. Instrum.* 63 (1992) 80.
11. R. Nyholm, S. Svensson, J. Nordgren and A. Flodström, *Nucl. Instrum. Methods* A 246 (1986) 267; J. N. Andersen, O. Brörneholm, A. Sandell, R. Nyholm, J. Farsell, L. Thånell, A. Nilsson and N. Mårtensson, *Synchrotron Radiation News* 4, No. 4 (1991) 18.

12. C. T. Chen and F. Sette, *Rev. Sci. Instrum.* 60 (1989) 1616.
13. P. A. Heimann, F. Senf, W. McKinney, M. Howells, R. D. van Zee, L. J. Medhurst, T. Lauritzen, J. Chin, J. Meneghetti, W. Gath, H. Hogrefe and D. A. Shirley, *Phys. Scripta* T 31 (1990) 127.
14. J. Schirmer, A. B. Trofimov, K. J. Randall, J. Feldhaus, A. M. Bradshaw, Y. Ma, C. T. Chen and F. Sette, to be published.
15. F. X. Gadea, H. Köppel, J Schirmer, L. S. Cederbaum, K. J. Randall, A. M. Bradshaw, Y. Ma, F. Sette and C. T. Chen, *Phys. Rev. Lett.* 66 (1992) 883.
16. K. J. Randall, H.-M. Köppe, A. L. D. Kilcoyne, J. Feldhaus, A. M. Bradshaw, F. X. Gadea, H. Köppel, J. Schirmer and A. M. Bradshaw, to be published.
17. D. Arvanitis, J. Singh, H. Rabus, T. Lederer and K. Baberschke, *Phys. Rev. B 45 (1992) 1518.*
18. L. Asplund, U. Gelius, S. Hedmann, K. Helenelund, K. Siegbahn and P. E. M. Siegbahn, *J. Phys. B* 18 (1985) 1569.
19. K. Randall, J. Feldhaus, W. Erlebach, A. M. Bradshaw, W. Eberhardt, Z. Xu, Y. Ma and P. D. Johnson, *Synchrotron Radiation News* 4, No. 6 (1991) 16.
20. J. D. Bosek, G. M. Bancroft, J. N. Cutler and K. H. Tan, *Phys. Rev. Lett.* 65 (1990) 2757.
21. K. J. Randall, A. L. D. Kilcoyne, H.-M. Köppe, J. Feldhaus, A. M. Bradshaw, Z. Xu and P. D. Johnson, Y. Ma, J.-E. Rubensson and W. Eberhardt, to be published.
22. L. J. Medhurst, P. A. Heimann, M. R. F. Siggel, D. A. Shirley, C. T. Chen, Y. Ma, S. Modesti and F. Sette, *Chem. Phys. Lett.* 193 (1992) 493.
23. J. N. Andersen, M. Qvarford, R. Nyholm, S. L. Sorensen and C. Wigren, *Phys. Rev. Lett.* 67 (1991) 2822.
24. J. N. Andersen, M. Qvarford, R. Nyholm and J. F. van Acker, Phys. Rev. Lett. 68 (1992) 94.
25. A. Nilsson, H. Tillborg and N. Mårtensson, *Phys. Rev. Lett.* 67 (1991) 1015.
26. W. Wurth, J. Stöhr, P. Feulner, X. Pan, K. R. Bauchspiess, Y. Baba, E. Hudel, G. Rocker and D. Menzel, *Phys. Rev. Lett.* 65 (1990) 2426.
27. C. T. Chen, L. H. Tjeng, J. Kwo, H. L. Kao, P. Rudolf, F. Sette and R. M. Fleming, *Phys. Rev. Lett.* 68 (1992) 2543.
28. C. T. Chen, L. H. Tjeng, P. Rudolf, G. Meigs, J. E. Rowe, J. Chen, J. P. McCanley Jr., A. B. Smith III, A. R. McGhie, W. J. Romanow and E. W. Plummer, *Nature* 352 (1991) 603.
29. A. F. Hebord, M. J. Rosseinsky, R. L. Hadden, D. W. Murphy, S. H. Glarum, T. T. M. Palstra, A. P. Ramirez and A. R. Kortan, *Nature* 350 (1991) 600.
30. H. Werner, W. Bensch, R. Schlögl, Th. Schedel-Niedrig, M. Keil, A. M. Bradshaw and J. Kirschner, to be published.
31. E. Weschke, C. Laubschat, T. Simmons, M. Domke, O. Strebel and G. Kaindl, *Phys. Rev. B* 44 (1991) 8304.

THE PHYSICS OF ELEMENTARY CLUSTERS

J. Friedel

Laboratoire de Physique des Solides, Laboratoire Associé au CNRS
Université Paris Sud, 91405 Orsay Cedex, France

ABSTRACT

A short review is made of our present knowledge of bonding in elementary clusters.

1. Introduction

Clusters are agregates of a few to a few thousands of atoms which constitute a domain intermediary between the gaseous and condensed phases. Classical examples comprise the whole of organic chemistry where carbon clusters have their broken bonds at the surface saturated with H or other atoms or reactants. Also, Guinier Preston zones, such as in aluminum solid solutions, can be considered as small metallic clusters imbedded in partial epitaxy in the surrounding aluminum matrix.

But, in metals, the study of clusters as such only started in earnest with Kubo's study of their electronic structure in 1962 and Ino's first production and structural studies in 1966. Then, activity developed in Europe, with international meetings in Lyon (1976), Lausanne (1980), Berlin (1984), Aix en Provence (1988), Constance (1990), together with many more local meetings The fact that the international meetings took place in the USA in 1992 and will be in Japan in 1994 show the world extension of a field where, however, Europe remains, I think, a leader.

I will restrict myself to *elementary clusters*, made of atoms all of the same element. But I should stress that the chemical resistivity of clusters and especially their potential use in catalysis are an increasingly active field. Also, the study of *clusters of clusters* is very active. Their production covers questions of dendritic growth, sintering, percolation. The study of their physical properties leads to interesting questions of optics and conduction, superconduction and magnetism

Clusters can be produced from the *vapor*. First, very small clusters can be characterized and studied in finite equilibrium concentrations, in dense enough vapors. But a more general way is to produce them in a jet where a vapor is suddenly cooled down. Conditions in jets are usually such that the size distribution of clusters is fairly narrow.

Clusters can also be produced from *condensed phases*. They can be produced, in neutral and ionized forms, by shooting fast ions or a laser beam on a solid target. They can also be extracted in an ionized form from the tip of a tungsten needle covered with a metallic liquid if a high enough electric field is applied.

In all these cases, one can deposit the clusters produced on a solid surface, or select them in mass and charge in an electromagnetic selector, ionize or neutralize them, and study their structure, in jets or as deposits, by electron diffraction.

2. A Simple Case: The Rare Gases

Here, the interatomic potentials can be considered, to a first approximation, as additive and central. The long range Van der Walls attraction and the short range electrostatic repulsion lead to an equilibrium distance a_o which is observed in the diatomic molecule, and also in the equilateral trimer or the regular tetrahedral tetramer. With five atoms, one atom sits on one of the faces of the tetrahedron made by the four other atoms. The long range attraction between the two summits of this triangular bipyramid, at second nearest neighbour distance, compresses the other bonds between nearest neighbors, so that most bonds are shorter than a_o: there is a compromise between long distance attractions and short distance repulsions.

This "frustration" effect increases with the number of atoms in the cluster, and leads to a progressive reduction of the average atomic volume when the number N of atoms increases further. For $N > 4$, the atomic structures are rather complex except for $N = 13$, where one observes an *icosahedronn*: one atom sits at the center of 12 other atoms arranged on a sphere so that each atom has five neighbours on the sphere. This is obviously more stable than nuclei of the face centred cubic or hexagonal close packed structures, where each of the surrounding atoms has only four neighbours on the sphere. But this increase in stability is paid by a (small) distortion, where the radial bonds are shortened with respect to the bonds along the spherical shell.

For $N \leq 13$, the stable atomic stuctures were studied already before the war by nuclear physicists, who thought, mistakenly, that nucleons in nuclei acted as rare gas atoms in a cluster. For larger values of N, studies started in earnest only about 20 years ago. Computer studies were stimulated by electron diffraction studies which were initiated in Orsay by Jean Farges.

Computer studied predicted, for $15 \leq N \leq 50$, complex structures with local icosahedral order, that looked somewhat like liquid or amorphous structures.

For $50 \leq N \leq 1000$, one obtained *"multilayer icosahedra"*, where a central icosahedron is covered by successive layers of close packed atoms. For large enough values of N, these clusters could also be considered as sectors of face centred cubic crystals multiply twinned along close packed faces.

Such multitwins leave an empty sector except if the crystals are stretched somewhat. The corresponding elastic energy increases as the volume of the cluster, while the stabilizing surface energy due to the possibility for all the crystalline sectors ot having close packed surfaces only increases as the surface One then expects a critical size $N_c \cong 1000$ above which a single FCC crystal of cuboctahedron form should be more stable.

This is indeed the succession of structures observed by electron diffraction on jets; and J. Farges' measurements were accurate enough to take into account size distributions and temperature effects. Further studies with mass selected clusters showed that some clusters were especially stable and therefore more numerous and less polarizable. These corresponded to values of N such that all the surfaces were atomically smooth. For multilayer icosahedra, the corresponding *atomic magic numbers* are:

$$N = 1/3\ (1(10n^3 - 15n^2 + 11n - 3) \qquad (1)$$

where n inters an integer. For cuboctahedra N varies also as n^3 for large n's with a similar slope. Near to the value N_c, there is indeed a transitory range of overlap between the atomic magic numbers of multilayer icosahedra and cuboctahedra.

In conclusion, it is clear that even in this simple case, the clusters provide a specific and complex domain, characterized by specific symmetries (such as five fold axes) which are not met either in the simple molecules or the macroscopic crystals. The field is dominated by the need to provide smooth surfaces. Finally, if the nature of bonding and atomic structure is well understood, practically all dynamical aspects have been neglected so far. Very little is known of atomic vibrations, diffusion processes, or martensitic phase changes. As a result, one knows next to nothing about the speed and ways by which a growing cluster can take successive atomic structures. Indeed it was a lucky chance that J. Farges' early diffraction pattern fitted so well computer studies, as the computations referred to equilibrium conditions while the jet experiments were made on dynamical conditions which might have been far from equilibrium !

3. Alkali-Metals

I will restrict myself to the best known and simpler case of alkali-metals, which show some similarities with rare gases.

When two atoms are brought at close distance, they exchange their valence electrons at a rate measured roughly by the transfer energy t. In alkali-metals where electron correlations can be safely neglected, the two atomic valence states $|a\rangle$ and $|b\rangle$ build a bonding state $|a\rangle + |b\rangle$ and an antibonding state $|a\rangle - |b\rangle$ with energies which deviate from the atomic energy by $\pm$ t. Only the bonding state is occupied, leading to a bonding energy t per atom. The length of the bond is obtained by equilibrating this attractive long range energy with a short range electrostatic repulsion, as in rare gases.

When $N > 2$ atoms are brought together, they will similarly exchange their valence electrons, thus build a large molecule. The atomic states broaden into an energy band of delocalized (molecular) states. Within the same approximations, this broadening is symmetrical in energy with respect to the atomic energy and only the lower half of the bond is occupied. Hence a bonding energy per electron (or per atom) given by the half width of the energy band.

When N is large, the distribution of states in the band is quasicontinuous and a good estimate of the width is given by the root square of the second moment of the energy distribution of states. This in turn is proportional in a close packed arrangement to nt^2, where n is the average number of nearest neighbours. Thus the bonding energy per atom is proportional to $n^{1/2}t$ while the repulsive short range one is proportional to n: bonding energies are *not* additive in metals.

A number of consequences follow, in the limit of large N's:

- The atoms must be at close distances, to optimize t.
- The structure must be close packed, to increase n in the volume of the cluster.
- The surface must be smooth and round, to increase n at the surface of the cluster.
- Finally, as bonding increases as $n^{1/2}$ while the short range repulsions increase as n, atoms on the surface have a tendency to have larger interatomic distances than in the volume of a cluster. In particular, the average atomic volume should increase when the size of the cluster increases .

The three first predictions are qualitatively similar to those for rare gases. The last one is just opposite. They are all characteristic of any metallic bonding.

To these general conclusion valid for large N's, one must add specific corrections characteristic of metallic bonds for smaller N's:

i) For *very small* values of N, one must discuss the energy distribution of molecular states case by case. Deviations from close packing can be observed and explained. Thus trimers are made of slightly oblate triangles, tetramers have a planar arrangement ...

ii) For values of N less than about 50, the energy difference between successive molecular states at the Fermi level is larger than k_BT at room temperature or below. As pointed out by Kubo, one should then distinguish clusters with an *odd* number of electrons, where the last occupied state is singly occupied, leading at low temperature to a Curie paramagnetism in T^{-1}, and those with an *even* number of electrons where all occupied states have two electrons of opposite spins, leading to a temperature independent dia-or Van Vleck-paramagnetism. These two tvpes of behaviors were clearly first characterised by Taupin and Charvolin in Orsay on Guisier Preston zones of Li developed in neutron irradiated LiF. Many other experiments on distribution of small clusters or on mass selected ones have confirmed this point, which can be considered as a signature of electron delocalization.

About the same time, Slodzian has also observed *even-odd alternation* in the rate of production of ionised Cu_N^+ cluster by secondary ionic emissions from Cu targets. These were confirmed and extended to many simple metals and alloys, notably by Joyes, who was one of the first to propose that this alternation is due to an alternation in stability, the clusters with full (doubly occupied) spin states being slightly more stable.

In fact the question is made somewhat complex for a number of reasons. From a theoretical side, Joyes himself showed that one has to take into account the change in atomic structure with N for any numerical fit with experiment; also the even-odd alternation in stability decreases and disappears for increasing electron correlations; but it should be observed in all monovalent clusters. From an experimental point of view, altemations in frequencv of appearance seemed more irregular or non existent in jets than in clusters produced by secondary ionic emissions or by electric extraction from needles. Later studies, notably by C. Brechignac in Orsay, have measured directly the energy even-odd alternation by measuring the rate of evaporation of atoms (and of dimers) from clusters as a function of their size, thus confirming Jones' interpretation.

The same kind of jet experiments, which can produce much larger clusters, have shown *electronic magic numbers* above wich the rate of production suddenly decreases and also the stability decreases. This effect, first observed by Knight, can be interpreted as due to the filling of molecular states with high degeneracies. For large enough values of N where clusters can be considered as approximately spherical, the molecular states can be classified by their orbital momentum. Using notations classical in nuclear physics, the states of maximum orbital momentum are $|1,\ell>$ states, i.e. states with no spherical node inside the cluster. The condition for such a state to be just filled is, for fairly large N's,

$$N \cong 1/5\ell^3 \qquad (2)$$

Such an electronic shell structure entends experimentally up to N = 2000, as shown by Martin's careful measurements of the susceptibility of mass selected clusters. Above that critical number, the law for atomic magic number (1) applies, showing that surface atomic smoothness becomes more relevant than the smoothness of electronic wave functions. Icosahedral structures are then observed for a while but the boundary cuboctahedral structures has not been studied in details yet.

When N is somewhat below an electronic magic number, one expects a spontaneous distorsion of the cluster orbital degeneracy that lift part of the orbital degeneracy so that the low energy states are fewer in numbers and can be completely filled. This Jahn-Teller distorsion into an ellipsoid can be followed by observing the plasmon absorption of the cluster, which is unique for a spherical shape but splits into two lines, one twice as strong as the other for ellipsoids of revolution. The gigantic plasmon absorptions characteristics of these metallic clusters have indeed been well known for ages as they provide the color of stain glasses or of an aluminum sheet used in architecture. The Jahn-Teller splitting characteristic of the approach to an electronic magic number is easily visible because of the narrow width of the plasmon lines.

Odd-even alternations, electronic magic numbers, plasmon excitations and Jahn-Teller distortions all show that electrons in metallic clusters act as delocalized fermions, in a way reminiscent of nucleons in nuclei. Indeed other reactions, such as fission of charged clusters or even fusion of clusters, can be produced and studied.

4. Covalent Structures

Electronic correlations, which have been neglected in alkali-metals, play here an increasing role. This can be understood qualitatively in terms of the intraatomic Coulomb interaction U which measures the charge fluctuations:

$$M + M \rightarrow M^{+} + M^{-} - U \qquad (3)$$

with U = I - A, where I is the first ionization potential of atom M, A its electron affinity. U increases systematically with valence in the periodic table, much as Pauling's electronegativity (which is proportional to I + A, with $A \ll I$).

Halides, oxygen, nitrogen, on the right of the table, have very sizeable electronic correlations. It can be shown that in such a case, the most stable arrangement is a loose *packing of dimers* M_2. This is indeed observed in clusters as in macroscopic condensed phases, for O_2 and N_2 as well as for Cl_2. Magic numbers due to multilayer icosahedral structures of molecules, are indeed observed for nitrogen, but also for CH_4 but not for $C0_2$, i.e., in all cases where well defined nearly spherical and weakly interacting *molecules* are produced.

Tetravalent elements present a transitional regime where three types of bonding can occur:

- metallic bonding in close packed (liquid) phases.
- sp^3 covalent (σ) bonding, in saturated structure with four bonds per atom.
- π bonding in unsaturated structures (with less than four bonds per atom).

Clusters of carbon have been most studied. Here, π bonding prevails, because at least partly of stronger electron correlations connected with a larger value of U than in Si or Ge. For N < 1, one observes *sticks* of equidistant carbon atoms, with two π bonds at right angles. For 8 < N < 40, the straight sticks bond into *closed circles*, where the elastic energy of bonding is expanded to saturate the broken bonds at the ends of the stick.

These two regimes appear clearly experimentally in two regimes of secondary ionic emission. For straight sticks, one observes an even-odd alternation of production rate, which is explained in the same way as in monovalent metals, taking into account that each successive C adds two π electrons to a doubly degenerate π band. For circles, one observes a periodicity 4 in production rate, which can be related to a supplementary degeneracy due to delocalization (electrons can turn along the circle in along opposite directions).

For larger value of N, one observes two dimensional clusters which prefigure the development of macroscopic graphitic planes. These *fullerenes* have various closed forms (spherical, ellipsoidal, cylindrical) so as to saturate all possible broken bonds. Their curvature is obtained by replacing in the graphitic plane a number of six fold C rings by fivefold ones. The best known of these is C_{60}, often called 'footballon', where C_6 and C_5 rings alternate as on a football.

If complex fullerenes are the usual constituent of suit, C_{60} and the sticks and circles have only been known for a fairly short time, because their production requires conditions of vacuum. Indeed a number of these structures have been observed and even understood from astrophysical measurements, and such structures probably play a significant role in the description of astronomical clust and of the analysis of possible life in space. If closed circles grow out of straight sticks, it is also probable that fullerenes grow out of closed circles.

5. Conclusion

I have purposedly not discussed the possible role of VUV in studying clusters. But many jet experiments use nowadays laser beam techniques; and many experiments on deposited clusters have used or could use the whole gamme of techniques of synchrotron radiation, from electronic excitations to EXAFS.

I prefer to stress that the new phases observed are very original and complex. Their dynamical properties are poorly known. And finally, this is typically an interdisciplinary field where atomic and molecular physicists and chemists can meet with condensed matter specialists, where astrophysicists talk with specialists of superconductivity.

6. References

1. International Conferences cited above were published by *J. Phys. (Paris)* for Lyon, and *Zeit. Phys. D, Atoms, Molecules and Clusters* for the following ones.
2. P. Joyes, *Les agrégats inorganiques elementaires*, Monographies de Physique, Ed. Phys. (Les Ulis) (1989).
3. J. Friedel, Atomic Bonding in Clusters, in *Elemental and molecular clusters and solids,* P. Martin and G. Pacchioni, Springer Series in Mater., Sciences 6, Springer Verlag, Berlin (1987).
4. J. Friedel, J. *Phys. (Paris)* **48**, 93 (1987); **51**, 2023 (1990); **2,** 959 (1992).

EXPERIMENTS WITH CIRCULARLY POLARIZED RADIATION

BERNT SCHMIEDESKAMP
Universität Bielefeld, Fakultät für Physik,
W-4800 Bielefeld, Germany

Abstract

Circularly polarized radiation is attracting increasing interest for research tasks in physics, chemistry and biology. This is due to the fact, that fundamental properties of the electronic structure as well as magnetic and chiral properties of the studied systems become accessible to experiment, since these properties induce in general a different response of matter to left and right handed circularly polarized radiation. Recent experimental work in this field is reviewed.

1 Introduction

Circularly polarized radiation (CPR) [1] is emitted in bending magnets of storage rings slightly above and below the ring planes [2]. It occurs thus very generally, but experiments with circularly polarized synchrotron radiation were until recently rather the exception than the rule, because monochromators in general do not transmit the high degree of circular polarization to the samples to be studied. Most of the experiments described in this work were done with the 6.5m normal incidence monochromator[3] at BESSY in Berlin, some at the SX700-3[4].

When the circularly polarized radiation interacts with a sample as displayed in Fig. 1 the interaction depends in general on the handedness. Experiments with circularly polarized radiation make use of these dependencies:

The *absorption A* and *photoelectron intensity I* in photoemission can depend on the helicity of the radiation, if the sample is magnetic or if it exhibits a chirality. The *spin polarization* $\vec{P}$ of electrons emitted in the photoionization of free atoms and molecules, in the photoemission from solids and adsorbates as well as in an Auger process is different for left and right handed circularly polarized radiation even for non-magnetic and non-chiral systems.

It will turn out that for all but two of the experiments described in this work spin polarized electrons due to optical orientation, i.e. a spin polarization transfer from the photon onto an electron [5–7] play a key role. We start therefore with a brief chapter about spin polarized electrons. After that topical experiments for each of the types given in Fig. 1 will be described: Electron Spin Polarization (ESP) analysis in the angle- and energy resolved photoionization of free atoms and molecules [8], Circular Magnetic X-ray Dichroism (CMXD) [9] and SPin resolved Extended X-ray Absorption Fine Structure (SPEXAFS) [10] in the absorption of σ-radiation, Magnetic Circular Dichroism (MCD) in the photoemission with σ-radiation [4], electron spin analysis in the photoemission from solids and adsorbates [11–14], Circular Dichroism (CD) [15] in the absorption of σ-radiation, Circular Dichroism in the angular distribution of photoelectrons (CDAD) [16] and finally the first experiments of the spin resolved Auger electron spectroscopy from non-magnetic solids and free atoms [17].

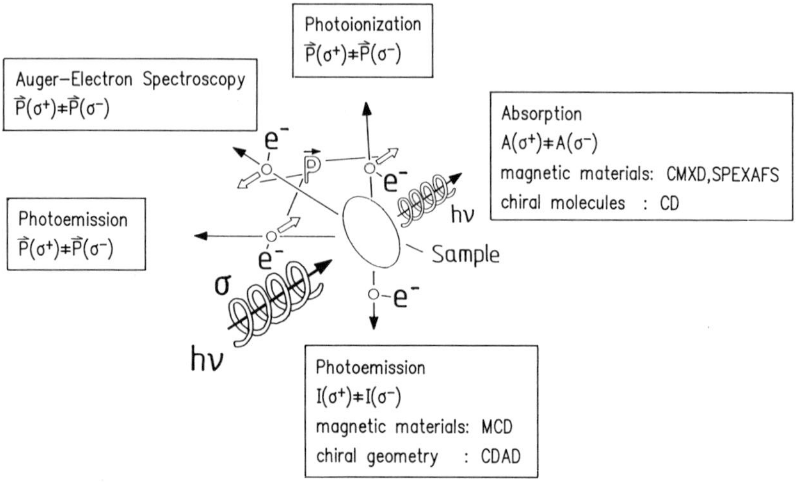

Fig. 1: The different types of experiment with circularly polarized radiation (abbreviations are explained in the text).

2 Experiments

2.1 *Spin polarized photoelectrons with unpolarized and circularly polarized radiation*

Since spin polarized electrons play a key role in most experiments with circularly polarized radiation it is worth remembering that spin polarized photoelectrons can already be obtained with *unpolarized* radiation from unpolarized matter and that this type of spin polarization is also always present in experiments with circularly polarized radiation). This has been well-known from the photoionization of free atoms and molecules for more than a decade [2,8,18–20]. Spin orbit interaction is required and the spin polarization is due to an interference of continuum wave functions. As illustrated in Fig. 2, symmetry considerations demand that the spin polarization vector $\vec{P}$ has only a non-zero component $P_\perp$ perpendicular to the scattering plane of the incident radiation and the emitted electron. $P_\perp$ is zero for $\theta = 0°$, $90°$, $180°$ and $-90°$ and its average over all electron emission directions in space vanishes also.

That a corresponding phase effect exists also for the photoemission from solid surfaces has been predicted [21] and shown experimentally [22] very recently. An example for the normal photoemission from a Pt(100) surface with unpolarized HeI radiation at 21.2eV is given in Fig. 3. The spin polarization vector has only a component $P_\perp$ perpendicular to the reaction plane; the size of $P_\perp$ is considerable as can be seen from the intensities $I_+^\perp = \frac{I}{2}(1 + P_\perp)$ and $I_-^\perp = \frac{I}{2}(1 - P_\perp)$ in the lower part of Fig. 3 and a corresponding spin polarization has been found for a number of other surfaces, materials and photon energies [22]. $P_\perp$ is independent of a rotation ϕ of the crystal about its surface normal (inset of Fig. 3) i.e. its direction is only determined by the direction of the incident photon and the emitted electron but not by crystal symmetries.

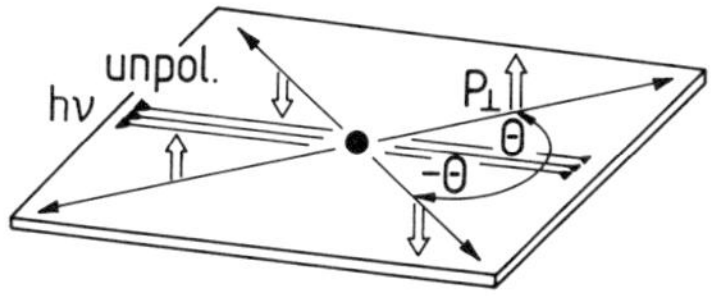

Fig. 2: Spin polarization in the photoionization with unpolarized radiation.

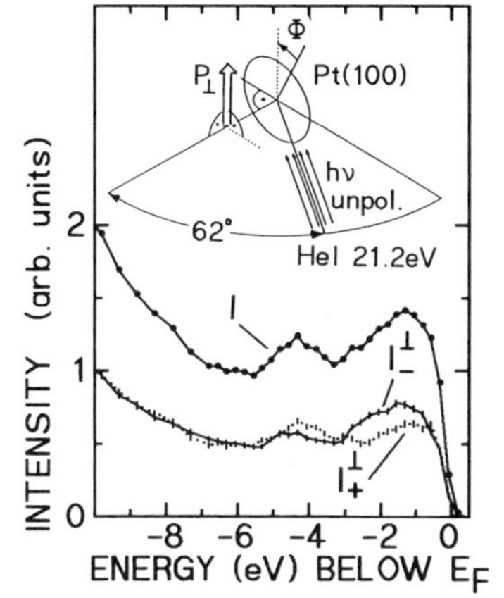

Fig. 3: Spin polarization with unpolarized radiation in the normal emission from Pt(100) (from Irmer et al, ref. 22).

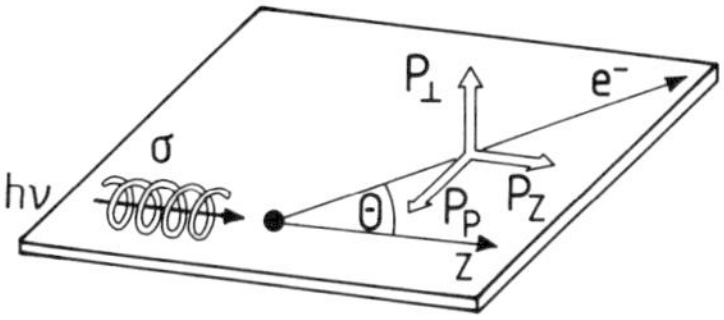

Fig. 4: Spin polarization in the photoionization with circularly polarized radiation.

In the photoionization with *circularly* polarized radiation additional spin polarization is induced by optical orientation, which requires also spin orbit interaction. This mechanism was first recognized by Fano [5] and experimentally verified by Heinzmann et al [6]. Theoretical treatments [19] and experiments [8,23] show that in this case $\vec{P}$ has in general also components P_p and P_z parallel to the reaction plane as displayed in Fig. 4. While $P_\perp$ is independent of the helicity of the radiation, P_p and P_z change their sign when the handedness of the radiation is switched from right to left handed [8,19,23]. Like $P_\perp$, P_p vanishes for $\theta = 0$, 90°, 180° and −90° and its average over all electron emission directions is also zero. P_z, however, is in general non-zero for almost all electron emission directions and has also a non-zero average value $\overline{P}_z$ if all photoelectrons produced are extracted regardless of their direction of emission.

2.2 *Electron spin polarization approaching unity in the photoionization of Tl*

Photoelectron spin polarization values close to unity are usually obtained only for special experimental conditions which require at least selection of a fixed electron emission angle θ. It is an absolute exception that a beam of almost completely polarized photoelectrons occurs for all θ simultaneously. In the photoionization of Tl this exception has been found even twice [24].

As displayed in the upper part of Fig. 5 the ground state of Tl is a $^2P_{1/2}$ state. Ionization of Tl with circularly polarized radiation yields an ionic $^1S_0$ state and a photoelectron which can leave as an $s_{1/2}$- or a $d_{3/2}$-continuum-wave. It can be derived that for two special combinations of the transition matrix elements D_s, D_d and the phase difference $\delta_s - \delta_d$ a photoelectron spin polarization $\vec{P} = 1$ can be achieved for all θ simultaneously:

The first case is shown in the left part of Fig. 5. If $D_d = 0$ the photoelectron is a pure s-wave resulting in $P_z = 1$ for all θ and $\overline{P}_z = 1$.

The second case is illustrated in the right part of Fig. 5. If $D_d = -2\sqrt{2}D_s$ and $\delta_s - \delta_d = 0$ the outgoing s- and d-wave interfere in such a manner that the spin polarization vector $\vec{P}$ remains in the reaction plane and is rotated by 2θ with respect to the light propagation direction for an electron emission angle θ.

That these ideal cases can indeed approximately be achieved in a photoionization experiment is shown by the experimental results in the lower left and right part of Fig. 5. Both results were obtained close to autoionization resonances, where matrix elements and phases show strong variations with radiation wavelength λ: At $\lambda = 148nm$ autoionization yields a drastic increase of the cross section σ for photoionization, which is exclusively due to s-waves and results in $\overline{P}_z = 1$ (bottom left). At $\lambda = 83nm$ P_z varies with Θ as expected for case 2 (bottom right).

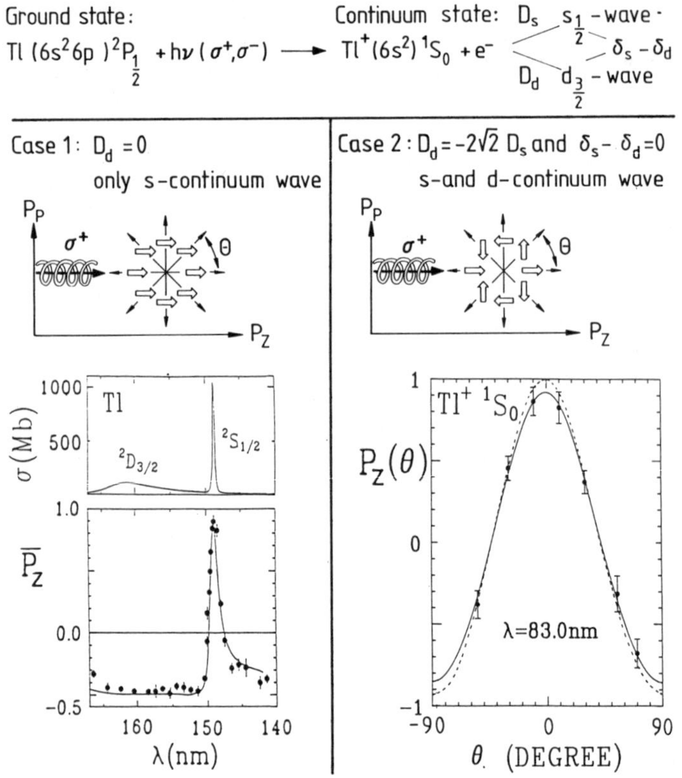

Fig. 5: Top: Reaction equation describing the photoionization of the Tl 6p level with σ-radiation. Bottom: The two cases for which $\overline{P} = 1$ can be achieved simultaneously for all Θ. Left: case 1, experimental result from Müller et al, ref. 24.
The experimental $\overline{P}_z$ data (dots with error bars) are compared with a calculation of Cherepkov (solid line).
Right: case 2, experimental result from Böwering et al, ref. 24.

2.3 *Probing magnetic properties of matter with circularly polarized light: CMXD, SPEXAFS in absorption and MCD in photoemission*

The techniques for probing magnetic samples displayed in Fig. 6 use circularly polarized radiation to produce a spin polarized electron by optical orientation in a transition from a p-level, i.e. exactly as in the photoionization of Tl.

In circular magnetic x-ray dichroism (CMXD)[9], the transitions occur into the unoccupied part of the exchange split valence bands only some eV above the Fermi level. Since the spin-up and spin-down-density of states are different and since left and right circularly polarized radiation yields spin polarized electrons with opposite signs the absorption of circularly polarized radiation is helicity-dependent. From the helicity dependence the local magnetic moments can be deduced.

Local magnetic moments for a number of heavy impurities in iron are given in the upper part of Fig. 7. When going from Os to Ir a change between the ferro-magnetic and anti-ferromagnetic coupling to the Fe lattice is observed [9,25]. Presently different groups [26,27] are working on a separation of magnetic moments into their spin and orbital parts, which yields

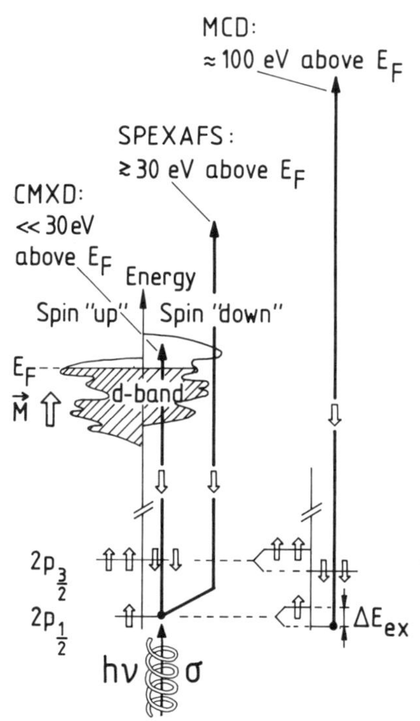

Fig. 6: Probing magnetic properties of matter with circularly polarized x-rays. Explanations are given in the text.

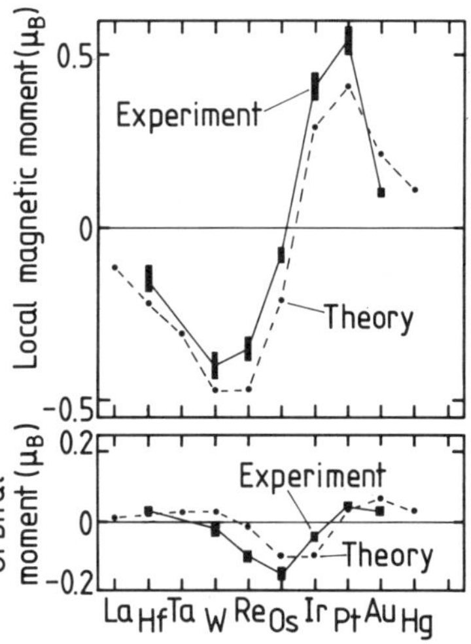

Fig. 7: Local magnetic moments for 5d-impurities in Fe determined from CMXD spectra.
Top: local magnetic moments
Bottom: estimation of orbital moments using the different absorption behaviour for the L_3- and L_2-edges (from Schütz et al [9,25,27].
Theoretical values from Ebert et al, ref. 28).

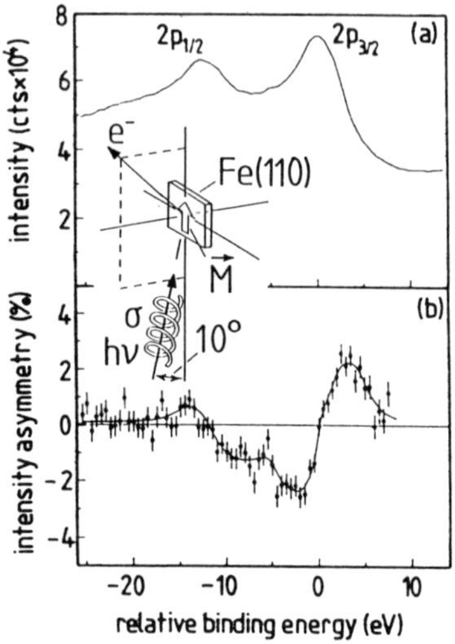

Fig. 8: MCD in the photoemission from the 2p core levels of iron.
Inset: experimental geometry.
Top: total photoelectron intensity I.
Bottom: Asymmetry given by $\frac{I(\sigma^+)-I(\sigma^-)}{I(\sigma^+)+I(\sigma^-)}$
(from Baumgarten et al, ref. 4).

a deeper insight into the interplay of Coulomb and spin orbit interactions, hybridization and crystal fields [26] and is required for a deeper understanding of magnetism.

The experimental [27] and theoretical [28] values of the orbital moments for impurities in iron are given in the lower part of Fig. 7. Orbital moments turn out to contribute considerably to the total magnetic moment and can become even dominant for Os, i.e. close to the point of change between ferro-magnetic and anti-ferromagnetic coupling.

Spin polarized extended x-ray absorption fine structure (SPEXAFS) [10] makes use of spin polarized electrons with kinetic energies larger than about $30eV$. Scattering processes with neighbouring atoms depend on the electron spin polarization and thus also on the handedness of the radiation. Information is gained about the magnetic short range order [10].

In the magnetic circular dichroism (MCD) photoemission experiments described below [4] the photoelectron energies are chosen about $100eV$ above the Fermi level. Dependencies of the photoelectron intensity on the handedness of the radiation can be expected, if - in addition to the spin orbit splitting - an exchange splitting ΔE_{ex} exists, which separates the levels for spin up and spin down electrons energetically. Due to the limited photoelectron range MCD in photoemission is a tool for the study of surface magnetism.

An example for MCD in the photoemission from core levels of iron [4] is displayed in Fig. 8. The sample magnetization $\vec{M}$ is parallel to the surface and the light incidence occurs at grazing angle of 10° with respect to the surface in order to have a large component of $\vec{M}$ aligned with the direction of light propagation. Switching the light helicity from σ^+ to σ^- results in a change of the photoelectron intensity I from $I(\sigma^+)$ to $I(\sigma^-)$ and yields the asymmetry $\frac{I(\sigma^+)-I(\sigma^-)}{I(\sigma^+)+I(\sigma^-)}$ in the lower part of Fig. 8. Reversing the sample magnetization $\vec{M}$ results in the same asymmetries.

Recent results [29] show that MCD in photoemission can even be observed if $\vec{M}$ is normal to the direction of light propagation due to final state selection effects of $\underline{angle - resolved}$ photoemission (see also contribution ThA2, this conference)

2.4 *Spin-, angle- and energy resolved photoemission from solids and adsorbate layers*

Spin-, angle-, and energy resolved photoemission with circularly polarized radiation from nonmagnetic matter has in the past several times been applied to the ideal cases of solid surfaces [11] and the weakly bound almost free atom-like rare gas adsorbates [11,30].

In the last three years a lot of work has been devoted to metal adsorbate layers and intermixed systems on different surfaces [31–40]. The studies include photoemission from d-core levels [31,32] which can be described like free atom photoionization, epitaxial layer systems for different coverages [33,35,37], systems which interact more or less strongly [33–38] with the underlying substrate as well as intermixed [38,39] and magnetic systems [40]. Some of these experiments are described. All results were obtained for the highly symmetrical setup of normal radiation incidence and normal electron emission.

For the photoemission from the d-core levels of a thick unordered thallium layer on Pt the photoemission spectrum in Fig. 9 has been obtained [32]. Due to a spin orbit splitting of $2.3eV$ into a $d_{3/2}$ and a $d_{5/2}$ level two peaks are visible. Spin analysis yields spin polarization values of +50% and −28% (with the background subtraction indicated by the dash dotted line below the peaks) for the photoemission from these two levels as can be seen from the separation of the spectrum into its parts I_+ with electron spin polarization vector $\vec{P}$ parallel and I_- with $\vec{P}$ antiparallel to the photon spin. These are almost exactly the values which are expected in a free atom photoionization experiment for the $5d$ levels of Tl (if spin orbit interaction of the final states is neglected) [32]. In addition a crystal field splitting for the $d_{5/2}$ level of about $0.2eV$ is detected.

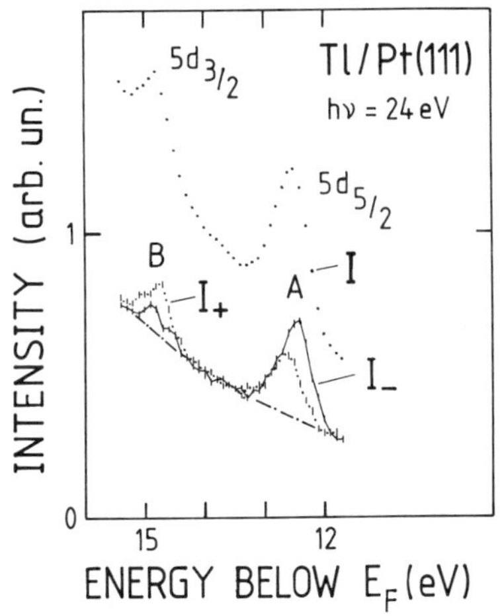

Fig. 9: Spin resolved spectrum for the photoemission from a Thallium adsorbate layer on Pt(111) (from ref. 32).

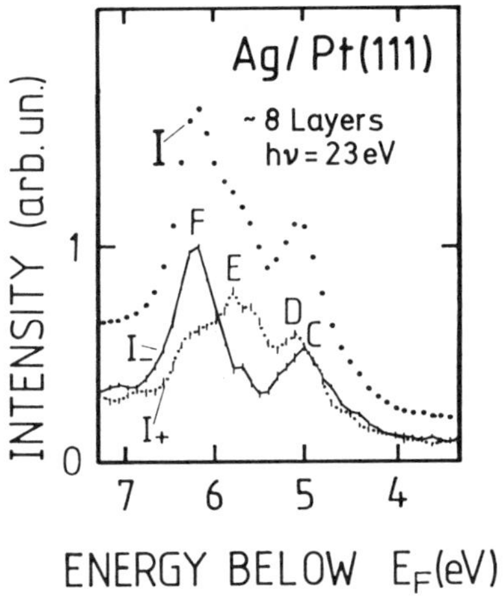

Fig. 10: Spin resolved spectrum for the photoemission from a thick epitaxial Ag-layer on Pt(111) (from ref. 33).

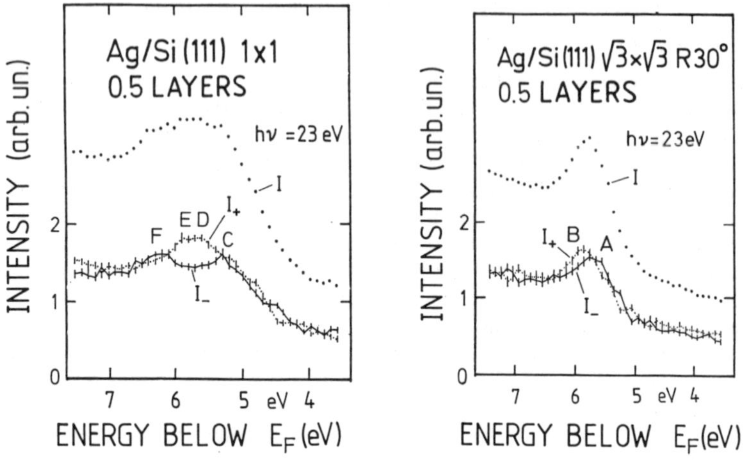

Fig. 11: Detection of Ag valence orbital overlap for Ag adsorbates on Si(111) by means of spin resolved photoemission (from Vogt et al, ref. 36).

Spin polarized photoemission with circularly polarized radiation from an 8-layer epitaxial Ag adsorbate on Pt(111) yields photoemission spectra like the one in Fig. 10 which are identical with those of Ag(111). Spin analysis reveals 4 peaks C,D,E,F with the spin polarization sign sequence minus-plus-plus-minus. They are due to the transitions from the valence bands of Ag. Compared to the photoionization from the d-levels of thallium we have now a large crystal field splitting due to Ag-Ag interaction in combination with a by far smaller spin orbit splitting, which results in the displayed spectrum with 4 peaks. Spin orbit interaction separates the valence band states into wave functions of Λ^3_{45}- and Λ^3_6–symmetry, which yield even completely spin polarized electrons for transitions with circularly polarized radiation into states with symmetry Λ^1_6 [33].

The photoemission from Ag adsorbate layers of different coverages and on different substrates has been studied [33,36] in the same manner as for $Ag(111)$[33,34] and yields almost always with up to now only one exception a photoemission spectrum with 4 peaks and the spin polarization sign sequence minus-plus-plus-minus as in Fig. 10. This indicates that in all of these cases Ag-d-electron overlap is present in combination with spin orbit splitting. The exception is a study of submonolayer Ag adsorbates on the (111) surface of silicon. Room temperature adsorption of 0.5 monolayers of Ag results in an adsorbate with (1 x 1) LEED pattern and the photoemission spectrum with the four Ag peaks, which show the presence of Ag-Ag interaction (left part of Fig. 11). Annealing of the sample at $500°C$ transfers the surface into a $\sqrt{3}$ x $\sqrt{3}R30°$ structure and alters the photoemission spectrum in such a manner that spin polarization analysis reveals only the spin orbit split peak (see Fig. 11, right) as in the photoionization of Tl in Fig. 9. No indication for a presence of AgAg interaction is found. The (mainly geometric) structure of this adsorbate system is still object of a long on-going debate in literature (several references are given in [36]). Structure models which require AgAg interaction can clearly be excluded by the described study.

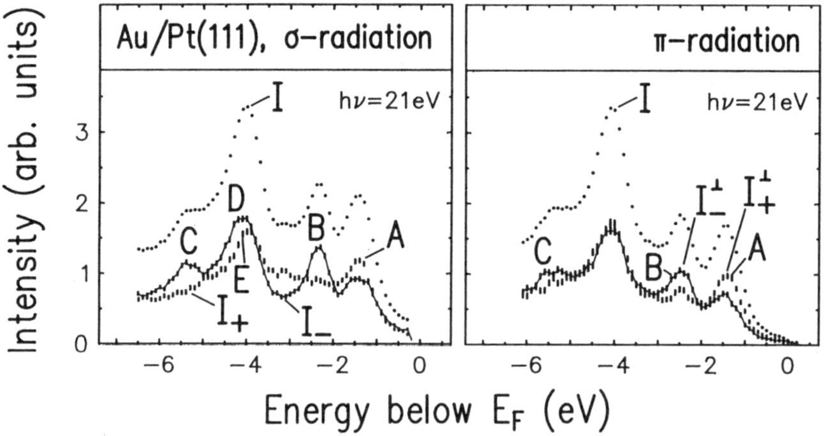

Fig. 12: The surface and interface electronic structure for a Au-monolayer on Pt(111) studied with spin resolved photoemission. Left: Photoemission with circularly polarized radiation. Right: Photoemission with linearly polarized light (from ref. 37). $I_+^\perp = \frac{I}{2}(1 + P_\perp), I_-^\perp = \frac{I}{2}(1 - P_\perp)$.

An understanding of symmetry properties of surface and interface states requires not only adsorbate systems which yield reproducible experimental results; it should also be ideal enough to be treatable theoretically. Au on Pt(111) can be grown epitaxially in a manner that the Au atoms are on lattice sites which are a direct continuation of the Pt lattice [41]. An experimental result for this system is given in the left part of Fig. 12. Spin analysis reveals several peaks which can be attributed to corresponding structures in a calculation of Tamura and Feder [41]. The spectrum consists of contributions from Pt-surface emission with symmetry Λ_{45}^3 [42] (peak A), a Au-induced surface state on Pt(111), which is located in the Au layer and the underlying first Pt layer [41] (peak B), states which are only located in the Au adsorbate layer (peaks C,E) and bulk states of the Pt-substrate (peak D).

The data contribute also to an understanding of a spin polarization effect with normally incident <u>*linearly*</u> polarized light in the normal photoemission from three-fold surfaces [41–43]. Spin polarized electrons of this type are not mentioned in this work up to now, since they occur in photoemission neither with circularly nor with unpolarized light. They require also the spin orbit interaction but they have no analogy in the photoionization of free atoms or the photoemission in 3-dimensional solids. They seem to be generated only in the photoemission from surfaces and interfaces and are predicted to require electronic states of the symmetry type Λ_{45}^3 [43]. An additional interesting property is the fact, that the spin polarization is sensitive to the geometrical structure of surfaces and adsorbate layers [41]. A photoemission spectrum obtained with linearly polarized radiation for the Au monolayer on Pt(111) is given in the right part of Fig. 12. Pronounced spin polarization with $\vec{P}$ perpendicular to the light direction (component $P_\perp, \vec{P}$ rotates by 3ϕ, if the sample is rotated by ϕ about the surface normal) is found in three of the four peaks. A comparison with the corresponding spectrum for σ light (Fig. 12, left) shows that spin polarization is indeed only obtained for photoemission from those states which have both surface or interface location <u>*and*</u> symmetry Λ_{45}^3.

2.5 *Spin polarized electrons due to optical orientation even for electronic states with small spin orbit interaction*

The generation of spin polarized electrons by optical orientation requires spin orbit interaction, which is large for matter with a high atomic number. Therefore most of the previous studies were done with materials of medium and large atomic number [11]. But even in the case of very small spin orbit interaction spin polarized electrons can be generated and yield valuable information about the electronic structure.

A first example is the photoemission from a Si(111) surface [44]. The photoemission spectrum in Fig. 13 shows a peak which is due to transitions from the p-valence bands into a free electron like final band. Spin analysis reveals a spin orbit splitting of the p-valence bands of about $30meV$ and allows a symmetry characterization of the valence bands.

The spin orbit splitting is even smaller in the Π^*-valence levels of NO. The NO $^2\Pi$-ground state is spin orbit split into a $^2\Pi_{3/2}$ and a $^2\Pi_{1/2}$ level with a value of only $15meV$ for the spin orbit splitting ΔE_{so}. At room temperature both fine structure states are significantly populated. Only a small spin polarization is therefore detected in an experiment with an effusive NO beam as can be seen in the left part of Fig. 14. A considerable degree of spin polarization up to more than 10% at $\Theta = 0$ degrees is, however, obtained when the same experiment is performed with an NO supersonic molecular beam (Fig. 14, right), since in this case the $^2\Pi_{1/2}$ ground state level is almost exclusively populated. From the Θ dependence of P_z spin polarization parameters can be determined, which is an important step towards a quantum mechanically complete characterization of the vibrationally resolved molecular photoionization of NO.

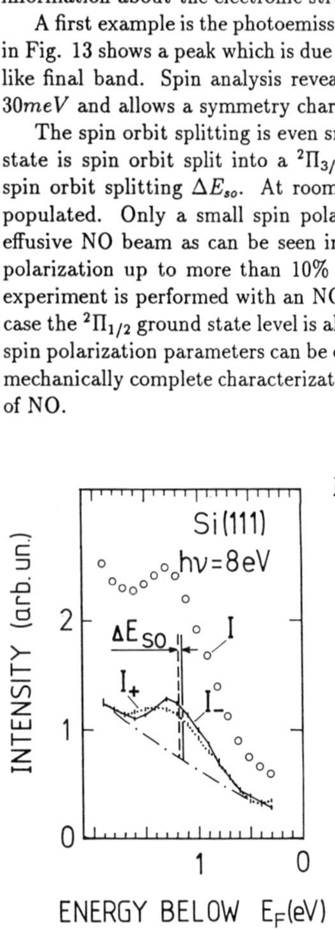

Fig. 13: Spin orbit splitting of the silicon valence band resolved with spin resolved photoemission with σ-light (from ref. 44).

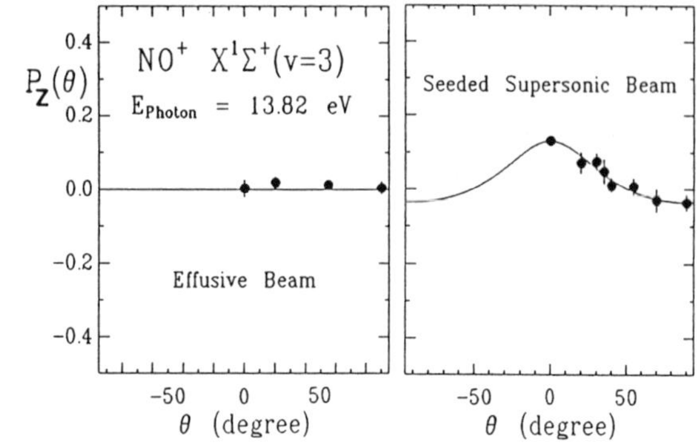

Fig. 14: Spin polarized electrons in the photoionization of NO. from left: an effusive beam, and right: a supersonic beam (from Salzmann et al, ref. 45).

2.6 *Circular dichroism in absorption (CD) and the angular distribution of photoelectrons (CDAD)*

In CD and CDAD the absorption and angular distribution of photoelectrons, respectively, depend on the light helicity, but for both techniques spin orbit interaction is not needed.

Circular dichroism in absorption (CD) has a long history in the visible [46]. It is based on the fact that the absorption of chiral molecules is different for σ^+- and σ^--radiation. The differences are usually small ($\leq 1\%$) since they arise from a combined action of electric and magnetic dipole transitions [46]. An example for the molecules fenchone and camphor is shown in Fig. 15. The two molecules are only different with respect to the position of the oxygen atom and two CH_3-groups. Such small differences can modify the biological and chemical properties of the molecules considerably. For camphor higher absorption A is obtained for σ^--radiation around $6.5eV$ photon energy, fenchone turns out to absorb σ^+-radiation more strongly.

While circular dichroism in absorption requires the handedness of the molecule itself, circular dichroism in the angular distribution of photoelectrons (CDAD) [16] can be observed already, if the whole system which is involved in the photoemission process (consisting of the incident radiation, the electronic orbital and the emitted photoelectron) exhibits a handedness. Differences in the photoelectron intensity for right and left handed circular polarized radiation are due to a phase shift of $\frac{\pi}{2}$ between outgoing continuum waves which is introduced by the σ-radiation in addition to the phase shifts which are already present in the photoemission with unpolarized radiation. This $\frac{\pi}{2}$ phase shift is opposite for σ^+- and σ^--radiation.

Until recently CDAD was experimentally found only for the photoemission from aligned orbitals as also expected from a simple model [16]. It was therefore not clear, whether CDAD can also exist for the photoemission from s-orbitals. A strong CDAD has, however, now been observed [48] in the carbon $1s$ core level photoemission from CO on Pd(111). A typical result is presented in Fig. 16. As shown in [49] a dependence of the photoelectron intensity on the light helicity can be explained even for s-orbitals, if the *final* states of the oriented CO molecule are properly described by symmetry adapted wave functions.

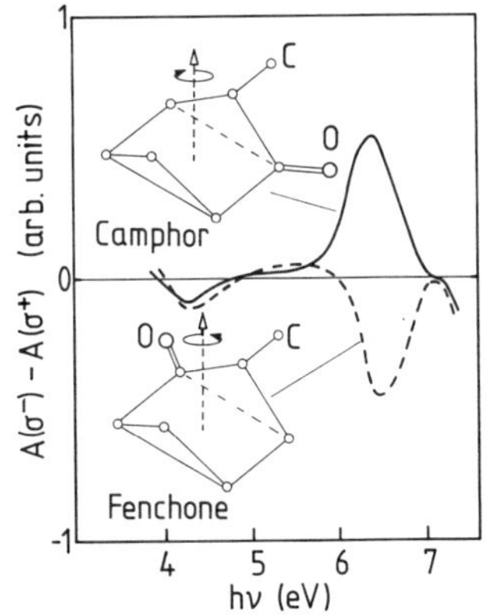

Fig. 15: Circular dichroism (CD) in the absorption of σ-light in chiral molecules (from Lagier, ref. 47).

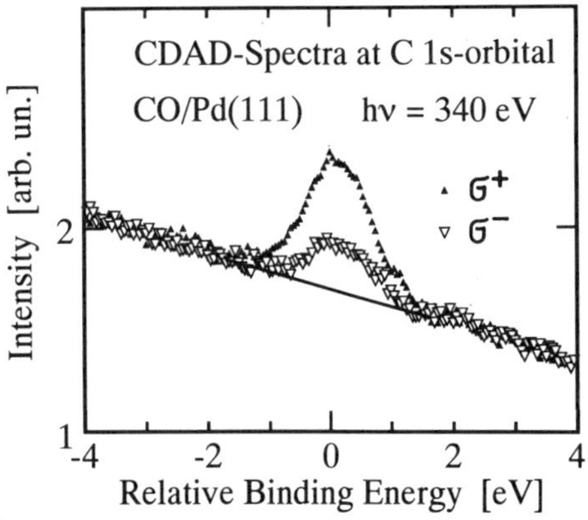

Fig. 16: CDAD for the photoemission from the C-k-core levels of CO adsorbed on Pd(111) (from Bansmann et al, ref. 48).

2.7 *Spin polarized Auger electrons after core hole generation with circularly polarized light*

The photoemission process with circularly polarized radiation produces in general spin polarized photoelectrons due to optical orientation. It generates, however, as illustrated for the case of a Rb metal layer in Fig. 17, also a hole, which can decay via emission of an Auger electron. One expects spin polarization for the hole left behind and the question arises whether and in which manner spin polarization is transferred from the hole state onto the Auger electron in the Auger decay.

The question has recently been addressed to alkali adlayers [17]. A high degree of spin polarization was found for the emitted Auger electrons. Results for the Rb metal layers are given in Fig. 18. A change of sign is observed for the spin polarization at a photon energy of $15.7eV$. This can be explained by the fact that the density of states above the Fermi level changes from dominantly s-like directly above E_F (indicated by A in Fig. 17) to dominantly d-like at higher energies (B in Fig. 17). Taking this into account the results can qualitatively be explained in a simple model: In step 1 a polarized $p_{3/2}$-hole is generated. Its spin polarization sign depends on whether the electron is transferred into s- or d-states. In step 2 the Auger decay is regarded in an atom-like model. Momentum conservation demands then that the spin polarization vector of the emitted Auger electron is opposite to the spin polarization vector of the electron emitted in step 1. Since the spin polarization of this electron depends on the symmetry of the final states also for energies between E_F and the vacuum level, spin polarized Auger electron spectroscopy can be used to deduce symmetry properties of states in this energy range and opens more generally a new method of studying excitations and unoccupied electronic states.

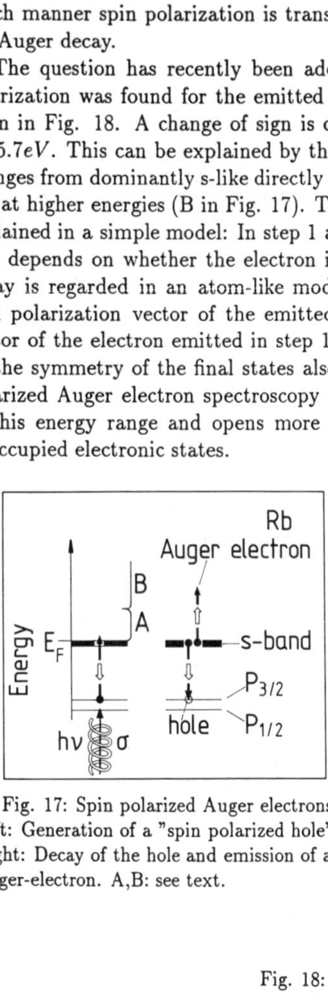

Fig. 17: Spin polarized Auger electrons from Rb layers. Left: Generation of a "spin polarized hole" with σ-radiation. Right: Decay of the hole and emission of a spin-polarized Auger-electron. A,B: see text.

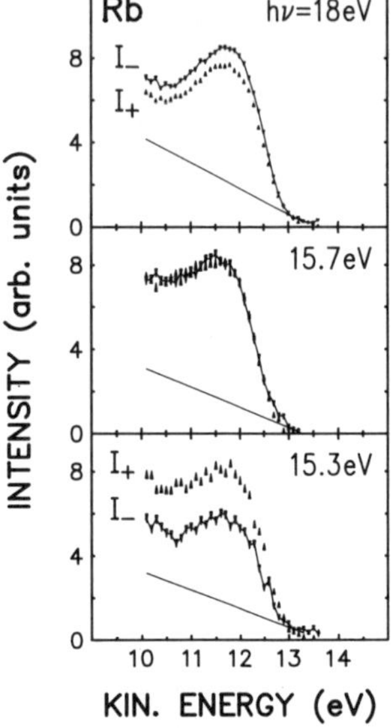

Fig. 18: Spin polarized Auger electrons from Rb layers, experimental result, (from Stoppmanns et al, ref. 17).

3 Future prospects

Future experiments with σ-radiation will certainly profit considerably from wiggler and undulator sources which are presently being put into operation in several synchrotron radiation laboratories and which yield higher fluxes of σ-radiation especially at higher photon energies. Together with further properties such as the tunability of the radiation wavelength and the degree of circular polarization, the ability of a fast change of the radiation helicity and a time structure which allows time resolved studies, one can expect considerable progress. Examples are photoemission and Auger processes involving core levels and even aligned and oriented molecules, time resolved studies of CD (in absorption and fluorescence) from chiral matter (in order to study the kinetics of structure changes), circular differential x-ray microscopy, MCD from large molecules or impurities and defects in solids, and light scattering from molecules.

4 Acknowledgements

I would like to thank Prof. U. Heinzmann for his continuous support and fruitful discussions. Several of the above-described experiments could not have been performed without the experimental collaboration with R. David, N. Irmer, B. Kessler, N. Müller, P. Stoppmanns and B. Vogt. Thanks are also due to the BESSY staff and colleagues of the Fritz-Haber-Institut der MPG. Cooperation with several further colleagues is gratefully acknowledged: N. Böwering, J. Bansmann, R. Feder, J. Hormes, J. Kirschner, M. Müller, M. Salzmann, C.M. Schneider, G. Schönhense, G. Schütz and E. Tamura. Several of the experiments were financially supported by BMFT and DFG.

5 References:

1. Circularly polarized radiation can alternatively be characterized by its handedness, helicity or photon spin. Left handed CPR has positive helicity, is also denoted as σ^+-radiation and its photon spin $(+\hbar)$ is parallel to the propagation direction. Right handed CPR has negative helicity (σ^--radiation) and the photon spin $(-\hbar)$ is antiparallel to the propagation direction of light
2. U. Heinzmann, J. Phys. B: Atom. Molec. Phys. 13 (1980) 4353
3. F. Schäfers, W. Peatman, A. Eyers, Ch. Heckenkamp, G. Schönhense and U. Heinzmann, Rev. Sci. Instr. 57 (1986) 1032
4. L. Baumgarten, C. M. Schneider, H. Petersen, F. Schäfers and J. Kirschner, Phys. Rev. Lett. 65 (1990) 492
5. U. Fano, Phys. Rev. 178 (1969), 131 and Phys. Rev 184 (1969), 250
6. U. Heinzmann, J. Kessler and J. Lorentz, Phys. Rev. Lett. 25 (1970), 1325
7. F. Meier, B. P. Zakharchenya (eds.): Optical Orientation, Modern Problems in Condensed Matter Sciences Vol. 8 (North Holland, Amsterdam 1984)
8. U. Heinzmann, Fundamental Processes in Atomic Collision Physics (Edited by H. Kleinpoppen, J. S. Briggs and H. O. Lutz) p. 269, Plenum (1985)
9. G. Schütz, W. Wagner, W. Wilhelm, P. Kienle, R. Zeller, R. Frahm and G. Materlik, Phys. Rev. Lett. 58 (1987) 737; G. Schütz, Phys. Bl. 46 (1990), 475
10. G. Schütz, R. Frahm, P. Mautner, R. Wienke, W. Wagner, W. Wilhelm and P. Kienle, Phys. Rev. Lett. 62 (1989), 2620
11. U. Heinzmann, Physica Scripta T17 (1987), 77 and in: M. Campagna and R. Rosei (ed.): Photoemission and Absorption Spectroscopy of Solids and Interfaces with Synchrotron Radiation (North Holland, Amsterdam 1990), 469
12. F. Meier, D. Pescia: In Ref. 7., Chap. 7
13. J. Kirschner: Polarized Electrons at Surfaces (Springer, Berlin, Heidelberg 1985)

14. R. Feder (ed.): Polarized Electrons in Surface Physics (World Scientific, Singapore 1985)
15. M. Carnell, S.D. Peyerimhoff, A. Breest, U.H. Gödderz, P. Ochmann and J. Hormes, Chem. Phys. Lett. 180 (1991), 427
16. G. Schönhense, Physica Scripta T31 (1990), 255
17. P. Stoppmanns, B. Schmiedeskamp, B. Vogt, N. Müller and U. Heinzmann, Physica Scripta T41 (1992), 190
18. U. Heinzmann, J. Phys. B: Atom. Molec. Phys. 13 (1980), 4367
19. N. A. Cherepkov, J. Phys. B: Atom. Molec. Phys. 12 (1979), 1279; K. N. Huang, Phys. Rev. A22 (1980), 223; H. Klar, J. Phys. B: Atom. Molec. Phys. 13 (1980), 3117
20. J. Kessler, Polarized Electrons, 2nd ed. (Springer, Berlin, 1985)
21. E. Tamura and R. Feder, Europhys. Lett. 16 (1991), 695
and Solid State Commun. 79 (1991), 989
22. B. Schmiedeskamp, N. Irmer, R. David and U. Heinzmann, Appl. Phys. A53 (1991), 418
N. Irmer. R. David, B. Schmiedeskamp and U. Heinzmann, Phys. Rev. B45 (1992), 3849
23. Ch. Heckenkamp, F. Schäfers, G. Schönhense, U. Heinzmann, Phys. Rev. Lett. 52 (1984), 421
24. N. Böwering, M. Salzmann, M. Müller, H.W. Klausing and U. Heinzmann, Physica Scripta 41 (1990), 429
M. Müller, N. Böwering. A. Svensson and U. Heinzmann, J. Phys. B. 23 (1990), 2267
25. R. Wienke, G. Schütz and H. Ebert, J. Appl. Phys. 69 (1991), 6147
26. B. T. Thole, P. Carra, F. Sette, G. van der Laan, Phys. Rev. Lett. 68 (1992), 1943
27. G. Schütz, private communication, November 1991
28. H. Ebert, R. Zeller, B. Drittler and P. Dederichs, J. Appl. Phys. 67 (1990), 4576
29. C. M. Schneider, D. Venus and J. Kirschner, Phys. Rev. B45 (1992), 5041
30. U. Heinzmann and G. Schönhense, chapter 11 of ref 14
G. Schönhense, Appl. Phys. A41 (1986), 39
31. B. Vogt, B. Schmiedeskamp and U. Heinzmann, Vacuum 41 (1990), 1118
32. B. Schmiedeskamp, B. Vogt and U. Heinzmann, submitted
33. B. Schmiedeskamp, B. Kessler, B. Vogt and U. Heinzmann, Surf. Sci. 223 (1989), 465
34. E. Tamura, R. Feder, B. Vogt, B. Schmiedeskamp and U. Heinzmann, Z. Phys. B77 (1989), 129
35. B. Vogt, B. Schmiedeskamp and U. Heinzmann, Z. Phys. B80 (1990), 359
36. B. Vogt, B. Schmiedeskamp and U. Heinzmann, Phys. Rev. B42 (1990), 9267
37. B. Schmiedeskamp, N. Irmer, P. Stoppmanns, B. Vogt and U. Heinzmann, submitted
38. B. Vogt, P. Stoppmanns, B. Schmiedeskamp and U. Heinzmann, Appl. Phys. A52 (1991), 323
39. C. M. Schneider, G. S. Sohal, P. Schuster and J. Kirschner, Vacuum 41 (1990), 511
C. M. Schneider, J. J. de Miguel, P. Bressler, P. Schuster, R. Miranda and J. Kirschner, J. Electron Spectr. and Rel. Phen. 51 (1990), 263
40. C. M. Schneider, P. Schuster, M. S. Hammond and J. Kirschner, Europhys. Lett. 16 (1991), 689
41. P. Stoppmanns, B. Heidemann, N. Irmer, N. Müller, B. Vogt, B. Schmiedeskamp, U. Heinzmann, E. Tamura and R. Feder, Phys. Rev. Lett. 66 (1991), 2645
42. B. Schmiedeskamp, B. Vogt and U. Heinzmann, Phys. Rev. Lett. 60 (1988), 651
43. E. Tamura, W. Piepke and R. Feder, Phys. Rev. Lett. 59 (1987), 934
44. B. Schmiedeskamp, B. Vogt and U. Heinzmann, Solid State Commun. 76 (1990), 1391
45. M. Salzmann, M. Müller, N. Böwering and U. Heinzmann in: I.E. McCarthy, W.R. MacGillivray, M. C. Standage (ed): Abstracts of contributed papers, XVII ICPEAC, Brisbane, Australia 1991 and to be published
46. E. Charney, The Molecular Basis of Optical Activity: Optical Rotary Dispersion and Circular Dichroism, Wiley, New York, 1979
47. H. Lagier, Diplomarbeit, Universität Bonn (1989), similar spectra are given in ref. 15
48. J. Bansmann, Ch. Ostertag, G. Schönhense, F. Fegel, C. Westphal, M. Getzlaff, F. Schäfers and H. Petersen, submitted to Phys. Rev. B.
49. N. Chandra, Phys. Rev. A39 (1989), 2256

II. SOURCES AND OPTICS

VERY LOW EMITTANCE STORAGE RINGS : GOALS AND PROBLEMS

Pierre MARIN
LURE, Centre Universitaire de Paris-Sud, 91405 Orsay Cedex, FRANCE

ABSTRACT

The derivation of the beam emittance in an electron storage ring is briefly reminded and its fundamental importance on the photon beam properties is stressed, as well as the relevance of several physical effects on its behavior. The other factors of merit of a synchrotron radiation source are reviewed in terms of the compromise to be reached with the present goal. Some tentative examples of synchrotron radiation sources with very low beam emittance are presented together with preliminary conclusions.

1. Introduction

A parameter of prime importance in experiments with synchrotron radiation sources is the spectral brightness defined as :

$$B = \frac{d^4 F}{d\sigma,\ d\Omega,\ d\ \lambda/\lambda,\ dt}$$

High intensity, small beam cross section, good focusing of the photon beam and high monochromaticity are looked for in order to maximize the useful photon flux. The special relevance of small electron beam emittance in both radial and vertical planes has been stressed many times in the past ten years[1] in particular for high energy sources equipped with undulators. However the trend towards reaching lower emittances meets other limitations on the beam performances either technical or fundamental which users and machine designers should bear in mind. I will shortly describe the origin of the beam emittance and discuss the various possibilities of decreasing it. There is a well known litterature on the subject[2] but it is worth reminding it shortly here. Later on, the other factors of merit of a synchrotron radiation source and the detrimental effects which can be brought on them by very low emittances will be discussed. I will end up by briefly stating the first steps towards going to what is described now as the fourth generation of synchrotron radiation sources.

2. Beam Emittance in an Electron Storage Ring

Emittance results from an equilibrium between the excitation of orbit jumps due to single photon emission and the damping of the oscillations by the accelerating voltage of the cavity. See Fig. 1 and Fig. 2.

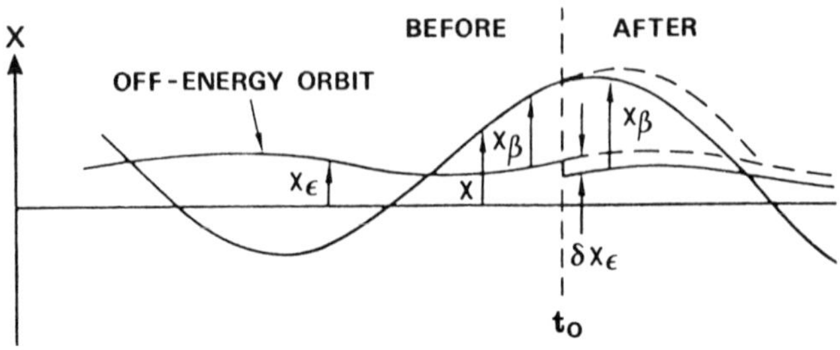

Fig. 1 : Effect of an energy change by radiation loss on betatron oscillations in a $D \neq 0$ section.

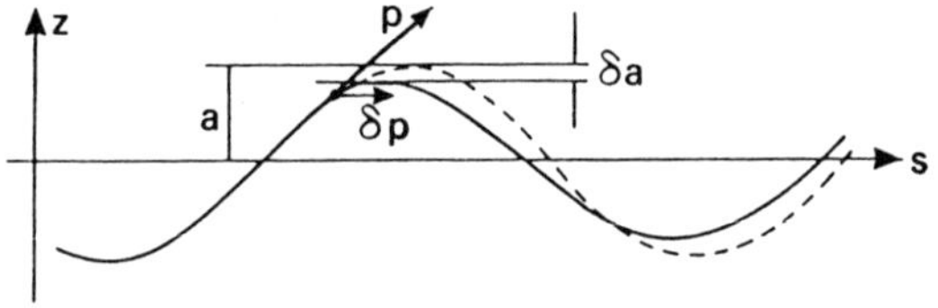

Fig. 2 : Effect of an energy change by RF acceleration on betatron oscillations.

More precisely, consider the Courant and Snyder invariant for a non radiating electron.

$$A_x^2 = \gamma_x(s)\, x^2 + 2\, \alpha_x(s)\, x\dot{x} + \beta_x\, \dot{x}^2 \tag{1}$$

α_x, β_x and γ_x are called the Twiss parameters. A_x^2 is independant of azimuth.

Suppose a photon energy u is emitted at a point where the dispersion function D is different from zero, then :

$x \Rightarrow x - D\frac{u}{E}$, and

$\dot{x} \Rightarrow \dot{x} - \dot{D}\frac{u}{E}$, where E is the beam energy.

The quadratic terms $\left(\frac{u}{E}\right)^2$ in (1) yield an increase of A_x^2

$$A_x^2 \Rightarrow A_x^2 + \delta A_x^2$$

and one can easily show that :

$$\langle \delta A_x^2 \rangle = \frac{\langle u^2 \rangle}{E^2} \mathcal{H}(s), \quad \text{where } \langle \ \rangle \text{ means average value}$$

with $\mathcal{H}(s) = \gamma_x D^2 + 2\,\alpha_x D\dot{D} + \beta_x \dot{D}^2$.

At equilibrium :

$$d\frac{\langle A_x^2 \rangle}{dt} = \frac{1}{2\pi R E^2} \underbrace{\int_L N \langle u^2 \rangle \mathcal{H}(s)\, ds}_{\text{excitation}} - \underbrace{2\frac{\langle A_x^2 \rangle}{\tau_x}}_{\text{damping}} = 0$$

and the emittance $\varepsilon_x = \frac{1}{2}\langle A_x^2 \rangle$ is given by

$$\boxed{\varepsilon_x = \frac{\tau_x}{4} \frac{1}{2\pi R E^2} N \langle u^2 \rangle \int_L \mathcal{H}(s)\, ds} \tag{2}$$

where :

τ_x is the horizontal damping time constant, $\tau_x \approx \frac{E}{P}$, P being the power loss

N is the number of photons emitted per second

$\langle u^2 \rangle$ is the average of the photon energy square

$\int_s \mathcal{H}(s)\, ds$ is taken only in that part of the circumference where photons are emitted, that is in the dipole magnets.

One can show that $N \langle u^2 \rangle \approx u_c P$, where u_c is the characteristic energy of the photons.

3. Lowering of the Emittance

Looking at the structure of formula (2) tells one immediately the three different possible ways of lowering the emittance of the beam :

- For a given machine by the decrease of the beam energy. In fact $\varepsilon_x \approx E^2$.

- For a given machine by increasing the radiated power P with the help of long wiggler magnets installed at locations where the dispersion D and its derivative $\dot{D}$ are zero. This creates no extra contribution to $\int_s \mathcal{H}(s)\, ds$, nor to $N \langle u^2 \rangle$, but lowers τ_x and therefore ε_x. This is the method of the so-called damping wigglers.

- Find lattices which make the integral $\int_s \mathcal{H}(s)\, ds$ small.

Several lattices have been proposed with in view to achieve a small beam emittance : FODO, C.G., DBA, TBA, QBA, etc[3].

All of them can be shown to produce a dependence of the minimum achievable emittance of the form :

$$\varepsilon_x = k_i \frac{Cq\, \theta^3 \gamma^2}{J_x} \tag{3}$$

where Cq is a constant, 3.8 10^{-13}, θ is the dipole bend, γ the energy, J_x the horizontal partition coefficient and k_i is a constant which is typical of the lattice.

Looking at (3), the question immediately arises how far can one go towards very small emittances. There is a limit which comes from the uncertainty principle, when applied to the transverse motion of the electron

$$\Delta x\, \Delta p_x \geq \frac{h}{2\pi}$$

with $\Delta p_x = \dot{x} \frac{h\nu}{c}$ being the transverse momentum change due to photon emission.

Therefore, the orbit position x and its derivative $\dot{x}$ are such that $x\dot{x} \geq \frac{1}{2\pi} \frac{c}{\nu}$, that is $\varepsilon_x \geq \lambda/2\pi$.

More precisely, one can compute the degree of coherence for the light emitted from an extended incoherent monochromatic source[4] which sets a limit $\varepsilon \approx \lambda/4\pi$, often refered to as the "coherent limit" for the emittance.

4. Factors of Merit of a Synchrotron Radiation Source and the Detrimental Effects of Lowering the Beam Emittance

Synchrotron radiation sources are expensive items to build and to run. They normally gather large crowds of users with different purposes. The spectral brightness, mostly determined by beam emittance and intensity, together with the properties of the dipole and insertion device sources is the most important factor of merit. However some of the users are interested mainly by the source dimensions σ_x, σ_z rather than the emittances ε_x, ε_z. This is the case for the so-called microphotography experiments. Others put a strong emphasis on achieving a short bunch length and a high intensity per bunch. This applies to the experiments making use of the temporal structure of the beam.

Other important aspects concern the beam lifetime which one would like to be as long as possible in order to avoid frequent refilling of the machine. In the last few years, strong emphasis has been put also on photon beam stability and reproducibility. With the achievement of smaller photon beam dimensions, this demand is more and more stringent for preventing the beam emittance being unnecessarily diluted. Generally speaking, synchrotron radiation sources are multipurpose machines aiming at different applications.

The most extreme cases are very flat beams characterized by a low aspect ratio σ_z/σ_x and the FEL experiments which are better matched with round shape beams.

The detrimental effects on low beam emittance are briefly discussed below together with possible cures :

4.1. Particle Beam Instabilities

These can occur in various ways.

In the single bunch case the chromaticities have to be substantially increased from slightly positive to much higher values to combat the so-called "fast head-tail" effect. In doing so, the dynamic aperture is usually reduced, thus decreasing the beam lifetime.

In the multibunch case, transverse and longitudinal instabilities can occur. The latest widen the horizontal beam size in the non zero dispersion part of the machine. The cure here, is to reduce the vacuum chamber impedance and to design "mono-mode" cavities. Considerable progresses have been achieved in the recent years with careful smooth designs of the vacuum systems. The most advanced proposal for a "mono-mode" cavity has been made at Cornell in a superconductive version[5], see Fig. 3. The design allows the higher modes to leave the cavity through wide openings, so as to be damped into special ferrites located outside the cryostat. High voltage gradients are secured using superconductivity, thus cutting down capital and running costs.

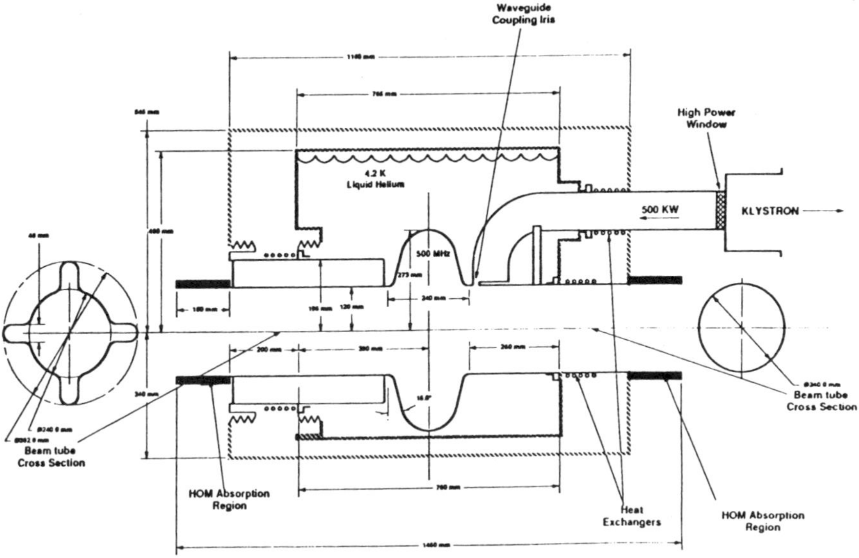

Fig. 3 : Monomode superconductive 500 MHz RF cavity proposed for the B-factory project at Cornell.

4.2. Ion and dust trapping in an electron beam

These may result in a variety of unwanted effects on beam lifetime, transitory increase of beam emittance together with small and fast beam current losses. However the use of positron filling in several machines, DCI and Super-ACO at Orsay, P.F. at Tsukuba has brought a definite cure to these.

4.3. Dynamic aperture and energy acceptance reductions

The strong focusing necessary for reaching low emittances reduces the dynamic aperture of the beam, thus decreasing the beam gas lifetime. Increasing the bunch intensity has a dramatic effect on the Touschek lifetime and leads to more and more frequent refillings, the smaller the beam emittance is. The use of many undulators and wigglers has similar adverse effects on both the dynamic aperture and the energy acceptance, with a further reduction of the beam lifetime.

4.4. Misalignment sensitivity

The strong focusing increases the machine sensitivity to small misalignements of its magnetic components, thus calling for more refined magnetic measurements and tighter mechanical tolerances.

4.5. Effect of ground movements and site vibrations

These are now thoroughly inspected for the machines under construction or assessed for those in the design stage. The smaller the photon beam size after demagnification by suitable optics, the most stringent the conditions are. Obtaining stabilities at the micrometer level is not easy for beam lines in the ten meter range and the proposed solutions are still far away from being properly assessed for the new generation of machines to come.

4.6. Heat load on optical systems

This problem was temporarely relieved by the use of undulator rather than bending magnet sources. It comes again into play through the higher prospected power densities.

5. Fourth Generation of Synchrotron Radiation Sources

I will simply summarize here the conclusions of a workshop held on the subject at Stanford at the beginning of the year[6].

5.1. First step towards decreasing the emittance of existing rings by a factor of 10. Two examples of high energy colliders operated a low energies can be quoted :

- PEP operated at 7.41 GeV which has a vertical emittance $\varepsilon_v = 2\ 10^{-10}$ m.rad.
- Tristan operated at 10 GeV together with wiggler magnets has an horizontal emittance $\varepsilon_h = 1.46\ 10^{-9}$ m.rad.

5.2. Preliminary designs of machines with a natural horizontal emittance in the 10^{-10} to 10^{-11} mrad range are proposed with a diffraction limit of 10 to 100 Å in x and 1 to 10 Å in z. This assumes that the correct horizontal to vertical coupling can be reached. These machines have basically large circumferences in the 1 km range in order to accomodate the large number of achromats and wigglers required, together with a high energy around 4 GeV in order to increase Toucheck lifetime.

Eventhough "Top up" of the beam intensity at regular intervals is envisaged to cope with the short Touschek lifetime. This however, is found to be insatisfactory owing to radiation safety problems and to perturbations of the experimental conditions for the users.

The problem of vibrations gets increasingly difficult at the submicron level and no satisfactory proposal is made in this area. Short bunches are also envisaged but the parasitic energy losses get enormous. To overcome this the bunch rotation scheme proposed by Orlow in a restricted part of the circumference can be envisaged. Finally positron filling is found to be compulsory.

6. Synchrotron Radiation from Protons

Unusual conditions for proton beams should be reached at CERN in the LHC project. Protons with an energy of 8 TeV, bent by 10 Tesla dipole fields will produce synchrotron radiation with a critical energy $h\nu_c = 70$ eV (182 Å). For a beam intensity of 0.85 A, the total radiated power should be 18.2 kW (13.5 W per dipole magnet of unit length 13 m). The beam emittance just after injection should be $\varepsilon_{x,z} = 4.5\ 10^{-10}$ m.rad, thus yielding a diffraction limit for $\lambda = 56$ Å. With a damping time for the proton beam around 1 day and with beam lifetimes of several days even lower emittances and diffraction limits can still further be expected.

7. Conclusions

We are still far from proposing well thought projects with very low emittances. However to pave the way towards this goal a number of recommandations can be made at this stage :

- Close attention to the Cornell efforts on superconductive cavities should be paid. Incorporating several of these items on existing high intensity rings would be of paramount importance.

- The experience of ESRF on the massive use of motorized jacks and of hydraulic levelling systems together with the concept of girders will be immensely valuable for the new very low emittance sources.

- Culture on site properties and building construction in relation with vibrations and stability of experimental systems should become familiar to machine physicists and users.

- More has to be understood on limitations of dynamic aperture by insertion devices.

Finally it appears that the only realistic possibility to build such expensive and fancy systems as very low emittance rings is to propose them, not as new projects, but as appendixes to existing facilities.

8. References

1. M. Cornacchia, Requirements and Limitations on Beam Quality in Synchrotron Radiation Sources, CAS, Synchrotron Radiation and Free Electron Lasers, CERN 90-03, 2 April 1990, p 53.

2. M. Sands, The Physics of Electron Storage Ring. An Introduction, Academic Press, New-York and London, p 373.

3. A. Ropert, High Brilliance Lattices and the Effect of Insertion Devices, CAS, Synchrotron Radiation and Free Electron Lasers, CERN 90-03, 2 April 1990, p 158.

4. Principles of Optics, Max Born and Emil Wolf, Pergamon Press, p 508.

5. CESR-B, Conceptual Design for a B-Factory Based on CESR, CLNS 91-1050.

6. M. Cornacchia and H. Winick, Workshop on Fourth Generation Light Sources, SSRL 92-02, Feb. 24-27, 1992.

HIGH RESOLUTION MONOCHROMATORS FOR UNDULATOR RADIATION

W.B. PEATMAN, F. SENF
Berlin Electron Storage Ring for Synchrotron Radiation (BESSY),
Lentzeallee 100, D-1000 Berlin 33, FRG

Abstract

The advent of synchrotron radiation sources in the 1960's and 1970's sparked a significant change from traditional light sources for the designers of monochromators and spectrometers: the source is intrinsically small and the radiation inherently directed in a very small cone. This has made it possible to have high collection efficiencies of the radiation on the one hand and, in principle, a high energy resolution on the other. Running counter to this is the need for monochromators in the soft x-ray portion of the electromagnetic spectrum where a significant transmission is only possible if very large angles of incidence of the radiation on the optical elements are employed. This requirement tends to reduce the optical acceptance of the system. Furthermore, the optical aberrations associated with large angles of incidence together with long optical surfaces reduce the resolution.

With the development of undulator sources on storage rings of the third generation the designers again have the advantage of a still smaller and still more collimated radiation source. In this paper we start with an analysis of the spatial characteristics of undulator radiation in order to determine the effective source size and source divergence for the wavelengths desired. This information is essential for dimensioning the monochromator. It is then seen that despite the small emittance of the source, it must still be demagnified in order to achieve a resolution of $E/\Delta E$ of 10^4 or more in the monochromator. The effectiveness of demagnifying systems is examined.

With this information the design for three different types of monochromator has been optimized for resolution for energies 100 - 800 eV:
a) a constant length Rowland circle spherical grating monochromator
b) a Petersen type plane grating monochromator
c) a Hettrick-Underwood type plane grating monochromator with a grating with variable line spacing.

These monochromators will be described and their anticipated performance compared with each other.

1) Introduction

The design of synchrotron radiation facilities of the third generation can be conveniently divided into three parts: the first and ultimately most important part involves the experiments and the associated equipment that would exploit a significantly higher spectral and spatial resolution than available to date. The other two parts involve the storage ring itself, which together with undulators is expected to produce an extremely stable, high brilliance source of soft x-ray radiation, and the monochromators which are required to deliver monochromatic radiation to the experiment spectrally pure and, if possible, with a microscopic focus. In this contribution, we have examined the factors that are essential for the generation of

a

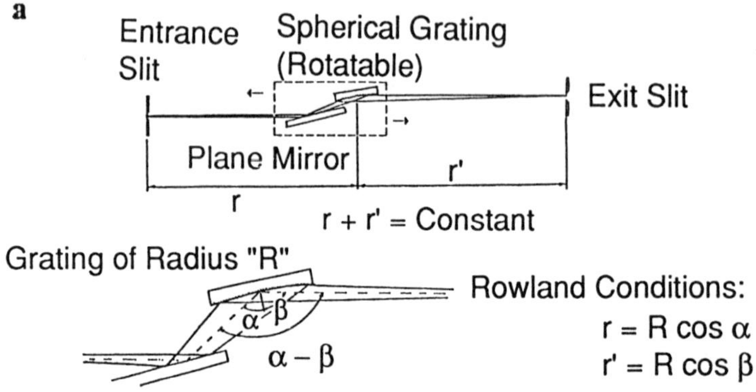

b

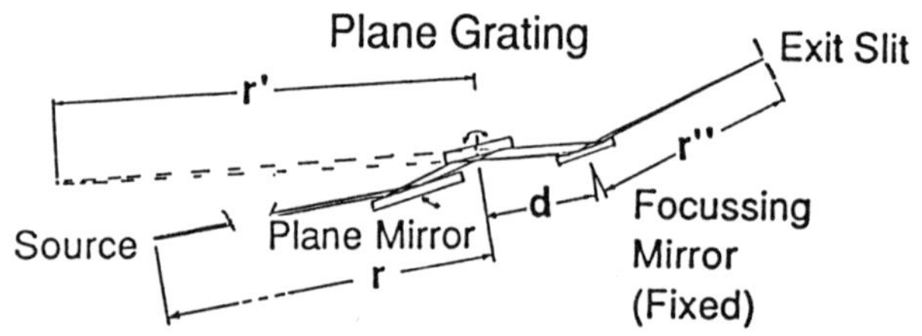

For Exact Focussing at all Wavelengths (Petersen [2])

$$-\frac{r'}{r} = \frac{\cos^2\beta}{\cos^2\alpha} = \text{Constant}$$

c

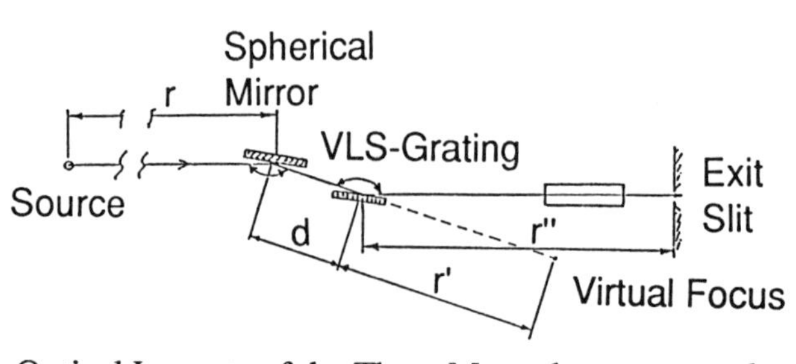

Fig. 1: The Optical Layouts of the Three Monochromators to be Compared:

a: Constant Length Spherical Grating Monochromator (CL-SGM)

b: Petersen Type Plane Grating Monochromator (P-PGM)

c: Hettrick-Underwood Type Variable Line Spacing Plane Grating Monochromator (HU-VLSPGM) highly monochromatic radiation from brilliant undulator sources for the energy range 100 - 800 eV and compare the anticipated performance of three types of monochromator especially suited for the task: Rowland circle spherical grating monochromators (RC-SGM)[1] , Petersen type plane grating monochromators (P-PGM)[2] and Hettrick-Underwood type of plane grating monochromators with variable line spacing gratings (HU-VLSPGM)[3]. A resolution of $E/\Delta E \gtrsim 10^4$ is sought. Only limited attention will be paid to the size of the image at the experiment.

The basic layouts of the three types of monochromator are shown in figure 1. All three types exist already, two of them at synchrotron radiation laboratories around the world[4, 5]. The HU-VLSPGM has been used with a plasma source in the near UV yielding a resolution of 35000[6]. The design of the RC-SGM addressed here is very simple: a vessel with a plane mirror and rotatable concave grating translates between fixed entrance and exit slits (Fig. 1a). By this means, the entrance and exit arm lengths are varied so as to fulfill the Rowland conditions. The entrance and exit optical axes are parallel to each other and are fixed as well[5, 7]. The most experience has been gained with the P-PGM (Fig. 1b), at least eight examples of which have been in operation for some years and the merits of the optical and mechanical design proven[4].

2) Demagnification of the source

Despite the very small emittance of the electron beam in the storage rings planned, the source is still too large for the resolution required for many of the experiments proposed. It should be mentioned here, that we assume a vertical dispersion plane in the monochromator in order to exploit the significantly smaller vertical source size of the electron beam. For the VUV storage rings of the third generation that are under construction or planned, the vertical source size to be expected is $\sigma_V \approx 40\ \mu m$. Furthermore, if one assumes a Gauss distribution, 95 % of the desired radiation will emanate from a source of $4 \times \sigma_V = 160\ \mu m$ in diameter. Considering the cross section alone, one determines that the source size limited resolution is about 6000 for an RC-SGM or about 18000 for a P-PGM without an entrance slit. Looking at the exit slit limited resolution and assuming for the moment that no demagnification of the source occurs in the beam line, one finds a limiting resolution of about 4000 for the RC-SGM and 1400 for the P-PGM. Clearly, the source must be reduced in size, before or within the monochromator if a resolution of 10^4 is to be achieved. For a quantitative estimate of the effective source size, the depth must also be included in the calculation. Finally, when tangent errors are also included one expects that demagnification of 10 or more is needed. We also require that the sum $\alpha - \beta$ decreases with decreasing photon energy in order to gain a partial suppression of higher order light.

The magnification, M, for the three types of monochromator considered here (fig. 1a, b, c) is given by the following expressions (definitions as shown in the figures):

$$M = \frac{\text{image size}}{\text{object size}} = \frac{r'}{r}\frac{\cos\alpha}{\cos\beta}$$

For the RC-SGM (fig. 1 a) $\quad M \equiv 1$!

For the P-PGM (fig. 1b) $\quad M = \frac{r''}{rc - d/c} \simeq \frac{r''}{rc} \simeq \frac{1}{20}$ typically.

Table 1: Comparison of Divergence of Power and Flux

Comparison of the total emitted power of an undulator with the flux of the first five harmonics as a function of the hard edge half opening angle, θ, of an aperture. The three undulators indicated are planned for BESSY II (1.7 GeV) [8].

	θ (μrad)	P+	F* 1.	2.	3.	4.	5.
	40	0.007	0.24	0.05	0.42	0.09	0.59
U-1	80	0.06	0.68	0.19	0.90	0.53	
(K = 2.0)	160	0.16		0.68			
λ_0 = 100 mm	320	0.50		0.95			
N = 40	640	0.92					
	40	0.007	0.34	0.07	0.50	0.16	0.51
U-2.5	80	0.06	0.84	0.25	0.95	0.53	0.96
(K = 2.0)	160	0.16		0.84		0.98	
λ_0 = 52 mm	320	0.50		0.95			
N = 80	640	0.92					
	40	0.008	0.40	0.08	0.42	0.18	0.49
U-3	80	0.07	0.83	0.29	0.92	0.53	0.95
(K = 1.6)	160	0.20		0.90		0.97	
λ_0 = 30 mm	320	0.59		0.95			
N = 110	640	0.95					

+ P = fraction of total power emitted

*F = fraction of the total flux of the harmonic. If F ≥ 0.99 no entry is made.

For the HU-VLSPGM (fig. 1c) $M = \frac{(r'+d)}{r} \cdot \frac{r''}{r'} \frac{\cos\alpha}{\cos\beta} \simeq \frac{1}{50}$ typically.

From these considerations it is evident that an RC-SGM requires a demagnifying optical system between the source and the entrance slit. The P-PGM and the HU-VLSPGM offer intrinsically a favorable, inherent demagnification which, however, as shown below, is not achievable in practice due to the three dimensional nature of the source and to the tangent errors of the optical surfaces involved.

Ray trace studies were made in order to determine what values of demagnification, 1/M, are realistic for a deep undulator source and mirrors with tangent errors. For this it is necessary to determine the vertical opening angle of the radiation emanating from an undulator on a storage ring of the third generation. In table 1 are shown the fractional intensities of the first five harmonics transmitted by various pinholes at a distance of 10 meters from the middle of the undulator[8]. This has been done for undulators with λ_0 = 100 mm, 52 mm and 30 mm for K-values of 2.0, 2.0 and 1.6 respectively. Below these values of K the emission angles decrease and present no limiting value to the acceptance of the beam line. Higher K-values have still larger angles of emission. As shown below, the angles required at K = 2 are already problematic for beam line designers. Values are shown including the even harmonics 2 and 4. By utilizing the even harmonics, whose peak intensities are less than but still within a factor of two of those of the third and fifth harmonics, values of $K \lesssim 2$ will lead to overlapping energy ranges for the first five harmonics. The values of F shown in table 1 were calculated using the code SMUT[9]. Also shown in the table are the fractional amounts of the total power (P) of the undulators transmitted by the various pinholes. Three conclusions can be drawn immediately: (1) all three types of undulators exhibit essentially the same behavior of F and P as a function of pinhole acceptance. (2) A pinhole acceptance of 2 x 160 μrad will accept effectively all of the desired radiation exept for some of the second harmonic. For 95 % transmission of a Gauss profile this corresponds to σ_V' = 80 μrad. This value is used in the following ray trace studies. (3) With this vertical acceptance, 84 % of the undesired radiative power of the undulator can be eliminated thus reducing heat load problems.

In figures 2 and 3 are shown the ray trace results for a nominal 1/M = 24 demagnifying system consisting of a spherical mirror and of a plane elliptical mirror respectively. We assume that sagittal focussing, not critical for the energy resolution being optimized here, is performed with a separate optical element in a Kirkpatrick-Baez arrangement[10]. The source parameters used for the calculations are given in table 2. The results for such ray trace studies for nominal demagnification factors of 10 to 30, and with rms tangent errors of 0, 1.0 and, for the plane ellipse, 2.0 arc seconds are shown in figure 4. As expected, the spherical mirror exhibits an asymmetric tail, which depends quadratically on the vertical acceptance. Since it is possible to obtain spherical optics with rms tangent errors of 0.1-0.2 arc seconds the main source of focussing problems stems from this tail. The situation is reversed for a plane elliptical mirror where tangent errors between 0.5 and 1 arc sec are possible but only at great cost. Smaller errors than 0.5 arc sec rms for this geometry are not known to the authors. Nevertheless, a plane elliptical mirror with 0.5 arc sec rms tangent error and a nominal demagnification of 25 or so would effectively reduce the vertical source size by a factor of about 7 enabling 95 % of the light from an undulator source with σ_V = 40 μm to pass through a 22 μm wide slit. For a

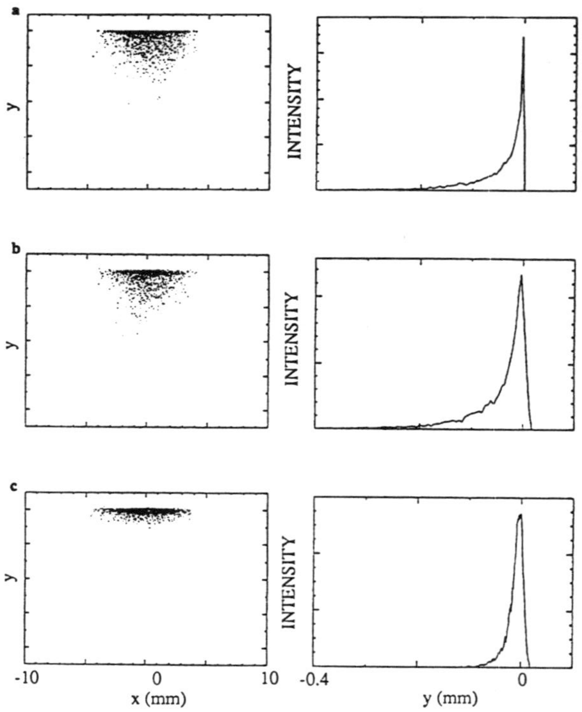

Fig. 2: Focussing Characteristics of a Spherical Mirror
To the left the spot diagram. To the right the integrated vertical profile. Angle of incidence $\theta = 87.5°$.
a) Demagnification = 24, $\sigma_{TE} = 0$, $\sigma'_{VSR} = 80\ \mu rad$
b) Demagnification = 24, $\sigma_{TE} = 1$ sec, $\sigma'_{VSR} = 80\ \mu rad$
c) Demagnification = 24, $\sigma_{TE} = 1$ sec, $\sigma'_{VSR} = 40\ \mu rad$

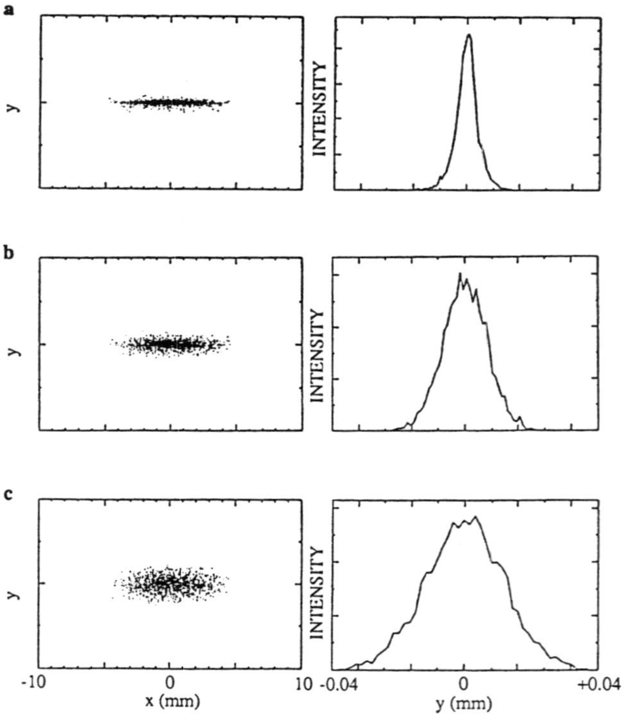

Fig. 3: Focussing Characteristics of a Plane Elliptical Mirror
To the left the spot diagram. To the right the integrated vertical profile. Angle of incidence $\theta = 87.5°$.
a) Demagnification = 24, $\sigma_{TE} = 0$, $\sigma'_{VSR} = 80\ \mu rad$
b) Demagnification = 24, $\sigma_{TE} = 1$ sec, $\sigma'_{VSR} = 80\ \mu rad$
c) Demagnification = 24, $\sigma_{TE} = 2$ sec, $\sigma'_{VSR} = 80\ \mu rad$

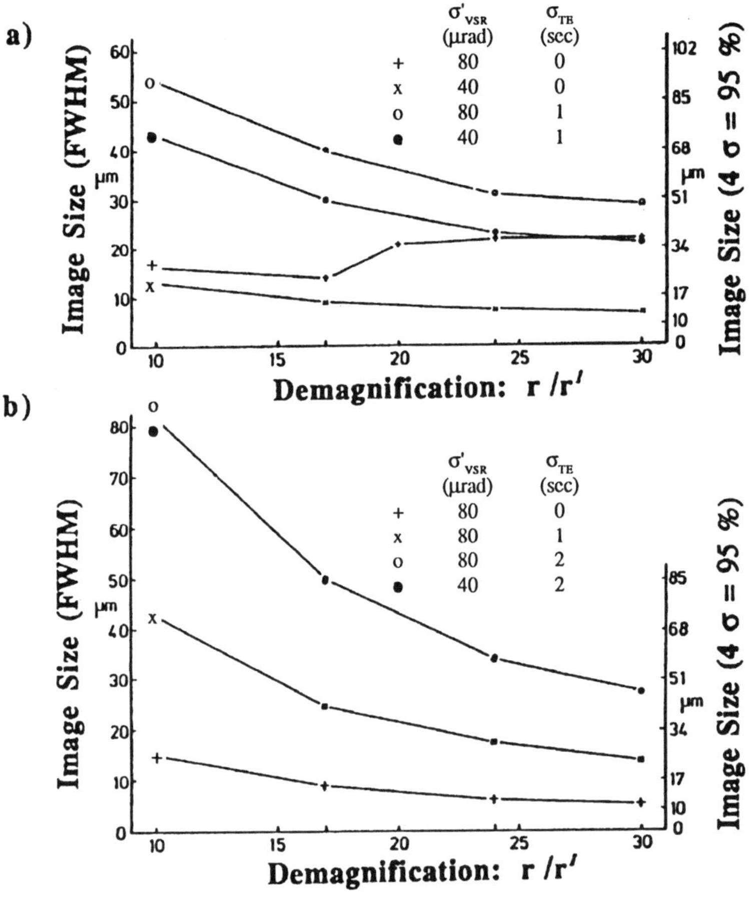

Fig. 4: Demagnification of an Undulator Source
Vertical image size as a function of demagnification for several tangent errors (σ_{TE}) and vertical acceptances (σ'_{VSR}).
(a) Spherical mirror. (b) Plane elliptical mirror.

higher resolution than permitted by a 22 μm wide slit, the vertical source size in the storage ring must be reduced.

From the foregoing arguments, it can be concluded that (1) an RC-SGM requires a demagnifying optical system before an entrance slit and that (2) the inherent demagnification of a P-PGM or HU-VLSPGM can only be realized if nearly perfect optical geometries are available.

3) Two comparisons of resolution for three types of monochromators

As shown in the standard treatises on the geometric theory of optical gratings[11] the largest contribution to the line width function is the defocus term F_{20}. For high resolution monochromators, it is essential that this term and primary coma are effectively zero over the working range of the monochromator. For the P-PGM this term is made equal to zero for all wavelengths by suitable selection of r, r', α, β.

$$F_{20} = \frac{\cos^2\alpha}{r} + \frac{\cos^2\beta}{r'} = 0,$$

and for the RC-SGM by additionally adjusting R:

$$F_{20} = \frac{\cos^2\alpha}{r} + \frac{\cos^2\beta}{r'} - \frac{\cos\alpha + \cos\beta}{R} = 0$$

In the P-PGM the zeroing of the defocussing equation is guaranteed by a focussing mirror (yielding a negative source distance) behind the plane grating. The same is accomplished with a focussing sphere in the HU-VLSPGM design (zeroing at zero order) in addition to the line density parameter, b_2, of the grating:

$$F_{20} = \frac{\cos^2\alpha}{r} + \frac{\cos^2\beta}{r'} + m \cdot N_o \cdot b_2 \cdot \lambda$$

and the line density equation $\frac{dn}{dy} = N_o\,(1+2b_2y + 3b_3y^2 + 4b_4y^3 + ...)$

where N_o is the groove density in the middle of the grating.

Thus, in contrast to most monochromators presently employed in SR-labs around the world, the P-PGM and RC-SGM are focussed exactly over their entire energy range. In addition, for the RC-SGM, coma and parts of the spherical errors are identically zero, also over the entire energy range. The HU-VLSPGM is able to fulfill these requirements to a great extent by adjusting all the parameters in the line density equation so, as to minimize the wavelength dependent defocussing, coma, and spherical errors respectively[3]. As shown in their original papers this can be accomplished to a high degree of success[3, 6].

We compare here these three types of monochromator. The boundary conditions for the comparison are given in table 2. The first comparison to be made is the achievable resolution for the same transmission. In this case, only the RC-SGM and the P-PGM will be

Table 2: Boundary Conditions: 90-800 eV High Resolution Monochromators

1)	Total length: middle of undulator to exit slit	26 m
2)	Length occupied by storage ring	9 m
3)	Undulator source characteristics	
	$E = 1.7$ GeV ;	$L = \lambda_0 N = 4100$ mm
	$\sigma_H'(e\ell) = 0.029$ mrad ;	$\sigma_H(e\ell) = 0.200$ mm
	$\sigma_V'(e\ell) = 0.015$ mrad ;	$\sigma_V(e\ell) = 0.040$ mm
		$\sigma_H'(SR) = 0.200$ mrad
		$\sigma_V'(SR) = 0.080$ mrad
4)	Lengths of optical components:	
	Vertical mirrors	≤ 300 mm
	Horizontal mirrors	≤ 1000 mm
	Gratings	≤ 160 mm
5)	Tangent errors assumed (rms)	
	Planes, Spheres	0.1 sec
	Ellipse	0.25 sec
	Horizontal elements	0.4 sec
6)	Slits	≥ 4 μm
7)	Grating line density	≤ 1600 ℓ/mm
8)	Types of monochromator considered:	
	Rowland Circle-SGM (RC-SGM)	
	Petersen-PGM (P-PGM)	
	Hettrick-Underwood Variable Line Spacing Grating - PGM (HU-VLS PGM)	

compared, the HU-VLSPGM having a smaller acceptance than either of the above. The ray trace results of this comparison are shown in figure 5 a. In the figure one sees that at the low energy end of the scale the P-PGM is slightly better than the RC-SGM. As one goes to higher energies, however, the dispersion of the P-PGM diminishes. In principle, the same occurs with a RC-SGM, but in this case another grating with a higher line number is taken showing a much better resolution than the P-PGM for equal transmission. The values of the transmission are given in the top row. If a plane elliptical mirror were to be used in the P-PGM in this comparison the results would be worse because of the larger rms tangent error assumed for such a mirror. The reason for the discontinuities in the P-PGM curve is that, in order to maintain a high resolution over the entire range 100 - 800 eV, the transmission for both monochromators was reduced twice, arbitrarily keeping the maximum of each portion of the RC-SGM at $E/\Delta E \simeq 15000$. Lest one forget, an important advantage of the P-PGM is that one grating can be used to continuously cover the energy range of the monochromator - in practice a range of 50 eV - 2000 eV, well beyond that of this comparison[4]. This large energy range can be covered by an RC-SGM in practice only if several gratings with corresponding line densities are used. With only one grating, a larger energy range can be covered only with large changes in the entrance and/or exit arm lengths. In addition, the low energy end is limited by the horizon wavelength - a problem which increases with increasing line density - the latter being required for high resolution!

In the second comparison, the achievable resolution regardless of the transmission is deter-mined with the help of ray traces for all three monochromators. The results are shown in figure 5 b. One sees that the RC-SGM curves are essentially the same as those in the first comparison (fig. 5 a).

In the case of the P-PGM, the acceptance has been significantly reduced (on the spherical mirror) and the dispersion increased by considering either a grating with 2400 ℓ/mm or the second order of a grating with 1200 ℓ/mm. In both cases, the efficiency is much lower than that of a 1200 ℓ/mm grating used in first order. Hence, this "trick" could not be employed in the first comparison. Nevertheless, the very high resolution over a large energy range demonstrates the capability of a P-PGM if equipped with a small but good spherical focussing mirror. A loss in the horizontal acceptance of a P-PGM is the consequence a design in which the effect of tangent errors is reduced by employing a horizontally deflecting focussing mirror[12].

The HU-VLSPGM resolution maximum at 280 eV, the optimization energy, shows the highest resolution of the three monochromator types. This is sufficiently encouraging to warrant building one for an SR light source with a small emittance[13]. In addition to the very high resolving power, the HU-VLSPGM also exhibits a strong demagnification (s. Section 2 above), if equipped with a horizontally focussing mirror as shown in figure 1c. Ray trace studies for a dipole source at BESSY I and including tangent errors yield a spot size of $5 \times 8\ \mu m^2$ (VxH) at the exit slit. This stigmatic image lends itself to a further reduction in size for microfocus studies.

One aspect of the HU-VLSPGM still has to be established: for a particular performance the groove spacings must correspond to the constants b_2, b_3 and b_4 to an extremely high degree. A systematic error in these constants would destroy the optimization of the grating i.e. the minimization of these three aberrations over the energy range desired.

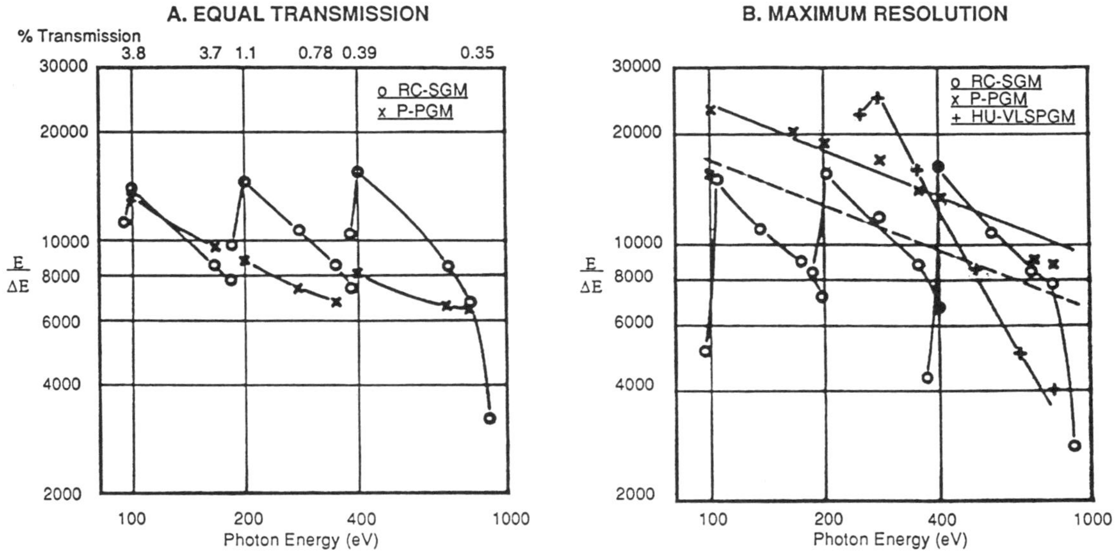

Fig. 5: Ray Trace Comparison: RC-SGM, P-PGM and HU-VLSPGM

A. Resolution for Equal Transmission (only RC-SGM, P-PGM). B. Resolution maximized at cost of Transmission.

1) RC-SGM: 10 µm slits; K-B preoptics of sphere and plane ellipse with 1/M = 6.4; Gratings (174°) 265 ℓ/mm, 530 ℓ/mm, 1060 ℓ/mm; Monochromator length = 8.0 m

2) P-PGM: 10 µm exit slit; spherical focussing mirror; Grating (174°) 1200 ℓ/mm, 1/M = 7.3 Monochromator length = 26.0 m. For "B" a 2400 ℓ/mm grating was used

3) HU-VLSPGM: 4 µm exit slit; spherical focussing mirror; VLS-Grating (174°) N_o = 1000 ℓ/mm, 1/M = 25 Monochromator length = 26.0 m

A non-systematic jitter in the groove spacings would also reduce the resolution dramatically. The published performance data[6] indicate that the latter error, jitter, can be mastered.

Conclusions: In order to achieve a resolution of $\gtrsim 10^4$ in the 100 - 800 eV range, monochromators which are inherently focussed and for which the other principle aberrations are effectively zero over this range, are required. A simple Rowland circle design fulfills these requirements and offers an angular acceptance which is very well scaled for undulators. A Petersen type PGM has a somewhat smaller acceptance for highest resolution but compensates this by enabling one grating to serve the entire energy scale. A Hettrick-Underwood VLSPGM also looks very promising. The small acceptance is compensated for by the very strong stigmatic demagnification of the source making this type especially suited for high resolution microfocus studies.

Acknowledgements:
Useful discussions with H. Petersen and J. Underwood and financial support from the Senate of Berlin are gratefully acknowledged.

References

1) a) H.A. Rowland, *Phil. Mag.* 16, 197 (1883)
b) ibid 16, 210 (1883)
c) H.G. Beutler. *J. Opt. Soc.* Am 35, 311 (1945)
2) H. Petersen, *Opt. Commun.* 40, 402 (1982)
3) M.C. Hettrick, J.H. Underwood, *AIP Corp. Proc.* 147, 237 (1986)
4) H. Petersen, *Nucl. Instr. Meth.* A 246, 260 (1986)
5) E. Ishiguro, M. Suzui, J. Yamazaki, E. Nakamura, K. Sakai, O. Matsudo, N. Mizutani, F. Fukui, M. Watanabe, *Rev. Sci. Inst.* 60, 2105 (1989)
6) M.C. Hettrick, J.H. Underwood, P.J. Batson, M.J. Eckart, *Appl. Opt.* 27, 200 (1988)
7) F. Senf, F. Eggenstein, W. Peatman, *Rev. Sci. Instrum.* 63, 1326 (1992)
8) "BESSY II, an Optimized Wiggler/Undulator Storage Ring Light Source for the VUV and XUV Spectral Regions" BESSY, Berlin (in German)
(a) Part 1, 1986 (b) Part 2, 1989
9) a) C. Jacobsen, "SMUT" Undulator spectrum program with finite electron beam dimensions. Lawrence Berkeley Laboratory, Center for x-ray optics
b) C. Jacobsen, H. Rarback, *SPIE* 582, 201-212 (1985)
10) P. Kirkpatrick, A.V. Baez, *J. Opt. Soc.* Am 38, 766 (1948)
11) See reference 1 c above. For a general treatment, see e.g. J.B. West, H.A. Padmore, "Optical Engineering" in *"Handbook on Synchrotron Radiation"*, Vol. 2, G.V. Marr, Editor, North Holland, Amsterdam, 1987
12) W. Jark, *Rev. Sci. Inst.* 63, 1241 (1992)
13) Joint project of the Center for X-Ray Optics, Lawrence Berkeley Laboratory and BESSY.

116 nm H_2 Laser Pumped by a Traveling-Wave Photoionization Electron Source

S. J. Benerofe, Guang-Yu Yin and S. E. Harris
Edward L. Ginzton Laboratory
Stanford University
Stanford, CA 94305 USA

We discuss the use of a photoionization electron source (PES) to pump a 116 nm laser in the Werner band ($C^1\Pi_u \rightarrow X^1\Sigma_g^+$) of molecular hydrogen. The laser is pumped by free electrons which are created by photoionizing molecular hydrogen with soft x-rays from a traveling-wave laser plasma. The pumping configuration presented has allowed 2 Hz operation and saturation of the 116 nm Werner band laser. Using PES pumping we have measured a small signal gain coefficient of 1.6 cm^{-1}, an improvement of more than a factor of 10 over previously reported e-beam and discharge pumped results. We also report results showing that the small signal gain coefficient can be increased by cooling the laser medium, thereby reducing the Doppler width of the laser transition. We show that even though the free electrons have an average temperature of ~10 eV, the lasing hydrogen molecules retain an ambient temperature of < 0.02 eV. We measure an extrapolated small signal gain of exp(43), with a 1064 nm pump energy of 580 mJ in 200 psec.

1. Overview

We describe the use of a traveling-wave photoionization electron source[1-3] (PES) to pump a 116 nm laser in the Werner band of H_2. The PES is constructed by using a grazing-incidence traveling-wave laser plasma[4] to make soft x-rays which in turn photoionize ambient hydrogen molecules. These are the energetic electrons which make-up the PES. We quantitatively demonstrate the advantages of PES over conventional electron pumping sources by generating saturated laser emission at 116 nm in the Werner band[5, 6] of H_2. We emphasize a special feature of this type of excitation, which is its ability to produce hot electrons while at the same time retaining an ambient (lasing medium) temperature which is comparatively cold. This is confirmed by measurements of the 116 nm gain as a function of the ambient H_2 temperature. In this work the free electrons have an average temperature of 10 eV while the lasing hydrogen molecules retain an ambient temperature of < 0.02 eV. This allows very high gain at modest pumping energy; here we obtain an extrapolated small-signal gain of exp(43), with a 1064 nm pumping energy of 580 mJ in a pulse width of 200 psec.

2. Photoionization Electron Source

A PES is produced when high energy photons pass through a medium and photoionize it. The photoionized electrons are left with a kinetic energy (KE) equal to the difference between the photon energy ($h\nu$) and the ionization potential (ϕ) of the electron donor medium

$$KE = h\nu - \phi.$$

Through a judicious choice of the electron donor medium for its ionization potential and photoionization cross section, a good overlap between the pump photon spectrum and the photoionization cross section can be achieved. In addition it may also be possible to achieve a selection which maximizes the overlap of the PES spectrum and the electron pumping cross section for the short wavelength laser scheme. We note that in PES pumping the electrons are produced in the lasing medium (if the electron donor medium and the lasing medium are intermixed or layered, for instance, in the case of a solid medium). This is in contrast to conventional electron pumping schemes where inductive and capacitive effects may limit how quickly a discharge pump can be delivered to the lasing medium. We also note that the photoionization event takes place quickly enough that the temporal characteristics of the PES will follow that of the photoionizing radiation source. This is advantageous in the pumping of short wavelength lasers. Using electrons to pump a laser scheme eases the restrictions put on the pump transitions since electrons can pump non-dipole allowed transitions and electron cross sections are usually much broader in energy than photon cross sections. This allows for efficient use of the broad band pumping sources used in short wavelength laser research.

In this work, the high energy photons are created using a laser-produced plasma. As shown schematically in Fig. 1, the high intensity pump laser is focussed onto a target. This creates a plasma which is then heated by the laser and emits soft x-rays. The soft x-rays then travel into the ambient H_2 gas surrounding the target and photoionize it. The high energy electrons created during this process are the PES. When we combine the PES with a traveling-wave pumping geometry we demonstrate its usefulness in pumping short-wavelength laser schemes.

3. Grazing Incidence Traveling-wave Geometry

Due to the rapid decay of short wavelength transitions, and the current lack of high quality mirrors, resonators for short-wavelength lasers are hard to build. Therefore, in order to observe the radiation of the short-wavelength lasers, amplified spontaneous emission is detected. To achieve the long gain lengths necessary we use a grazing incidence geometry as shown in Fig. 2.

The pump laser is focussed (in the plane of the figure) to a line on the target creating a long plasma which acts as the pump source for the PES. As we can see from Fig. 2a, as the angle of incidence (θ) is increased, the length of the plasma is increased by $1/\cos(\theta)$. This allows for long gain lengths but would lead to a reduction of the intensity of the pump laser on the target, and therefore a reduction of the conversion efficiency of the pump laser energy to soft x-ray energy. As shown in fig. 2b this problem is avoided by threading the target like a screw to bring the local angle of incidence of the pump beam (θ-α) to be near normal to the target. This preserves the projected area of the pump beam on the target and hence the conversion efficiency of the laser plasma. The tips of each of the grooves pick off the pump beam creating a plasma. The soft x-ray emission distribution of the individual point plasmas merges together at a distance from the target on the order of the groove spacing, creating a uniform line source of soft x-rays.

From fig. 2a we can also see that the angled pump beam produces traveling-wave pumping as the pump pulse front sweeps along the target. As has been suggested by several authors (see ref. (7) for example), the group velocity of a pulse in an amplifying

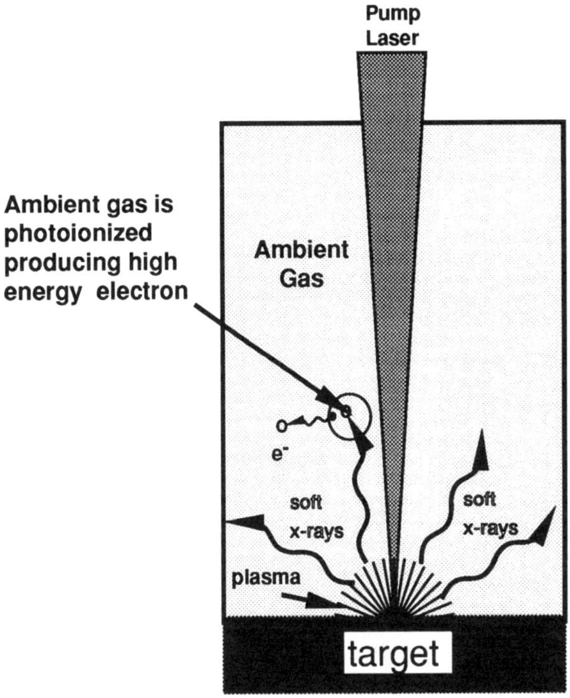

Figure 1: The high intensity pump laser is focussed on a metal target creating a plasma. The plasma emits soft x-rays into the ambient gas surrounding the target. The soft x-rays photoionize the gas producing a high energy electrons. These are the electrons which make up the PES.

medium may be less than (in some cases, significantly less than) the speed of light. To match the speed of the plasma excitation to the group velocity of the ASE pulse being amplified, the suggestion of Zs. Bor[8] could be used: Replace one mirror of the pump laser beam line by a grating used in first order to create a delay across the full width of the pump pulse equal to the propagation time of the short wavelength signal through the amplifying medium. In this way nearly perfect traveling-wave pumping can be achieved. For more details on the grooved target geometry see refs. (3,9,10).

4. Experiment

For the experiments discussed in this paper the H_2 gas is both the electron donor and the lasing medium. In a preliminary set of experiments[3] we were able to measure gain

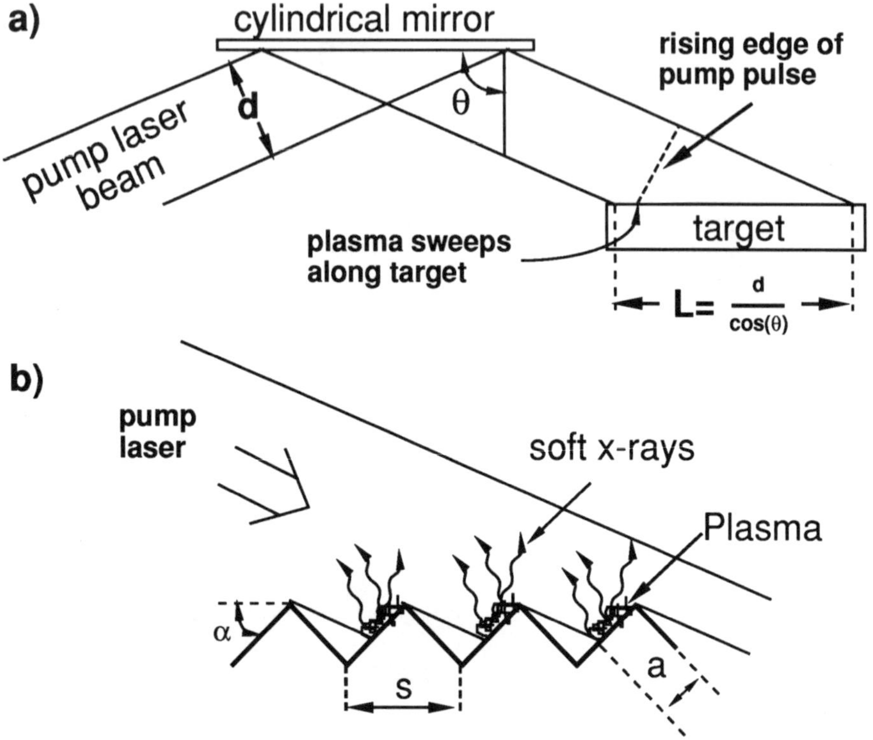

Figure 2: Grazing incidence traveling-wave geometry: a) oblique incidence geometry and traveling-wave pumping as pump laser sweeps along the target and b) the grooved target surface.

on many transitions in the Lyman and Werner bands of H_2. Concentrating on the 116 nm Werner band line, we optimized the experimental setup and in the experiments discussed below we show that even though the PES is a high energy pump source, the lasing medium is pumped and lases before it has significantly increased in temperature. We set out to show this by experimentally confirming that the 116 nm transition is Doppler broadened and that the gain cross section can be increased by decreasing the ambient H_2 temperature

The experimental setup is shown in fig. 3. For this experiment, the 580 mJ, 170 psec, 1064 nm pump pulse, initially a 3.5 cm x 1.2 cm oval beam, is reflected off of the aluminum coated cylindrical mirror at 82.6° from normal incidence with the 3.5 cm axis of the beam in the plane of incidence. The mirror focuses the beam to a line on the target with a transverse spot size of ~100 μm and a length of 27 cm, yielding an intensity on the target of 8×10^{10} W/cm^2. The target was a stainless steel tube threaded at 317.5 μm/groove (80 pitch) with $\alpha = 60°$; the density of hydrogen molecules surrounding the target was fixed for all temperatures at 17 torr (5.6 x 10^{17} molecules/cm^3). The output signal of the H_2 laser was detected using a microchannel plate detector (Chevron configuration) mounted on a 1 meter normal incidence MacPherson VUV spectrometer (model number 225). The entrance slits of the spectrometer were 75 cm away from the end of the plasma. The spectrometer and the hydrogen cell were separated by a 1 mm thick LiF window. The cell was cooled by filling thin plastic bags with liquid nitrogen or solid CO_2 (dry ice) and placing them in a styrofoam container around the experiment cell.

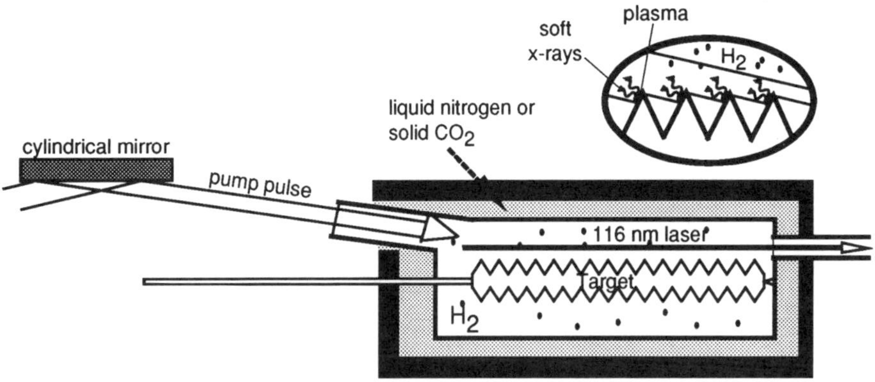

Figure 3: A schematic of the setup use for the 116 nm H2 experiment

The temperature of the hydrogen gas inside the cell was measured using two separate methods: 1) a K-type thermocouple gauge was suspended inside the cell, and 2) we measured the pressure of the H_2 inside the cell, which for a fixed density and volume determines the temperature. Both of these methods agreed to within 10 K during the two colder-than-room-temperature experiments. Given the measured gain coefficient of the 116 nm laser at room temperature, and assuming Doppler broadening of the lasing transition, we can calculate the gain expected at other temperatures due to the change in the linewidth of the laser transition and the redistribution of the population for the ground state of the hydrogen molecule subject to the known selection rules. We will outline this calculation in the next sections.

Figure 4 shows two representative gain measurements from this experiment. The change in the gain coefficient clearly shows that by simply cooling the ambient H_2 gas we can dramatically increase not only the gain coefficient, but also the length of saturated gain and hence the power output of the laser. Figure 5 shows a plot of the 116 nm gain coefficient vs. temperature. The measured data points and the theory (as discussed below) agree to better than 10%. This support both the Doppler broadening hypothesis and the assumption that the H_2 remains essentially unheated by the surrounding high energy electrons during the pumping and lasing process. Each data point represents the average of 10 to 20 separate gain measurements. Each gain measurement consists of approximately 15 equally spaced measurements of the output signal versus the length of the plasma on target. Furthermore, each data point in each of the gain measurements represents 10 separate measurements of the output signal at the given length.

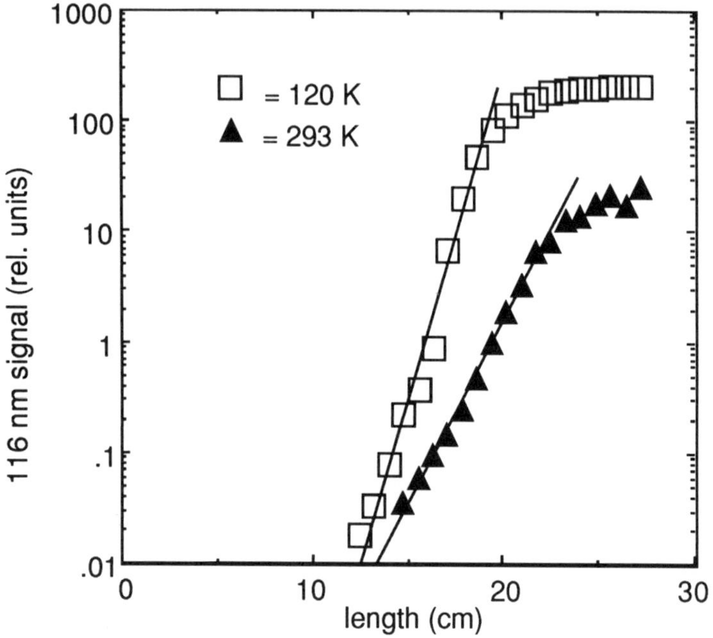

Figure 4: Two representative signal vs. length measurements for the 116 nm transition. ▲ denotes signal measured with an ambient H_2 temperature of 293 K, and ❑ denotes signal measured with an ambient H_2 temperature of 120 K. The gain was measured to be 0.9 cm-1 at 293 K and 1.6 cm-1 at 120 K.

5. A PES Calculation

Using a simple model for the PES excitation mechanism we will now outline a rough calculation of the expected gain on the 116 nm transition. Molecular hydrogen has been extensively studied and as a result most of the important quantities have been measured and/or calculated and are available in the literature. The PES calculation proceeds as follows (see Fig. 6): First the spectral distribution and intensity of the soft x-rays emitted from the laser-produced plasma must be ascertained (Fig. 6a). These variables commonly can be roughly estimated with a knowledge of the target material and the pump laser wavelength and intensity. Next, the pertinent photoionization cross sections have to be obtained (Fig. 6b). These two distributions can then be "overlapped" spectrally to obtain an electron density distribution vs. energy (Fig. 6c). It is important to note that the

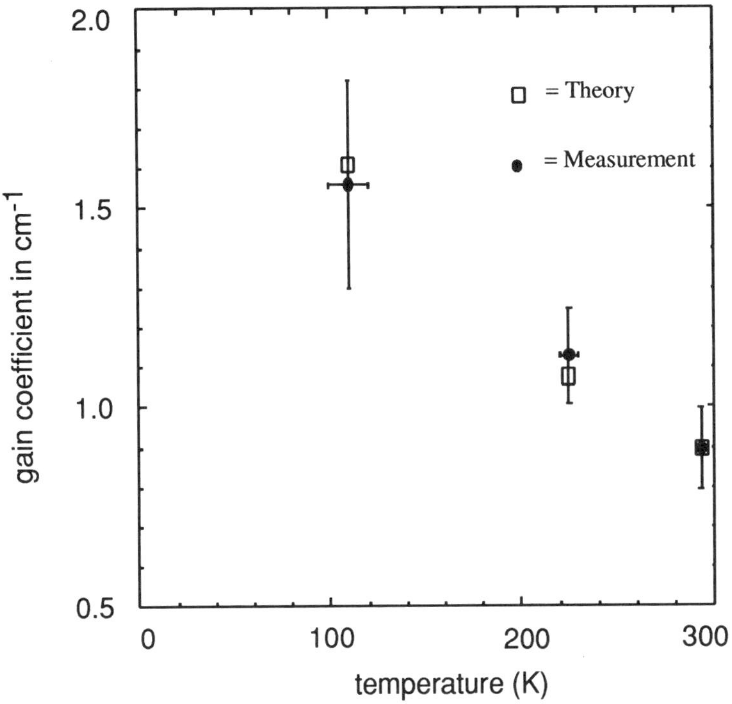

Figure 5: 116 nm gain coefficient vs. ambient hydrogen pressure. ❑ denotes the calculated gain, • denotes the average gain. The error bars in gain on the graph are the standard deviation of the measured gain for the given temperature. The error bars in temperature represent the uncertainty in the temperature measurement of the ambient H_2 gas because of the discrepancy between the two temperature measurement methods used.

kinetics of this step will often be complicated and may depend strongly on the parameters of the calculation such as the pump pulse length and the density and composition of the ambient medium[11]. (This rough calculation ignores the dynamical collision cooling effects implicit in this step.) The population of the upper laser level is then determined by overlapping the electron density distribution and the electron excitation cross section.

We now proceed with the details of the 116 nm laser calculation. First we calculate the soft x-ray pump flux (Fig. 6a). The pump laser intensity on target was calculated to be ~10^{11} W/cm^2. From measurements performed under similar conditions by the author, the laser plasma conversion efficiency into the energy range of interest at this pump laser intensity is estimated to be approximately 2%. A convenient, but not necessarily accurate way to represent the emission spectrum of a laser-produced plasma is by a blackbody. Using the parameters of this experiment we calculate a blackbody temperature of 12 eV.

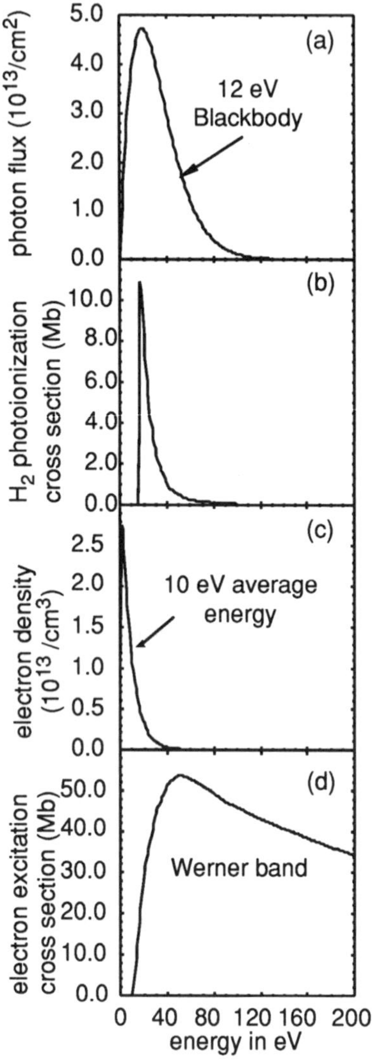

Figure 6: A graphical illustration of the PES pumping process: a) Photon flux vs. energy for a 12 eV blackbody, b) H_2 photoionization cross section vs. energy, c) Calculated electron density vs. energy distribution; the average electron energy is 10 eV, d) Werner band electron excitation cross section vs. energy.

The soft x-rays for the plasma travel from the target through the surrounding gas, photoionizing it. This "creates" the electrons which are the PES. To obtain the density of free electrons, n_e, the H_2 photoionization cross section[12] (Fig. 6.1b) is spectrally overlapped with the blackbody distribution where a uniform 60° (half) cone angle of emission has been assumed (~π steradians). Since we are calculating the gain of the laser at a specific distance from the target, measured using an aperture for spatial resolution, we attenuate the photon flux by propagating it through the intervening H_2 gas using Beer's Law. We thus obtain an electron density distribution (Fig. 6.1c). For this calculation we obtain a total electron density of $\sim 10^{15}$ cm^{-3} and an average electron energy of 10 eV. The kinetic energy of the electrons is set by the difference in energy between the photoionizing soft x-rays and the ionization potential of H_2. The electron density vs. energy distribution can be used to calculate an equivalent discharge current j which can be compared to currents produced by conventional discharges:

$$j = \langle n_e v e \rangle$$

where: v = electron velocity, e = the charge on an electron, $\langle \, \rangle$ denotes averaging over the given distribution. We thus obtain a discharge current of $\sim 4 \times 10^4$ A/cm^2, which we note has a temporal profile which closely follows that of the laser pulse[11]. As a result, this large discharge current has a sharp rise time and a fast fall time, which makes it especially useful for diagnostics, probing and pumping short wavelength lasers.

This electron density distribution then excites the upper level of the laser through electron collisions with neutral H_2. The electron excitation cross section for the Werner band of H_2 (σ_{e-}) has been calculated by Gerhart[13]. The cross section for electron pumping the 116 nm line is then calculated using refs. (14,15). We can now calculate an electron pumping rate for the upper laser level of the 116 nm transition($C^1\Pi_u$, υ'=1, J'=1):

$$N_{upper} = N_{H2}(J{=}1) \langle e v n_e \sigma_{e-} \rangle$$

where: $N_{H2}(J{=}1)$ is the number of H_2 molecules in the J=1 ground state. If we assume that the duration of the PES is the same as the pump laser pulse we obtain $N_{upper} = 2 \times 10^{13}$ cm^{-3} This yields a calculated gain on the 116 nm transition of 1.5 cm^{-1} at room temperature (293 K).

6. Cold Cell Calculation

We now want to calculate the change in the gain on the 116 nm transition as the ambient H_2 molecules are cooled. For a Doppler broadened transition, the gain cross section is inversely proportional to the Doppler width . The Doppler width of a transition is proportional to the square root of the temperature of the lasing medium as shown in the equation below:

$$\Delta\nu_D = \nu_a \sqrt{\frac{(8\ln 2)kT}{Mc^2}}$$

Where: $\Delta\nu_D$ is the Doppler width, k is Boltzmann's constant, T is the temperature of the lasing medium, M is the mass of the lasing molecule and c is the speed of light. For the three temperatures the experiment was conducted at, the calculated increase in the gain coefficient due to the narrowing of the Doppler width and the redistribution of the ground state H_2 molecules is shown in table 1 This result was used to compute the 'theory' data points in fig. 5. From that figure we observe better than 10% agreement between theory and experiment.

Temp. (K)	$\sqrt{\frac{T_{rt}}{T_{expt}}}$	$\frac{N_{rt}J=1}{N_{expt}J=1}$	$\frac{\alpha_{rt}}{\alpha_{expt}}$
293	1.00	1.00	1.00
225	1.14	1.08	1.23
120	1.56	1.14	1.78

Table 1: Calculated increase (relative to room temperature) in the 116 nm gain coefficient due to lowering the ambient temperature. Column 1 shows the three temperatures at which the experiment was conducted, column 2 shows the calculated relative increase of the gain cross section at each of these temperatures due to a narrowing of the Doppler width. Column 3 shows the calculated relative increase of the ground state population of the lasing medium at each of the three temperatures. Column 4 shows the overall calculated increase in the gain coefficient relative to the room temperature value.

7. Summary

We have shown how a traveling-wave source of laser produced x-rays may be used to produce an electron source of high average energy, high equivalent current, and very fast rise time. We note the dramatic improvement of the gain on the 116 nm transition as compared to conventional discharge pumping. We have also demonstrated that this high current pumping source will not significantly heat the lasing medium before it has lased, and that by simply cooling the lasing medium we can significantly increase the gain on the 116 nm transition and extract more energy from the system.

8. Acknowledgements

This work was jointly supported by the Strategic Defence Initiative Organization, the U. S. Office of Naval Research, and the Air Force Office of Scientific Research.

9. References

1. Wang, J.C., Caro, R.G. & Harris, S.E. *Phys. Rev. Lett.* **51**, 767-770 (1983).
2. Benerofe, S.J., Yin, G.-Y., Barty, C.P.J., Young, J.F. & Harris, S.E. *Phys. Rev. Lett.* **66**, 3136-3139 (1991).
3. Benerofe, S.J. *Methods for Efficient Pumping of Vacuum Ultraviolet Lasers* (Stanford University, 1992).
4. Sher, M.H., Macklin, J.J., Young, J.F. & Harris, S.E. *Opt. Lett.* **12**, 891-893 (1987).
5. Waynant, R.W. *Phys. Rev. Lett.* **28**, 533-535 (1972).
6. Hodson, R.T. & Dreyfus, R.W. *Phys. Rev. Lett.* **28**, 536-539 (1972).
7. Casperson, L. & Yariv, A. *Phys. Rev. Lett.* **26**, 293-295 (1971).

8. Bor, Z., Szatmari, S. & Muller, A. *Appl. Phys. B* **32**, 101-104 (1983).
9. Sher, M.H., Benerofe, S.J., Young, J.F. & Harris, S.E. *J. Opt. Soc. Am. B* **8**, 114-116 (1991).
10. Sher, M.H. *Techniques For Short Wavelength Photoionization Lasers: A 2 Hz 109 nm Laser* (Stanford University, 1989).
11. Wang, J.C. *Excitation of Metastable Levels of Li and Na Using a Laser-Produced Plasma X-Ray Source* (Stanford University, 1984).
12. Samson, J.A.R., Haddad, G.N. & Gardner, J.L. *Ionization Processes in Molecules* 1-157-158 (Pergamon, Viewig, Braunschweig, Hamburg, 1974).
13. Gerhart, D.E. *J. Chem. Phys.* **62**, 821-832 (1975).
14. Herzberg, G. *Spectra of Diatomic Molecules* 1-660 (Robert E. Krieger Publishing Co. Inc., Malabar, 1989).
15. Spindler, R.J., Jr. *J. Quant. Spectrosc. Radiat. Transfer.* **9**, 627-642 (1969).

COHERENT XUV ~ SOFT X-RAY GENERATION BY SUBPICOSECOND VISIBLE AND ULTRAVIOLET LASERS

K.Miyazaki, H.Sakai, H.Takada, and G.U.Kim*

Laser Section, Optoelectronics Division, Electrotechnical Laboratory
1-1-4, Umezono, Tsukuba, Ibaraki 305, Japan

ABSTRACT

High-order harmonic generation in He, Ne, Ar, Kr and Xe has been studied by ultrashort-pulse dye (616 nm) and XeCl (308 nm) lasers at intensities of 2 x 10^{14} and 3 x 10^{15} W/cm^2, respectively. Based on the experimental results, discussion is made on the characteristic harmonic distributions including a broad plateau, the role of ac Stark effect in the plateau formation, the fundamental-wavelength dependence of harmonic generation, and the influence of multiphoton ionization followed by dielectric breakdown of the nonlinear medium.

1. Introduction

Recent developments of intense ultrashort-pulse lasers have made it possible to produce light intensities of 10^{14} - 10^{18} W/cm^2, and much attention has been focused on the high-order harmonic generation in a gaseous medium by the extremely high-brightness lasers. Several experiments of the harmonic generation have been made using Nd:YAG[1] and glass,[2,3] KrF,[4,5] dye,[6] and Ti:sapphire lasers[7], and the highest order (shortest wavelength) of harmonic radiation generated to date is the 109th (7.4 nm) in Ne with fsec Ti:sapphire laser (806 nm) pulses.[7] The most characteristic behavior in the harmonic generation is a *plateau* where the intensities of successive harmonics are in a range of an order of magnitude. Typical harmonic distributions including a broad plateau were observed by Saclay group with psec Nd:YAG and glass laser pulses,[1,2] and also by the present authors with sub-psec dye laser pulses.[6] The plateau formation has been studied theoretically, and much effort has been devoted to the new domain of nonlinear optics at high intensities.[8-12] The high-order harmonic generation provides a promising approach to develop (tunable) coherent radiation sources in the extreme ultraviolet (XUV) and soft X-ray spectral regions, and the plateau behavior of harmonics suggests possible extension of the coherent radiation to a shorter-wavelength region. For these, there are still subjects to be studied with respect to the detailed nonlinear interaction of an intense radiation field with atomic gases.

In this paper we report a series of experimental results of the high-order harmonic generation in rare gases with sub-psec dye (616 nm) and XeCl (308 nm) laser pulses, and discuss the role of ac Stark effect in the plateau formation process, the fundamental wavelength dependence of harmonic generation, and the influence of strong multiphoton ionization (MPI) followed by dielectric breakdown of the nonlinear medium.

2. Experimental

The sub-psec 616-nm radiation was produced from a linear-cavity dye laser pumped by frequency-doubled, mode-locked Nd:YAG laser pulses. The dye laser pulse is amplified at 10 Hz by three- stage amplifiers up to the energy of 2 mJ in ~ 800 fsec. For

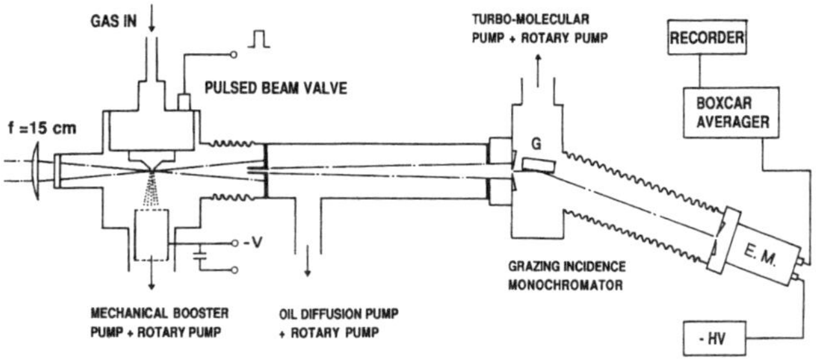

Fig. 1. Experimental arrangement for the harmonic genration.

the generation of sub-psec UV pulses, the dye laser output was frequency doubled in a 0.5 mm thick BBO crystal. The 308-nm pulse of about 100 μJ was amplified up to ~ 30 mJ by two XeCl laser amplifiers.[13] The amplified UV laser pulse widths were reliably measured to be 380 fsec by the background-free autocorrelation method using the the 3rd- and 5th-harmonic generation in a rare gas jet, of which technique for the ultrashort UV pulse measurement has been developed by the present authors.[14] The experimental arrangement for the harmonic generation[15] is shown in Fig.1. The linearly-polarized dye or XeCl laser output is focused with a 15-cm focal length CaF_2 lens into a supersonic atomic beam. The confocal parameter of the focused beam was 8.2 mm for the 616-nm radiation and 3.7 - 4.5 mm for the 308 nm, depending on the beam divergence, which are much longer than the interaction length of 1 mm. The maximum focused intensities in the nonlinear medium were 2 x 10^{14} W/cm^2 at 616 nm and ~ 10^{16} W/cm^2 at 308 nm. The gas density in the interaction region was fixed throughout the experiments and estimated to be 4.0 x 10^{18} cm^{-3} for He, 2.7 x 10^{18} cm^{-3} for Ne, and 4.3 x 10^{17} cm^{-3} for Ar, Kr and Xe. In the downstream of atomic beam, an electrode is placed to detect ions produced through multiphoton ionization (MPI) and dielectric breakdown of the nonlinear medium.

3. Harmonic distributions with the 616-nm laser

Figure 2 shows an example of the high-order harmonic spectrum obtained in He. The highest-order harmonics (shortest wavelengths) observed were the 41st (15.0 nm) in He, the 37th (16.6 nm) in Ne, the 19th (32.4 nm) in Ar, the 17th (36.2 nm) in Kr and the 13th (47.4 nm) in Xe. The detection limit of small harmonic signal was governed by the

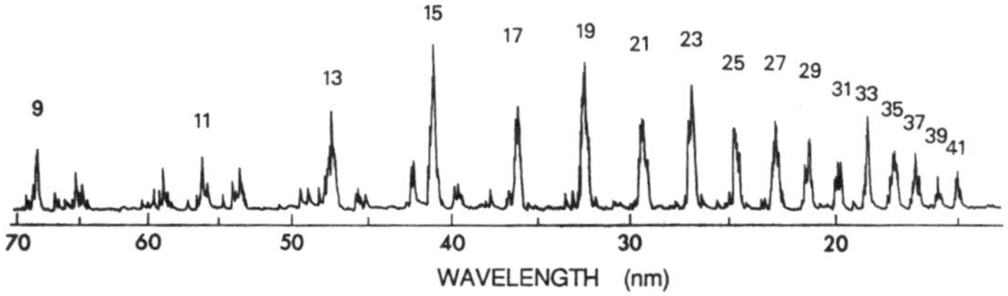

Fig. 2. Example of the harmonic spectrum observed in He with the 616-nm laser. The spectral peaks are labelled by he harmonic order, and the unmarked peaks are the second-order diffractions from the grating.

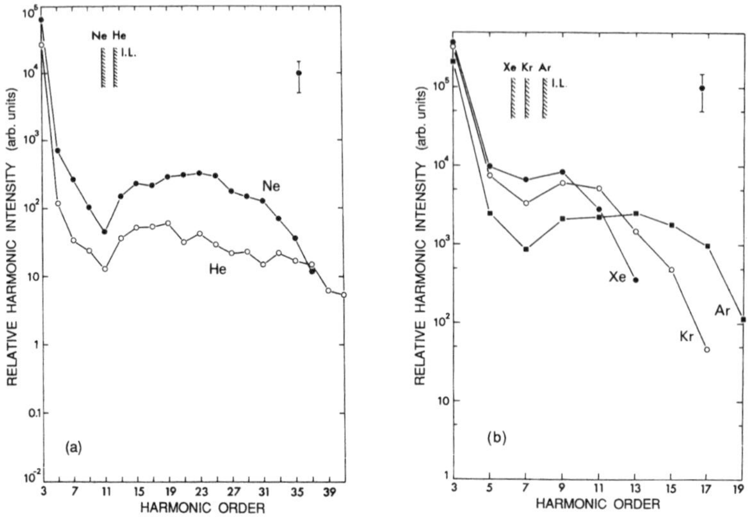

Fig. 3. Harmonic distributions in (a) He and Ne, and in (b) Ar, Kr, and Xe. I = 1.4, 1.9, 1.9, 1.5, and 2.0 x 10^{14} W/cm^2 for He, Ne, Ar, Kr, and Xe, respectively. For comparison, the harmonic intensities in (a) and (b) are plotted on the same relative scale.

random noise due most probably to MPI or subsequent dielectric breakdown. At the highest fundamental intensity, the ion signal was detected in all the rare gases employed. In Xe, visible dielectric breakdown was observed and accompanied with relatively large random noise. The weakest harmonic signal could be detected in He.

Figure 3 shows the calibrated harmonic distributions (a) in He and Ne, and (b) in Ar, Kr, and Xe. In He and Ne, the 3rd-order harmonic is most intense, and after a decrease from the 3rd to the 9th, an intensity *minimum* appears at the 11th order in the energy region near the ionization limit (IL). Then the higher-order harmonics increase again and represent a broad *plateau* from the 13th to the 37th in He and to the 33rd in Ne where the harmonic signals are within about an order of magnitude. The harmonic distributions observed in Ar, Kr, and Xe are similar to those in He and Ne: The harmonic signal decreases from the 3rd, and the plateau which is preceded by an intensity minimum at the 7th harmonic develops from the energy region near the IL.

It is clearly seen that there is a systematic trend in the harmonic distributions observed. The intensity of low-order (3rd to 9th) harmonics increases with atomic number from He to Xe. The highest order of harmonics and the widths of plateau region decrease with increasing the atomic number. (The detectable order of harmonics for the heavier atoms was partly limited by the MPI and/or dielectric breakdown.) These results strongly suggest that the harmonic radiation is mainly generated by the nonlinear response of neutral atoms. In this respect, the high-order harmonic generation, as well as the characteristic harmonic distributions, should be understood as the intrinsic nature of an atomic single-electron dipole induced by the intense radiation field, as suggested by several theoretical studies.[8-12] However, the experimental results obtained here include some characteristic features that have not been expected and observed so far.

4. Formation of a plateau

We note in Fig.3 that the plateau in all the rare gases is preceded by an intensity minimum and start to be formed around the region near the IL. To see this in more detail, we measured the fundamental intensity I dependence of the harmonic spectrum generated near the IL of Ar, and the result is shown in Fig.4. The intensity of each-order harmonic never follows the simple power law of I in the weak field limit. At the lowest value of I = 1.7 x 10^{13} W/cm^2, neither intensity minimum nor plateau of harmonics is observed. By increasing I, the 9th harmonic just above the IL increases rapidly to overcome the 7th. With further increases in I, the 9th harmonic signal decreases the I-dependent growth rate, and the 11th harmonic and subsequently the 13th increase rapidly to form a plateau. This result shows that theplateau is induced by the incident strong field and results from the nonperturbative nonlinear response of atomic electrons to the intense laser pulse.

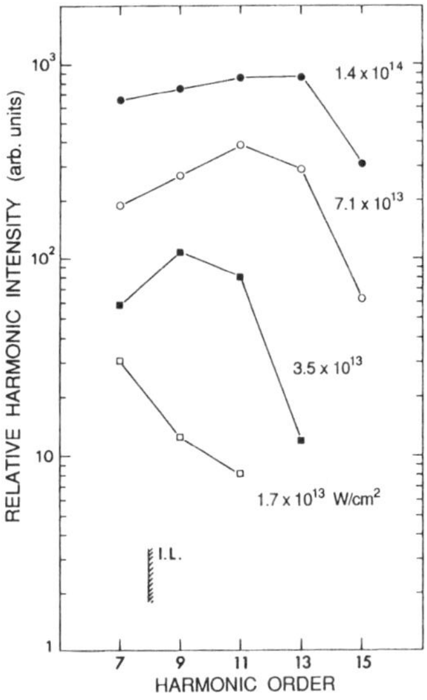

Fig. 4. Harmonic distributions in Ar at several fundamental intensities.

To see the effect of single-atom response for increasing I on the harmonic distribution, we made a nonperturbative calculation using the atomic model proposed by Becker *et al.*[11] where no excited state was incorporated. In spite of the simple model, the calculated results for He, Ne and Ar provided a very gross feature of the plateau formation or the I-dependent increase in the high-order harmonics. This has been shown already by the detailed nonperturbative calculations for Ne at 616 nm by Kulander *et al.*[16] and for Xe and model atoms at 1.064 μm by several authors.[8-12] The qualitative agreements between the theoretical and experimental results lead to a conclusion that the plateau itself is due mainly to a general property of the nonperturbative nonlinear response of atoms to an intense radiation field. However, the detail of the characteristic harmonic distributions observed here, especially the minimum at a specific harmonic order and the characteristic I-dependent increase in each-order harmonic, have never been reproduced completely by the calculations, suggesting some other important effect on the plateau formation.

Recently, L'Huillier *et al.*[17] have shown that the importance of conventional phase matching is considerably reduced in the nonperturbative regime, compared with that in the weak field limit. This conclusion was based on the theoretical result that the nonlinear polarization proportional to the single-atom response varies much less rapidly with the incident strong field, and the single-atom emission has a weak intensity dependence at the harmonic frequencies. This weak I-dependence of high-order harmonics appears to be supported by the result presented in Fig.4. In order to know the detailed I-dependence and the characteristic nature of each-order harmonic generated from the energy region of IL, we measured the harmonic intensity as a function of I. The results for the 7th, 9th,

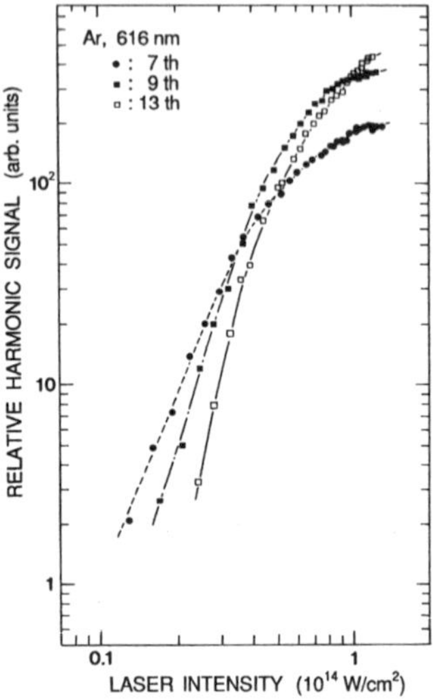

Fig. 5 Harmonic intensity as a function of the laser intensity.

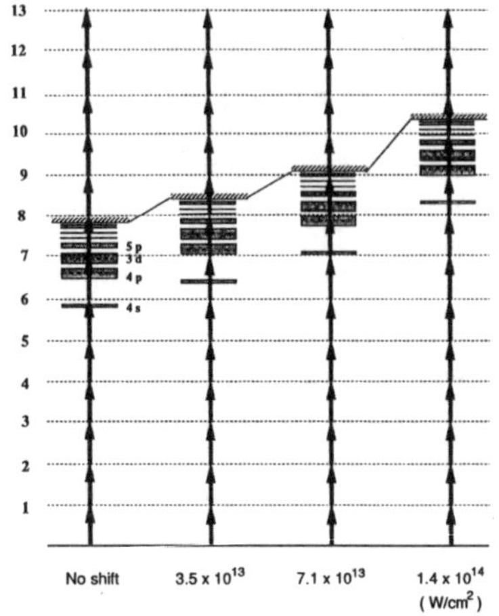

Fig. 6. ac Stark shifts of excited states and ionization limit of Ar with the increase in laser intensitiy.

and 13th harmonics in Ar are shown in Fig.5, which are in consistent with those in Fig.4. In the region of $I < \sim 3 \times 10^{13}$ W/cm², the higher-order harmonic is weaker and increases more rapidly with an increase in I. With the increase in I, the 7th-order harmonic decreases rapidly the growth rate and begins to saturate first for $I > \sim 3 \times 10^{13}$ W/cm² to form the minimum shown in Fig.3(b). The 9th harmonic overcomes the 7th at $I \sim 3 \times 10^{13}$ W/cm² and also tends to saturate around $I \sim 7 \times 10^{13}$ W/cm². In contrast to these, the 13th harmonic continues to increase rather rapidly with increasing I up to a level comparable to (or higher than) the 9th harmonic, but it also goes into the saturation region at the highest intensities. It is noted in Fig.5 that the harmonic increases in a way characteristic to the each order with the increase in I, whereas the different order of harmonics represents a slow increase at the highest intensities. We observed the similar I-dependences of harmonics in Ne having the IL much higher than that of Ar, where the 11th harmonic representing the minimum in Fig.3(a) decreased most rapidly its growth rate with an increase in I, and the 13th overcome the 11th to form a plateau at the highest intensities.

5. The role of ac Stark effect

From these results, we note the importance of atomic structure of each nonlinear medium. It has been pointed out by several authors that the harmonic generation at high intensity is closely associated with the above threshold ionization (ATI) in the interaction of an intense laser pulse with atoms.[9,18] The plateau formation process observed here may be illustrated by referring to the ATI results, especially by taking into account the ac Stark effect. In the intense radiation field, the excited states and IL of rare gas atoms are well known to be Stark shifted toward higher energy approximately by the quiver or ponderomotive energy $E_P \sim e^2E_0^2/4m\omega^2$, where E_0 is the peak electric field of the laser at frequency ω, and m and e are the electron mass and charge.[19] In the present experiment, E_P or the shift becomes equal to the incident photon energy of 2.0 eV at $I = 5.7 \times 10^{13}$ W/cm^2. For example, the shifting IL of Ar exceeds the 9th- and 11th-harmonic energy at $I \sim 6.6 \times 10^{13}$ and 1.8×10^{14} W/cm^2, respectively. The shifts of excited states and IL of Ar for I in Fig.4 are schematically shown in Fig.6.

The temporal shifts of *a lot of* excited states and IL in *a single* laser pulse would provide two significant effects on the harmonic generation process concerned. One of them is the temporal and spatial phase change in the harmonic radiation that are generated from the energy region of shifting excited states. When an excited state is crossing a harmonic photon energy, the harmonic radiation experiences a rapid and large change in the refractive index due to anomalous dispersion. The rapid and multiple refractive index changes due to *many* shifting excited states induce large temporal and spatial phase mismatches between the harmonic radiation and the nonlinear polarization induced by the fundamental, and should suppress the rapid growth of the harmonic generated from this (resonance) region. As seen in Fig.6, at low intensities the 9th harmonic in Ar is generated from the region above the IL. With the temporal increase in the shifts of excited states and IL, the photon energy of the 9th harmonic falls into the resonance region at more than $I \sim 6.6 \times 10^{13}$ W/cm^2. This I corresponds approximately to that at which the I-dependent growth rate of the 9th harmonic decreases rapidly in Figs.4 and 5. For the 7th harmonic, the refractive index changes would be fairly large even at lower I, because it is originally produced from the resonance region. Thus, the 7th cannot be generated efficiently. This would lead to the characteristic I-dependence of the 7th harmonic shown in Figs.4 and 5, resulting in the formation of a minimum. In addition to the strong field effect discussed by L'Huillier *et al.*,[17,20] the refractive index changes due to the Stark shifts should be taken into account in the phase matching, especially for the harmonics produced from the resonance region, and in the resulting formation of the characteristic distributions.

The other effect of the ac Stark shifts is possible resonance enhancement of the higher-order harmonics. For example, at $I \sim 2 \times 10^{14}$ W/cm^2 the shifting excited states of Ar are in resonance with a multiple number (7 to 11) of fundamental photon during the course of evolution of a laser pulse. The *dynamic resonances* can contribute to the efficient enhancement of the higher-order harmonics generated above the shifted IL where the refractive index change due to the shifting excited states would be small. In Fig.4, the 9th harmonic in Ar can be enhanced by the 7 and 8 photon resonances in a range of $I \approx 1.7 \sim 3.5 \times 10^{13}$ W/cm^2, and the enhancement of the 11th harmonic is expected by the 7 to 10 photon resonances at higher intensities.

The formation of characteristic harmonic distributions in Kr and Xe is possibly illustrated as in Ar. For He and Ne, the formation of a minimum at the 11th harmonic should be due to the the same mechanism as for the 7th in Ar. In addition, comparing the harmonic distributions shown in Fig.3(a), the enhancement of higher-order harmonics above the 11th is larger in Ne than in He. This may suggest the importance of a lot of shifting excited states of Ne for the efficient enhancement of the harmonics produced above the shifted IL. At the peak values of I in Fig.3 (a), the 15th harmonic in He and Ne is the *lowest* order locating in the continuum above the *shifted* IL, and then the 11th harmonic representing the intensity minimum is the *highest* order below the shifted resonance region. Similarly, in Fig.3 (b) the 13th in Ar and the 11th in Kr and Xe are the lowest above the shifted IL, and the 7th is the highest below the resonance region. In these heavier atoms having a relatively low IL, especially in Xe and Kr, however, the MPI or the dielectric breakdown would limit the formation of a broader plateau, as discussed later.

It is clear that the harmonic order which represents a minimum or starts to form the plateau should depend on I, as seen in Fig.4. Thus, we may conclude that at high intensities the minimum in the distribution appears around the harmonic order which experiences the most significant refractive index change due to the shifting excited states in a laser pulse, and the plateau starts to be formed from the region near the IL by additional enhancement due to the dynamic resonances. This scheme for the minimum and plateau formations should reconcile with the theoretical models given so far.

6. Harmonic generation with the 308-nm laser

We made the same experiments with sub-psec XeCl (308 nm) laser pulses to see the fundamental wavelength dependence of the harmonic generation. An example of the harmonic spectrum observed in He is shown in Fig.7, and Fig.8 shows the calibrated harmonic distributions in He and Ne at $I \sim 3 \times 10^{15}$ W/cm^2. The highest-order harmonics (wavelength) observed with the 308-nm radiation is the 15th (20.5 nm) in He, the 13th (23.7 nm) in Ne, and 9th (34.2 nm) in Ar, Kr and Xe. The high-order harmonics were generated much more efficiently than with the visible laser pulses, but strong dielectric breakdown was observed in all the rare gases at the highest intensity used. It is noted in Fig.8 that the most intense harmonic signal in He was observed at the 7th which is generated near the energy region of the IL, and a plateau-like distribution ranges from the 3rd to the 11th order, while the higher-order harmonics decrease monotonically with an increase in the harmonic order. A similar distribution is observed in Ne, where the 7th-order harmonic is fairly enhanced. In contrast to those observed with the visible dye laser, however, no broad plateau is formed around the IL. For Ar, Kr and Xe, a monotonic decrease in the higher-order harmonics was observed at $I \sim 3 \times 10^{15}$ W/cm^2.

Figure 9 shows the harmonic distributions for He at several values of I. Only the 3rd to 7th-order harmonics are observed at less than $I \sim 2 \times 10^{14}$ W/cm^2. With an increase in I, the 7th harmonic increases rapidly to overcome the 5th. However, the harmonic distribution having a peak at the 7th does not change with a further increase in I up to 2.5 x 10^{15} W/cm^2, while the each-order of harmonics increases appreciably with the increase in I. The similar behavior of harmonic radiation was also observed in Ne. As discussed in Sec.5, at the UV intensity of $I \sim 2.5 \times 10^{15}$ W/cm^2 one may expect the ac Stark shift of ~ 22 eV. However, the results shown in Fig.9 do not suggest such a large shift, in contrast

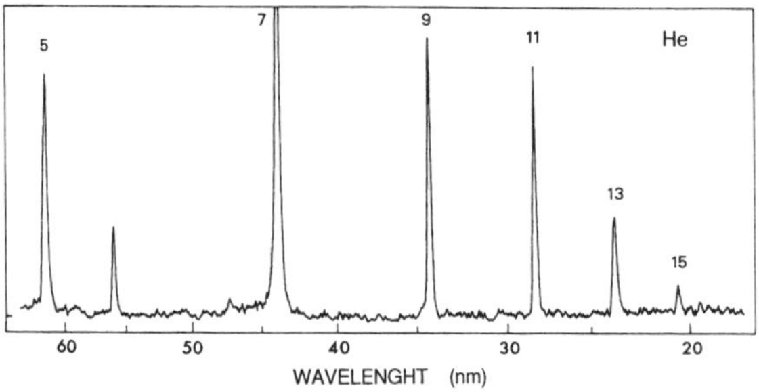

Fig. 7. Example of the harmonic spectrum observed in He with the 308-nm laser. The spectral peaks are labelled by the harmonic order.

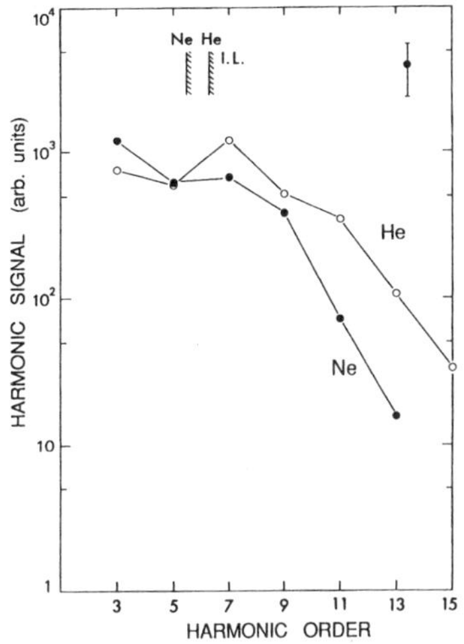

Fig. 8. Harmonic distributions observed in He and Ne with the 308-nm laser. I = 2.5 and 2.8 x 10^{15} W/cm^2 for He and Ne, respectively.

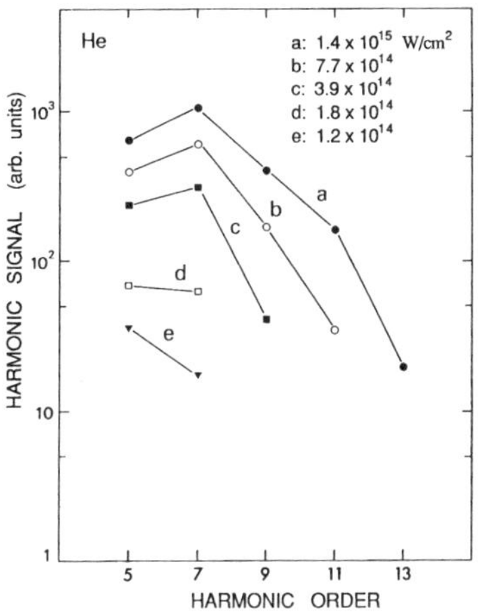

Fig. 9. Harmonic distributions in He with the increase in laser intensity.

to those obtained with the visible dye laser. This is certainly due to the dielectric breakdown of nonlinear medium which is the avalanche increase in electron density initiated by the strong MPI. This has been confirmed by the simultaneous measurement of harmonic and ion signals as a function of the position of laser focus, and an example of the results for He is shown in Fig.10. At I = 1.1 x 10^{14} W/cm^2, the 5th harmonic signal is maximized when the UV laser is focused into the central region of He beam, and the beam center of He may be identified to the position of peak ion signal. However, at I = 3.4 x 10^{15} W/cm^2, where the visible strong dielectric breakdown was observed in the medium, the UV radiation focused into the He beam generates the harmonic less efficiently than that focused apart from the central part of He beam. With the 616-nm radiation, the similar

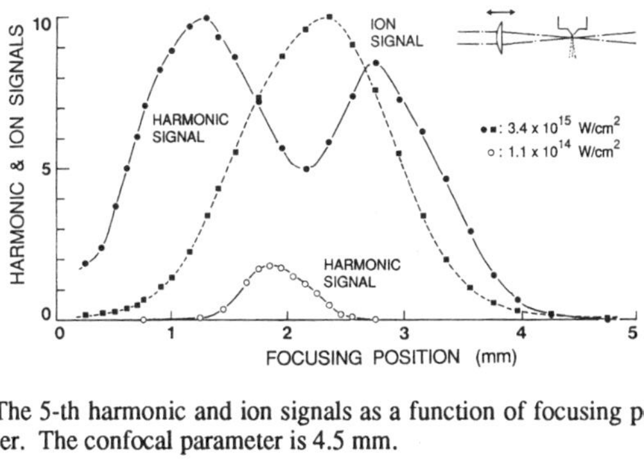

Fig. 10. The 5-th harmonic and ion signals as a function of focusing position of the 308-nm laser. The confocal parameter is 4.5 mm.

results were observed for Kr and Xe, but for He and Ne the visible laser did not produce the double peaks of harmonic signal.

These results show that the strong MPI followed by the dielectric breakdown limits I in the nonlinear medium to a value much lower than that used or expected, and prevents a large shift for efficient enhancement of the higher-order harmonics. The temporal evolution of dielectric breakdown usually destroys the coherence of the incident field or the coherent interaction process. In Fig.9, one can see that the effective I to form the harmonic distribution in He should be approximately less than $\sim 4 \times 10^{14}$ W/cm^2, and then a shift of $3 \sim 4$ eV is expected before the onset of dielectric breakdown even at higher intensities. This shift can enhance the 7th harmonic through the 5- and/or 6-photon resonance in He and Ne but is too small to generate efficiently the higher-order harmonics by the dynamic resonances. This leads to the harmonic distribution shown in Fig.8. The same limiting process due to the dielectric breakdown would certainly take place in the harmonic generation with the intense KrF laser pulses,[4,5] where very high intensities of 10^{16} -10^{18} W/cm^2 were used for the experiments, but no broad plateau formed around the IL has been observed in the harmonic distributions.

Recently, Krause *et al.*[21] have pointed out a possible large contribution of ions to the high-order harmonic generation, because they found some disagreement between the experimental results and the theoretical predictions concerning the highest-order of harmonic to be observed. However, the present results obtained at 308 nm show that the high-order harmonics are mainly generated by *neutral atoms*, and the ions produced in the evolution of a laser pulse do not contribute appreciably to the harmonic generation. The discrepancy between the theory and experiment may be corrected by incorporating the enhancement of harmonics due to the dynamic resonances.

7. Conclusion

We have studied the high-order harmonic generation with intense sub-psec dye and XeCl laser pulses. The characteristic harmonic distributions obtained by the visible laser represent a broad plateau that is preceded by an intensity minimum at a specific harmonic order. The results strongly support that the ac Stark shifts of excited states and IL play an important role in the formation of the characteristic harmonic distributions. The UV laser pulse induced strong MPI followed by dielectric breakdown of the nonlinear medium, which limited I to a much lower value than that used and restricted efficient higher-order harmonic generation. No evidence was observed of the ion contribution to the high-order harmonic generation, so far as the ions were produced in the evolution of a single laser pulse. The plateau formation process discussed here suggests possible extension of the (tunable) coherent radiation to a shorter-wavelength region. For this, the present results suggest advantages of longer-wavelength, ultrashort pulse lasers that can induce the dynamic resonances effective for the higher-order harmonic generation and reduce the influence of medium ionization. Furthermore, because of the ω^{-2} dependence of E_p, the longer-wavelength radiation alleviates I necessary to induce a shift of excited states and IL for the resonance enhancement of the high-order harmonics. Recently, Macklin *et al.*[7] have observed the high-order harmonic radiation up to the 109th (7.4 nm) in Ne with fsec Ti:sapphire laser pulses, and we have also obtained a preliminary result of the high-order harmonic generation up to the 67th (11.7 nm) with fsec Ti:sapphire laser pulses. These observations are consistent with the present conclusions.

* On leave from Korea Research Institute of Science and Technology.

1. M.Ferray, A.L'Huillier, X.F.Li, L.A.Lompré, G.Mainfray, and C.Manus, *J. Phys. B: At. Mol. Opt. Phys.* **21** (1988) L31; X.F.Li, A.L'Huillier, M.Ferray, L.A.Lompré, and G.Mainfray, *Phys. Rev. A* **39** (1989) 5751.
2. A.L'Huillier, L.A.Lompré, G.Mainfray, and C.Manus: in *Proc. 5th Int. Conf. on Multiphoton Processes*, eds. G.Mainfray and P. Agostini (CEA, Paris, 1991) p.45; Ph.Balcou, C.Cornaggia, A.S.L.Gomes, L.A.Lompré, and A.L'Huillier, to be published.
3. J.K.Crane, M.D.Perry, R.W.Falcone, S.M.Herman, K.S.Budil, and D.Stickland, in *Quantum Electronics and Laser Science Conference 1992, Technical Digest Series Vo.13* (OSA, Washington, DC,1992) p.276.
4. A.McPherson, G.Gibson, H.Jara, U.Johann, T.S.Luk, I.A.McIntyre, K.Boyer, and C.K.Rhodes, *J. Opt. Soc. Am. B* **4** (1987) 595.
5. N.Sarukura, K.Hata, T.Adachi, R.Nodomi, M.Watanabe, and S.Watanabe, *Phys. Rev. A* **43** (1991) 1669.
6. K.Miyazaki and H.Sakai, *J. Phys. B: At. Mol. Opt. Phys.* **25** (1992) L83.
7. J.J.Macklin, J.D.Kmetec, and C.L.Gordon III, in *Quantum Electronics and Laser Science Conference 1992, Postdeadline Papers* (OSA, Washington, DC,1992) p.18.
8. K.C.Kulander and B.W.Shore, *Phys. Rev. Lett.* 62 (1989) 524.
9. J.H.Eberly and Q.Su, and J.Javanaien, *Phys. Rev. Lett.* **62** (1989) 881; *J. Opt. Soc. Am. B* **6** (1989) 1289.
10. R.M.Potvliege and R.Shakeshaft: *Phys. Rev. A* **40** (1989) 3061.
11. W.Becker, S.Long, and J.K.McIver, *Phys. Rev. A* **41** (1990) 4112.
12. J.L.Krause, K.J.Schafer, and K.C.Kulander, *Phys. Rev. A* **45** (1992) 4998.
13. K.Miyazaki, T.Fukatsu, I.Yamashita, T.Hasama, K.Yamada, and T.Sato, *Appl. Phys. B* **52** (1991) 1.
14. H.Sakai, K.Miyazaki, and G.U.Kim, to be published in *J. Opt. Soc. Am. B*.
15. K.Miyazaki, H.Sakai, and T.Sato, *Appl. Opt.* **28** (1989) 699.
16. K.C.Kulander, K.J.Schafer, and J.Krause, in *Proc. 5th Int. Conf. on Multiphoton Processes*, eds. G.Mainfray and P. Agostini (CEA, Paris, 1991) p.119.
17. A.L'Huillier, K.J.Schafer, and K.C.Kulander, *Phys. Rev. Lett.* **66** (1991) 2200; *J. Phys. B: At. Mol. Opt. Phys.* **24** (1991) 3315.
18. M.D.Perry, A.Szoke, and K.C.Kulander, *Phys. Rev. Lett.* **63** (1989) 1058.
19. R.R.Freeman and P.H.Bucksbaum, *J. Phys. B: At. Mol. Opt. Phys.* **24** (1991) 325.
20. A.L'Huillier, P.Balcou, and L.A.Lompré, *Phys. Rev. Lett.* **68** (1992) 166.
21. J.K.Krause, K.J.Schafer, and K.C.Kulander, *Phys. Rev. Lett.* **68** (1992) 3535.

GERMANIUM LASER AT 236 Å

P. JAEGLE[1], A CARILLON[1], H Z CHEN[2], P DHEZ[1], L. DWIVEDI[4], J JACOBY[2], G JAMELOT[1], JIE ZHANG[5], M H KEY[2], A KIDD[2], A KLISNICK[1], R KODAMA[2], J KRISHNAN[4], C L S LEWIS[3], D NEELY[3], P NORREYS[2], D O NEILL[3], G J PERT[6], S A RAMSDEN[6,7], J P RAUCOURT[1], G J TALLENTS[4], J UHOMOIBHI[3].

1 Laboratoire de Spectroscopie Atomique et Ionique, Université Paris-Sud, France
2 Central Laser Facility, Rutherford Appleton Laboratory, Chilton, England
3 Department of Pure and Applied Physics, Quenns University, Belfast, N Ireland
4 Department of Physics, University of Essex, Colchester, England
5 Clarendon Laboratory, University of Oxford, England
6 Department of Computational Physics, University of York, England
7 Department of Physics, University of Hull, England

ABSTRACT

A half-cavity using a multilayer mirror has been set up close to the end of a large gain germanium-target system. An unambiguous saturation behaviour of the 232 Å and 236 Å lines has been observed. The role of the cavity mirror in this result is discussed. Suggestions are made for practical applications of an XUV laser in this wavelength range.

1. Introduction

A half-cavity, consisting of a multilayer mirror perpendicular to the axis of an amplifying plasma column, is expected to strongly increase the the XUV laser gain, even with a small reflection coefficient of the mirror [1]. In the case of a collisionally pumped X-ray laser, such a half cavity makes it possible to increase the effective path length of radiation much beyond the maximum length allowed by refraction. This is especially important when the plasma is produced from massive targets because then the free-electron density gradient is generally steep, and causes a fast refractive ejection of the beam out of the amplifying region.

Furthermore germanium slab targets irradiated by 1.05 μ pulsed laser have proved to yield neon-like ion lasing lines at 232 Å and 236 Å [2]. Later on an important gain optimization for these lines has been achieved at the R.A.L.with a doubled target system with which a partial compensation of refractive deviation was obtained [3] Therefore to bring together a half-cavity and the double target system could lead to a very large gain, hopefully larger than the saturation threshold. This has been effectively obtained [4]

The present paper discusses the part of the half-cavity in the achievement of the saturation intensity . Furthermore the characteristics of the X-ray laser pulse produced by this laser are given in view of future applications.

2. Half-cavity

The scheme of the experimental device is shown in figure 1. The two targets have their laser-irradiated surfaces opposite each other [3]. The concave mirror is located at 22 mm from the edge of the first target. It is protected by a thin metal shield. A 1mm-diameter aperture is punched for the X-ray laser beam passing through. The first target length is of 22 mm while the second can take values of 5 mm, 9 mm and 14 mm.The mirror curvature radius is of 130 mm, so that the focus is very near the exit end of the laser.The flight time for a double pass along the first plasma alone is≈ 400 ps.

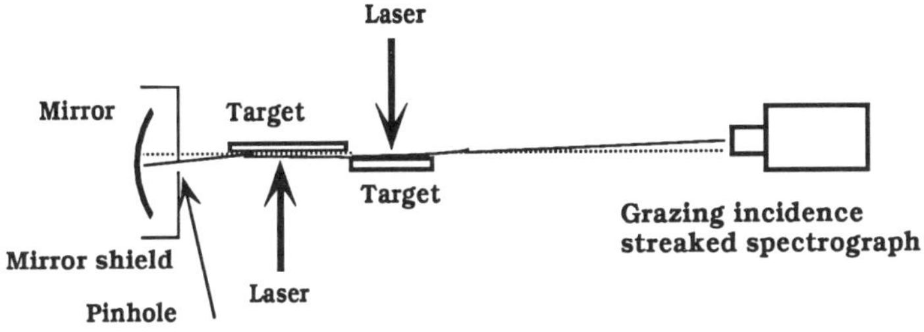

Fig. 1: Sketch of the germanium laser.

The multilayer mirror was made of 40 pairs of Mo-Si layers deposited by sputtering on a super-polished glass support*. Figure 2 displays the curve of the spectral dependence of the mirror reflectivity**. The reflection coefficients are of 24% and 28% at 232Å and 236 Å respectively . We will see that the mirror operates with an effective reflection coefficient much less than these values.

The half-cavity carried out with this mirror will improve greatly X-ray laser operation owing to two main properties:

a) A half-cavity provides a large gain increasing even though the mirror reflection coefficient may be small. This is because the effective gain-length product resulting from radiation double-pass in plasma is approximately given by the expression:

$$(Gl)_{eff} = 2Gl + \mathrm{Ln}\, R \qquad (1)$$

where l is the plasma length and R the reflection coefficient. Ln R ranges between - 5 and -1. Thus, in a large gain medium, a small R gives only a small negative contribution to the doubling of Gl coming from double-pass. For instance, for a plasma column of Gl = 10, a mirror of 5% reflection will give an effective gain-length of 17 which is not much smaller than 20 for 100% reflection.

b) From (1) one can define an effective length, l_{eff}, travelled by radiation when the cavity is working:

---

* The mirror was coated by C M Falco, J M Slaugther and D Schultze, Laboratory for X-Ray Optics, Optical Science Center, University of Arizona, Tucson

** The mirror was coated by C M Falco, J M Slaugther and D Schultze, Laboratory for X-Ray Optics, Optical Science Center, University of Arizona, Tucson

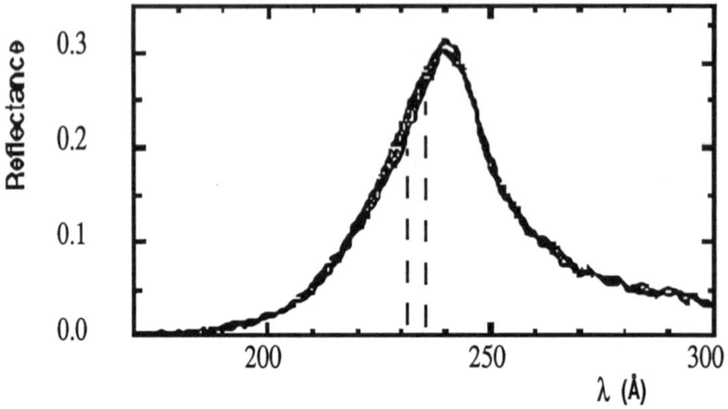

Fig. 2: Spectral band of the multilayer mirror; the dashed lines give the location of the germanium lasing lines. The figure shows the quasi-perfect overlapping of three curves recorded at three different places of the mirror.

$$l_{eff} = 2l + \frac{1}{G} \mathrm{Ln}\, R \qquad (2)$$

Contrary to what occurs for a physical lengthening of plasma, the increasing from l to l_{eff} is not limited by refraction since, in a stationary plasma, the curvature of the light trajectory is the same before and after reflection. This is illustrated by the figure 3. The plasma expansion between the first and the second pass tends to reduce the steepness of the density gradient, and also the beam refractive deviation. This small effect has been observed in our experiments and led us to make small corrections to the detector position. But it has no significant consequence on l_{eff} as far as the target length matches the refractive deviation. If the target length was greater than the useful one, this effect could increase slightly the apparent saturation threshold. Since this is opposite to the refraction effect it cannot bring about confusion between saturation and refraction at large gain.

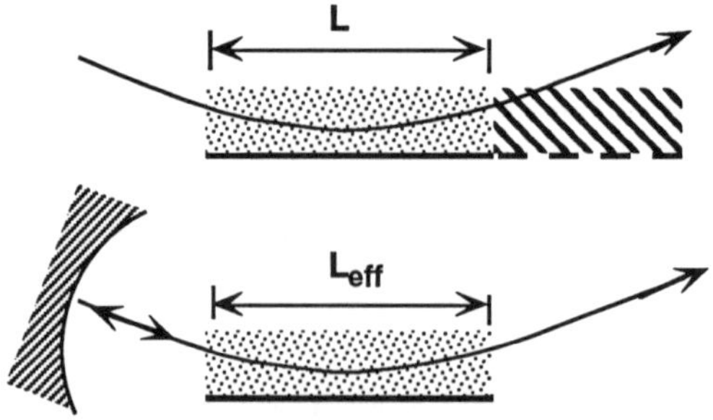

Fig. 3: The upper diagram illustrates the refractive limitation of the gain-length product; below, one sees that the mirror increases the effective laser length without additional refractive effect.

The fact that the mirror does generate a dramatic increase of the intensity at 236 Å is clearly shown in figure 4. This figure displays the time-resolved signal without and with mirror for the first target alone. The typical increase is a factor 30. Moreover the time duration of the signal is not changed by double-passing which was not the case in a similar experiment [5]. This shows the mirror is not destroyed from the beginning of the interaction but keeps its efficiency, at least partially, during the main part of the gain life-time.

At this point there is still no saturation of intensity at all. The gain coefficient of the plasma, measured without mirror, is near 4 cm^{-1}. With the two targets, an exponential intensity increase is observed up to $G \approx 13$. The increase by a factor 30, produced by the mirror, means that the target length of 2.2 cm is to be replaced by an effective length, $l_{eff} \cong 3.1$ cm, or in other words,that the effective gain-length is about 12. This is still in the range of the exponential intensity behaviour.

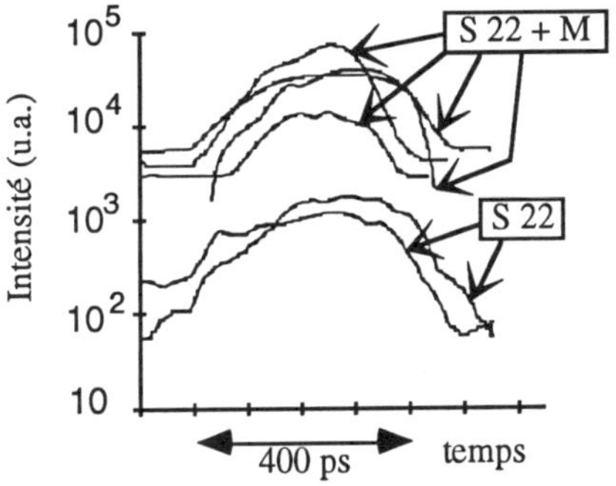

Fig. 4: Time dependant output from the first target (l = 22 mm) without (lower curves) and with (upper curves) multilayer mirror. The typical increasing factor is about 30.

At the same time, this result shows that the mirror doesn't work with the reflection coefficient exhibited in figure 2. Otherwise, from the relation (1), the effective gain-length should be of 15 instead of 12. In fact there is an "effective reflection coefficient" which is of the order of 2.10^{-2}. This is partially due to the fact that not all the reflected radiation falls into the amplifying region of the plasma during the second pass. But it is also possible that the heating of the mirror by the X ray laser beam leads to destructive pertubations at the thin layer interfaces.

3. Saturation

When the the intensity reaches a large value, such that the rate of new photon creation becomes larger than the rate of emitter creation, the density of population inversion starts to decrease. At given gain coefficient this happens when the plasma length (or effective length) is above some threshold which is discussed below. Then the gain coefficient starts also to decrease and the intensity ceases to behave exponentially as a function of the plasma length.

In order to estimate the saturation threshold let us consider a two level system, i and j. Let us call n_i and n_j the reduced level populations, g_i, the statistical wight of level i , A_{ij}, the spontaneous emission probability, $\phi_{sp}(\nu)$, the line profile function for spontaneous emission, Θ, the solid angle. From the radiative transfer equation as applied to an uniform plasma column, the relation between the photon density, n(ν,l), and the length l can be written [6]:

$$\frac{dn(\nu,l)}{dt} = g_i n_i \phi_{sp}(\nu) \frac{A_{ij}\Theta}{4\pi} \exp(Gl) \tag{3}$$

Regarding the emitter creation, let us call τ the inversion population recovery time. In order that the photon creation rate remains smaller than the average velocity of emitter creation we must have:

$$\frac{dn(\nu,l)}{dt} \leq \frac{g_i(n_i-n_j)}{\tau} \phi_{ASE}(\nu) \tag{4}$$

where ϕ_{ASE} (ν) is a new profile function accounting for line narrowing. In setting:

$$\alpha = \frac{\phi_{sp}(\nu)}{\phi_{ASE}(\nu)} \text{ and } \eta = 1 - \frac{n_j}{n_i} \tag{5}$$

from (3) and (5) we see that saturation will take place as soon as:

$$Gl \geq Ln \frac{4\pi\eta\alpha}{A_{ij}\Theta\tau} \tag{6}$$

For Gaussian lines $\alpha \cong \sqrt{Gl}$. For the 236 Å line of Ge^{22+} one has [7]: $\tau \cong 1/270\ A_{ij}$, $\eta \cong 0.13$ and $\Theta \cong 10^{-5}$.. Furthermore $\alpha \cong 5$. This gives the saturation threshold at:

$$Gl \cong 19 \tag{7}$$

A more detailed study of the inversion population shows that a small decrease of gain coefficient may already be observable at Gl = 17 - 18.

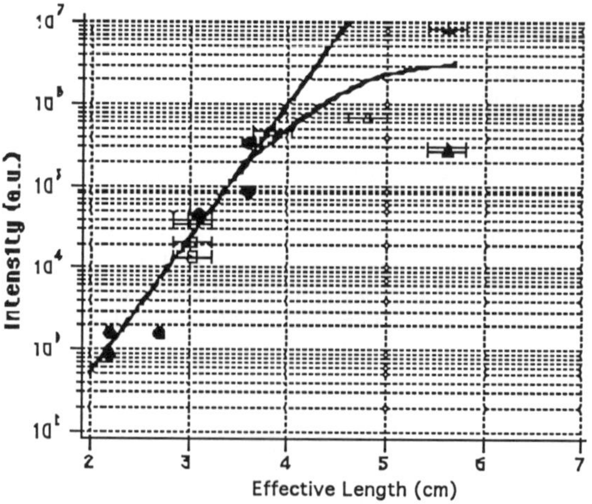

Fig. 5: 236 Å line intensity as a function of the effective path length in the germanium plasma. For l < 4 cm one observe an exponential increasing with G = 4 cm^{-1}. Saturation is seen to push down the gain coefficient once l ≅ 4.5 cm, i.e. Gl ≅ 18, and above.

If now we look at the set of experimental data for the 236 Å-line reported in figure 5, we see an important change to occur in the line-to-length behaviour just above l = 4 cm. Up to this (effective) length the intensity obtained with or without mirror exhibits an exponential increase with the gain coefficient of 4 cm^{-1} previously mentioned. Then, for $l_{eff} > 4.5$ cm, i.e $(Gl)_{eff} > 18$, appears an obvious reduction of the gain coefficient, the largest gain-length being near 22. This gives the evidence of saturation. The observed threshold is in good agreement with the above calculated value.

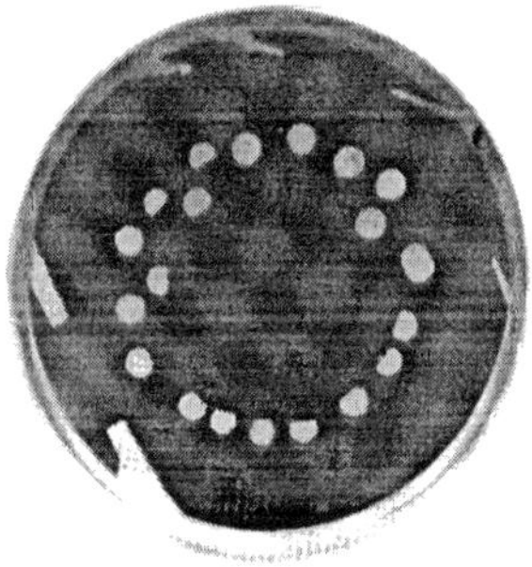

Fig. 6: Photography of the mirror showing the impact spots of the X-ray laser beam. The diameter of the spots is near 1 mm.

If the mirror reflectivity was as good as expected from synchrotron radiation measurements the gain-length could reach a value of 25 instead of the observed 22. The fact the the damage brought about by the X-ray laser beam itself may contribute to the reduction of R is supported by the photography of figure 6 which shows the mirror after being used in the experiment. The impact spots are visible on the mirror surface. Each shot destroys completely the non-protected small area of the multilayer which is crossed by the X-ray laser axis. Reflectivity controls performed after using the mirror have shown the reflection coefficient not to be significantly altered outside the impact spots. That is why the same mirror can be used for many shots provided that a fresh zone each time is put on the laser axis

4. What can we do with a laser at 236 Å?

Most of X-ray laser applications, which have been considered up to now, need wavelengths in the water window (23 Å $\leq \lambda \leq$ 44 Å). As regards the laser at 236 Å, its possible applications are settled by the main characteristics of the source, especially by the fact that the radiation penetration depth is very small. It is about 1000 Å in aluminium and silicon but only a few tens of angströms in most elements. Before suggesting a field where this peculiarity should be rather beneficial, we summarize the main characteristics of the laser pulse.

4.1 Main characteristics of the X-ray laser pulse

The brightness of the laser is such that a shot supplies 2 - 8 10^{12} photons in about 300 ps and 10^{-5} steradian. This brightness is larger by several order of magnitude than in any other laboratory source, including undulators. On another side, it is very interesting that the cooperation of M. Carter et al.* using an experimental technique developed at L.L.N.L. [8], enabled us to make a preliminary investigation of the transverse coherence of the beam. Some of the results, corresponding to the mirror described above, are shown in the table I.

MIRROR	L_c (z = 150cm)	Incoh. source size	L_c (z= 0)	Divergence
NO	140 μ	40 μ	3.4 μ	5.5 mr
YES (R=130 mm)	390 μ	14 μ	14.5 μ	1.4 mr

Table I: Coherence of the X-ray laser;
- column 2: transverse coherence length measured at 150 cm from the plasma end;
- column 3: incoherent source size which would correspond to the value of column 2
- column 4: transverse coherence length at the plasma end deduced from the values of column 2 (ref.9).

These results (column 2 and 4 in table I) point out a new advantage of the half-cavity, that is a large enhancement of beam spatial coherence

4.2 Study of very thin layer growth:

High brightness, significant spatial coherence and short penetration depth, these three characteristics turn the research effort to surface and interface chemical-physics. More especially workers in new optics, semi-conductor manufacturing and new magnetic materials are interested in the quality of very thin layers deposited on crystal surfaces. Now defects are introduced in thin layers by the formation of aggregates which break the layer uniformity during the growth and modify its physical properties.

With current techniques it is difficult to get evidence of the possible cluster formation in the vapour jet used for material deposition. The exceptional brightness of the X-ray laser may allow us to develop investigation methods making use of photoelectron spectroscopy, or even UV-fluorescence, in order to detect and characterize clusters in the vapour jet.

---

* M Carter, S. Mrowka, J. Trebes, K. Nugent

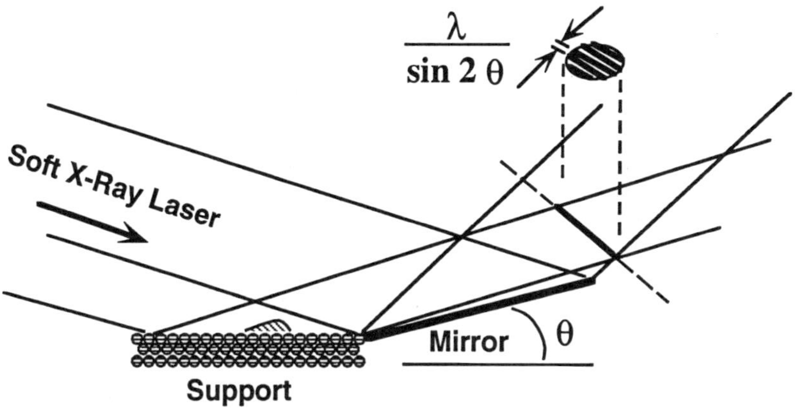

Fig 7: X-ray laser interferometric study of aggregate formation during thin layer deposition.

At the same time, the soft X-ray laser beam coherence would be used for carrying out an interferometric diagnostic of surface uniformity during the layer growth. This is illustrated by the sketch of figure 7. Up to the present time there was no source allowing this type of experiment with high resolution on large surfaces.

That is why we believe that a specific field of applications of economical interest is open at the wavelengths of the germanium laser.

References:

1. Dhez P ,. Jamelot G , Carillon A , Jaeglé P , Pardo P , Naccache 1984 D .AIP Conference Proceedings N°119, Subser. N° 5, ed. by S.E. Harris and T.B. Lucatorto, Amer. Inst. of Phys, New York pp 199.
2. Lee T N , McLean E A , Elton R C , 1987 Phys Rev Lett. **59** pp 1185
3.Neely D , Lewis C L S , O'Neill D M , Uhomoibhi J , Key M H, Rose S J , Tallents G J , Ramsden S A , 1992 Opt. Commun. **87** pp 231
4.Carillon A , Chen H Z , Dhez P , Dwidedi L , Jaeglé P , Jamelot G , Zhang J , Key M H, Kidd A , Klisnick A , Kodama R , Krishnan J , Lewis C L S , Neely D , Norreys P , O'Neill D M , Pert G J , Ramsden S A , Raucourt J P , Tallents G J , Uhomoibhi J O , 1992 Phys. Rev. Letters **68** pp 2917-20.
5. Kato Y, Daido H, Shiraga H, Yamanaka M ,Azuma H, Murai K, Miura E , Yuan G , Ohmi M ,Tanaka K , Kanabe T , Takagi M , Nakai S , Lewis C S L , O'Neill D M , Neely D , Shinoara K , Niibe M , Fukuda Y , 1991 Proceedings of the SPIE, Suckewer ed., 1551 pp 56
6. Sureau A , Proceedings of the 3rd International Colloqium on X Ray Lasers, Schliersee, Germany, May 1992, IOP Publishing, in press.
7.Whitten B L , London R A , Walling R S , 1988 J.O.S A **B5**, pp 2537
8.Walling R S , Shimkaveg G M , Carter M R , Steward R E , Trebes J E , London R A , Osterheld A L , Ratowsky R P , Craxton R S , Feit M D , Da Silva L B , MacGowan B J , Matthews D L , same volume as ref. 6.
9.Raucourt J. P , Thèse de Doctorat de l'Université Paris VI, 6 Avril 1992

VUV RADIATION PRODUCED BY HARMONIC GENERATION IN RARE GASES

Anne L'HUILLIER, Ph. BALCOU and L.A. LOMPRE
Service des Photons, Atomes et Molécules
Centre d'Etudes de Saclay, 91191 Gif-sur-Yvette, France

Abstract : The recent advances concerning high-order harmonic generation processes in strong laser fields are discussed. We report the observation of the 135th harmonic (160 eV) of a 1053 nm Nd-Glass laser focused into a 40 Torr gas jet of neon. There are two aspects to the physics of these highly nonlinear processes, the single atom response, which determines in particular how many harmonics can be produced, and the collective response from the nonlinear medium (phase matching), which allows one to optimize the emission. We also discuss the perspectives for these processes to become an useful XUV coherent source.

1. Experimental method and results

As an introduction, we briefly present the experimental method and the results obtained recently in Saclay[1]. Then, we discuss the physics of these processes, starting with the microscopic properties of these high-order harmonic generation processes. We examine how the number of harmonics produced varies with the parameters that affect the response of the individual atoms to the radiation field. We consider the influence of phase matching effects on the conversion efficiency. In conclusion, we briefly discuss the possibility for these processes to lead to an useful XUV source and what would be its specifications.

The experiment consists in focusing a laser into an atomic gas jet and in analyzing the radiation emitted on axis. The laser required must be of short pulse duration (typically from 100 fs to a few tens of picoseconds) and of high power (the intensity at best focus must be higher than 10^{13} W/cm^2). Experiments have been performed with excimer, Ti:Sapphire, Nd-Glass and dye lasers, with wavelengths ranging from 248 nm to 1064 nm, all being nevertheless laboratory-scale type of systems. The experimental setup used in our laboratory is shown in Fig.1.

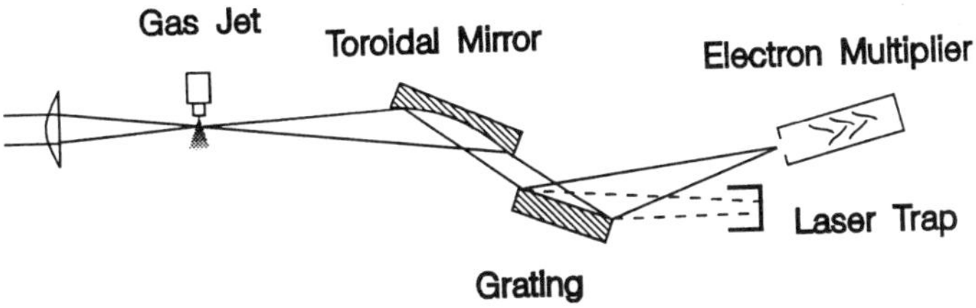

Fig. 1. Experimental set up

The laser is focused just below the nozzle of a pulsed gas jet. The light is analyzed on axis by a monochromator consisting of a toroidal mirror and a plane grating (Jobin Yvon) both coated with gold. The mirror, placed at 1 meter from the laser focal spot to minimize the risks of damaging the optics with the incident laser light, refocuses the radiation onto a 100 μm output slit. The laser light reflected by the grating is trapped before being too much focused in order to avoid plasma generation on the walls of the chamber. The photons are detected by an electron multiplier. The spectral range covered by this system is approximately 7-60 nm. There is no entrance slit and no diaphragms, so that this spectrometer has a good detection efficiency.

Fig. 2 (a) shows a spectrum obtained in neon at an intensity of about 10^{15} W/cm^2. Each point is an average of about 3 to 5 laser shots, with a strict selection in pulse energy and duration. The positions of the harmonics are indicated by the straight lines at the top of the figure. The width of the harmonics is constant, equal to the resolution of the monochromator (0.5 Å). Below 13 nm, the harmonics start overlapping, which results in an increase of the background level. The reflectance of gold with which both the mirror and the grating are coated, decreases by more than one order of magnitude from 10 to 7 nm for the 11-degrees incidence angle on the mirror and grating.

Fig. 2(b) shows the region from 12 to 8 nm, where we have divided the signal by the monochromator's response and taken into account our detector's gain. There is no apparent cut off, merely a steady slight decrease. The harmonics can be resolved until about the 135th harmonic (the highest order reported to date).

2. Physics of high-order harmonic generation processes

2.1 Microscopic properties

The first condition required for the observation of high order harmonics is

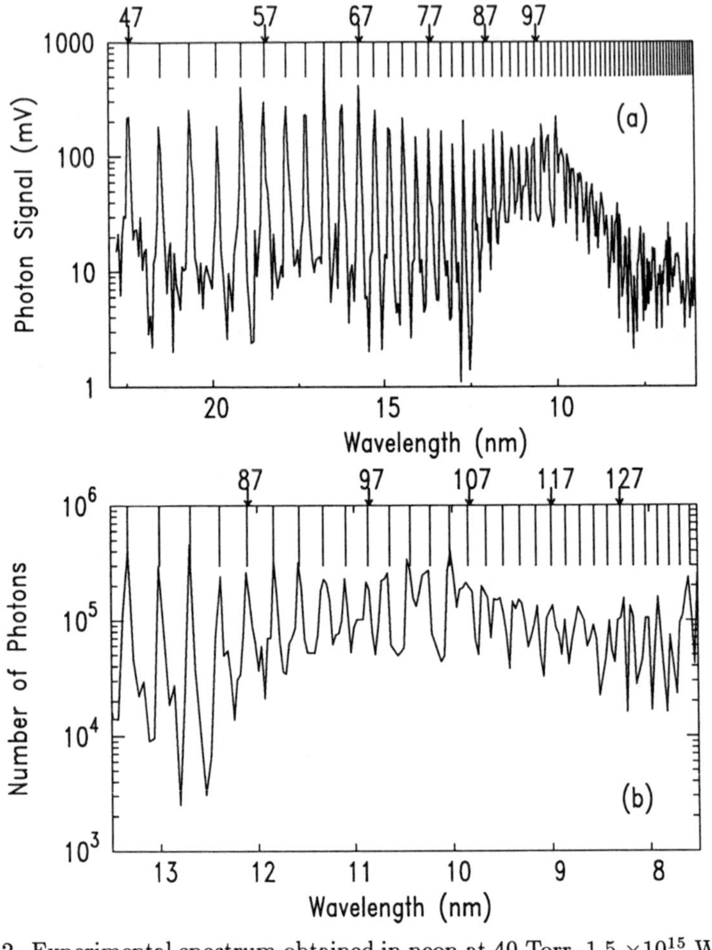

Fig. 2. Experimental spectrum obtained in neon at 40 Torr, 1.5 $\times 10^{15}$ W/cm^2; (a) raw data; (b) data corrected from the spectrometer's response over 12-8 nm (from Ref.1).

that each atom be strongly perturbed by the incident laser field and consequently radiate high order harmonics. The single atom response actually determines the harmonic intensity distribution. Instead of rapidly decreasing as expected within lowest-order perturbation theory, the intensity of the harmonics in general is found to be approximately constant (slightly decreasing), from the 5th or 7th order up to a very high order. This behavior is characteristic of a strong (non-perturbative) laser-atom interaction.

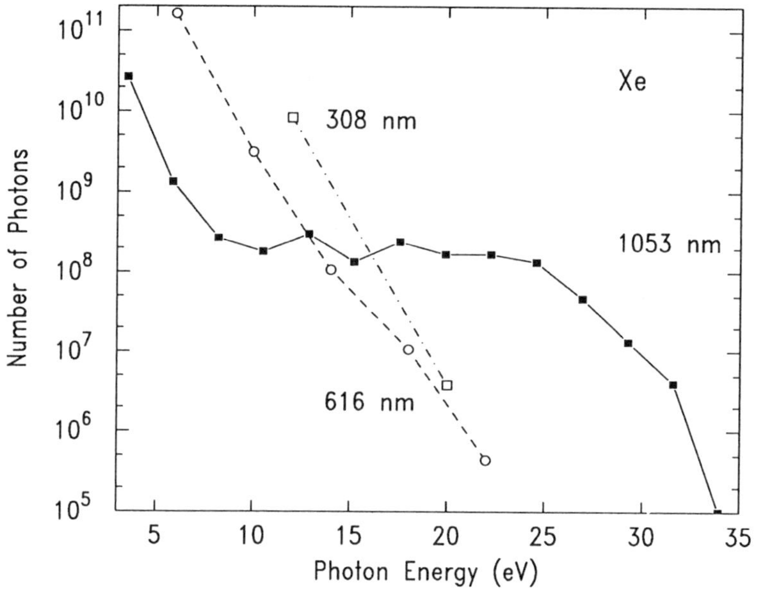

Fig. 3. Comparison between results obtained in xenon at 1053 nm, 616 nm and at 308 nm.

Figure 3 illustrates the influence of the wavelength on harmonic generation[2]. Here, we compare results obtained at 1053 nm (full squares), 616 nm (open circles) and 308 nm (open squares) using short pulse lasers (1.2, 2 and 1.4 ps respectively). All of the results have been obtained at intensities above saturation, between 5 $\times 10^{13}$ W/cm^2 and 10^{14} W/cm^2. The comparison between the three results shows that (i) the plateau behavior, which is remarkable at 1053 nm, disappears at shorter excitation wavelengths (the harmonic intensities decrease more rapidly with increasing order for shorter wavelengths) (ii) the conversion efficiency is higher for low energy photons and lower for high energy photons when using short incident wavelengths. The laser-atom interaction regime departs more strongly from the perturbative limit for longer incident wavelengths. The plateau is more pronounced and high harmonics can be produced more efficiently.

Although most of the experiments so far have been carried out at intensities below 10^{14} W/cm^2 (Ref.3), results obtained at higher intensities begin to appear. Sarukura et al. reported observation of the 25th harmonic in He and Ne using a 500 fs 248 nm laser[4], a result interpreted[5,6] as harmonic emission from the ions. Miyazaki and Sakai detected the 41st harmonic in He with a 800 fs 616 nm dye laser[7]. More recently, Macklin et al. observed the generation of up to the 109th harmonic in Ne, using a 150 fs 806 nm Ti:Sapphire laser[8] (170 eV), Crane et al., the generation of the 45th harmonic in He with the second harmonic (527 nm) of a 600 fs Nd-Glass laser[9].

In Fig. 4, we compare the results[1] obtained in xenon, argon, neon and helium at an intensity at best focus equal to 1.5×10^{15} W/cm^2. In xenon, the intensity of the harmonics is constant from the 5th harmonic to the 23rd and then falls off abruptly over four orders of magnitude. In argon, there seems to be two successive plateaus, one up to the 27th harmonic, the other from the 39th to the 49th harmonic. In neon and helium, we see very long and quite similar plateaus, whose limit (around 160 eV) is experimental, due to our monochromator's lack of resolution.

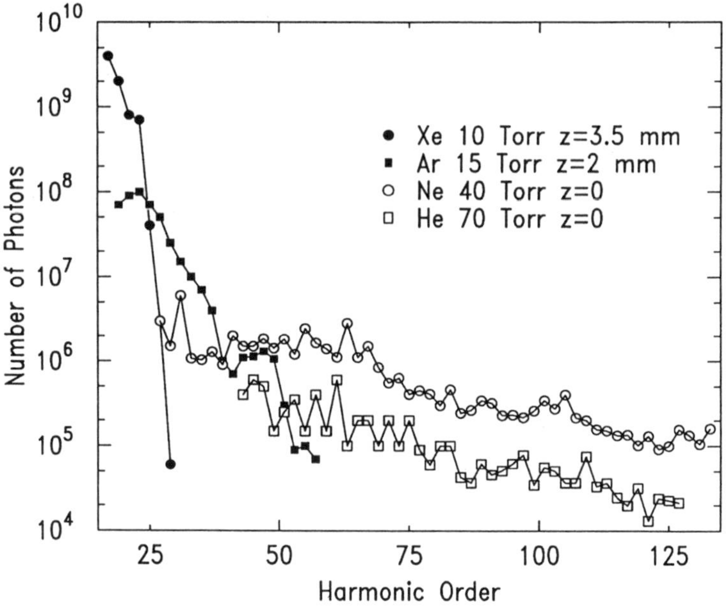

Figure 4. Number of photons obtained per laser pulse in xenon, argon, neon and helium as a function of the harmonic order.

Ions which are more difficult to ionize than neutrals can produce even more harmonics and with a comparable efficiency. Thus, Krause et al.[5] predict generation of photons of more than 400 eV energy in He^+ irradiated by the second harmonic of a Nd-Glass laser at an intensity of 5 $\times 10^{15}$ W/cm^2. These calculations do not include propagation effects which may alter this prediction, possibly reducing the conversion efficiency for ions. We now discuss the influence of these effects for neutral media.

2.2 Phase Matching

The second condition for harmonics to be efficiently generated is that the atomic dipole moments add in phase over a macroscopic distance in the nonlinear medium. The best signature of a coherent process is that the signal depend quadratically on the atomic density. This was indeed the case in our experiments over the (limited) range investigated. The coherence length over which the qth harmonic field E_q can be built up coherently[10] is in general imposed by the focusing of the incident radiation. It can be approximated by $L_{coh} = b\tan[\pi/(q-1)]/2$. If the medium's length is larger than L_{coh} and if the amplitude of the polarization field is not too much concentrated at the best focus site where the intensity is maximum (as is the case in a perturbative picture), the harmonic fields build up through a series of interferences. We show in Figure 5(a) the interference fringes corresponding to the generation of the 13th harmonic in a 2 mm long uniform nonlinear medium.

These phase matching oscillations have been observed experimentally by varying the position of the focus in the gas jet[11]. The result for the 13th harmonic is shown in Fig. 5(b). The period of these oscillations are close to twice the coherence length owing to focusing, thus confirming our interpretation.

Increasing the coherence length by using weak focusing optics provides a simple means for improving the conversion efficiency, as shown in Fig. 6. The number of photons for a few harmonics in xenon is plotted as a function of the laser confocal parameter b (see Ref.2). The strength of the harmonic field outside the medium depends on the length over which it can be built up coherently, i.e. on b, so that the intensity of the harmonic field leaving the medium is approximately proportional to b^2. This yields a b^3 scaling for the integrated intensity, since the focal section of the harmonic beam depends on b. This is valid as long as $L_{coh} \leq L$. In the case of a collimated geometry where $L_{coh} \geq L$, the harmonic field does not depend on b, and the integrated intensity (the number of harmonic photons) varies linearly with the confocal parameter b.

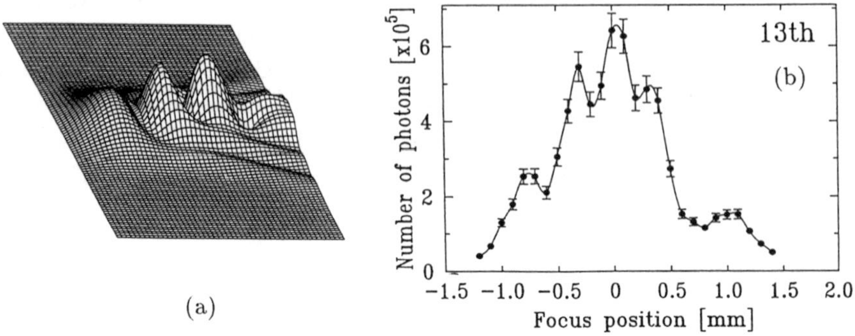

Fig. 5. (a) Graphical representation of $|E_q(r,z)|$ ($q = 13$) in the nonlinear medium. The laser propagates from the left to the right. The polarization is assumed to follow a I^5 power law (simulating a strong field regime); (b) Experimental results showing the13th harmonic signal as a function of the position of the laser focus in the nonlinear medium.

If the intensity at best focus is higher than that needed to ionize the gas (and simultaneously to produce high harmonics), one can also defocus the beam relative to the gas jet (the coherence length increases away from the focus)[1,2]. This allowed us to optimize the conversion efficiency in xenon up to 10^{-6}, with numbers of photons per pulse as high as 5×10^9 at 20 eV (see Fig.4).

3. Harmonic generation user's guide

In conclusion, we would like to give a description as exhaustive as possible of the properties of this XUV source, in the hope of interesting a broad community.

The number of photons we obtain at 20 eV (17th harmonic in xenon) is almost 10^{10} photons[1]. Since the pulse duration of the harmonics follows that of the pump pulse (1 ps), this represents an instantaneous photon flux of 10^{22} photons/s. Even the average photon flux reaches a rather high value equal to 10^9 photons/s, in spite of the very low repetition rate of our laser, 1 shot/10 seconds (10 Hz lasers could be employed and would yield 100 times more flux). The instantaneous power is 30 kW, a value that is several orders of magnitude higher than that reached by more conventional sources. The number of photons obtained around 100-150 eV, remains much lower in these experiments, about 10^6 photons per pulse but could be optimized by using more powerful and weakly focused lasers.

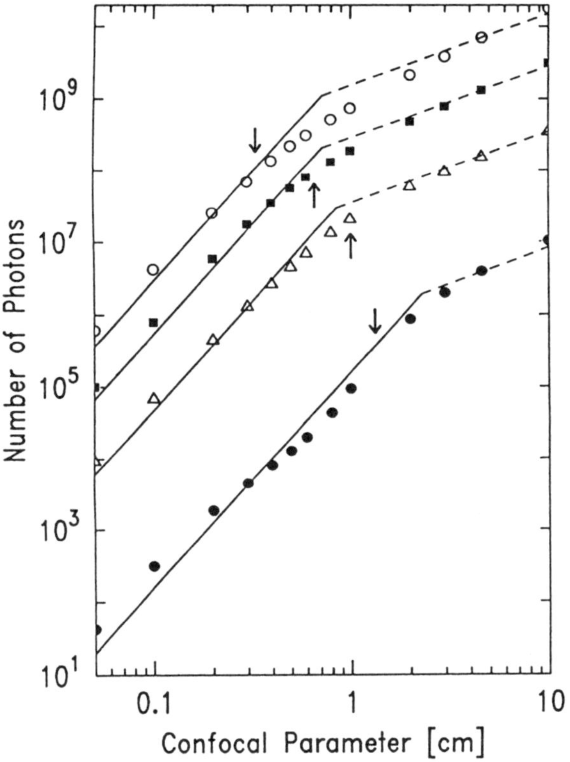

Fig. 6. Number of photons in xenon at 5 $\times 10^{13}$ W/cm^2 for the 7th, 13th 19th and 25th harmonic as a function of the laser confocal parameter.

The most interesting characteristics of this radiation is its short pulse duration. 100-fs lasers (e.g. Ti:Sapphire systems) lead to XUV pulses shorter than 100 fs (typically a factor of 2 less). It enables the possibility to implement pump-probe techniques in the XUV range, by splitting the incident laser in two arms, with a variable delay between the two. Note that the spectral width of the XUV radiation is not narrow, precisely because of the short pulse duration. Typical finesses are of the order of 10^3 with a 1 ps laser, around 20 eV (it gets better at higher energies). However, the radiation is coherent, which means that even though the spectral resolution might not be good, no information is lost. This aspect of the harmonic emission should be extremely useful in molecular spectroscopy studies.

Finally, the harmonic emission is directional, spatially coherent and would be well refocused. It can be tunable over a broad energy range (3 to at least 150 eV), if tunable lasers (e.g. dye or Ti:Sapphire) are used. We think that the investigation of these processes is worth pursuing, not only for its basic interest as a probe of the behavior of atoms in strong laser fields but also because it might lead to short pulse, bright and compact sources of coherent radiation in the XUV range.

References

1. A. L'Huillier and P. Balcou, to be published.
2. P. Balcou, A. S. L. Gomes, C. Cornaggia, L. A. Lompré and A. L'Huillier, J. Phys. B, in press.
3. A. L'Huillier, L. A. Lompré, G. Mainfray and C. Manus, in "Atoms in Intense Laser Fields", Ed. M. Gavrila, Academic Press, Orlando, Florida, in press.
4. N. Sarakura, K. Hata, T. Adachi, R. Nodomi, M. Watanabe and S. Watanabe, Phys. Rev. A **43** 1669 (1991).
5. J. L. Krause, K. J. Schafer and K. C. Kulander, Phys. Rev. Lett. **68**, 3535 (1992).
6. H. Xu, X. Tang and P. Lambropoulos, Phys. Rev. A, in press.
7. K. Miyazaki and H. Sakai, J. Phys. B **25**, L83 (1992)
8. J. J. Macklin, J. D. Kmetec and G. L. Gordon III, to be published.
9. J. K. Crane, M. D. Perry, S. Herman and R. W. Falcone, to be published.
10. A. L'Huillier, K. J. Schafer and K. C. Kulander, Phys. Rev. Lett. **66**, 2200 (1991) ; J. Phys. B, **24** 3315 (1991); A. L'Huillier, Ph. Balcou, S. Candel, K. J. Schafer and K. C. Kulander, Phys. Rev. A, 1 September 1992.
11. A. L'Huillier, P. Balcou and L. A. Lompré, Phys. Rev. Lett. **68**, 166 (1992).

III. ATOMS

PRESENT TRENDS IN THE THEORY OF ATOMIC PHOTOIONIZATION

CHRIS H. GREENE and FRANCIS ROBICHEAUX

Department of Physics and Joint Institute for Laboratory Astrophysics, University of Colorado, Boulder, Colorado 80309

ABSTRACT

Nonperturbative electron correlation effects on photoionization spectra are discussed for a variety of different systems. Recent efforts have demonstrated that an accurate theoretical description of complex perturbed multichannel Rydberg spectra can be obtained from a small-scale calculation adopting quantum defect and R-matrix viewpoints, even for atoms as heavy as barium. This will be further documented by new calculations of autoionizing resonance spectra in p-subshell photoionization of all the halogen atoms, and in s-subshell photoionization of chlorine, which reproduce striking regularities observed experimentally. Photoionization of excited states is another topic gaining increasing experimental and theoretical attention. Photoionization of excited barium has shown recently how atomic hyperfine structure plays an unexpectedly strong role in breaking down selection rules, and will be particularly important to incorporate in the description of any anisotropic process. Lastly, we discuss a new experimental and theoretical study of the complicated angular distribution of photoelectrons ejected from aligned excited states of atomic magnesium.

1. Overview

The theoretical description of single-electron photoionization processes in complicated atoms has improved in dramatic fashion in recent years. The past few decades showed great progress[1] in treating photoionization processes having basically only a single participating electron. More recently the focus has been increasingly directed by theory and experiment toward an understanding of photoexcitation processes in which two or more electrons interact in a closely correlated manner. Even in energy ranges where only single-electron escape is energetically possible, resonant two- or three-electron atomic states frequently dominate the spectrum.

Meanwhile, *ab initio* theoretical methods have enjoyed much success for describing photoabsorption spectra in the lighter atoms of the periodic table.[2] Attempts to extend these methods to heavier atoms have found it increasingly difficult or even impossible to attain spectroscopically-useful calculations for atoms heavier than Ar. This is largely a result of the fact that d- and f-wave atomic orbitals are unusually sensitive to the details (such as the barrier height) of the atomic "effective potential", so sensitive in fact that this potential cannot be predicted *ab initio* to the accuracy needed. To get around this limitation of current *ab initio* descriptions, an alternative attack has been to use some empirical information to build into such potentials a few critical constraints. This basic approach dates back to the early use of model potentials in atomic calculations by Laughlin and Victor.[3] More recently, such potentials were used by Bartschat and Scott[4] to treat Ba photoionization.

The most extensive and systematic use of semiempirical model potentials has occurred more recently, in calculations of alkaline earth atom valence spectra by Aymar, Greene, and coworkers.[5] The approach to these alkaline earth atom calculations can be summarized as follows:

(i) Find a model potential whose Schrödinger eigenenergies accurately reproduce experimental levels of the alkaline earth ion with one valence electron moving outside of closed shells.

(ii) Assume that each of the "valence electrons" in the atom or ion of interest moves in this same independent electron potential, and in addition each pair of valence electrons experiences the full Coulomb potential $1/r_{ij}$ between them.

(iii) Solve the multidimensional time-independent Schrödinger equation variationally, more or less by brute force, within a radial hypercube representing a reaction volume of configuration space in which each valence electron is confined within a radius r_0 from the nucleus. A convenient and rapidly converging algorithm which has been adopted for the calculations discussed in this review is the eigenchannel R-matrix method recast in its noniterative[6] and streamlined[7] form.

(iv) Match the variational solution on the boundary of the reaction volume to a linear combination of regular and irregular Coulomb wave functions for the outermost electron, multiplied by channel functions in the remaining degrees of freedom of the atomic valence electrons. This gives the usual smooth, short-range reaction matrix of multichannel quantum defect theory(MQDT)[8] which includes some channels which are energetically open and some which are closed asymptotically.

(v) Solve the comparatively small-dimension set of linear equations of MQDT, which amounts to imposing boundary conditions at $r \to \infty$ analytically using the known properties of regular and irregular Coulombic radial solutions. In many applications the reaction matrix is initially calculated from the fully nonrelativistic Schrödinger equation, ignoring fine structure and other relativistic effects. Nevertheless many of these effects can be very important. Fortunately, the frame transformation ideas developed by Fano *et al.*[9] permit an accurate description of these effects in many energy ranges. (Alternatively, spin-orbit terms can be included explicitly in the Hamiltonian, which is preferable for the very heaviest atoms, as discussed in Ref. 5.)

This procedure has proven effective for describing a very large number of diverse spectra for the alkaline earth atoms over the past five years. Many examples are discussed in Ref. 5 and in the references cited there. To show one comparatively recent example, Fig. 1 displays the calculated photoionization spectrum[10] of the excited $6s6p\,^1P^o$ state of barium. Also shown in the figure is the relative experimental measurement of Lange *et al.*[11] In this experiment, a linearly polarized laser produces the excited $6s6p$ level in only its $M_J = 0$ magnetic substate, choosing the quantization axis to coincide with the laser polarization vector. The experimental spectrum of Fig. 1 was obtained by scanning a second pulsed dye laser which reaches the ionization continuum between the $Ba^+(6s)$ and $5d$ thresholds and it is rich with autoionizing

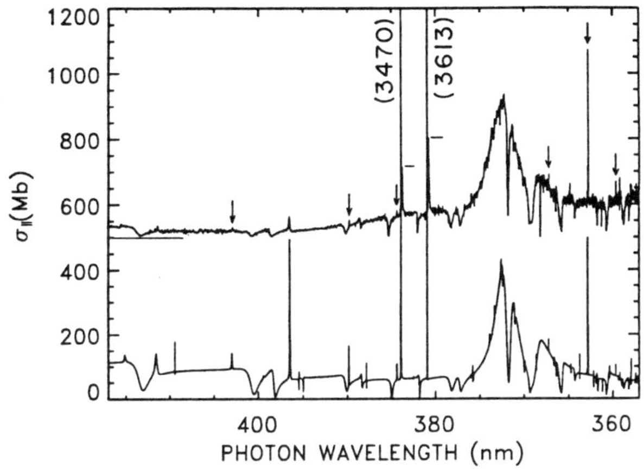

Fig. 1. Photoionization cross section of Ba $6s6p\ ^1P_1^o$ state as a function of wavelength of the second laser, between the Ba^+ $6S$ and $5d_{3/2}$ thresholds for parallel polarizaion of the two lasers. The hyperfine depolarization is included in the calculation.

resonances, most of which are Rydberg levels attached to the 5d level of the barium ion (i.e. $5dns, 5dnd, 5dng$). Because the second laser has polarization vector parallel to the first, selection rules readily show that only $J = 0$ and $J = 2$ final states are reached by this arrangement.

The first thing noticeable about Fig.1 is that theory[10] and experiment[11] are in generally good agreement. This is not a trivial matter in itself, as a competing R-matrix calculation of Bartschat and McLaughlin[12] fails to give the prominent $6p^2\ ^1S^e$ broad "complex resonance" which peaks around 373 nm. The first comparisons of theory and experiment for this spectrum showed that forbidden lines associated with the electronic angular momentum $J = 1$ are present in the experimental data. (These are the lines marked with arrows in Fig. 1.) After some analysis discussed at length in Ref. 10, the appearance of these levels was found to be caused by hyperfine-induced mixing of the $6s6p\ ^3P_1$ the electronic states. This can be pictured qualitatively as the first dye laser exciting the $M = 0$ at time $t = 0$, after which the Ba isotopes with nonzero spin undergo a precession of $\vec{J}$ as a result of the hyperfine interaction. Even though 82% of Ba isotopes have spin $I = 0$, the remaining 18% precess rapidly and cause the forbidden lines to appear at nearly the 10% level. This shows how the inclusion of hyperfine interaction in the theoretical description of valence-shell photoionization can be quite important.

2. Atoms with more than Two Valence Electrons

The generalization of these ideas to more complex atoms is simple in principle, but nontrivial in practice. For a two-electron atom such as barium, the spatial portion of configuration space which must be treated explicitly in the eigenchannel R-matrix calculation has only six dimensions. After making a spherical harmonic expansion of the angular degrees of freedom, there are really only nontrivial basis functions in two nonseparable radial degrees of freedom. On the other hand to treat the photoionization of the outermost *s*- and *p*-subshells of the halogen atoms, for instance, this requires treating 7 valence electrons with 7 radial degrees of freedom to be expanded. Fortunately, the remarkable efficiency of the independent-electron approximation (i.e. Hartree-Fock)[13] leads to far more efficient basis functions than expected.

The main elements which require special consideration for treating several valence electrons in an eigenchannel R-matrix calculation can be summarized as follows. First, exact ionic channel states are not available at the outset. When many valence electrons are present, a configuration-interaction calculation is needed just to find the ionic energy levels and wavefunctions to adequate accuracy. Next, we have found that convergence is rapidly improved by using the so-called "natural orbitals" of Löwdin[14] for the *d*- and *f*-orbitals of the basis rather than ordinary Hartree-Fock orbitals. These natural orbitals penetrate into the core and therefore overlap the core electrons much more strongly than do the Hartree-Fock orbitals. A last new element for the halogen calculations, which should be generally necessary in calculations for many valence electrons, is the importance of intermediate coupling in the ionic channel states. For instance, the $5s^2 5p^4\ {}^3P_2$ and ${}^1D_2$ levels of I^+ are mixed by the spin-orbit interaction, as the LS-designation is only approximate. This rotation of the LS-coupled channels is important, as it allows some states to autoionize which would be unable to do so if the core levels were strictly LS-coupled. For a more detailed description of this additional rotation, see Ref. 15.

The ionization thresholds of the halogen atoms arranged in order of increasing energy are $ns^2np^4\ {}^3P_2, {}^3P_1, {}^3P_0, {}^1D_2, {}^1S_0$, and Rydberg series are present with an outer *s*- or *d*-electron attached to each of these. The symmetries which are expected to be excited in the spectrum from the $ns^2np^5\ {}^2P^o$ state, in LS-coupling, are ${}^2S^e$, ${}^2P^e$, and ${}^2D^e$. States of other symmetries are coupled in by the frame transformation as well, namely ${}^2F^e$, ${}^4P^e$, ${}^4D^e$, ${}^4F^e$. Each of these seven LS-symmetries requires a separate final state eigenchannel R-matrix calculation to give an LS-coupled reaction matrix K^{LS}. The basis set size ranges from a few hundred up to a maximum of about 700 many-electron basis functions for the largest of these R-matrix calculations. The resulting LS-coupled reaction matrices are then combined into a jj-coupled reaction matrix for each allowed final state total angular momentum J, after which the quantum defect calculation is performed at a large number of energy mesh points (typically 20,000 points in the spectra shown below), followed by convolution with the experimental resolution. Despite the large number of energy mesh points, and the fact that seven separate LS-coupled R-matrix calculations must be performed in addition to the ground state diagonalization, the entire calculation for one atom takes about 2.5 hours of CPU time on a Decstation 5000/200 desktop computer.

Figure 2 shows the photoionization spectrum of the ground state $^2P_{3/2}$ level, for each of the heavy halogens, in the energy range below the $^1D^e$ ionization threshold. The dominant features for each of these atoms are broad $^1D^e$ md autoionizing Rydberg states, for which the ^2P and ^2D overlap. These broad sawtooth-shaped autoionizing states are reminiscent of those for the rare gas atoms Ar, Kr, and Xe, which look very similar in the energy range between the ionic ground state fine structure thresholds. The much narrower resonances for Cl and Br have character $^1D^e$ ms ^2D and $^1D^e$ md ^2S. (The latter Rydberg series would be forbidden to autoionize in LS-coupling, and accordingly they are missing from the calculations of Brown, Carter, and Kelly.[16] By including the departures from LS-coupling through a frame transformation, the slow decay of these states is built into the present calculation.) The similarity between the Cl and Br spectra is striking evidence of the identical character of the valence shells of these atoms.

Iodine has a somewhat different appearance because of the greater importance of spin-orbit coupling. Agreement between theory and experiment is good for Cl and Br, but there is at best only qualitative agreement for I, probably indicating the limitations of the frame transformation approach for Rydberg series converging to ionic states with very large spin-orbit splittings. The calculation for iodine is sufficiently accurate to reproduce qualitatively the large "complex d-resonance" centered nearly right on top of the $^3P_0$ ionization threshold close to 1100 Å. This causes an intensity modulation of the entire Rydberg series converging to this threshold. Another aspect which is particularly noticeable for iodine is the presence of additional features on the broad md-resonances. These are the "dark" Rydberg series of LS-forbidden symmetry, such as the $^1D^e$ md ^2F Rydberg levels which cause small dips in the photoionization cross section just to the right of each sawtooth peak, e.g. very close to 1075 Å. Closer inspection of the Cl and Br spectra show a similar small dip to be present in these atoms also, a consequence of the intermediate-coupling frame transformation discussed above. For Cl and Br there is no experimental evidence of this LS-forbidden autoionizing state, but it seems to be observed in I. A nontrivial prediction of the Ref.15 calculations is that higher resolution spectra will eventually observe this narrow dip in Br.

More recently, the ns-subshell photoionization processes in the halogen atoms have begun to receive attention in synchrotron radiation experiments of van der Meulen *et al.*[17] The measurements range from around 21 eV photon energy up to around 25 eV, which reaches singly-excited autoionizing states of the type $nsnp^5$ $^{1,3}P^o$ mp. The calculation in this energy range requires some additional effort because of the larger number of channels in the problem, and because the lower channels have more nodes within the reaction volume owing to their higher kinetic energy. Our calculated total and partial cross sections[18] are in good agreement with the experimental results.[17]

(a)

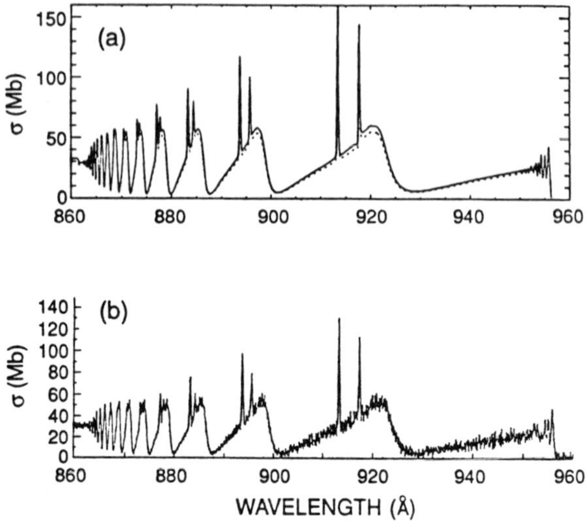

(b)

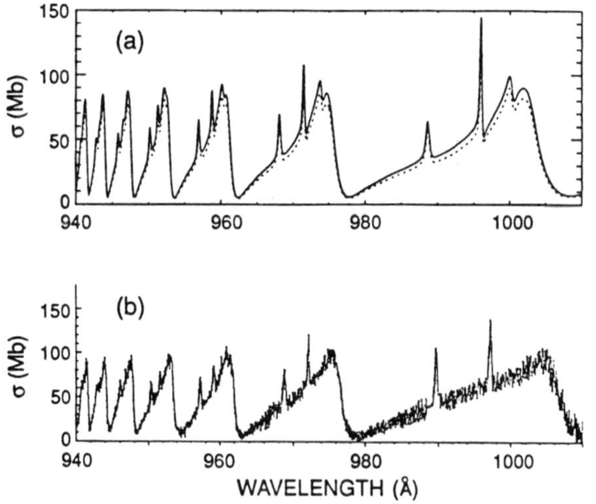

Fig. 2. Photoionization from the halogen $ns^2np^5\ {}^2P_{3/2}$ ground state between the ns^2np^4 ${}^3P$ and ${}^1D$ thresholds: a) Cl, experiment by B. Ruščić and J. Berkowitz [Phys. Rev. Lett. **50**, 675 (1983)]; b) Br, experiment by B. Ruščić, J.P. Greene and J. Berkowitz [J. Phys. B **17**, 1503 (1984)]; c) (following page) I, experiment by J. Berkowitz, C.H. Batson, and G.L. Goodman [Phys. Rev. A **24**, 149 (1981)].

(c)

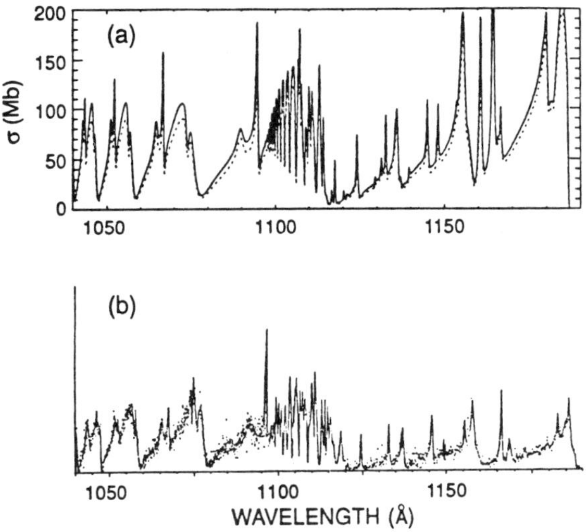

3. Angular Distribution of Photoelectrons Ejected from an Aligned Excited State

Gallagher and coworkers have studied magnesium photoionization in a series of recent papers, using multistep laser excitation to reach a bound Rydberg state $3sn\ell$ below the ionization threshold, followed by its photoionization by the isolated core excitation (ICE) method.[19–21] The ICE basically excites the inner electron from the $3s$ to the $3p$ state, after which the $3pn\ell$ state can autoionize. They have previously done experiments in which two lasers excite a $3sns\ {}^1S^e$ level with $n \approx 10-18$, measuring the total photoionization cross section and also the resulting photoelectron angular distributions. These angular distributions are characterized by a single (second rank, $k=2$) asymmetry parameter β, as is well known, because the $3sns$ state is isotropic. Their complicated resonant spectra have been found in Refs. 19b and 20 to agree well with MQDT calculations conducted using reaction matrices calculated previously by Greene,[22] using an LS-coupled eigenchannel R-matrix calculation of the same basic type discussed above in Sec. 1. As in Sec. 2, the fine-structure effects were incorporated through a frame transformation procedure.

The most recent efforts of Lindsay *et al.*[21] have treated a similar system, namely two-step laser excitation of a $3snd\ {}^1D^e$ level of Mg followed by photoionization through the ICE mechanism. While seemingly identical, this adds some nontrivial complexities and tests new physics which was not probed by the earlier photoionization of Rydberg $3sns$ states. Firstly, just the detailed photoionization spectroscopy reaches higher angular momentum J-states and tests the interactions of $3pns, 3pnd$, and $3png$ Rydberg autoionizing levels. Secondly, and perhaps more fundamentally, the photoelectron angular distributions become intrinsically richer and more complicated. Because the two-step laser excitation of the $3snd$-state causes it to be aligned, with up to rank $k = 4$ irreducible multipoles, the addition of yet one more (ionizing) photon causes the photoelectron angular distribution to have multipole moments up to rank $k = 6$. Thus the final state energies reached (relative to the ground state) are up in an energy range which would be normally reached in a synchrotron radiation experiment. But because three photons in all are absorbed, the photoelectron angular distributions carry new dynamical information which could not be accessed by a one-photon experiment.

To see this in detail, consider the simplest possible experimental geometry, in which all three laser photons are linearly polarized along a common axis $\hat{z}$. In this case the angular distribution has cylindrical symmetry about this axis, and in particular it has the form

$$\frac{d\sigma}{d\Omega} = \frac{\sigma}{4\pi}\left(1 + \beta P_2(\cos\theta) + \gamma P_4(\cos\theta) + \epsilon P_6(\cos\theta)\right),$$

in terms of Legendre polynomials P_k and now *three* asymmetry parameters β, γ, ϵ. The combined use of Wigner-Racah algebra techniques, along with the previously-calculated reaction matrices K^{LS} in LS-coupling and the ICE dipole matrix element is described in detail in Ref. 21. The net result is that expressions are found for the three asymmetry parameters in addition to the total photoionization cross section for the aligned Rydberg state, σ. All of these are rapidly-varying functions of energy, as they sharply reflect the autoionization dynamics into the multiple available continua.

Figure 3 shows these parameters as functions of the third (ionizing) photon energy, for photoionization of the $3s10d\ ^1D^e$ state of magnesium. Again the agreement between theory and experiment is very good overall, although perhaps somewhat poorer than had been reported previously in Ref. 19b for the total photoionization cross sections. Poorer agreement is perhaps expected since the angular distributions are notoriously sensitive to approximations made in the description of the atomic dynamics, in contrast to the total cross section. This sensitivity derives from the fact that differential cross sections are sensitive to additional interferences, such as between the $3sEp$ and $3sEf$ continuum phases, which cannot interfere in the total cross section. Of the three asymmetry parameters, β and γ are dependent on the p-f phase interference, while ϵ is not, as is readily seen from angular momentum coupling considerations. The agreement between theory and experiment is highly encouraging; the use of model potentials along with the eigenchannel R-matrix approach and quantum defect methods describes complicated electronic dynamics with high efficiency.

4. Conclusions

The above examples show how theoretical capabilities have improved in recent years far beyond a primarily *qualitative* description of complex atomic spectra. The capabilities have existed for some years to calculate properties of first-row atoms to

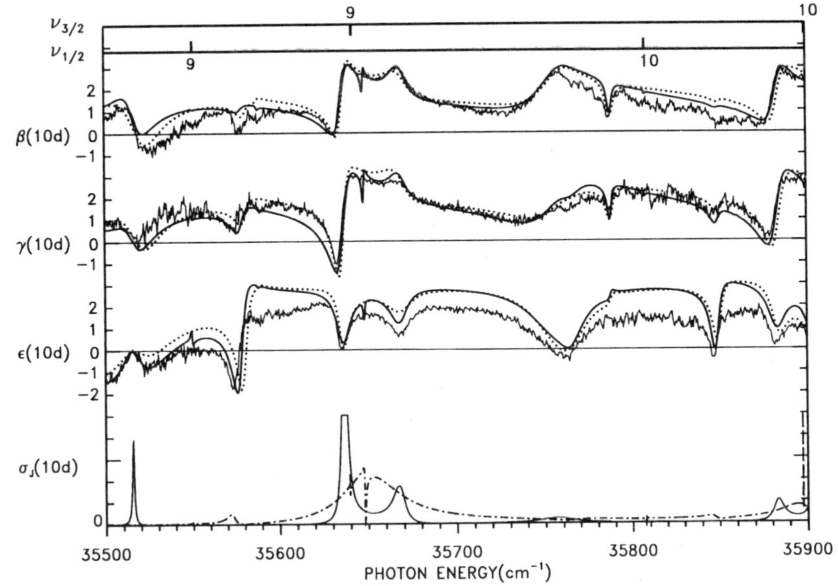

Fig. 3. The angular distribution parameters β, γ, and ε and photoionization cross section for the *3s10d→ 3p10d* transition in Mg from Ref. 21. The experimental results for the angular distribution parameters are shown with the jagged solid line. The predictions of two MQDT calculation are shown with smooth lines.

approximately the level of accuracy shown here, but the algorithms need to be modified to describe spectra of the type discussed here — particularly for heavy atoms. In addition to bolstering our confidence in the present level of theoretical descriptions of complicated autoionizing physics, this permits us to focus on new problems which are not efficiently described by these techniques at present. The most glaring of these is the excitation of both electrons to very high-lying bound and continuum states. Such high double-excitations and also the threshold law for two-electron ejection remain inaccessible because of the limiting assumption that both electrons can interact strongly only within a small region of configuration space. Overturning these limitations is the next major goal in sight for theory.

5. Acknowledgment

Discussions with Robert Wood and Darrell Armstrong on the physics of barium are gratefully acknowledged, as are conversations with Mark Lindsay and Tom Gallagher on the physics of magnesium. Most of the work discussed here was supported by the Division of Chemical Sciences, Office of Basic Energy Sciences, U. S. Department of Energy.

6. References

1. A.F. Starace, *Handbuch der Physik* **31**, S. Flügge and W. Mehlhorn,eds. (Springer-Verlag, Berlin and New York, 1982).
2. P.G. Burke and W.D. Robb, Adv. At. Mol. Phys. **11**, 143 (1975).
3. C. Laughlin and G.A. Victor, Adv. At. Molec. Phys. **25**, 163 (1988), and references therein.
4. K. Bartschat, M.R.H. Rudge, and M.P. Scott, J. Phys. B **19** 2469 (1986).
5. C.H. Greene and M. Aymar, Phys. Rev. A **44**, 1773 (1991) and references therein.
6. C.H. Greene, Phys. Rev. A **28** 2209 (1983); H. Le Rouzo and G. Raseev Phys. Rev. A **29** 1214 (1984).
7. C.H. Greene and L. Kim, Phys. Rev. A **38**, 5953 (1988).
8. U. Fano and A.R.P. Rau, *Atomic Collisions and Spectra* (Academic, Orlando, 1986) and M.J. Seaton, Rep. Prog. Phys. **46**, 167 (1983).
9. U. Fano, Phys. Rev. A **2**, 353 (1970); C.M. Lee and K.T. Lu, Phys. Rev. A **8**, 1241 (1973); U. Fano, J. Opt. Soc. Am. **65**, 979 (1975).
10. R.P. Wood, C.H. Greene, and D. Armstrong, submitted to Phys. Rev. A.
11. V. Lange, U. Eichmann, and W. Sandner, Phys. Rev. A **44**, 4737 (1991).
12. K. Bartschat and B.M. McLaughlin, J. Phys. B **23**, L439 (1990).
13. C. Froese-Fischer, *The Hartree-Fock Method for Atoms* (Wiley, New York, 1977).
14. P.O. Löwdin, Phys. Rev. **97**, 1474 (1955); P.O. Löwdin and H. Shull, Phys. Rev. **101**, 1730 (1956).
15. F. Robicheaux and C.H. Greene, Phys. Rev. A in press (1992).
16. E.R. Brown, S.L. Carter, and H.P. Kelly, Phys. Rev. A **21**, 1237 (1980).
17. P. van der Meulen, M.O. Krause, C.D. Caldwell, S.B. Whitfield, and C.A. de Lange, Phys. Rev. A in press (1992); P. van der Meulen, M.O. Krause, C.D. Caldwell, S.B. Whitfield, and C.A. de Lange, J. Phys. B **24**, L573 (1991); P. van der Meulen, M.O. Krause and C.A. de Lange, J. Phys. B **25**, 97 (1992).
18. F. Robicheaux and C.H. Greene, submitted to Phys. Rev. A.
19. a) W.E. Cooke, T.F. Gallagher, S.A. Edelstein, and R.M. Hill, Phys. Rev. Lett. **40**, 178 (1978); b) G.W. Schinn, C.-J. Dai, and T.F. Gallagher, Phys. Rev. A **43**, 2316 (1991).
20. M.D. Lindsay, L.-T. Cai, G.W. Schinn, C.-J. Dai, and T.F. Gallagher, Phys. Rev. A **45**, 231 (1992).
21. M.D. Lindsay, C.-J. Dai, L.-T. Cai, T.F. Gallagher, F. Robicheaux, and C.H. Greene, Phys. Rev. A in press (1992).
22. C.H. Greene, unpublished calculation 1990.

MANY-ELECTRON INTERACTIONS IN SINGLE AND MULTIPLE PHOTOIONIZATION OF METALLIC VAPORS

MATHIAS RICHTER
Institut für Strahlungs- und Kernphysik, Technische Universität Berlin, Hardenbergstraße 36, W-1000 Berlin 12, FRG

ABSTRACT

Photoionization studies on free open-shell metal atoms play an important role for the investigation of many-body systems. Examples for recent photoabsorption, photoion-yield, and photoelectron spectroscopy results are presented and discussed. Topics are electron angular distribution on Li and Be, direct double photoionization of Na, Ag, and Pt, and wavefunction collapse and giant resonances on 3*d* and 4*f* elements.

1. Introduction

Electron systems of free atoms are well suited for theoretical and experimental investigation of many-body effects since the physical forces are well-known and no further disturbance as in solid or molecular systems must be taken into account. With the use of synchrotron radiation for spectroscopy experiments photoionization of free atoms out of inner shells has been developed therefore to an extensive field of research during the last thirty years.[1-5]

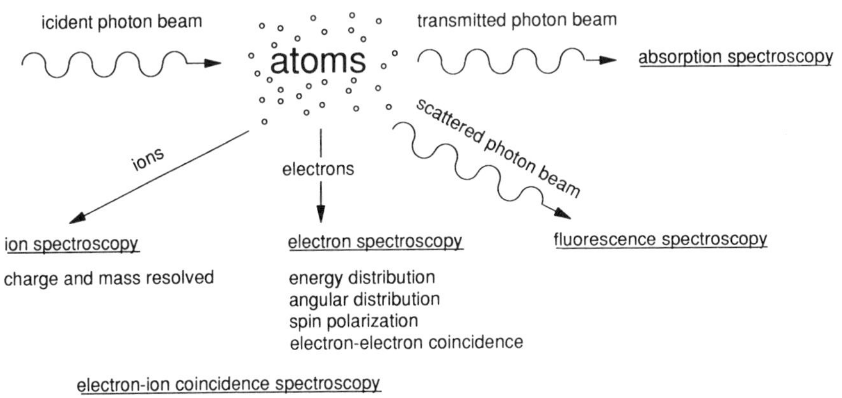

FIG. 1. Representation of the interaction between VUV radiation and free atoms.

Fig. 1 summarizes the situation of the interaction between VUV radiation and free atoms. After impact of a photon decay processes finally lead to the creation of a single or multiple charged ion and one or more photo- and/or Auger electrons. Fluorescence can also be a part of the decay cascade. Hence absorption, ion, electron, and fluorescence spectroscopy are corresponding experimental methods to investigate many-electron decay after inner-shell excitation.

Besides measurements on closed-shell rare gas atoms which are easier to carry out, studies on open-shell metal atoms play an important role in this context. Interest in strong and manyfold coupling of partially filled outer subshells to inner-hole states gives many reasons to solve technical problems in producing metallic vapors of high density for photoionization experiments.[5]

During the last years optimized VUV light sources and improved oven technics made a series of detailled photoionization studies on metallic vapors feasible.[5] Here examples for recent results of photoabsorption, photoion-yield, and photoelectron-spectroscopy measurements with emphasis to excitation by synchrotron radiation are presented.

2. Electron Angular Distribution of Photo- and Auger Emission on Atomic Lithium and Beryllium

Lithium is the simplest open-shell many-electron system of the periodic table and represents an ideal object for many-body studies. Fig. 2 shows the photoelectron spectrum of atomic Li taken at 75.5eV photon energy in the range of the $1s^{-1}$ photoionization.[6] Besides the two 'main lines' ($1s\ 2s\ ^{1,3}S$) at 64.41eV and 66.31eV ionization energy two satellites ($1s\ 2p\ ^{1,3}P$) appear at 66.67eV and 67.61eV. They are due to a so-called conjugate shakeup process where the residual ion is excited by a $1s2s \rightarrow 1s2p$ transition during $1s^{-1}$ photoionization. This implies in addition to transfer of energy between ion and photoelectron exchange of angular momentum. The ion remains in a P state whereas the photoelectron has S symmetry ($1s\ 2p\ ^{1,3}P\ \varepsilon s$) which is opposite to the main process ($1s\ 2s\ ^{1,3}S\ \varepsilon p$).

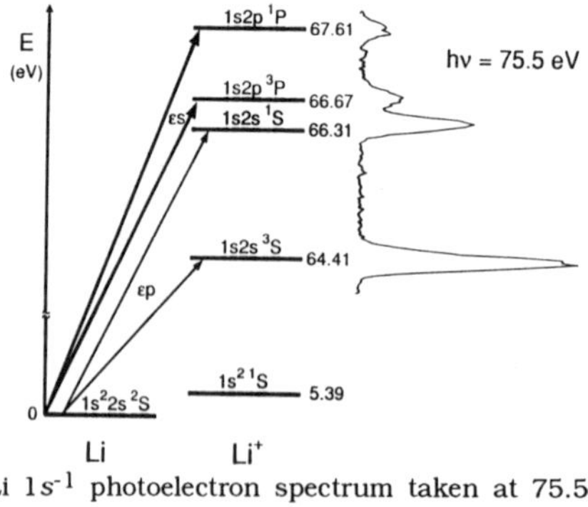

FIG. 2. Li $1s^{-1}$ photoelectron spectrum taken at 75.5eV photon energy together with the corresponding ionization energy diagram.[6]

Fig. 3 shows experimental results of the angular-distribution asymmetry parameter β for the Li $1s^{-1}$ photolines as a function of photon energy.[6] From the threshold behavior of β the symmetry character of the residual ion can be directly read off. Confirming qualitative predictions the β value for the main process remains constant whereas for the conjugate shakeup process it decreases towards threshold approaching from higher photon energies.

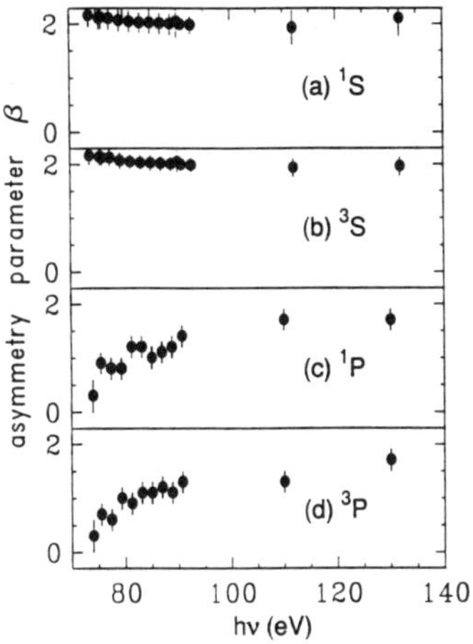

FIG. 3. Angular-distribution asymmetry parameter β for the $1s^{-1}$ photolines of atomic Li.[6]

On Beryllium the same type of processes has been studied.[7] In contrast to Li the residual Be^+ ion is able to decay via Auger into the Be^{2+} ground state:

$$\text{Be: } 1s^2\,2s^2 \;\rightarrow\; \begin{matrix} 1s\,(2s^2\;{}^1S)\;\varepsilon p \\ \\ 1s\,(2s\,2p\;{}^{1,3}P)\;\varepsilon(s,d) \end{matrix} \;\rightarrow\; 1s^2\,\varepsilon l\;\varepsilon_{Au}\,l_{Au}. \tag{1}$$

Information about the symmetry character of the photoelectron can be deduced here from angular distribution of Auger emission since the intermediate single ionic state of Be created by polarized radiation is aligned. Calculating the alignment A_2 which is directly proportional to the asymmetry parameter β_{Au} for Auger emission it has been found for the conjugate shakeup photoelectron near threshold an εd portion of 0.39 + 0.08.

Investigations as described point directly to the field of 'complete experiments'. Angle resolved detection of photo- and Auger electrons in coincidence represents one possibility to obtain a sufficient set of experimental parameters to describe a photoionization process completely.[8] Alternate methods are to measure angular distribution as well as spin polarization of photoelectrons[9] or angular distribution of photoelectrons on atoms which have been aligned by a polarized laser field.[10] For details the reader is refered to the contribution by V. Schmidt in this volume.

3. Direct Double Photoionization of Atomic Sodium, Silver, and Platinum

In a satellite process as discussed in the last chapter an electron of the residual ion is excited into a discrete state during photoionization. If now this excited state becomes a continuum state two electrons leave the atom simultaneously. This is the situation of a direct double photoionization process, also called shakeoff, which has attracted increasing interest during the last years.[11]

Unfortunately shakeoff manifests itself as unstructured background in a photoelectron spectrum since the available kinetic electron energy is portioned arbitraryly to both electrons. Therefore it is often helpful to study direct double photoionization by photoion-yield spectroscopy.

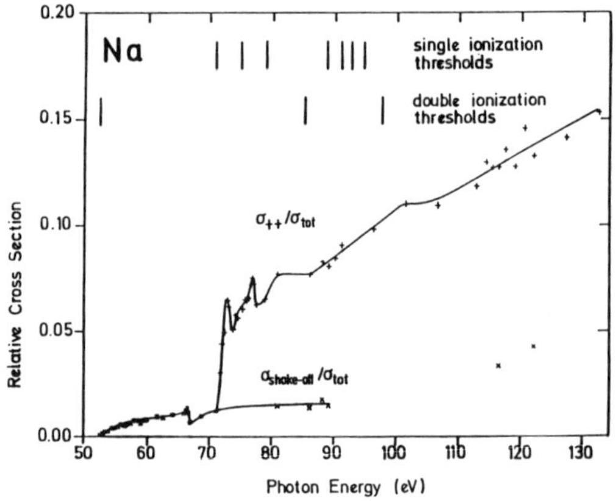

FIG. 4. Double photoionization cross section and direct double photoionization cross section (shakeoff) normalized to the total photoionization cross section for atomic Na.[12]

Fig. 4 shows the relative cross section for double photoionization of atomic Sodium as a function of photon energy.[12] Below 52.4eV which is the first threshold for double photoionization[13] no Na^{2+} signal is observed. Between 52.4eV and 71.0eV photon energy only direct double photoionization (shakeoff) Na: $1s^2\ 2s^2\ 2p^6\ 3s \rightarrow 1s^2\ 2s^2\ 2p^5\ \varepsilon l\ \varepsilon' l'$ contributes to the creation of Na^{2+} ions since other open photoionization channels do not allow any Auger decay and end up in Na^+ states.[13] The cross section of shakeoff between 52.4eV and 71.0eV therefore can be read off directly from Fig. 4 to increase from 0% to at least ~1% of the total photoionization cross section. The weak Fano type resonance structure at ~66eV indicates coupling of the shakeoff process to the corresponding $2s \rightarrow 3p$ resonance.[14]

At 71.0eV photoionization out of the $2s$ shell starts leading now via Auger decay also to a Na^{2+} final state[15] (Na: $1s^2\ 2s^2\ 2p^6\ 3s$ –> $1s^2\ 2s\ 2p^6\ 3s\ \varepsilon l$ –> $1s^2\ 2s^2\ 2p^5\ \varepsilon l\ \varepsilon_{Au} l_{Au}$) manifested by a sudden increase of the Na^{2+} signal. The photon energy range above 71.0eV is characterized by a large number of additional thresholds for single photoionization leading via Auger decay to Na^{2+}.

With the help of photoelectron spectroscopy data of the latter processes[15-18] data points for relative shakeoff intensity could be deduced also for a few photon energy values above 71.0eV as shown in Fig. 4. The relative intensity of direct double photoionization seems to converge to ~1.5% below 90eV photon energy whereas at ~120eV the obtained values are higher (~3.5%). An explanation for that might be the fact that at 97.8eV direct double photoionization Na: $1s^2\ 2s^2\ 2p^6\ 3s$ –> $1s^2\ 2s^2\ 2p^4\ 3s\ \varepsilon l\ \varepsilon' l'$ opens which involves two $2p$ electrons.

Corresponding measurements on atomic Silver resulted in still higher values, between 10% and 20%, for the relative direct double photoionization cross section.[19] One reason for that could be the large number of different direct double photoionization channels which open within a small energy range above the first threshold at 29eV and finally all contribute to the Ag^{2+} signal:

$$\begin{aligned} \text{Ag: [Kr]}\ 4d^{10}\ 5s \quad &\text{–>} \quad \text{[Kr]}\ 4d^9\ \varepsilon l\ \varepsilon' l' \\ &\text{–>} \quad \text{[Kr]}\ 4d^8\ 5s\ \varepsilon l\ \varepsilon' l' \\ &\text{–>} \quad \text{[Kr]}\ 4d^8\ 5p\ \varepsilon l\ \varepsilon' l'. \end{aligned} \tag{2}$$

For direct double photoionization of atomic Platinum the situation is just as complicated and shakeoff accounts for ~15% of the total photoionization:[20]

$$\begin{aligned} \text{Pt: [Xe]}\ 4f^{14}\ 5d^9\ 6s \quad &\text{–>} \quad \text{[Xe]}\ 4f^{14}\ 5d^8\ \varepsilon l\ \varepsilon' l' \\ &\text{–>} \quad \text{[Xe]}\ 4f^{14}\ 5d^7\ 6s\ \varepsilon l\ \varepsilon' l' \\ &\text{–>} \quad \text{[Xe]}\ 4f^{14}\ 5d^7\ nl''\ \varepsilon l\ \varepsilon' l'. \end{aligned} \tag{3}$$

A second reason for different relative shakeoff intensity could be that on Ag and Pt one or two d electrons are involved which may offer more oscillator strength than s or p electrons in Na. An exact explanation for the different relative shakeoff intensities can only be obtained by many-electron calculations.

A small detail of the photoion-yield spectrum in Fig. 4 should still be mentioned: resonance structures at ~73eV and ~77eV photon energy. They can be ascribed to Na: $1s^2\ 2s^2\ 2p^6\ 3s$ –> $1s^2\ 2s\ 2p^6\ (3d,4s)\ 3p$ and $1s^2\ 2s^2\ 2p^6\ 3s$ –> $1s^2\ 2s\ 2p^6\ (3d,4s)\ 4p$ double excitations which are not observed in absorption.[14] Obviously they couple to the $2s^{-1}$ photoionization and finally lead to a Na^{2+} state. The corresponding decay cascade $1s^2\ 2s\ 2p^6\ (3d,4s)\ (3p,4p)$ –> $1s^2\ 2s\ 2p^6\ 3s\ \varepsilon l$ –> $1s^2\ 2s^2\ 2p^5\ \varepsilon l\ \varepsilon' l'$ implies two Auger processes from which the first is of special interest because the inner $2s^{-1}$-hole state here remains as a 'spectator'. In the resonances therefore three different processes contribute to the Na^{2+} signal:

1) one-step process via shakeoff,
2) two-step process via $2s^{-1}$ photoionization and Auger decay,
3) three-step process via double excitation and two-step Auger decay.

4. Wavefunction Collapse and Giant Resonances on 3*d* and 4*f* Elements

The idea of a 'spectator' of an Auger process as mentioned in the last chapter has originally been used not for electron holes in inner shells but for excited electrons in outer states, for example to describe resonant Auger decay on Argon:[21] Ar: [Be] $2p^6\ 3s^2\ 3p^6$ –> [Be] $2p^5\ 3s^2\ 3p^6\ 3d$ –> [Be] $2p^6\ 3s^2\ 3p^4\ \varepsilon l$. About the passive role of a spectator for the 3*d* electron a controversial discussion was held and on Calcium (Ca: [Ar] $4s^2$) the 'breakdown of the spectator model' for decay of $2p^{-1}3d$ excited states was experimentally and theoretically demonstrated.[22] Responsible for the more active role of a 'participator' for the excited 3*d* electron in the Calcium Auger decay is the collapse of the 3*d* wavefunction in connection with strong overlap and coupling to inner shells.

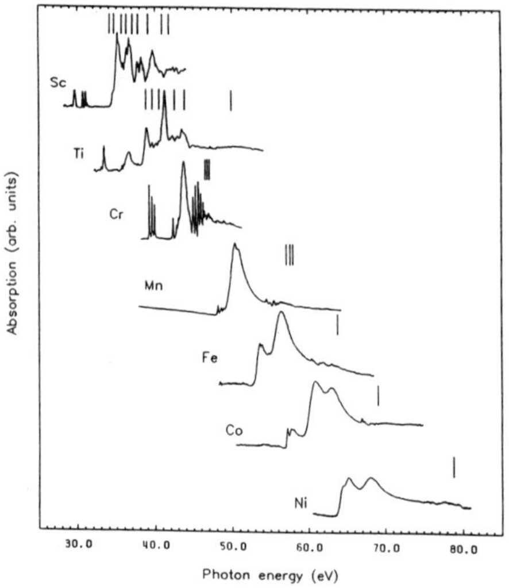

FIG. 5. Photoabsorption spectra of 3*d* elements. The bars mark the energy positions of the $3p^{-1}$ ionization thresholds.[5]

For 3*d* elements ([Mg] $3p^6$ $(4s,3d)^n$) which directly follow Calcium in the periodic table wavefunction collapse of 3*d* electrons influences strongly 3*p* –> 3*d* excitations as demonstrated in Fig. 5. Wavefunction collapse has found to be 'term dependent'.[23,24] With increasing atomic number more and more terms of [Mg] $3p^5$ $(4s,3d)^{n+1}$ excited states imply collapsed 3*d* orbitals. 3*p*-3*d* overlap increases as well as 3*p*-3*d* coupling. This results finally in complicated broad and overlapping multiplett structures of strong $3p \rightarrow 3d$ excitations which cannot be resolved experimentally and form a so-called atomic giant resonance.

Atomic giant resonances have been studied intensely on $3d$ elements most of all on Manganese with a half filled $3d$ shell which makes many-body calculations easier to carry out. An overview of references is given in Ref. 4 and Ref. 5.

A second important group of elements in this context are Xe, Cs, Ba, La, and lanthanides.[4,5] Their $4d \rightarrow (4,\varepsilon)f$ giant resonances are partially continuum resonances.[25-28] Only stepwise the resonances shift from an energy region in the $4d^{-1}$ continuum for Xenon ($4d \rightarrow \varepsilon f$ resonance) into a region below the $4d^{-1}$ thresholds for Gadolinium ($4d \rightarrow 4f$ resonance) with increasing atomic number. Elements between show intermediate $4d \rightarrow (4,\varepsilon)f$ resonance character.[28]

Responsible for the character of $4d^{-1}(4,\varepsilon)f$ excited states is again the degree of wavefunction collapse - here for excited $4f$ electrons - which depends sensitively on the form of the corresponding effective radial potential consisting of a potential well in the inner region of the atom separated from the outer region by a potential barrier.[4] A low barrier compared to the depth of the inner well allows a $4f$ wavefunction collapse. The corresponding strong $4d$-$4f$ overlap results in a $4d \rightarrow 4f$ giant resonance in analogy to $3p \rightarrow 3d$ resonances of $3d$ elements. A high potential barrier prevents the collapse and leads to small $4d$-$4f$ overlap. $4d$ oscillator strength is then transfered into the $4d^{-1}\varepsilon f$ continuum. A high potential barrier also for the excited εf electron results in an atomic $4d \rightarrow \varepsilon f$ shape resonance with delayed onset.

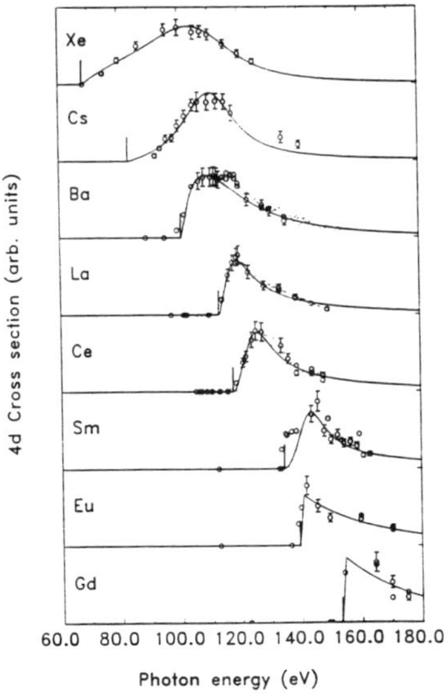

FIG. 6. Partial $4d \rightarrow \varepsilon f$ cross section of atomic Xe and following elements[25-28] together with fit curves according to Eq. 8. The vertical bars mark the centers of gravity for experimental $4d^{-1}$ multipletts.

The latter is demonstrated in Fig. 6 where $4d \rightarrow \varepsilon f$ cross sections of atomic Xe,[25] Cs,[26] Ba, La, Ce,[27] Sm, Eu, and Gd[28] are displayed. The vertical bars mark the centers of gravity for $4d^{-1}$ multipletts. Up to Sm the cross section at threshold accounts for ~0. The slow increase of intensity with photon energy is due to increase of amplitude of the εf continuum wave in the inner region of the atom which can be interpreted as tunneling through the barrier.[4] This results finally in strong $4d$-εf overlap and gives rise to a cross section maximum some eV above threshold.

If the effective radial potential with barrier of an excited continuum electron is approximated by a square potential of the form:

$$V_{\rm eff}(r) = \begin{cases} -I, & 0 \leq r < L_1 \\ V_0, & L_1 \leq r < L_1 + L_2 \\ 0, & L_1 + L_2 \leq r, \end{cases} \tag{4}$$

where I denotes the ionization energy, V_0 (≥ 0) the height of the barrier, and L_1 and L_2 the widths of inner potential well and barrier, the square of the amplitude $|A|^2$ of the continuum wave in the inner region of the atom can be easily calculated:

$$|A|^2 = \begin{cases} [(\cos k_2L_2 \sin k_1L_1 + k_1/k_2 \sin k_2L_2 \cos k_1L_1)^2 \\ \quad + (k_2/k_3 \sin k_2L_2 \sin k_1L_1 - k_1/k_3 \cos k_2L_2 \cos k_1L_1)^2]^{-1}, & h\nu > I+V_0, \\ 1, & h\nu = I+V_0, \\ [(\operatorname{ch}\kappa_2L_2 \sin k_1L_1 + k_1/\kappa_2 \operatorname{sh}\kappa_2L_2 \cos k_1L_1)^2 \\ \quad + (\kappa_2/k_3 \operatorname{sh}\kappa_2L_2 \sin k_1L_1 + k_1/k_3 \operatorname{ch}\kappa_2L_2 \cos k_1L_1)^2]^{-1}, & I \leq h\nu < I+V_0, \end{cases} \tag{5}$$

with:

$$k_1 = [2m\, h\nu]^{1/2}/\hbar; \quad k_2 = [2m\,(h\nu - I - V_0)]^{1/2}/\hbar; \quad \kappa_2 = ik_2; \quad k_3 = [2m\,(h\nu - I)]^{1/2}/\hbar, \tag{6}$$

and the restriction:

$$L_1 = (2n+1)(\pi/2)[\hbar/(2m(I+V_0))^{1/2}], \quad (n = 0, 1, 2, \ldots) \tag{7}$$

The cross section for excitation from an inner shell into the continuum state is proportional to $|A|^2$ and must decrease with $h\nu^{-l-7/2}$ for high photon energies where l denotes the angular momentum of the initial electron state.[29] Since the expression for $|A|^2$ in Eq. 5 converges to 1 for high photon energies an expression for the cross section in the framework of the present model is given by:

$$\sigma(h\nu) = \sigma_0\,(I/h\nu)^{l+7/2}\,|A|^2. \tag{8}$$

The experimental $4d \rightarrow \varepsilon f$ cross sections in Fig. 6 have been fitted by resonance profiles of Eq. 8. For I the experimental centers of gravity for $4d^{-1}$ multipletts have been applied. 'n' in Eq. 7 was set to 0, 'l' in Eq. 8 to 2.0. Since the experimental cross sections are only measured on a relative scale the parameter σ_0 in Eq. 8 has no meaning for the fits. Height V_0 and width L_2 of the potential barrier are the only remaining free fit parameters which define the cross section profile.

The agreement between data points and fits in Fig. 6 is fairly well. Moreover for Xenon and Barium height and width of the barrier derived from the fits agree excellently with corresponding many electron calculations.[30] Therefore fits by the formula in Eq. 8 give information to effective radial potential barriers.

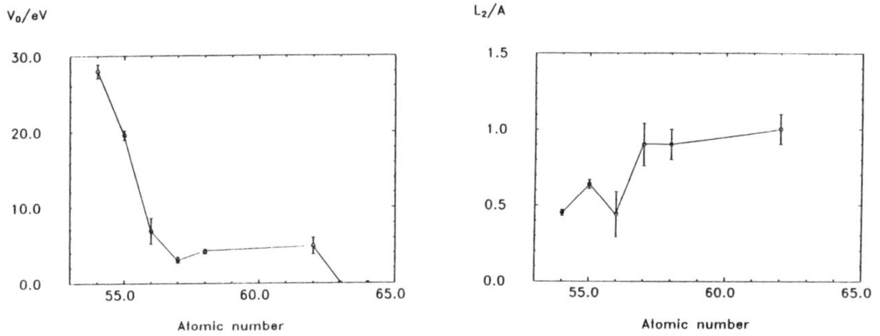

FIG. 7. Height and width of the potential barrier derived from fits to the data of Fig. 6 by the formula in Eq. 8.

The values for barrier height and width derived from the fits in Fig. 6 are displayed in Fig. 7. Summarizing the results the potential barrier disappears between Xe and Gd in two steps, at Ba (Z=56) and Eu (Z=63), elements with filled or half filled subshells. For Gadolinium missing potential barriers for excited f electrons result in a $4d \rightarrow \varepsilon f$ cross section without any delayed onset and with only a small contribution to the giant resonance. The latter is dominated by $4d \rightarrow 4f$ excitations which mainly autoionize into $4f^{-1}$ continua.[28]

1. R. P. Madden and K. Codling, *Phys. Rev. Lett.* **10** (1963) 516.
2. *Photoionization and Other Probes of Many-Electron Interactions*, ed. F. J. Wuilleumier (Plenum Press, New York - London, 1976).
3. *Handbuch der Physik* **XXXI**, ed. W. Mehlhorn (Springer, Berlin - Heidelberg - New York, 1982).
4. *Giant Resonances in Atoms, Molecules, and Solids*, eds. J. P. Connerade, J. M. Esteva, and R. C. Karnatak (Plenum Press, New York - London, 1987).
5. B. Sonntag and P. Zimmermann, *Rep. Prog. Phys.* (1992) in press.

6. B. Langer, J. Viefhaus, O. Hemmers, A. Menzel, R. Wehlitz, and U. Becker, *Phys. Rev.* **A43** (1991) 1652.
7. M. O. Krause and C. D. Caldwell, *Phys. Rev. Lett.* **59** (1987) 2736.
8. A. Hausmann, B. Kämmerling, H. Kossmann, and V. Schmidt, *Phys. Rev. Lett.* **61** (1988) 2669.
9. U. Heinzmann, *J. Phys.* **B13** (1980) 4353.
10. M. Pahler, C. Lorenz, E. v. Raven, J. Rüder, B. Sonntag, S. Baier, B. R. Müller, M. Schulze, H. Staiger, and P. Zimmermann, *Phys. Rev. Lett.* **68** (1992) 2285.
11. S. D. Price and J. H. D. Eland, *J. Phys.* **B23** (1990) 2269.
12. J. M. Bizau, D. Cubaynes, L. Journel, B. Rouvellou, F. J. Wuilleumier, M. Richter, K. H. Selbmann, P. Sladeczek, and P. Zimmermann, *BESSY Annual Report 1991*, p. 101
13. Ch. E. Moore, *Atomic Energy Levels* I (US Government Printing Office, Washington, 1949).
14. H. W. Wolff, K. Radler, B. Sonntag, and R. Haensel, *Z. Phys.* **257** (1972) 353.
15. M. Richter, J. M. Bizau, D. Cubaynes, T. Menzel, F. J. Wuilleumier, and B. Carre, *Europhys. Lett.* **12** (1990) 35.
16. D. Cubaynes, J. M. Bizau, F. J. Wuilleumier, B. Carre, and I. Gounand, *Phys. Rev. Lett.* **63** (1989) 2460.
17. D. Cubaynes, J. M. Bizau, M. Richter, and F. J. Wuilleumier, *Europhys. Lett.* **14** (1991) 747.
18. D. Cubaynes, J. M. Bizau, M. Richter, and F. J. Wuilleumier, to be published.
19. M. Y. Adam, L. Hellner, G. Dujardin, A. Svensson, P. Martin, and F. Combet Farnoux, *J. Phys.* **B22** (1989) 2141.
20. M. Martins, M. Richter, K. H. Selbmann, P. Sladeczek, and P. Zimmermann, to be published.
21. H. Aksela, S. Aksela, H. Pulkkinen, G. M. Bancroft, and K. H. Tan, *Phys. Rev.* **A37** (1988) 1798.
22. M. Meyer, E. v. Raven, M. Richter, B. Sonntag, R. D. Cowan, and J. E. Hansen, *Phys. Rev.* **A39** (1989) 4319.
23. K. T. Cheng and C. Froese Fischer, *Phys. Rev.* **A28** (1983) 2811.
24. K. T. Cheng and W. R. Johnson, *Phys. Rev.* **A28** (1983) 2820.
25. B. Kämmerling, H. Kossmann, and V. Schmidt, *J. Phys.* **B22** (1989) 841.
26. Th. Prescher, M. Richter, E. Schmidt, B. Sonntag, and H. E. Wetzel, *J. Phys.* **B19** (1986) 1645.
27. M. Richter, M. Meyer, M. Pahler, T. Prescher, E. v. Raven, B. Sonntag, and H. E. Wetzel, *Phys. Rev.* **A39** (1989) 5666.
28. M. Richter, M. Meyer, M. Pahler, T. Prescher, E. v. Raven, B. Sonntag, and H. E. Wetzel, *Phys. Rev.* **A40** (1989) 7007.
29. H. A. Bethe and E. E. Salpeter, in *Handbuch der Physik* **XXXV**, ed. S. Flügge (Springer, Berlin, 1957), p. 88.
30. Z. Crljen and G. Wendin, *Phys. Rev.* **A35** (1987) 1571.

THRESHOLD BEHAVIOUR OF PHOTODOUBLE IONIZATION IN ATOMS

GEORGE C. KING
Department of Physics, University of Manchester
Manchester M13 9PL, UK

ABSTRACT

A threshold electron spectrometer has been developed for the detection of near-zero kinetic energy photoelectrons produced in threshold photoionization of atoms and molecules by synchrotron radiation. This spectrometer has been used to study various aspects of threshold photoionization processes in the rare gases and also small molecules. These studies have provided information on the dynamics of threshold photodouble ionization and also on the spectroscopy of singly and doubly charged ionic states.

1. Introduction

Single photon double ionization has become a subject of considerable interest since it is a means of studying the dynamics of two electrons in the field of a doubly charged ion. In the threshold region these two electrons have only a small amount of kinetic energy to share (< 1 eV) and are expected to be highly correlated. Precise theoretical predictions are available about the behaviour of this electron-correlation dominated system. The theory of Wannier[1], for example, is based upon the dynamics of two electrons initially opposite to each other on either side of the ionic core in a configuration known as the Wannier geometry. Wannier predicted precise behaviours for the shape of the cross section near threshold and for the energy and angular distributions of the two outgoing electrons. Since that time, much theoretical effort has gone into developing quantal and more general descriptions of the photodouble ionization process.

On the experimental side, the goal has been to obtain precise information in order to test the theoretical predictions. However, because the direct single photodouble ionization process occurs only through electron-electron correlations, the cross section is low and experimental observations are difficult. In the present experimental studies it has been possible to investigate various aspects of the photodouble ionization process by the use of a threshold electron spectrometer that features very high detection efficiency as well as very high energy resolution.

2. Experimental Technique

The apparatus consists of a threshold electron analyser used in conjunction with the Synchrotron Radiation Source at Daresbury Laboratory, UK. The photon beam has an adjustable energy over the range 10-120 eV and an energy spread that is typically 3-75 meV, depending upon the optical monochromator used. Of note is the long lifetime of the circulating electron beam between injections ~ 30 hours, which considerably facilitates the present studies of processes with low cross sections.

The threshold analyser has been described in detail elsewhere[2] and is based upon the penetrating field technique. The analyser provides very high collection efficiency for low energy photoelectrons (essentially 4π collection) and very high threshold energy resolution (< 0.003 eV). The performance of the analyser is illustrated in figure 1. This shows a threshold photoelectron spectrum of the $^2P_{3/2,1/2}$ states of Ar^+ and was obtained by measuring the yield of threshold electrons as the photon energy was varied across the region containing the $^2P_{3/2,1/2}$ ionic states. Of note is the width of the peaks in the spectrum (~ 3 meV) and the excellent signal to noise ratio. The threshold photoelectron spectrum is superimposed on the total ion yield spectrum which was obtained by simply reversing the polarity of the potentials applied to the analyser.

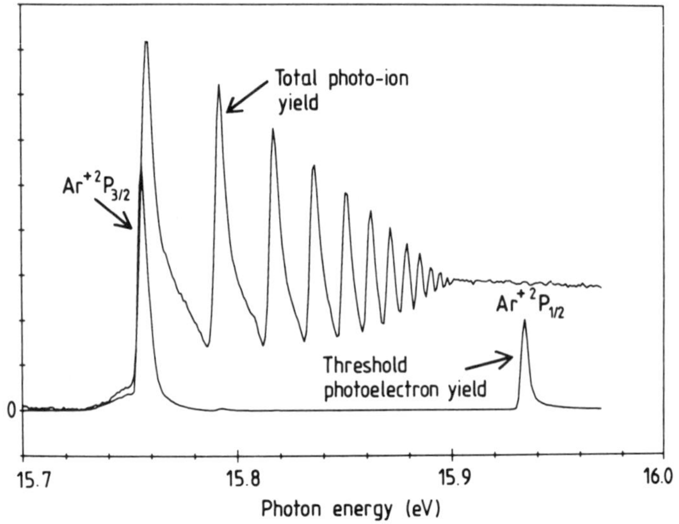

Figure 1. Threshold photoelectron spectrum superimposed on a photoion spectrum obtained in argon.

3. Threshold Photodouble Ionization in Helium and Electron-electron Correlations

A threshold photoelectron spectrum of helium obtained in the region of the double ionization potential[3] is shown in figure 2. Satellite states of He^+ up to n = 12 are clearly visible. Higher levels merge into an almost flat shoulder which falls off rapidly reaching a sharp minimum at the double ionization potential. This singularity which has the form of a strongly asymmetric cusp can be seen more clearly in the inset. Since the yield of threshold electrons (of energy between zero and ΔE) is being

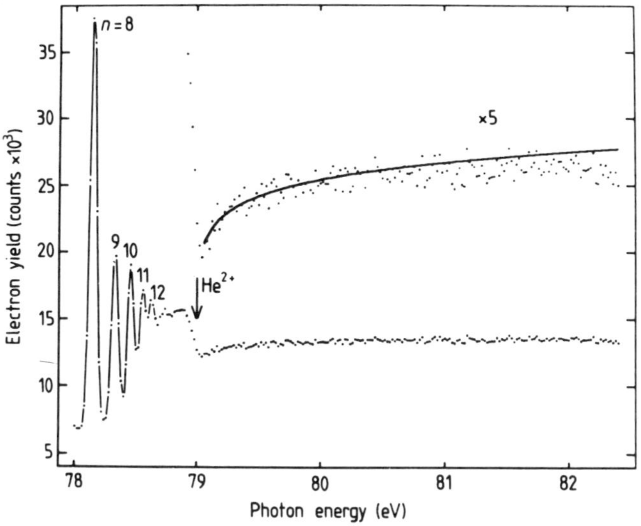

Figure 2. Threshold photoelectron spectrum of helium over an energy range in the region of the double ionization potential.

measured this corresponds to a partial cross section which may be written

$$\sigma_{\text{partial}} \propto E^{n-1}$$

where E is the excess energy available to the two electrons and n is predicted to be 1.056 by Wannier theory. Since n-1 rather than n is being measured this is a very sensitive test of the theory. The full curve shown in figure 2 is a fit of the data over the energy range $0.08 < E < 0.48$ for the Wannier exponent $n = 1.056$ and then extrapolated to higher energies. The data are consistent with this value of n and give the range of validity of the threshold law of about 2 eV.

Information on the angular behaviour of the photoelectrons was obtained by tuning the electron analyser to a particular value of photoelectron energy (eg 0.14 eV) and measuring the yield of photoelectrons at angles of 0° and 90° with respect to the electric field vector of the photon beam. β parameters for He^{+} satellite states obtained in this way[3] are shown in figure 3 together with results from Heimann et al[4]. Greene[5] has predicted that in the case of helium, β tends to -1 at the threshold of the satellite states as n tends to infinity, ie at the double IP. In contrast the present results tend to level off at about n = 7.

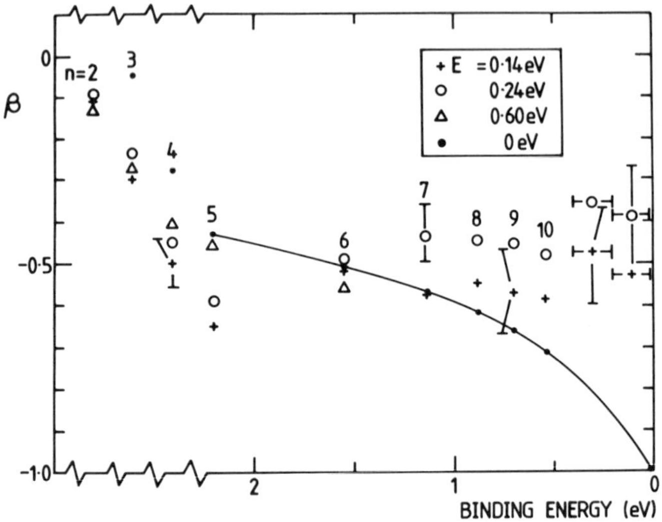

Figure 3. Assymetry parameters β plotted against binding energy for $He^{+*}(n)$ satellite states for varous values of photoelectron energy. The triangles are the results of Heimann et al[4].

4. Threshold Photoelectron/Photoion Coincidence Measurements

4.1. Helium

Coincidence techniques can be very powerful in the unravelling of the various decay paths in photoionization processes. In the case of photoionization in helium for example it is possible to separate the He^{++} contribution from the He^{+} contribution. This is done by detecting the threshold photoelectrons in coincidence with either the yield of He^{+} ions, below the double ionization potential (corresponding to excitation of satellite states) or the yield of He^{++} ions, above the double ionization potential (corresponding to excitation of continuum states). For these measurements[6], a time of flight analyser was added to the threshold electron analyser. This TOF analyser also featured a penetrating field stage and again had very high detection efficiency (~ 75%). This TOF analyser enabled the He^{+} and He^{++} ions to be separated and so coincidences could be measured between either of these species and the threshold electrons. The measured TPEPICO (He^{++}) and (He^{+}) spectra are shown in figure 4A and 4B respectively, together with the threshold photoelectron spectrum (figure 4C). As well as allowing the discrete and continuum regions to be separated and compared, the technique also allowed the elimination of the contribution of spurious background electrons. In this way, Hall et al[6] were able to improve their earlier measurements of the Wannier exponent n.

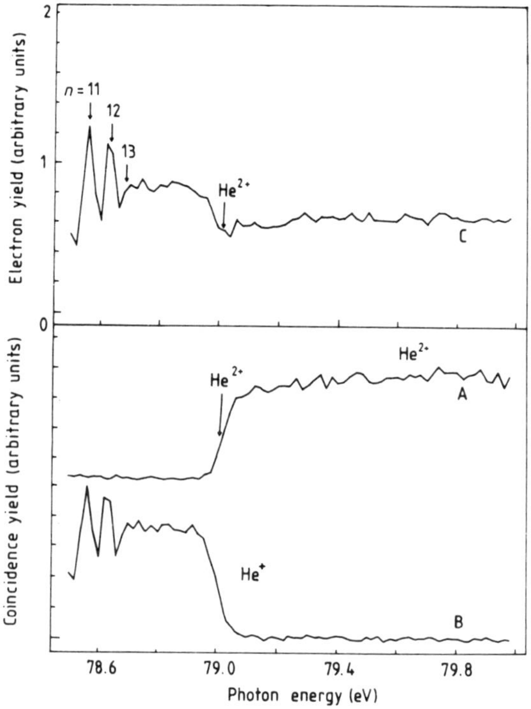

Figure 4. TPEPICO (He^{++})(A), TPEPICO (He^{+})(B) and threshold photoelectron spectrum in the region of the double I.P.

4.2. The Other Rare Gases

Hall et al[7] used the above coincidence technique to study the other rare gases Ne, Ar, Kr and Xe. Examples of their data are shown in figures 5 and 6 . Figure 5 shows a threshold photoelectron spectrum of krypton in the region of the double IP while figure 6a and 6b show the TPEPICO spectra for Kr^{++} and Kr^{+} respectively. The most striking feature of the coincidence spectra is that once the Kr^{++} channel is open no further signal is detected in the Kr^{+} channel. Another interesting feature is the very low cross section at the double ion onset itself. In fact the double ionization cross section only becomes intense once the satellite states are excited and the measurements show that this indirect process is the dominant contribution to the double ionization cross section.

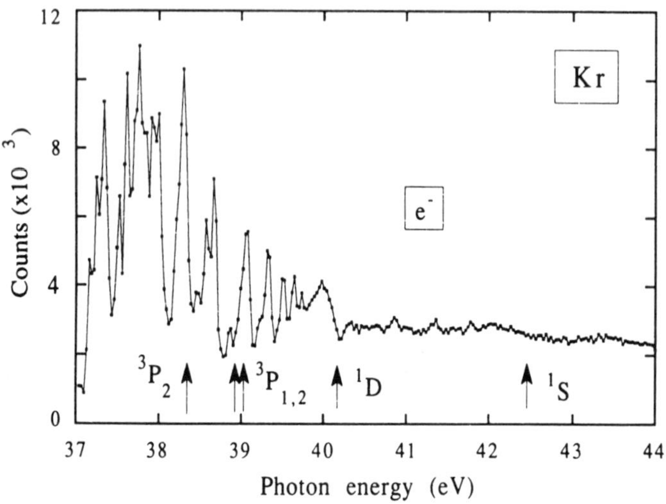

Figure 5. Threshold photoelectron spectrum of krypton.

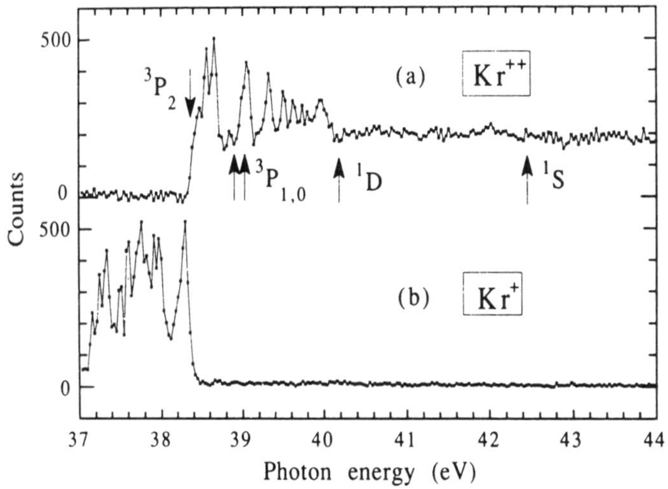

Figure 6. TPEPICO spectra recorded for singly and doubly charged ions of krypton.

5. Threshold Photoelectron/Photoelectron Coincidence Measurements

In the above coincidence studies the threshold photoelectron was measured in coincidence with one of the ionic products. In threshold photoelectrons coincidence spectroscopy two near zero energy photoelectrons are measured in coincidence, again as the energy of the incident photons is varied. Clearly two such low energy photoelectrons will only be produced at the threshold of photodouble ionization. Thus a spectrum of threshold electron coincidences against photon energy will map out the double ion state(s) of the target atom.

The apparatus for this technique[8] consists of two of the threshold analysers discussed above, situated on opposite sides of the interaction region where the tunable synchrotron radiation intersects the target gas beam. In the photodouble ionization process one photoelectron may travel in the direction of one of the threshold analysers and the other photoelectron in the direction of the second analyser. A TPESCO spectrum, obtained in argon[8] is shown in figure 7. Theory indicates that certain final states of the double charged ion, characterised by the symmetry of their two-electron wave functions should be kinematically disallowed and hence of low cross section[9,10]. The rare gases are ideal for a study of this selectivity since they have several doubly charged states of differing symmetry with which to test theory.The most important point of note in the spectrum is that all the possible ionic states are observed. Furthermore, the unfavoured ones are excited with intensities that are not very different from the favoured ones.

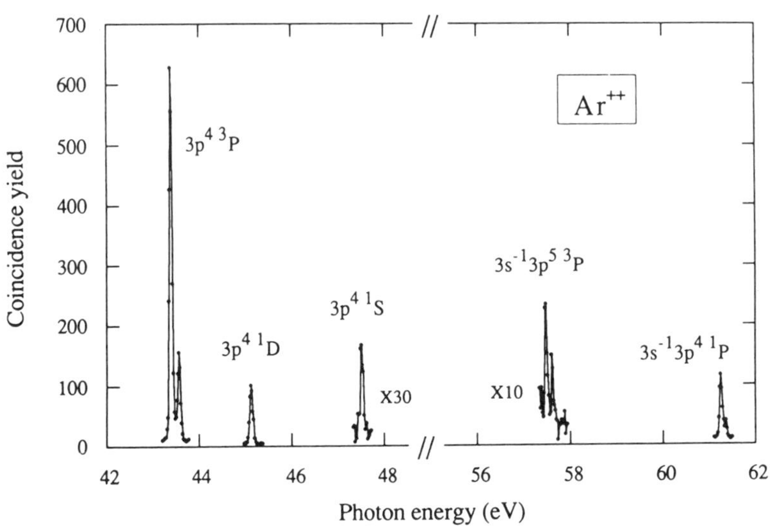

Figure 7. A threshold photoelectrons spectrum in argon showing the various double ion states.

5.1. Molecules

The TPESCO technique can be readily extended to molecules. Strictly speaking doubly charged molecular ions are not stable since the Coulomb repulsion between ions leads to a system whose potential energy is everywhere greater than the dissociation energy. However, the binding force produced by the exchange interaction between the electrons can produce a local potential minimum which can support vibrational levels. For example, the ground state of O_2^{++} is predicted to be several eV deep and to contain 18 vibrational levels. A TPESCO spectrum obtained in O_2 is given in figure 8, and shows a long vibrational progression which is assigned to the O_2^{++} ground state[11]. Nineteen levels are observed and this has allowed the molecular parameters of this state to be deduced for the first time. The measurements have also enabled the double ionization potential off O_2 to be located at 36.13± 0.02 eV. Interestingly this is nearly 0.5 eV lower than the previously accepted value. This difference is explainable by the relatively low intensity of the first two vibrational levels but which are observed in the present work.

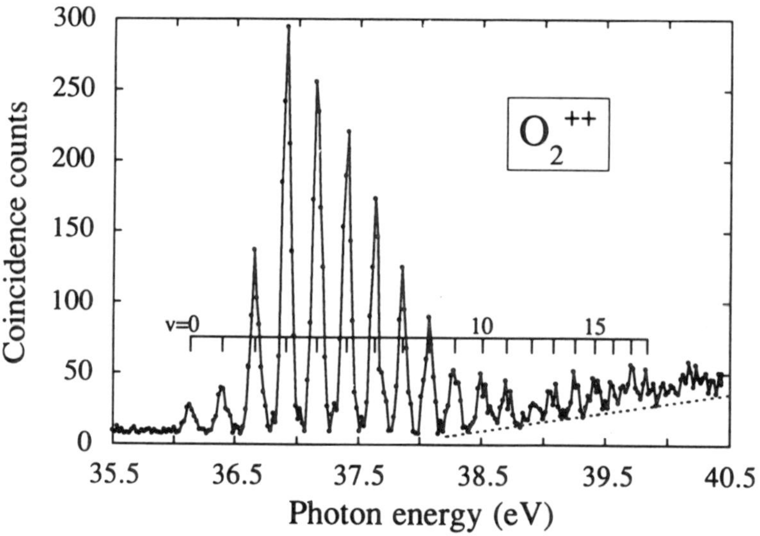

Figure 8. A TPESCO spectrum obtained in O2 showing the vibrational strucure of the O_2

6. Conclusions

Photodouble ionization may have contributions from direct and indirect processes. Direct single photon double ionization is dominated by electron-electron correlations and consequently its cross section is low. On the other hand, indirect processes such as photodouble ionization via the decay of satellite states can have much larger cross sections and are usually the dominant contribution to photodouble ionization. Threshold photoelectron spectroscopy can be a very useful tool in the study of the dynamics of the photo double ionization process. This is especially true when the technique is allied to the methods of coincidence counting.

7. Acknowledgements

The majority of the work described here was undertaken by Drs R I Hall, G Dawber, A McConkey, K Ellis, L Avaldi, M A MacDonald and G C King.

8. References

1. G. H. Wannier, *Phys. Rev.* **90** (1953) 817.

2. R. I. Hall, A. McConkey, K. Ellis, G. Dawber, L. Avaldi, M. A. MacDonald and G.C. King, *Meas. Sci. Technol.* **3** (1992) 316..

3. R.I. Hall, L. Avaldi, G. Dawber, K. Ellis and G. C. King, *J. Phys. B.* **24** (1991) 115.

4. P. A. Heimann, U. Becker, H. G. Kerkhoff, B. Langer, D. Szostak, R. Wehlitz, D. W. Lindle, T. A. Ferrett and D. A. Shirley, *Phys. Rev.* **A34** (1986) 3782.

5. C. H. Greene, *Phys. Rev. Lett.* **44** (1968) 441.

6. R. I. Hall, A. McConkey, L Avaldi, K. Ellis, M. A. MacDonald, G. Dawber and G. C. King, *J. Phys. B.* **25** (1992) 1195.

7. R. I. Hall, K. Ellis, A. McConkey, G. Dawber, L. Avaldi, M. A. MacDonald and G. C. King, *J. Phys. B.* **25** (1992) 377.

8. R. I. Hall, G. Dawber, A. McConkey, M. A. MacDonald and G. C. King, *Z.Phys.D.* **23** (1992) 377.

9. C. H. Greene and A. R. P. Rau, *Phys. Rev. Lett.* **48** (1982) 533.

10. A. D. Stauffer, *Phys. Lett.* **91A** (1982) 114.

11. R. I. Hall, G. Dawber, A. McConkey, M. A. MacDonald and G. C. King, *Phys. Rev. Letts.* **68** (1992) 275.

COMPLETE FRAGMENTATION PATTERNS IN TWO-STEP DOUBLE PHOTOIONIZATION OF ATOMS

VOLKER SCHMIDT
Fakultät für Physik, Universität Freiburg, D7800 Freiburg/Germany

ABSTRACT

Energy-, angle- and spin-resolved fragmentation patterns for photon-induced double ionization with special emphasis on two-step processes (photoionization with subsequent Auger decay) are discussed with respect to the parametrization of the manifold differential cross section, the measurement of fragmentation patterns, and the example of a complete fragmentation pattern for $4d_{5/2}$ photoionization in xenon with subsequent N_5-$O_{23}O_{23}$ $^1S_0$ Auger decay.

1. Introduction

Photon-induced double ionization processes in atoms lead to rather complicated fragmentation patterns which describe the spatial properties of two-electron emission. Without observing the polarization of the remaining ion, a complete fragmentation pattern reflects the energy-, angle- and spin-dependence of two-electron emission as given by the corresponding manifold differential cross section. In the most general way, it is a 6-fold differential cross section,

$$\frac{d^6\sigma}{d\Omega_1\, d\Omega_2\, dE_1\, dE_2}(\hat{x}_1,\hat{x}_2;m_{s_1},m_{s_2})\Big|_{\mathbf{S}}\ \delta(E_{ph}-E_1-E_2-E_B{}^{++})$$

which shows separately the dependences on different parameters: on the electron emission into the solid angles $d\Omega_1$ and $d\Omega_2$ at the directions $\hat{x}_1$ and $\hat{x}_2$, respectively (defined in a collision frame where the z-axis points into the direction of the incident photon beam and the x-axis coincides with the major axis of the polarization ellipse), on the energies E_1 and E_2 for each of the two electrons whereby the δ-function ensures energy conservation, on the spin projections m_{s_1} and m_{s_2} of both electrons as measured in the individual detector frames, and on the polarization of the incident light, which is described by the Stokes parameters $\{S_1,S_2,S_3\} = \mathbf{S}$. Of course, in addition to these parameters the differential cross section also depends on the inherent properties of the photoprocess, i.e., on the matrix elements responsible for photon-induced double ionization. Hence, it is of general interest to parametrize the manifold differential cross section in terms of "geometrical" (directions of electron emission and spin projections including the polarization of the incident light) and of "dynamical" (energy sharing between the electrons and the energy-dependence of the matrix elements) quantities, because such a parametrization provides the desired information on the photon-atom interaction in the case of double ionization as well as a view on the complete fragmentation pattern.

2. General Parametrizations

In order to simplify the discussion of a parametrized expression for the manifold differential cross section of two-electron emission, the spin dependence shall be omitted and completely linearly polarized light shall be assumed. For randomly oriented atoms and after integration over one of the two energies ($E_1 = E$, $E_2 = E_{ph} - E_B^{++} - E_1$) one obtains[1] (note some slightly different definitions; compare also parametrizations for direct double photoionization[2,3], for two-step double ionization,[4-6] and for continuous double Auger decay[7] which can be adapted also to direct double photoionization):

$$\left.\frac{d^3\sigma}{d\Omega_1 d\Omega_2 dE}\right|_{lin.pol.} = 1 + \sum_k A_{kk0}\, P_k(\cos\Theta_{12}) + \sum_{k_1,k_2} A_{k_1k_22}\, B_{20}^{k_1k_2}(\alpha_1,\varphi_1;\alpha_2\varphi_2)$$

with the bipolar harmonic function

$$B_{km}^{k_1k_2}(\hat{\varkappa}_1,\hat{\varkappa}_2) = \sum_{m_1,m_2} (k_1m_1\, k_2m_2 \,|\, km)\; Y_{k_1m_1}(\hat{\varkappa}_1)\; Y_{k_2m_2}(\hat{\varkappa}_2)\,,$$

the Clebsch-Gordan coefficient $(k_1m_1\, k_2m_2|km)$

and the spherical harmonics $Y_{km}(\hat{\varkappa})$.

In the given formula the A-coefficients contain the dynamical parameters of the double ionization process, i.e., the matrix elements with relative phases. In particular they depend on the energies E_1 and E_2 of the ejected electrons. The "geometry" is described now explicitly by the angular functions where the polar and azimuthal angles, α_i and φ_i comprised in $\hat{\varkappa}_i$, refer to the direction of the electric field vector as reference axis. The terms associated with the Legendre polynominal $P_k(\cos\Theta_{12})$ represent the angular correlation between the escaping electrons which is independent of the light beam, and the terms associated with the bipolar harmonics are due to the multipole expansion of the interactions of the electrons with the electrical field vector. The summation indices extend to infinity with the restriction $\{k_1 + k_2 = \text{even}\}$ and $\{k_i = 0, 1, 2, \ldots, 2\ell_i+1\}$ with ℓ_i being the orbital angular momentum of the respective continuum electron.

The formula applies for any photon-induced double ionization process, i.e., in particular for the two interesting and different cases of direct double photoionization and two-step double ionization. However, photon-induced two-step double ionization brings in several advantages which make the application of the general formula tractable and allow finally the establishment of a complete fragmentation pattern. The reason for these essential simplifications is the formulation of two-electron emission in two steps which fixes the kinetic energies $E_1=E_{ph}-E_B{}^+$ and $E_2=E_B{}^+-E_B{}^{++}$ of the ejected electrons and their orbital angular momenta ℓ_1 and ℓ_2. Hence, there is no energy sharing between the two continuum electrons, the summation indices k_1 and k_2 are restricted to only a few terms, and the condition of $\{k_1 + k_2 = \text{even}\}$ reduces to the condition that each k_i value must be even.

As an application of the foregoing parametrization, the case of 2p photoionization in magnesium with its subsequent L_{23}-M_1M_1 Auger decay shall be considered. Because Russell-Saunders coupling applies well, photoionization of a 2p electron is described by two dipole amplitudes, D_s and D_d, which correspond to the partial waves $\varepsilon_1 s$ and $\varepsilon_2 d$ of the photoelectron. In addition, the simple structure of the system allows only one partial wave, $\varepsilon_2 p$, for the Auger electron. With these orbital momenta the summation in the general expression for the parametrized cross section is restricted to only a few terms, i.e., the coefficients A_{220}, A_{022}, A_{222} and A_{422}. The differential cross section obtained in this way is identical with the parametrization result that follows from the more elaborate treatment of photon-induced two-step double photoionization on grounds of statistical and efficiency tensors[4-6] which has the advantage that it makes the factorization into quantities attached to the first and second step of the two-step ionization process transparent and it yields explicitly the dependences of the A-coefficients on the matrix elements including their relative phases (for the magnesium example see Ref.8). Hence, with known values for the matrix elements involved, complete fragmentation patterns can be calculated for any experimental set-up. As an example, spatial views for two energy- and angle-resolved fragmentation patterns are shown in the left hand part of Fig.1. For comparison, the corresponding patterns for direct double photoionization in helium are shown in the right hand part of Fig.1. Two fundamental features can be noted. First, only if one electron is emitted in the direction of the oscillating electric field vector of the incident light does rotational symmetry appear. Second, in the process of direct double photoionization in helium the system avoids electron emission in the same directions (which is a consequence of interplay between angular and correlation factors in the parametrized expression of Ref.3). This is not the case for the two-step process in magnesium (which is a consequence of the symmetry properties of the alignment tensor of the intermediate state).

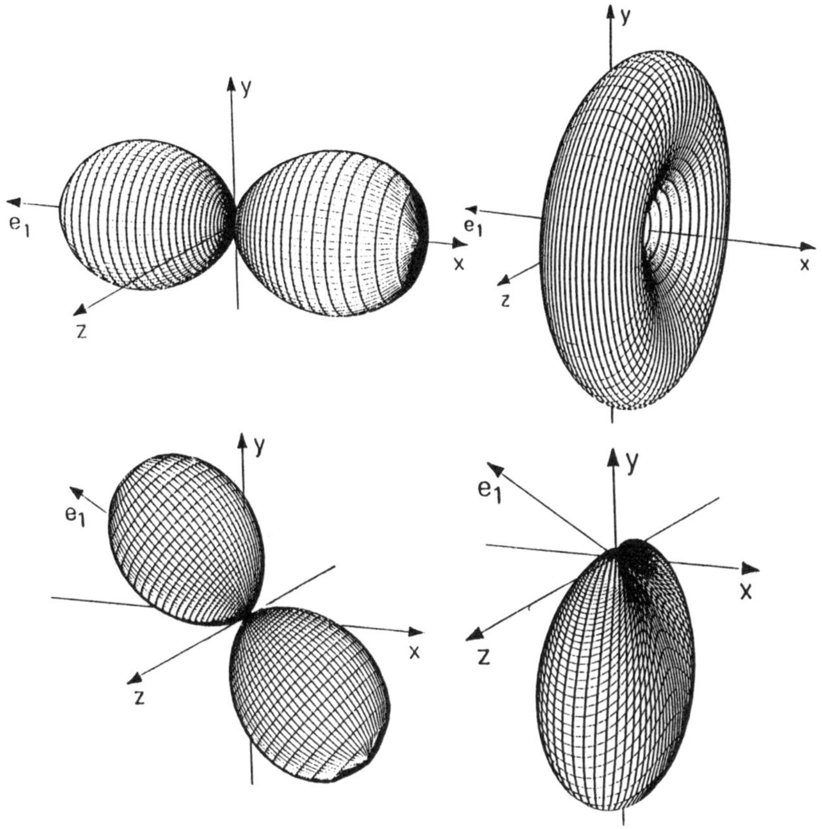

Fig.1. Predicted spatial views of angle- and energy-resolved fragmentation patterns for completely linearly polarized incident light (electric field vector along x-axis). Left: two-step double ionization in magnesium at 80 eV photon energy ($2p_{3/2}$ photoelectron e_1 with the L_3-M_1M_1 Auger electron e_2; data from Ref.9); right: direct double photoionization in helium at 80 eV photon energy (equal energies for both ejected electrons; data from Ref.3 with $\Theta_0=91^0$). Two cases are selected for the polar angle of the electron e_1 with fixed direction, $\alpha_1 = 180^0$ and $\alpha_1 = 150^0$.

3. Measurement of Fragmentation Patterns

The measurement of fragmentation patterns requires the coincident detection of both electrons, and such experiments are difficult to perform due to the low cross section for photon-induced double ionization and the special constraints when working with synchrotron radiation. Therefore, the main aspects to be considered for succesful experiments shall be listed.

First, the coincident angular correlation depends on the linear and circular polarization of the monochromatized synchrotron radiation. Though the linear polarization (given by the Stokes parameters S_1 and S_2) can be measured easily, this is not the case for circular polarization (given by the Stokes parameter S_3) in the VUV region. Hence, the quantitative interpretation of an observed fragmentation pattern relies on the correct information on the light polarization, and strong preference is given to experiments using monochromatized undulator radiation if the first harmonics can be tuned to the desired photon energy, because this radiation is practically completely linearly polarized ($S_1 \approx 0.99$).

Second, the experimental set-up for electron spectrometry has to be optimized for the measurement of coincidences.[10,11] Important aspects are a small coincidence resolving time (which is, however, determined essentially by the time spread from different geometrical paths of the electrons in the spectrometers), large solid angles accepted by each of the spectrometers and the limitation of the source volume for simulating a point-like source (possibly with a system of diaphragms). The last item then enables both spectrometers to view the same source volume and it simplifies the necessary correction for the influence of large solid angles.

Third, for the energy selection of both kinds of electrons the respective spectrometer voltages have to be set at the corresponding desired nominal energy which correspond to optimum transmission, and the height of the transmission function is a measure of intensity. However, from non-coincident electron spectrometry it is known that -depending on the angle-setting- there might be some shift in the line position and, therefore, in the height. For direct double photoionization with its continuous energy distribution such a "line" shift will have little effect on the measured coincident intensity, but in the case of two-step double ionization with fixed nominal energy values it might cause problems. Hence, in the latter case it is necessary to scan the respective spectrometer voltages over the whole photo- and Auger lines. For isolated electron lines such a time-consuming procedure can be avoided by registering each electron line at once using large-scale detectors (channelplates instead of channeltrons).

Fourth, the usually extremely low coincidence rates imply long times of data accumulation in order to achieve sufficient statistics. Therefore, high demands are put onto stable experimental conditions. When working with monochromatized synchrotron radiation this means in particular stable energy and polarization and a careful calibration for the unavoidable intensity decrease after each injection. Because true and accidental coincidences depend differently on the source strength, the change of photon intensity then requires the simultaneous registration of total and accidental coincidences.

In Fig.2 the first measured fragmentation pattern for photon-induced two-step double ionization is shown. It concerns $4d_{5/2}$ photoionization in xenon at 94.5 eV with subsequent N_5-$O_{23}O_{23}$ $^1S_0$ Auger decay[12] (for first observed fragmentation patterns in direct double photoionization see Refs.13,14). The angular correlation was measured in a plane perpendicular to the photon beam direction, for a fixed position of the emitted photoelectrons (azimuthal angle $\Phi_1=150^0$ in a coordinate frame with the z-axis pointing into the direction of the photon beam) and for several angles Φ_2 of the coincident Auger electrons. The experimental values are given as points with error bars, and the solid curve represents a least-squares fit of the theoretical angle-dependence to the experimental data. In the present case this angle-dependence requires

$$N_{true} = A_0 + A_2 \cos 2\Phi_2 + A_4 \cos 4\Phi_2 + B_2 \sin 2\Phi_2 + B_4 \sin 4\Phi_2,$$

where all three Stokes parameters are allowed to contribute, and from which the fit yields numerical values for the "observables" A_i/A_0 and B_i/A_0.

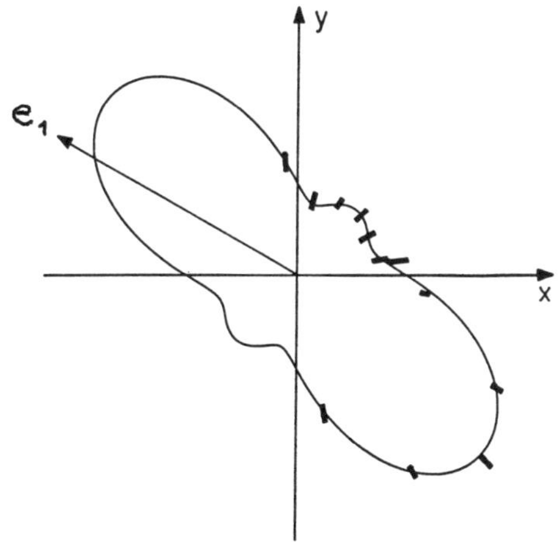

Fig.2. Fragmentation pattern for photon-induced two-step double photoionization at 94.5 eV photon energy for $4d_{5/2}$ photoionization (electron e_1) and subsequent N_5-$O_{23}O_{23}$ $^1S_0$ Auger decay (electron e_2) in xenon measured in a plane perpendicular to the photon beam direction.[12]

4. Complete Fragmentation Patterns for Two-Step Double Photoionization in Xenon

The selected process of $4d_{5/2}$ photoionization in xenon with subsequent N_5-$O_{23}O_{23}$ $^1S_0$ Auger transition represents a special case of photon-induced two-electron emission for two reasons: First, the condition for the two-step formulation is fulfilled (a small disturbance by post-collision interaction is taken into account) and, second, only one partial wave, $\varepsilon_2 d_{5/2}$, contributes to the selected Auger transition. The latter property makes the dynamics dependent only on the photoionization matrix elements (the energy-independent Auger matrix element solely governs the overall intensity, and the coupling of angular momenta involved in the Auger transition brings in some simple known numerical factors). Therefore, a measurement of the angular correlation between these electrons gives additional information on the photoionization process (within the dipole approximation, which is well fulfilled here) and provides (together with data on the partial cross section $\sigma(4d_{5/2})$, on the angular distribution parameter $\beta(4d_{5/2})$, and on the alignment parameter $\mathcal{A}_{20}(4d_{5/2})$) a sufficient number of observables for the evaluation of the dipole matrix elements d_γ and their relative phases $\delta_{\gamma\mu}$ (there are three involved matrix elements, corresponding to the channels $4d_{5/2} \to \varepsilon f_{7/2}$, $4d_{5/2} \to \varepsilon f_{5/2}$ and $4d_{5/2} \to \varepsilon p_{3/2}$, respectively; hence, one must determine three values $d_\gamma = \{d_+, d_0, d_-\}$ and two relative phases $\delta_{\gamma\mu} = \{\delta_0 - \delta_+, \delta_0 - \delta_-\}$). In other words, a complete experiment can be performed which then completely determines the fragmentation patterns for any selected experimental set-up.

The "observables" A_i/A_0 and B_i/A_0 from the fragmentation pattern of Fig.2 depend on the matrix elements d_γ and their relative phases $\delta_{\gamma\mu}$ (which are to be determined), on the unknown Stokes parameter S_3 for the circular polarization of the incident light (only the second harmonic of the monochromatized undulator radiation with S_1=0.957(5), S_2=0.0 could be tuned to the desired photon energy), and on known attenuation factors from the finite solid-angle effect. In order to extract the desired quantities d_γ, $\delta_{\gamma\mu}$ and S_3 from all the observables of the selected photoionization process, a system of non-linear equations was solved taking into account certain constraints for the unknowns (e.g., 2π range of the phases; each $|d_\gamma|^2$ multiplied with the known prefactors cannot be larger than the cross section; $S_3{}^2$ has to be smaller than $1-(S_1{}^2+S_2{}^2)$). The values obtained and published for d_γ and $\delta_{\gamma\mu}$ yielded a surprisingly large value for the relative phase $\delta_{\varepsilon f5/2} - \delta_{\varepsilon f7/2}$.[12] This was ascribed to strong spin-orbit effects in the continuum and seemed to be in accord with the results of a complete experiment on grounds of a spin-polarization measurement of the photoelectrons (5p photoionization in xenon[15]). However, the explicit calculation of d_γ and $\delta_{\gamma\mu}$ for $4d_{5/2}$ photoionization in xenon within the relativistic random phase approximation including intershell couplings[16,17] brought the result $\delta_{\varepsilon f5/2}-\delta_{\varepsilon f7/2} \approx 0$, which is in strong contradiction to the experimental value. Therefore, the root-finding procedure for solving the non-linear equations was reconsidered and it was found that the relative phases are strongly related to the correct value of S_3. For a better understanding of this connection a different procedure was worked out by decoupling the unknowns d_γ and $\delta_{\gamma\mu}$ from S_3 and evaluating their values for preselected S_3 values within the allowed range. This approach brought the result that the d_γ values are not much influenced by S_3, but without accurate information on S_3 large error bars have to be ascribed to the relative phases (in particular to $\varepsilon f_{5/2}-\varepsilon f_{7/2}$). Hence, within these error bars no disagreement exists between experimental and theoretical values (for details see Ref.6).

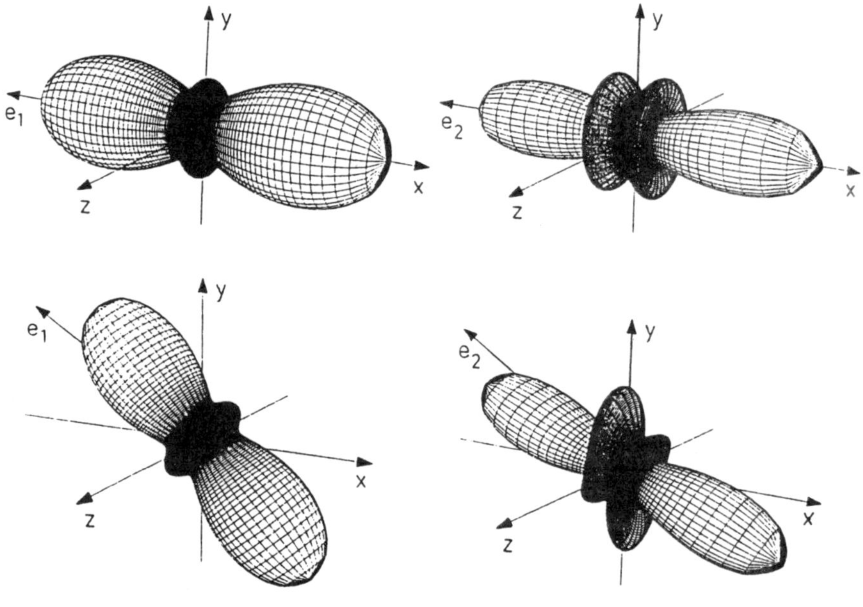

Fig.3. Spatial views of angle- and energy-resolved fragmentation patterns for completely linearly polarized incident light (94.5 eV, electric field vector along x-axis) and $4d_{5/2}$ photoionization with subsequent $N_5-O_{23}O_{23}$ $^1S_0$ Auger decay in xenon (data from Ref.6). The fixed direction for the emission of one of the electrons is specified as e_1 for the photoelectron and e_2 for the Auger electron, respectively; two cases are selected for the polar angle of the electron with fixed direction, $\alpha_1 = 180^0$ and $\alpha_2 = 150^0$.

Despite of the not-yet completely solved problem of the correct relative phases of the involved photoionization matrix elements, the mean values of the experimental data set can be used to establish complete fragmentation patterns for any desired geometry. Spatial views for the angular correlation pattern following photoionization by completely linearly polarized light and for fixed directions of the $4d_{5/2}$ photo- or N_5-$O_{23}O_{23}$ $^1S_0$ Auger electron, respectively, and ejection of the coincident other electron into the whole space are shown in Fig.3. In comparison to the fragmentation patterns of magnesium shown in Fig.1 there appears more structure, which is due to the higher orbital angular momenta of the electron whose angle-dependence is observed. The same argument applies when the fragmentation patterns for fixed direction of the Auger or photoelectron are compared.

Finally, the completeness of the experimental data set allows also the inclusion of a spin measurement for both electrons, thus yielding the full energy-, angle- and spin-resolved fragmentation pattern. As example, a sectional view in the plane perpendicular to the photon beam direction is shown in the left hand part of Fig.4 for longitudinal spin components. The essential changes caused by the additional spin observation can be understood qualitatively, if all spin-

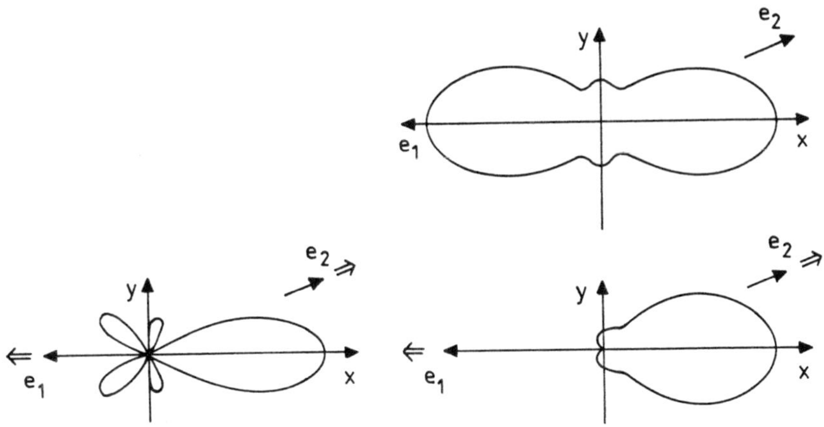

Fig.4. Sectional view of angle-, energy- and spin-resolved fragmentation pattern in the plane perpendicular to the photon beam direction, for the experimental conditions listed in the caption of Fig.3 and for longitudinal spin components of both electrons as indicated by the double arrows (left part). Right part: top, sectional view without spin-observation; bottom, spin-dependent pattern evaluated on grounds of a precessing spin in a singlet state only.

orbit effects are neglected and the two-electron pair wave function is assumed to be in a singlet state. With these prerequisites the spin decouples from the angular correlation pattern and can be visualized for the underlying double Stern-Gerlach experiment as follows:[18] Once the photoelectron's spin is measured along the given direction, the Auger electron's spin is fixed in space, too, and can precess only around the axis opposite to the photoelectron's spin. A measurement of the longitudinal spin component of the Auger electron then requires a projection of its precessing spin onto the direction of electron emission. Obviously, the projection gives either the maximum value or zero if both electrons are emitted into opposite or alike directions, respectively. Therefore, in these directions the fragmentation pattern which includes the observation of longitudinal components of the spin will be identical to the spin-independent pattern or zero, respectively, but at the other angles the result will be between these extreme values. The quantitative treatment of this model leads to the lower curve in the right hand part of Fig.4. The differences to the correct pattern (lower curve in the left hand part of Fig.4) are due to spin-orbit effects, in particular to the fine-structure splitting in the $4d_j$ hole-state and to contributions from triplet components in the final state; for a full discussion of these aspects see Refs.6,19).

5. Outlook

Experimental and theoretical studies on fragmentation patterns for two-electron emission are still only just beginning. However, they represent the natural extension of the simple and widely applied case where only one electron is observed.[20] Due to the progress in experimental coincidence techniques and due to the theoretical interest on a deep understanding of those electron correlations which are responsible for photon-induced two-electron emissions and show up most distinctly in the manifold differential cross section, it can be expected that such studies will become a fast growing field in the future.

6. Acknowledgements

I would like to thank the members of my research group, in particular B. Kämmerling, for their hard work and many ideas for deriving the results described, and the German Federal Minister for Research and Technology (BMFT, Contract no. 05472 AAI and 055VF AAI) for financial support of the research projects.

7. References

1. H. Klar and M. Fehr, *Z. Physik* **D** (1992) in press.
2. P. Selles, J. Mazeau and A. Huetz, *J. Phys.* **B20** (1987) 5183.
3. A. Huetz, P. Selles, D. Waymel and J. Mazeau, *J. Phys.* **B24** (1991) 1917.
4. E.G. Berezhko and N.M. Kabachnik, *J. Phys.* **B12** (1979) 2993.
5. N.M. Kabachnik, submitted to *J. Phys.* **B** (1992).
6. B. Kämmerling and V. Schmidt, *to be published.*
7. M.Ya. Amusia, I.S. Lee and V.A. Kilian, *Phys. Rev.* **A45** (1992) 4576.
8. V. Schmidt, *X. Intern. Conf. X-Ray and Inner-Shell Physics, AIP Conf. Proc.* **215** (1990) 559.
9. A. Hausmann, B. Kämmerling, H. Kossmann and V. Schmidt, *Phys. Rev. Lett.* **61** (1988) 2669.
10. A. H. Wapstra, in: *Alpha-, Beta- and Gamma-Ray Spectroscopy, Vol.1,* ed. K. Siegbahn (1966) North-Holland, 539.
11. M. Völkel and W. Sandner, *J. Phys.* **E16** (1983) 456.
12. B. Kämmerling and V. Schmidt, *Phys. Rev. Lett.* **67** (1991) 1848.
13. J. Mazeau, P. Selles, D. Waymel and A. Huetz, *Phys. Rev. Lett.* **67** (1991) 820.
14. B. Krässig and V. Schmidt, *X. Intern. Confer. VUV Radiat. Physics/Paris* (1992) contributed papers.
15. Ch. Heckenkamp, F. Schäfers, G. Schönhense and U Heinzmann, *Z. Physik* **D2** (1966) 257.
16. W.R. Johnson and K.T. Cheng (1992), private communication and to be published.
17. Z.W. Lin and H.P. Kelly (1992), private communication.
18. A. Bohm, *Quantum Mechanics: Foundations and Applications,* 2nd ed. (1986), Springer.
19. B. Kämmerling and V. Schmidt, *X. Intern. Confer. VUV Radiat. Physics/ Paris* (1992) contributed papers.
20. C.N. Yang, *Phys. Rev.* **74** (1948) 764.

RECENT DEVELOPMENTS IN PHOTOIONISATION OF EXCITED ATOMS AND IONS

J.M. BIZAU, D. CUBAYNES and F.J. WUILLEUMIER
Laboratoire de Spectroscopie Atomique et Ionique and L.U.R.E., Bât. 350, Université Paris-Sud, 91405 Orsay Cedex, France

ABSTRACT

Recent results obtained at LURE on photoionisation of laser excited atoms and ions are presented. An accurate test of theoretical predictions on the behaviour of photoionisation cross section and asymmetry parameter for the first inner shell of excited sodium and potassium atoms has been performed using electron spectrometry and synchrotron radiation. We also present the first experiment demonstrating the feasibility to use electron spectrometry to study the photoionisation of positively charged atomic ions using the high flux of photons emitted from an undulator of Super ACO.

1. Introduction

Photoionisation experiments combining the use of synchrotron and laser radiations are now more than 10 years old[1]. With the advent of a new generation of synchrotron radiation sources, this field has already expanded and will reach soon new boundaries. Using an efficient technique like electron spectrometry to analyse the photoionisation processes, it is now possible to measure with good accuracy relative partial cross sections and asymmetry parameters for the inner shell photoionisation of atoms prepared in an excited state. As for the photoionisation of positively charged ions, the situation is quite different: they are very few experiments on ions using synchrotron radiation, and the feasibility of measurements by electron spectrometry was demonstrated only one year ago[2].

The kind of information that can be obtained from the study of photoionisation processes on atomic species can be divided into two categories: spectroscopic information on the structure of the atoms, and information on the dynamics of the photon-atom interaction. Since these experiments deal with a many-body system as soon as the atom or the ion under study is not hydrogen-like, the modelisation of both classes of processes is made difficult by the coulombic interaction between the electrons of the atom, the so-called "electron correlations". The importance of electron correlations was revealed by the experiments performed in the early 60's using synchrotron radiation on ground state atoms. Only through a constant interplay between higher accuracy

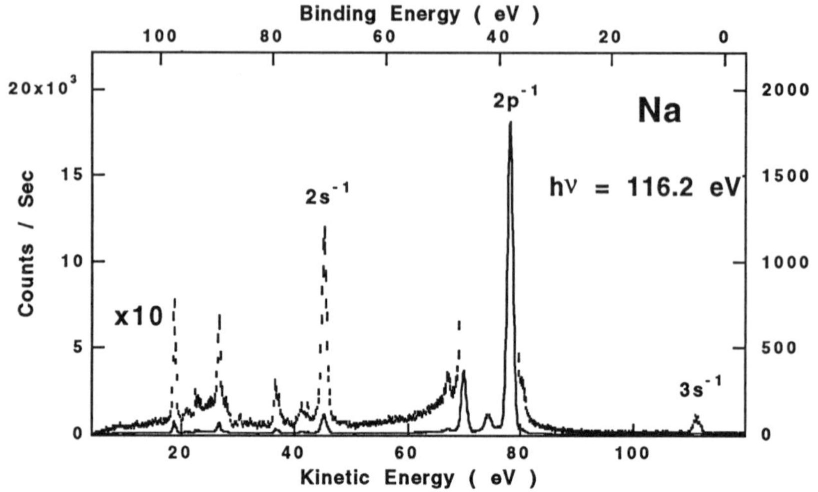

Fig. 1 - Complete photoelectron spectrum resulting from the photoionisation of Na atoms with photons of 116.2 eV, showing many electron correlation features.

experimental results, and more elaborated theoretical predictions, a satisfactory understanding of photoionisation processes was reached.

Among the various experimental techniques offered to the physicists to explore these processes, electron spectrometry is by far the most performant. The wealth of informations it gives on electron correlations is quite obvious when one considers an electron spectrum, such as the one presented in Fig. 1. This spectrum results from the photoionisation of sodium atoms in the ground state by photons of 116.2 eV photon energy. We observe several groups of lines. The most intense line of each group corresponds to single photoionisation in the various subshells of the Na atom: the line labeled $3s^{-1}$ at 5.1 eV binding energy corresponds to the photoionisation of the 3s subshell, the line labeled $2p^{-1}$ at 38 eV corresponds to the photoionisation of the 2p subshell, and the one labeled $2s^{-1}$ at 71 eV corresponds to the photoionisation of the 2s subshell. These lines are the only ones expected from the simplest model based on non-interacting electrons. All the other lines are a manifestation of electron correlations: i) the photoionisation of an inner subshell electron accompanied by the simultaneous excitation of the valence electron, gives rise to the so-called "satellite" lines. These lines are between 40 eV and 52 eV binding energy in the case of the 2p subshell, and between 75 eV and 95 eV in the case of the 2s subshell; ii) the Auger process, corresponding to the decay of the holes created in the photoionisation of the 2s subshell, is responsible here for the lines of kinetic energy between 17 eV and 32 eV; iii) the double photoionisation, or "shake off", (simultaneous ejection into the continuum of two electrons) creates a continuous distribution of electrons, which rises at the first double ionisation threshold (52eV binding energy) and is visible with a significant intensity over more than 10 eV.

In counterpart, the main difficulty in performing an electron spectrometry experiment, is the necessity to work with a low density of atoms, below 10^{13} atoms/cm^3. This limits the accuracy of the measurements of photoionisation processes with low cross sections. The situation is even worse for the studies of excited atoms and of ions, since only a small proportion of the atoms can be prepared in the excited state in the first case, and the currently available ion sources deliver low currents of rather fast ions in the second case. The only way to overcome this difficulty is to use a high flux of ionising photons. Only synchrotron radiation has the intense and clean energy spectrum, extending over a large enough photon energy range in the XUV, required to allow a useful comparison between measured and calculated innershell photoionisation cross sections. In addition, it is possible to take advantage of the polarization of the synchrotron radiation to measure the angular distribution of the emitted electrons. This parameter, which is related to the angular structure of the continuum wave functions of the ejected electrons, gives often a better sensitive test of the theories than the measurement of only the photoionisation cross sections, at least near thresholds and in the resonance regions.

In the following, we will present some of the most significant experimental results recently obtained on the photoionisation of laser excited atoms and ions, using electron spectrometry and synchrotron radiation.

2. Photoionisation of Laser Excited Atoms

Photoionisation of excited atoms is an efficient tool to explore electron correlations: using a laser to excite the atom, a well defined initial state for the photoionisation process can be prepared. Because the excited orbitals extend over a larger distance from the nucleus than the ground state orbitals, the nuclear potential can be tested at higher radial distance. The effect of the valence electron on an inner shell can be carefully studied in pumping, with lasers, the outer electron to more and more excited orbitals, and probing the inner shell electron with the synchrotron radiation.

The first experiments on excited atoms used electron impact as a pump, and a laser as a probe[3]. Because of the narrow tunability range of the laser available up to now, only photoionisation in the outer shell of excited atoms could be measured over a range of a few eV above the first ionisation threshold. Later on, pulsed lasers tuned to the energy of resonant transitions of the neutral atom were used, in synchronisation with a pulsed source of VUV continuum, allowing to begin photoabsorption experiments in inner shells of mainly alkali- and alkaline-earth atoms[4]. With the use of synchrotron radiation, resonant photoionisation was studied first using photoelectron spectrometry[5-8]. For example the oscillator strengths of the 2p -> 3s even parity resonances in Na were measured, with the valence electron on the first empty 3p orbital[5], or on higher excited orbitals using stepwise laser excitation[7,8]. The access to such even parity resonances, which is allowed in the two photon experiments, is forbidden when starting from the ground state of the atom. Direct photoionisation of an excited orbital could be measured over a large energy range only in one particular case, the 5d subshell of barium[9]. Because the Ba* 6s5d $^{1,3}D$ orbitals are metastable, a high proportion of the atoms (more than 90 %) could be excited.

Because the laser radiation is polarized, it is possible to aligne the excited atoms. Experiments measuring first the angular dependence[10], then the angular distribution[11] of electrons emitted in the resonant photoionisation of aligned lithium atoms as a function of laser polarisation have effectively been performed. This kind of experiments must be carried out with a low density of atoms (below 10^{11} atoms/cm^3), because radiation trapping occurs at higher atom density and destroys the initial alignment of the atoms. All the experiments presented below have been performed with a higher density of atoms, close to 10^{13} atoms/cm^3, in such a way that the excited atoms in the vapor can be considered as being non-aligned.

It is worthwhile to note also that, for the experiments combining laser and synchrotron radiations, only continuous wave lasers have been used up to now. This is due to the pulsed structure of synchrotron radiation with a high repetition rate, giving a small number of photons per pulse, and preventing to take advantage of the high power delivered by a pulsed laser synchronised with the synchrotron radiation.

Super ACO, which is the first ring of the third generation, came into operation in 1988. With the enhancement of photon flux it delivers, a new step could be reached: the measurement, with a good accuracy, of angularly integrated and resolved partial cross sections for direct inner-shell photoionisation of excited atoms .

2.1 Angularly Integrated Measurements

For the measurement of angle integrated cross sections, the electrons emitted around the magic angle were energy analysed with a cylindrical mirror analyser (CMA) whose axis is colinear to synchrotron radiation. As for the ground state Na atoms, the electron spectra obtained in the photoionisation of the 2p subshell of Na atoms laser excited in the $2p^6\ 3p\ ^2P_{3/2}$ state, show intense

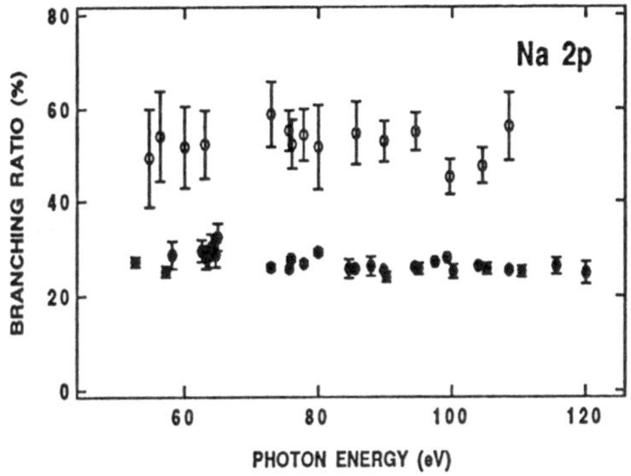

Fig. 2 - Variation of the branching ratio between photoionisation with excitation of the valence electron and single photoionisation in 2p subshell of Na atoms in the ground state (•) and in the 3p excited state[12] (o). We note the large enhancement of the relative intensity of correlation satellites in the excited atom.

satellite lines emphasizing the important role of electron correlations[12]. The most intense satellites correspond to "shake up" transitions of the valence electron, leaving the ion in the $2p^5$ 4p Na^+ state, and a "conjugate shake up" transition leaving the ion in the $2p^5$ 3d Na^+ state .

Fig. 2 compares the variation, as a function of photon energy, of the relative intensity of the sum of these correlation satellites, normalised to the intensity of the main line, for the ground state and the laser-excited sodium atoms[12]. One clearly observes a strong enhancement, by almost a factor of two, of the relative intensity of the satellites in the excited atom. One theoretical calculation has been performed, in the MBPT approximation[14], predicting that the variation, as a function of the photon energy, of the partial photoionisation cross sections of the 2p inner-shell electrons of sodium and 3p shell of potassium is almost independent of the orbital occupied by the valence electron. In order to verify this prediction, one compares in Fig. 3 the variation of the ratio between the 2p cross section for the atom in the excited state and in the ground state[12]. Fig. 3a, where the intensity of only the main line is considered, seems to refute the theoretical prediction, since we find this cross section is 20 % lower than the one of the same subshell for atoms in the ground state. This disagreement can certainly be accounted for by the fact that correlation satellites are not explicitly taken into account in the calculation. The experimental ratio of the cross sections, when the contribution of the satellites is added to the intensity of the main line, is close to 1 as shown in Fig. 3b, in much better agreement with the

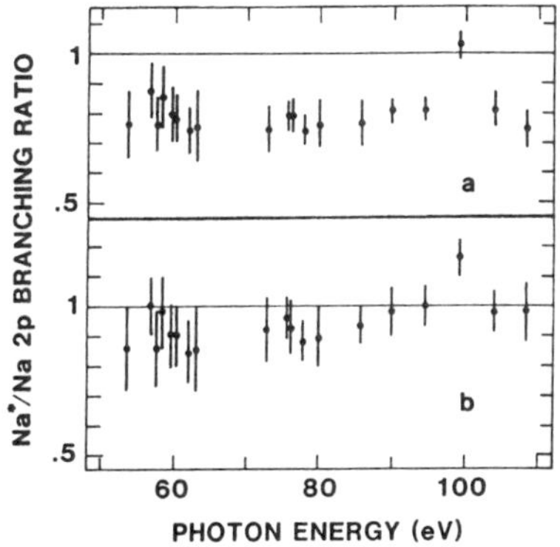

Fig. 3 - Variation, as a function of photon energy, of the branching ratio between the 2p photoionisation cross section of excited Na atoms and ground state Na atoms[12]: a) 2p main lines, b) 2p main lines + satellites.

theoretical prediction.

The sodium presents the special feature for metal atoms that absolute partial photoionisation cross sections can be determined using electron spectrometry and making the partition of the total photoabsorption cross section of Na atoms in the ground state, one of the very few absolute cross sections that has been experimentally measured in a metallic vapor[15]. A completely independant test of theoretical calculations is then allowed. Fig. 4 displays the variation, as a function of photon energy, of the partial single photoionisation cross section in 2p subshell of excited sodium atoms[13], as well as the cross section for the production of the correlation satellites, in comparison with the 2p cross section MBPT calculation[14]. The relative uncertainty in our absolute values, around 45 %, is for a large part due to the error on the absorption measurement, i.e. between 20 and 25 %. The calculated cross section seems too high compared to the measured 2p main line cross section, and too low compared to the total 2p cross section (main line plus satellites), although it stays within the error bars on this measurements.

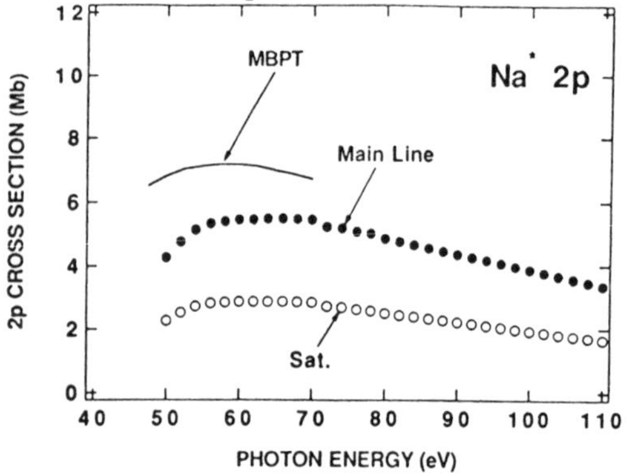

Fig. 4 - Variation, as a function of the photon energy, of the photoionisation cross sections in 2p subshell of Na atoms in the 3p excited state: ● main line, o satellites[9]; Theoretical results: —— MBPT[14].

The important transfer of oscillator strength towards the correlation satellites we have observed in the excited Na atom was not predicted by the MBPT calculation. This effect, discovered for the first time in the case of excited Na atoms[12,16], seems to be a general feature of inner shell photoionisation of excited alkaline atoms, since we have also observed it later for the 2s subshell of sodium[16], as well as for the photoionisation in the 3p subshell of potassium[17]. Fig. 5 shows the variation of the relative intensity of the sum of the correlation satellites for the $3p^6$ 4s $^2S_{1/2}$ ground state of K atom (left panel) and for the $3p^6$ 4p $^2P_{3/2}$ excited state (right panel). The average intensity of the satellites relative to the intensity of the main line is about 45 % in the excited atom, compared to 25 % for the ground state atom. The enhancement of the relative intensity of the correlation satellites in the excited atoms was attributed[12,16] to the fact that relaxation of the ionic core following inner shell ionisation leads to an overlap between the wave function of the shaken electron in the core-ionised final ionic

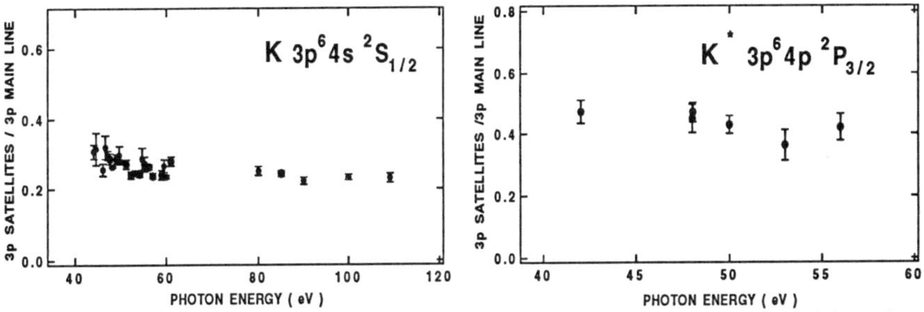

Fig. 5 - Variation of the branching ratio between photoionisation with excitation of the valence electron and single photoionisation in 3p subshell of K atoms in the ground state (left panel) and in the 4p excited state[17] (right panel).

state and the wave function of the valence electron in the initial state which is larger for the atom in the excited state than in the ground state. The same behaviour was recently predicted in a calculation of the photoionisation cross section for 1s electrons of lithium atoms in various $1s^2$ np highly excited state[18] ($n \geq 3$), cases which will be unfortunatly difficult to check experimentally for some time.

2.2 Angularly Resolved Measurements

The non-sensitivity of the inner-shell photoionisation cross sections to the orbital occupied by the valence electron, was also predicted for the behaviour of the asymmetry parameter for the 2p-subshell electrons of Na and the 3p-subshell electrons of K. Our experimental results[19] on the asymmetry parameter are displayed in Fig. 6, for Na atoms in the $2p^6$ 3s $^2S_{1/2}$ ground state (left panel) and in the $2p^6$ 3p $^2P_{3/2}$ excited state, together with theoretical results[14,20-22]. To perform this angularly resolved photoemission experiment, six channeltrons have been placed in the focal plane of the same CMA as the one previously used for the angularly integrated measurements. Six electron spectra are then simultaneously recorded, corresponding to electrons emitted at six different polar angles, taking as a reference the main axis of the ellipse of polarisation of the synchrotron radiation. The advantages of this method, where the CMA is kept fixed, compared to the classical one where the electron spectrometer is rotated, are the capability to avoid all the calibrations necessary when one has to take care of the variations, with time, of the photon flux and of the density of atoms. The experimental points presented in Fig. 6 are the result of our measurements for the intensity of the 2p main line at the six angles. Contrary to what was observed for the partial cross section, the MBPT prediction[14] seems here to be confirmed, since it is not possible to distinguish, within our error bars,

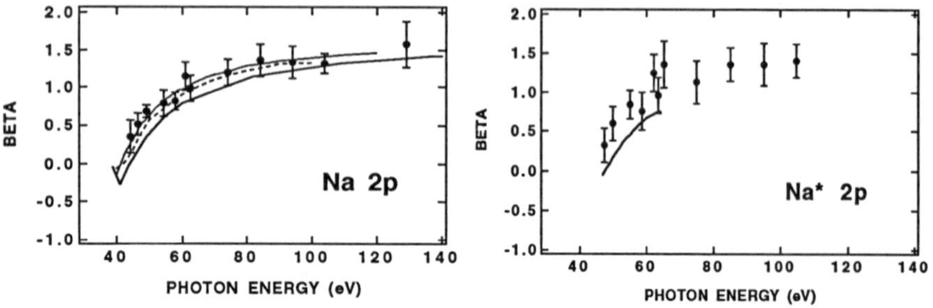

Fig. 6 - Variation as a function of the photon energy of the 2p asymmetry parameter of Na atoms in the ground state and in the 3p excited state: • experiment; theoretical results for the ground state: - - - HS[20], ... HF+relaxation[24], —— MBPT[14]; for the excited state : —— MBPT[21].

between the results for the ground state and for the excited atom. This better agreement could be explained if we remember that, within the model of Cooper and Zare[23], the asymmetry parameter is given by the ratio of radial matrix elements. The effect of even a substantial change in the cross section can then be largely cancelled out. This behaviour is well known already in the case of the 2p subshell of neon[24], the isoelectronic atom of Na^+. It is interesting to note that the variation, as a function of kinetic energy of the ejected electron, of the asymmetry parameter for the 2p subshell of Na atoms is the same, within 10%, as the variation of the asymmetry parameter for the Ne 2p subshell[25]. A complete comparaison with the theoretical results would require the measurement of the asymmetry parameter for the various satellites in the ground state and in the excited state. To perform these difficult experiments, the use of an undulator will be necessary, especially in the case of the excited atoms.

3. Photoionisation of Positively Charged Ions

Up to now, very little is known experimentally on the photoionisation or photoexcitation of ionic species. Most of our knowledge is obtained from theoretical predictions, based mainly on the simplest, non-interacting electron model[26]. The conclusion of all these calculations is that photoionisation cross sections in an inner subshell are almost not modified by the successive disappearance of outer electrons, except for a shift of the ionisation thresholds to higher energies[27]. The interest to have experimental data available to test theoretical calculations is quite obvious from the previously presented results, which show the inaccuracy of similar predictions for the photoionisation of excited atoms. Studies on positively charged atomic ions are a natural continuation to the studies on excited atoms, corresponding to the case where the valence electron is excited into the continuum. In addition, they also give the opportunity to test the variation of electron correlations along isoelectronic or isonuclear sequences.

The first experiments on ion species have used the same photoabsorption techniques developed for the studies on excited atoms: a pulsed laser tuned to the resonant transition energies of the atoms in a dense vapor produces, after some delay, a complete ionisation of the vapor. With a pulsed source of VUV continuum synchronised to the laser, photoabsorption spectra of some singly or doubly charged alkali- and alkaline-earth ions have been recorded[28]. Another method, known as the dual laser produced plasma method, consists in generating two plasmas using intense pulsed lasers focused on thick targets. One laser beam produces the ionised medium and the other one produces the photon continuum. With this method, almost complete stripping of the atoms of the first target can be achieved. Photoabsorption spectra of light ions in various charge states have been obtained with this technique[28,29].

The first successful photon-ion experiment involving synchrotron radiation was a merged beam experiment using ionic detection. Singly charged ions were produced in an ion source and accelerated. After focusing, the ion beam was merged on a length of about 10 cm with the monochromatised synchrotron radiation beam. Using ion spectrometry, absolute cross sections for resonant photoionisation of some singly charged ions were measured[30].

Some attempts to combine a non-monochromatised beam of synchrotron radiation with an ion trap have been carried out to produce very low-energy multiply-charged ions. One of these experiments allowed to analyse the charge distribution of ions produced in the K shell photoionisation of Ar^{2+} ions with white synchrotron radiation[31].

To perform electron spectrometry study on the photoionisation of ionic species, only a beam technique can be used. The great difficulty of such an experiment arises from the extremely low density of the ionised medium it is possible to focuse into the source volume of an electron spectrometer. In the best cases this density is around 10^8 ions/cm^3, that is to say almost 4 or 5 orders of magnitude lower than the density used for the electron spectrometry experiments on atomic species. To accomplish this experiment, two conditions must be fulfilled, i.e. having a high flux of photons available and studying a process with a high cross section.

In order to demonstrate the feasibility of electron spectrometry experiments using synchrotron radiation, we combined the synchrotron radiation beam produced by an undulator of the Super ACO storage ring, with an ion beam of singly charged ions, both beams being focused in the source volume of the CMA previously used for the experiments on excited atoms. To choose a good candidate, we have made use of the informations previously obtained in photoabsorption[32] and in ion spectrometry[33] experiments, which

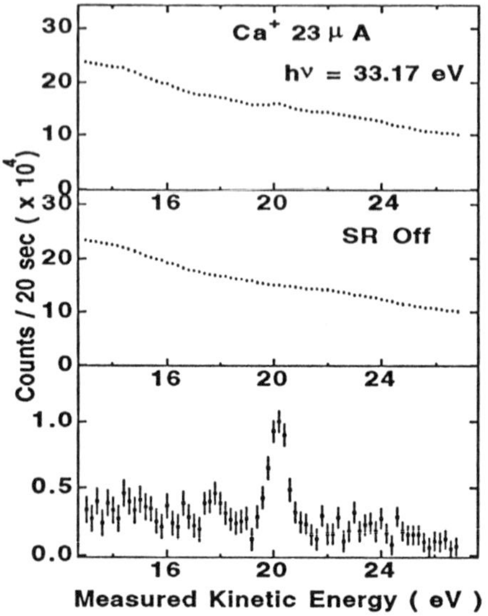

Fig. 7 - Photoelectron spectra produced in the resonant photoionisation of a Ca^+ ion beam by photons of 33.17 eV energy[2].

have shown the existence of a very intense resonance (more than 2000 Mb) in the photoionisation cross section of Ca^+ ions at 33.17 eV photon energy. This resonance was identified as a "giant resonance" corresponding to the transition 3p -> 3d in Ca^+. The Ca^+ ions produced in a Nier-Bernas type ion source are extracted at 20 kV. After mass and charge selection with a first magnet, the trajectory of the ions is bent by a toroidal electrostatic deflector to be parallel to the synchrotron radiation beam. The Ca^+ beam is then focused into the source volume of the CMA before being deflected by a second magnet. The photons are produced in the SU6 undulator of SuperACO. After monochromatisation, the photon beam is also focused into the source volume of the CMA. A chopper can be introduced after the monochromator to modulate the photon beam.

The electron spectra[2] obtained when the monochromator is tuned to 33.2 eV photon energy are shown in Fig. 7. On the two upper spectra, recorded with and without the synchrotron radiation interacting with the ions, we observe mainly a very intense background. It is produced by collisional ionisation of the residual gas in the experimental chamber (the pressure was typically 1×10^{-7} Torr) with the ion beam, involving direct ionisation and charge exchange processes. Using the chopper, this background (middle spectrum) can be instantaneously substracted from the spectrum obtained with the synchrotron radiation (upper spectrum), in order to obtain the pure signal produced by the photons (lower spectrum). We can now clearly observe, on this spectrum, a line which can be identified from its kinetic energy as being due to the expected resonant photoionisation process, i.e. to electrons produced in the autoionisation of the giant resonance in Ca^+. The smaller structures at lower kinetic energy are due to direct photoionisation of H_2O molecules, the main component of the residual gas. When the monochromator energy is slightly shifted from 33.2 eV, the intense line desappears, because the cross section for direct photoionisation into the continuum of the 4s electron in $3p^6\ 4s\ ^2S$ Ca^+ ion is very weak in this photon energy range.

4. Concluding Remarks

We have summarized the latest studies performed on excited atoms, in association with the fine analysis offered by electron spectrometry and the broad energy spectrum of synchrotron radiation. This gives a sensitive test of theoretical predictions in probing electron correlations. For the future developments of experiments combining the use of laser and synchrotron radiations for the study of atoms or molecules, some attempts have been made to synchronise a pulsed laser with synchrotron radiation[34]. Also the possibility to use a free electron laser, lasing in the infrared, in conjunction with synchrotron radiation is in project to study vibrationally excited molecules[35].

The experiments on atomic ions using electron spectrometry and synchrotron radiation are still at an early stage of feasibility. In order to achieve actual studies on resonant photoionisation of singly charged positive ions, the first improvement is to modify the experimental set-up presented here. We need to analyse the electrons under ultra-high vacuum conditions in order to supress the high background of electrons emitted in collisional ionisation. As far as the studies on multicharged ions are concerned, they will be possible under the condition of using an ECR ion source (the best ion source available at this time) in conjunction with an undulator or a multipole wiggler giving an extended photon spectrum. Such studies will be of great use for many other fields of physics, for exemple in inertial fusion where the photoionisation of low charged ions is considered as a heavy ion injector to heat and to confine the matter[36].

5. Acknowledgements

The authors want to thank N. Berrah-Mansour, B. Carré, E. Källne, C. Marinelli, T.J. Morgan, J. Novak, J. Obert, M. Pahler, J.C. Putaux, M. Richter, B. Rouvellou, S. Sorensen and F. Schlahter for their help in taking some of the data presented here.

6. References

1. J.M. Bizau, J.L. Le Gouët, D. Ederer, P. Koch, F.J. Wuilleumier, J.L. Picqué and P. Dhez, XII ICPEAC, *Book of Abstracts*, Post deadline papers (1981) p.1.
2. J.M. Bizau, D. Cubaynes, M. Richter, F.J. Wuilleumier,J. Obert, J.C. Putaux, T.J. Morgan, E. Källne, S. Sorensen and A. Damany, *Phys. Rev. Lett.* **67** (1991) 576.
3. R.F. Stebbing, F.B. Dunning, F.K. Tittel and R.D. Rundel, *Phys. Rev. Lett.* **30** (1973) 815.
4. T.B. Lucatorto and T.J. McIlrath, *Phys. Rev. Lett.* **A 37** (1976) 428.

5. J.M. Bizau, F. Wuilleumier, D. Ederer, J.C. Keller, J.L. Le Gouët, J.L. Picqué, B. Carré and P. Koch, *Phys. Rev. Lett.* **55** (1985) 1281.
6. A. Nunnemann, T. Prescher, M. Richter, M. Schmidt, B. Sonntag, H.E. Wetzel and P. Zimmermann, *J. Phys.* **B18** (1985) L337.
7. M. Ferray, F. Gounand, P. D'Oliveira, P.R. Fournier, D. Cubaynes, J.M. Bizau, T.J. Morgan and F. Wuilleumier, *Phys. Rev. Lett.* **59** (1987) 2040.
8. B. Carré, P. d'Oliveira, M. Ferray, P. Fournier, F. Gounand, D. Cubaynes, J.M. Bizau and F.J. Wuilleumier, *Z. Phys.* **D15** (1990) 177.
9. J.M. Bizau, D.Cubaynes, P. Gérard, F.J. Wuilleumier, J.L. Picqué, D.L. Ederer, B. Carré and G. Wendin, *Phys. Rev. Lett.* **57** (1986) 306.
10. M. Meyer, B. Müller, A. Nunnemann, T. Prescher, E. v. Raven, M. Richter, M. Schmidt, B. Sonntag and P. Zimmermann, *Phys. Rev. Lett.* **59** (1987) 2963.
11. M. Pahler, C. Lorenz, E. von Raven, J. Rüder, B. Sonntag, S. Baier, B.R. Müller, M. Schulze, H. Staiger, P. Zimmermann and N. Kabachnik, *Phys Rev. Lett.* **68** (1992) 2285.
12. D. Cubaynes, J.M. Bizau, F. Wuilleumier, B. Carré and F. Gounand, *Phys. Rev. Lett.* **63** (1989) 2460.
13. D. Cubaynes, J.M. Bizau, M. Richter and F. Wuilleumier, Tenth Int. Conf. on Vacuum Ultraviolet Radiation Phys. (1992) Book of abstracts, p. Mo71.
14. T.N. Chang and Y.S. Kim, *AIP Conf. Proc.* **94**, ed. B. Crasemann (New York: Amer. Inst. Phys., 1982) pp. 633-643.
15. K. Codling, J.R. Hamley and J.B. West, *J. Phys.* **B10** (1977) 2797.
16. M. Richter, J.M. Bizau, D. Cubaynes, T. Menzel, F. Wuilleumier and B. Carré, *Europhys. Lett.* **12** (1990) 35.
17. D. Cubaynes, J.M. Bizau, T.J. Morgan, B. Carré and F. Wuilleumier, Tenth Int. Conf. on Vacuum Ultraviolet Radiation Phys. (1992) Book of abstracts, p. Tu79.
18. Z. Felfi and S.T. Manson, *Phys. Rev. Lett.* **68** (1992) 1687.
19. D. Cubaynes, J.M. Bizau, C. Marinelli, T.J. Morgan, J. Novak, M. Pahler, B. Rouvellou, F. Schlachter, N. Berrah Mansour and F. Wuilleumier, Tenth Int. Conf. on Vacuum Ultraviolet Radiation Phys. (1992) Book of abstracts, p. Tu80.
20. C. Theodosiou and W. Fielder, *J. Phys.* **B15** (1982) 4113.
21. M. Isenberg, S. Carter, H.P. Kelly and S. Salomonson, *Phys. Rev.* **A32** (1985) 1472.
22. B. Craig and F.S. Larkins, *J. Phys.* **B18** (1985) 3569; ibid. (1985) 3713.
23. J. Cooper and R.N. Zare, in *Lectures on Theoritical Physics XIc Atomic Collision Processes*, ed. S. Geltman, K.T. Makanthappa and W.E. Brittin (New-York, London, Gordon & Breach Science Pub., 1969), p.317.
24. D.J. Kennedy and S.T. Manson, *Phys. Rev.* **A5** (1972) 227.

25. K. Codling, R.G. Houlgate, J.B. West and P.R. Woodruff, *J. Phys.* **B9** (1976) L83.
26. R.F.Reilman and S.T.Manson, *Astrophys. J. Supp.* **40** (1989) 815.
27. S.T. Manson, *AIP Conf. Proc.* **205**, ed. A. Delgarno, R.S. Freund, P.M. Koch, M.S. Lubell and T.B. Lucatorto (Amer. Inst. Phys. New-York, 1989), p.189 and ref. therein.
28. For a review of these studies, see F.J. Wuilleumier, in *Proc. of the ESRF-LURE Workshop on Photon-ion Interactions*, eds. F. Wuilleumier and E. Källne (ESRF Grenoble, 1989), p.231
29. P. Nicolosi, E. Jannitti and G. Tondello, *J. Phys.* II **1** (1991) C1-89.and ref. therein.
30. J.B. West, *Z Phys.* **D5** (1987) 265 and ref. therein.
31. S.D. Kravis, D.A. Church, B.M. Johnson, M. Meron, K.W. Jones, J. Levin, I.A. Sellin, Y. Azuma, N. Berrah Mansour, H.G. Berry and M. Druetta, 1991 *Phys. Rev. Lett.* **66** (1991) 2956.
32. B.F. Sonntag, C.L. Cromer, J.M. Bridgeis, T.J. McIlrath and T.B. Lucatorto, in Short Wavelength Coherent Radiation: Generation and Applications, *AIP Conf Proc.* **147** (New-York Am. Inst. Phys. 1986), p.412.
33. I.C. Lyon, B. Peart, K. Dolder and J.B. West, *J. Phys.* **B20** (1987) 1471.
34. D.L. Ederer, J.E. Rubensson, D.R. Mueller, R. Shuker, W.L. O'Brien, J. Jai, Q.Y. Dong, T.A.Callcott, G.L. Carr, G.P. Williams, C.J. Hirschmugl, S. Etemad, A. Inam and D.B. Tanner, *Nuclear Inst. and Method* **A319** (1992) 250.
35. "*Combustion Dynamics Facility*", Lawrence Berkeley Laboratory and Sandia National Laboratories (Berkeley, 1990), PUB-5284.
36. C. Rubbia, in *Proc. of the Third Eur. Particule Accelerator Conf.*, eds. H. Henke, H. Homeyer and Ch. Petit-Jean-Genaz (Ed. Frontieres, Gif sur Yvette, 1992) Vol. 1, p. 353.

PHOTOIONISATION OF POSITIVE AND NEGATIVE IONS

Vadim K. IVANOV
St-Petersburg Technical University
Polytekhnicheskaya 29, St-Petersburg, 195251 Russia

ABSTRACT

The application of many-body methods to the calculations of positive and negative ion photoionisation processes is discussed. Most of the results presented concern resonant effects in photoionisation cross sections and photoelectron angular distributions.

1. Introduction

The photoionisation process in positive and negative ions is now beginning to attract more attention from of both experimentalists and theoreticians because of its importance to the understanding of ion structure. There are large difficulties in the experimental investigations of ions, especially to measure the absolute cross sections, due to the fact that the absorbing sample density is rather small. So in fact, only a few absolute measurements on positive ions exist[1,2]. More absolute experimental data are available for negative ion photodetachment using laser radiation (see for a example [3-6]). Theoretical studies of the photoionisation processes in ions have on the other hand been carried out more intensely. For positive ions most calculations are performed using the R- matrix method, as in the case of the Stellar Opacity Project [7,8]. There are some other methods, which have recently been used for calculation of the resonant structure in the positive ion photoionisation spectrum: Configuration Interaction (CI)[9], different versions of the Random Phase Approximation (RPAE)[10-13] and Relativistic Many Body Perturbation Theory (MBPT)[14]. For negative ions up to the end of the eighties most of the calculations were based on semi-empirical methods[15,16]. But at the present time the photodetachment calculations have been performed using *ab initio* methods: RPAE[17,18], RRPA[19], CI[20], Dyson equation method[21,22].

The aim of this report is to discuss the application of many-body theory, in particular the RPAE, to the photoionisation process in positive and negative ions. The main attention is paid to the investigation of many electron correlations, in particular resonant effects, in photoionisation cross sections and photoelectron angular distributions. It should be noted that many electron correlations play a different role in the photoprocesses for positive and negative ions. For positive ions the photoionisation spectrum is in general determined by the discrete resonant structure. So the interactions between the different resonances and between discrete transitions and continuum spectra are the most significant and should be taken into account. In the case of negative ions, where the main contribution is given by the transitions into continuum spectra, there are additional correlation effects. Besides the strong intrashell and intershell interaction, the effects of rearrangement and the influence of the core polarisation are more significant than for neutral atoms.

Our many-body calculations are based on the use of the Hartree-Fock (HF) approximation as a zeroth order approximation for the following account of many-electron correlations within the frame of different modifications of the RPAE. Many body perturbation theory is used for a better description of energy level positions and energy transitions and for calculation of the fine structure. The polarisation potential is taken into account for the photodetachment from some negative ions by the Dyson equation method.

We consider the following photoionisation processes

$$\begin{aligned} \mathbf{h}\omega + \mathbf{A}^{\Delta z} &\rightarrow \mathbf{A}^{\Delta z+1} + \mathbf{e}^- \\ &\searrow \mathbf{A}^{(\Delta z+1)*} + \mathbf{e}^-, \qquad \Delta z = -1,\ 0,\ +1, +2, \ldots, \end{aligned}$$

which results in the ejection of one photoelectron, leaving the atom or ion in either the ground or an excited state. As a result of our calculations we obtain the dipole amplitudes, the partial and total cross sections, the angular distribution and the parameters for discrete and autoionizing resonances.

2. Photoionisation of positive ions

2.1. Continuum cross section

The RPAE and some its modifications have previously been applied with much success to neutral atoms[10]. So there is no doubt that the RPAE method should give a resonable description of the continuum spectrum of positive ion photoionisation, at least its smooth part, because of weaker many electron correlations. During recent years we have calculated the partial and total photoionisation cross sections of outer and intermediate subshells of the following ions: La^+, La^{+2}, La^{+3}, Ce^{+4}, Mg^+, C^+, Ca^+ among others.

For recent calculations[13,23] of Mg^+ and Ca^+ we used the Spin-Polarized versions of the HF (SPHF) and the RPAE (SP RPAE), which permit us to apply these methods, created for atoms or ions with filled subshells, to the ions with half-filled subshells. In this case the electronic structure, for example, of Mg^+ is the following:

$$1s\uparrow\ 1s\downarrow\ 2s\uparrow\ 2s\downarrow\ 2p^3\uparrow\ 2p^3\downarrow\ 3s\uparrow$$

where the arrows denote the electron spin projections.The results of the SP RPAE calculations of photoionisation cross sections for the 3s electron of Mg^+ and for the 3p electrons of Ca^+ are presented in fig.1. The influence of the inner 2p subshell on the 3s subshell in Mg^+ is rather small, excluding the resonance region near the 2p threshold. For Ca^+ the intershell interaction between the outer 4s and inner 3p subshells is more significant, as are the intrashell correlations for the 3p electrons, seen in fig.1b. There are no experimental results for the smooth part of the continuum spectrum in Mg^+ and Ca^+.

2.2. Autoionizing resonances

The photoionisation cross section may be measured more easily near resonances due to its larger value. Particulary it is concerned with the so-named giant dipole autoionizing resonances which appear due to the interaction between a strong discrete transition with large oscillator strength and the continuum spectrum. There are many calculations for well-known 4d → 4f resonances in atoms and ions of the lanthanides. For a example, the results

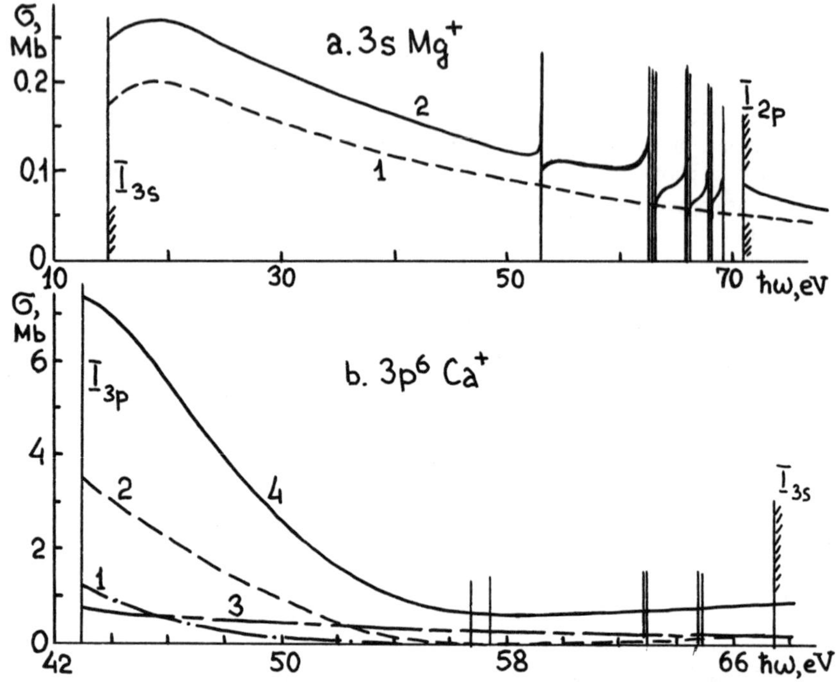

Figure 1. Photoionisation cross sections of the 3s Mg^+ (a) and the 3p Ca^+ (b). The vertical lines show the resonance positions. a: 1 - SPHF calculation, 2 - SP RPAE; b: 1 and 2 - SPHF and SP RPAE partial $3p\downarrow \to \varepsilon d\downarrow$ cross section correspondingly, 3 - $3p\uparrow \to \varepsilon s\uparrow$ cross sections, 4 - total $3p^6$ cross section.

of recent RPAE calculations for the Ce^{+4} ion cross section in the vicinity of this giant resonance are presented in Fig.2. Here are shown the different contributions of the outer $5p^6$ and $5s^2$ subshells in the total cross section and the results of calculations with fine structure splitting included. The calculated spin-orbit splitting value is 3.23eV, is comparable with the autoionizing width Γ= 1.77eV of these resonances.

A few years ago the absolute cross section for the 4s subshell in Ca^+ was obtained in the vicinity of the inner 3p threshold in the 20-43 eV photon energy range[2]. The main feature of the 4s Ca^+ photoionisation spectra is the giant autoionizing resonance due to a discrete $3p \to 3d$ transition. Figure 3 shows the results of our SP RPAE calculations[13] superimposed on the experimental data[2] and represents two different theoretical and experimental scales incorporating a shift of ~ 1eV. The R-matrix calculations[24] give better agreement for the energy position (E= 33.23eV), but are not so good for the resonance width (Γ=0.051eV). The very strong interaction of the $3p \to 3d$ transitions (in the SP RPAE method we have two transitions with different spin projections) with the 4s continuum leads to a significant energy shift of the resonance position from the SPHF values of 30.81eV and 31.02eV to 34.13eV. In addition we obtain only one resonance instead of the two peaks predicted by the SPHF calculation. Obviously this interaction

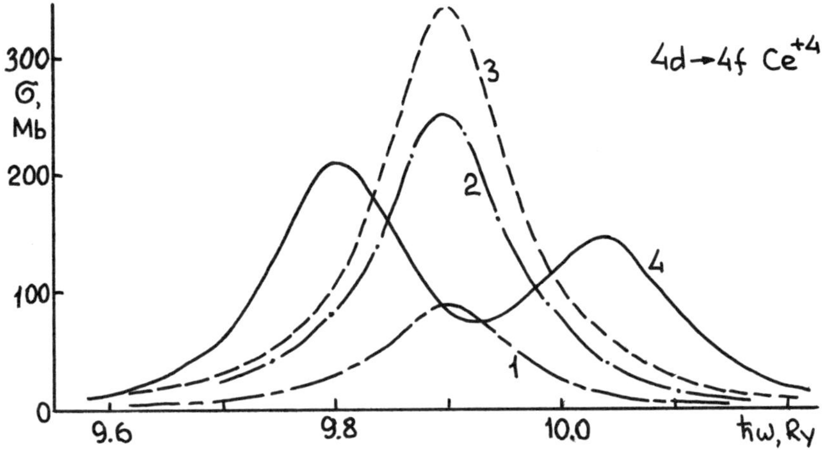

Figure 2. The 4d → 4f autoionizing resonance in Ce^{+4}. 1 and 2 - partial $5s^2$ and $5p^6$ cross sections correspondingly; 3, 4 - total cross sections without and with account of 4d fine structure.

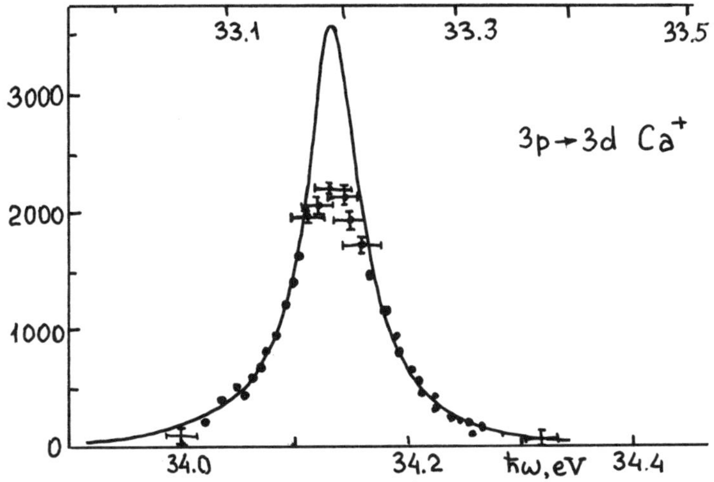

Figure 3. The comparison between experimental data[2] and the SP RPAE theory for the 4s Ca^+ cross section in the vicinity of the 3p→3d autoionizing resonance. The scale at the top refers to the experimental data.

results in the merging of the fine structure for the 3p → 3d transition, which is equal to 0.345eV for the 3p subshell from perturbation theory calculations[13].

So our calculations of giant resonances and weaker structure[13,23] showed that the RPAE and its modifications together with fine structure calculations describe reasonably well positive ion photoionisation spectra in the vicinities of single electron excitations. The inclusion of interaction between the discrete transitions themselves, and between them and transitions into the continuum spectrum permits us not only to identify this structure but to obtain the parameters of these single electron resonances. As far as the more complicated structure is concerned the RPAE method is able to help in the definition of some resonances[23] but it would be only tentative identification since only single electron excitation are included. The strong interaction between double and single electron excitations directly or via the continuum may change the resonance positions significantly, and redistribute their oscillator strengths. Thus, in such cases, to identify features in the photoionisation spectrum it is necessary to go beyond the RPAE method and to include in the calculations spin orbit interaction and double electron excitations together with single electron transitions, within the RPAE method.

3. Photodetachment from negative ions

3.1. Intrashell and intershell correlation effects

Many body calculations of negative ion photodetachment have been under way since the middle of the eighties and are now applied to ions with filled and half-filled subshells[17-19,25], and with one electron above closed subshells[21,22,25].

It should be noted that the use of the HF method as an initial approximation is not sufficient for accurate representation of the ground states of most negative ions, since the total HF energy of the ion is often the same or higher than that for the neutral atom. So such ions appear to be unstable within the HF method. Nevertheless, for ions with more than one electron in the outer shell, the single electron states are bound and their energies agree satisfactorily with electron affinity[17-19]. Therefore a basis, formed from the HF single electron states is suitable for many body calculations, in particular the RPAE.

Starting from the HF wave functions , we have obtained the RPAE photodetachment cross sections and photoelectron angular distributions for a series of the negative ions with filled and half-filled outer subshells[17,18,25]: Na^-, Ag^-, Cu^-, I^-, C^-, Si^-, Ge^-. It was shown the intrashell and intershell correlations are sometimes more significant in negative ions than in neutral atoms.

The most interesting resonance features may occur in photodetachment processes for ions with open outer subshells. For neutral atoms with the same electronic structure there are discrete transitions from the inner subshells into the unoccupied states in the outer subshell. A sign of such transitions appear for negative ions as resonances in the continuum spectra of the inner subshells[18].

In fig.4 the results of the SP RPAE calculations for the $2s\downarrow$ partial cross section in C^- ($...2s\uparrow 2s\downarrow 2p^3\uparrow$) are presented. The shape resonance, with a width of 0.94eV and 60 Mb in magnitude, in the single-channel SPHF photodetachment cross section is attributed to the transition to the "$2p\downarrow$" quasi-bound state. Taking into account intershell correlations, the interaction of the $2s\downarrow \rightarrow$"$2p\downarrow$" resonance with $2p\uparrow \rightarrow \varepsilon s\uparrow, \varepsilon d\uparrow$ channels leads to sharp interference structure with a deep minimum in the $\sigma_{2p\uparrow}$ cross section (see fig.5a) and the asymmetry parameter $\beta_{2p\uparrow}$ (fig.5b) as well. In fig.4 the reverse influence of the $2p^3\uparrow$ subshell on the $2s\downarrow \rightarrow \varepsilon p\downarrow$ shape resonance is shown. Such energy dependence of the outer subshell cross sections in C^-, Si^-,Ge^- [18] is similar to a Fano profile for autoionizing resonances, but here it appears due to strong intershell interaction between two continuum

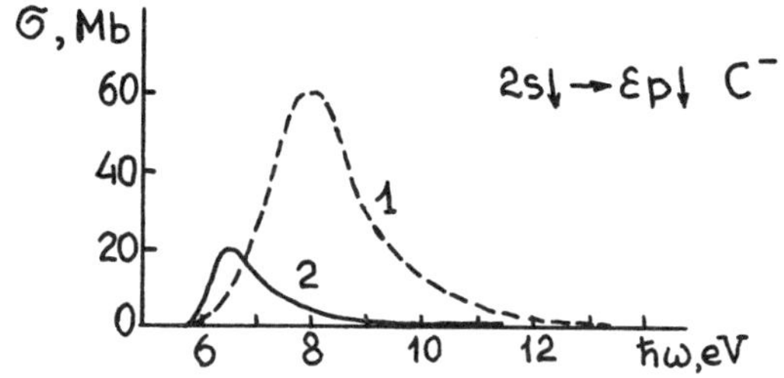

Figure 4. The $2s\downarrow \rightarrow \varepsilon p\downarrow$ partial cross section in C^-. 1- SPHF, 2 - SP RPAE.

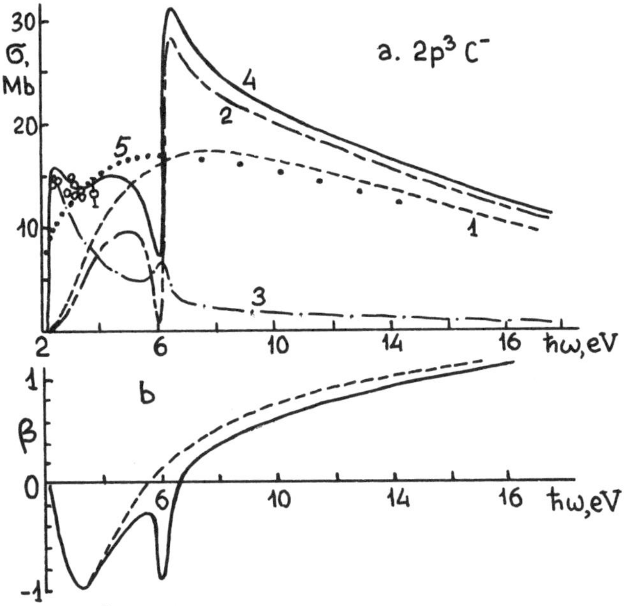

Figure 5. The $2p^3$ C^- cross section (a) and asymmetry parameter (b). a: 1,2 - $2p\uparrow \rightarrow \varepsilon d\uparrow$ transition with account of intrashell only and intershell correlations correspondingly, 3 - $2p\uparrow \rightarrow \varepsilon s\uparrow$ transition, 4 - total $2p^3$ SPRPAE cross section, 5 - model calculation[15], - experimental data[26]. b: 1,2 - without and with account of $2s\downarrow \rightarrow \varepsilon p\downarrow$ transition.

spectra. Our calculations are limited in this application of many body theory to ions with half-filled subshells but it is obvious that analagous effects can be observed for the other open-shell negative ions.

3.2. Rearrangement effects

It is well known that the rearrangement of the outer subshells can significantly influence inner shell photoionisation processes in neutral atoms[12]. It is also obvious that rearrangement effects in negative ions have to be more important than for neutral atoms due to strong changes in the average field in which photoelectron leaves the core. Our preliminary calculations confirm this proposition.

We have calculated the photodetachment cross sections from the inner subshells of the Ge^- ion[27] (...$3d^5\uparrow 3d^5\downarrow 4s\uparrow 4s\downarrow 4p^3\uparrow$) in an effort to find continuum spectrum resonances due to transitions into the unoccupied states of the 4p subshell. However the partial $3d\downarrow \rightarrow \varepsilon p\downarrow$ cross section, obtained by the SP RPAE method, doesn't reveal the resonance behaviour expected (see Fig.6). On the contrary, the SP RPAE partial $3d\uparrow \rightarrow \varepsilon p\uparrow$ cross section has a narrow, strong resonance near threshold, surprising at first. The analysis shows that, due to use of the frozen-core field approximation within the RPAE, the $3d\downarrow \rightarrow \varepsilon p\downarrow$ transition has a real discrete excitation, lying below the $3d\downarrow$ threshold, so the expected shape resonance moves to the real discrete spectrum and the continuum cross section becomes smooth.

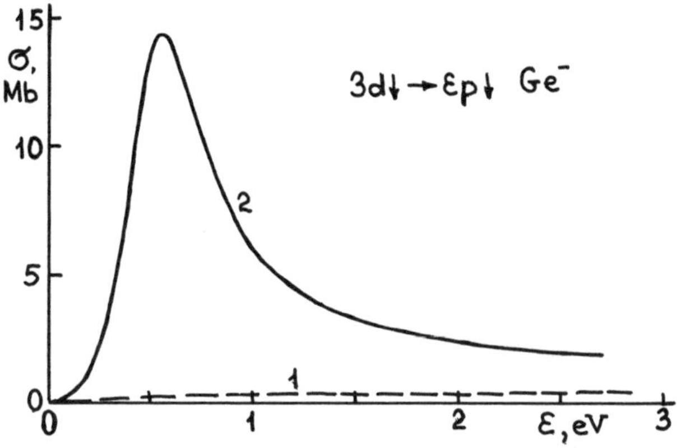

Figure 6. The $3d\downarrow \rightarrow \varepsilon p\downarrow$ Ge^- partial cross section, obtained within SP RPAE (1) and GRPAE (2).

It is clear, however, that in reality the outgoing photoelectron moves in the field of the other electrons, which rearrange during the photodetachment process. To take into account the dynamics of this process is rather difficult, so we have performed the calculations using the generalized RPAE method (GRPAE)[12]. The latter takes into account the static rearrangement effects, when the photoelectron moves in the field of the completely rearranged core. This rearrangement leads to dramatic changes in the energy dependence of the photodetachment cross section: for the $3d\downarrow \rightarrow \varepsilon p\downarrow$ transition we obtain a shape resonance at threshold (fig.6) and for the $3d\uparrow \rightarrow \varepsilon p\uparrow$ transition the strong peak disappears. It is very likely, in comparison with the rearrangement effects in neutral atoms, that the GRPAE gives a more realistic description of inner shell photodetachment. Moreover, the rearrangement effects may be significant for photodetachment processes from many electron outer subshells of negative ions also.

3.3. Polarisation effects

As was mentioned above, the ground state of negative ions with one electron outside closed shells cannot be obtained using the HF approximation. To bind an extra electron to an atom with closed shells, starting from the HF basis, it is necessary to include the polarisation potential, attracting the electron to the neutral core[21,22]. Only a few atoms, with large polarisability and a strong polarisation potential can form negative ions. These are: Pd^-, Ca^-, St^-, Ba^-, Yt^-, which were discovered in the recent years.

In our calculations to obtain the ground state of these ions we used the Dyson equation method[21,22], which takes into account the non-local energy-dependent polarisation potential. At first the photoelectron wave functions are calculated in the HF approximation in the field of the neutral atom. Using these wave functions we have obtained the photodetachment cross sections from the outer subshells of the above negative ions. In fig.7 the results of our calculations for the 4p Ca^- and 6p Ba^- are presented[28] and

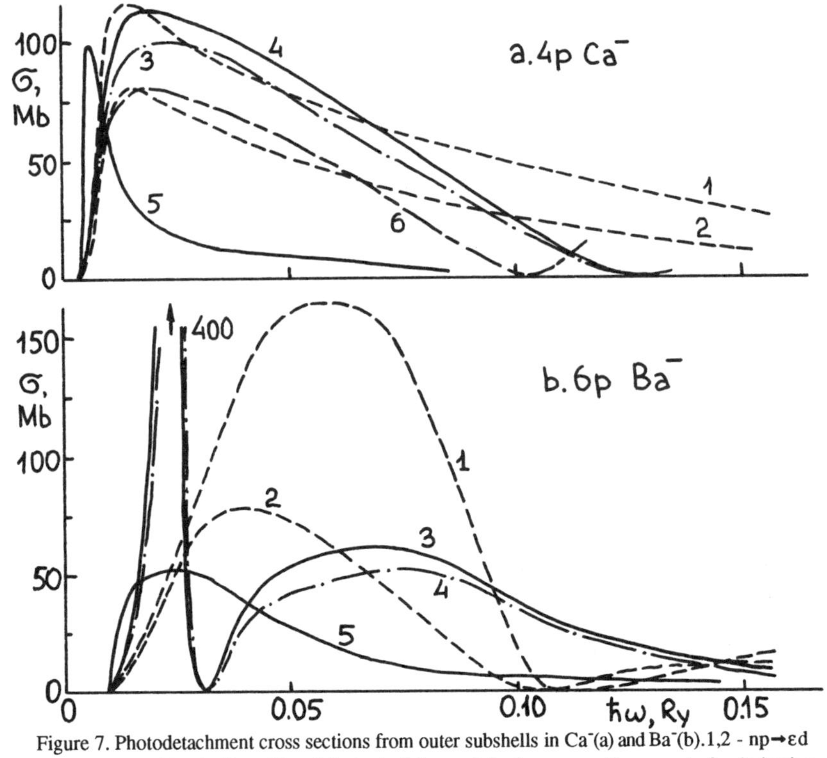

Figure 7. Photodetachment cross sections from outer subshells in Ca^-(a) and Ba^-(b). 1,2 - np→εd transition in "length" and "velocity" forms; 3,4 - the same with account of polarisation potential. 5 - np→εs transition (length form). 6 - CI calculation[20] for 4p→εd Ca^-.

we can see the usual energy dependence of the cross section and the differences between the "length" and "velocity" forms of the calculations.

But it is clear that a strong polarisation potential can play a very important role, not only for bound electrons, but also for outgoing photoelectron, because it moves in the field of the neutral atom which has a large polarisability. Also it is well known that there are low energy resonances in elastic electron scattering from alkaline earth atoms, and these are very sensitive to the polarisation potential. So, using the same Dyson equation method, we included the polarisation potential effects in the calculations of the photoelectron wave functions and the photodetachment cross sections (Fig.7)[25]. As a result we obtained for the 4p Ca^- electron photodetachment a noticeable shift in the Couper minimum position, to smaller energies and better agreement between different forms of calculations. But the most dramatic changes happen for the Ba^- ion. Due to a strong resonance in the elastic d-wave electron scattering by the neutral Ba atom we obtain a giant resonance in the $6p \rightarrow \varepsilon d$ partial photodetachment cross section and a very large energy shift of the minimum position.

So the results obtained demonstrate the very important role of polarisation effects for small photoelectron energies and it is obvious that they should be taken into account in photodetachment calculations for many other negative ions.

4.Summary

The application of many body methods, in particular the RPAE and its modifications, has been successful in the description of photoionisation processes in atoms and ions. But now it is clear that for a better description of the complicated resonance structure of positive ions by these methods we need to go beyond the RPAE and include fine structure interaction and double electron excitations. At the moment there is no significant dicrepancy between currently experimental data for negative ions and many body calculations. Nevertheless the results of our calculations show that the correlation effects for them are more diverse than for neutral atoms and positive ions. For a proper description of negative ion photodetachment one has to consider interchannel interaction, rearrangement effects and the influence of polarisation potential together.

5. Acknowledgements

I am very grateful to my colleagues: Dr. G.F.Gribakin, Dr. V.A.Kupchenko, Dr.. A.A.Gribakina and Dr. L.P.Krukovskaya from St-Petersburg Technical University, with whom the majority of results presented here were obtained. I would like to express my gratitude to Dr. J.B.West for his initiative and support for our positive ion calculations. I also thank Prof. M.Ya. Amusia, Dr. L.V. Chernysheva, Dr. M.Yu. Kuchiev and Dr. B.V. Gul'tsev for their collaboration and numerous discussions.

6. References

1. I. C. Lyon, B. Peart, J. B. West and K. Dolder, *J. Phys. B: At. Mol. Phys.* **19** (1986) 4137.
2. I. C. Lyon, B. Peart, K. Dolder and J. B. West, *J. Phys. B: At. Mol. Phys.* **20** (1987) 1471.

3. J. R. Peterson, Y. K. Bae and D. L. Huestis, *Phys. Rev. Lett..* **55** (1985) 692.
4. D. J. Pegg, J. S. Thompson, J. Dellwo, R. N. Compton and G. D. Alton, *Phys. Rev. Lett..* **64** (1990) 278.
5. C. W. Walter and J. R. Peterson, *Phys.Rev.Lett..* **68** (1992) 2281.
6. C. Brink, P. Balling, T. Andersen and H. Haugen, in *Atomic and Molecular Physics* (IV ECAMP) , *Abstracts, part 1* (Riga, 1992), p.184.
7. M. J. Seaton, *J. Phys. B: At. Mol. Phys.* **20** (1987) 6363.
8. K. A. Berrington, P. G. Burke, K.Butler, M. J. Seaton, P. J. Storey, K. T. Taylor and Yu Yan, *J. Phys. B: At. Mol. Phys.* **20** (1987) 6379.
9. K. T. Cheng and C. Froese Fischer, *Phys. Rev.* **A28** (1983) 2811.
10. M. Ya. Amusia and N. A. Cherepkov, *Case Stud. At. Phys.* **5** (1975) 47.
11 M. Ya. Amusia, L. V. Chernysheva, V. K. Ivanov and V. A. Kupchenko, *Z. Phys.* **D14** (1989) 219.
12. M. Ya. Amusia, V. K. Ivanov and L. V. Chernysheva, *Phys. Lett.* **59A** (1976) 191; M. Ya. Amusia,V. K. Ivanov, S. A. Sheinerman and S. I. Sheftel, *Zh. Exp. Teor. Fiz. (Soviet Physics JETP)* **78** (1980) 910.
13. V. K. Ivanov and J. B. West, (1992) to be published.
14. C. Guet and W. R. Johnson, *Phys. Rev.* **A44** (1991) 1531.
15. E. J. Robinson and S. Geltman, *Phys. Rev.* **153** (1967) 4.
16. D. L. Moores and D. W. Norcross, *Phys.Rev.* **A10** (1974) 1646.
17. M. Ya. Amusia, G. F. Gribakin, V. K. Ivanov and L. V. Chernysheva, *Izv. Akad. Nauk SSSR Ser. Fiz.* **50** (1986) 1274; *J. Phys. B: At. Mol. Opt. Phys.* **23** (1990) 385.
18. G. F. Gribakin, A. A. Gribakina, B.V. Gul'tsev and V. K. Ivanov, *J. Phys B: At. Mol. Opt. Phys.* **25** (1992) 1757.
19. V. Radojevic, H. P. Kelly and W. R. Johnson, *Phys.Rev.* **A35** (1987) 2027.
20. C. Froese Fischer and J. E. Hansen, *Phys. Rev.* **A44** (1991) 1559.
21. L. V. Chernysheva, G. F. Gribakin, V. K. Ivanov and M. Yu. Kuchiev, *J. Phys. B: At. Mol. Opt. Phys.* **21** (1988) L419.
22. G. F. Gribakin, B. V. Gul'tsev, V. K. Ivanov and M. Yu. Kuchiev, *J. Phys. B: At. Mol. Opt. Phys.* **23** (1990) 4505.
23. V. K. Ivanov, J. B. West, G .F. Gribakin and A. A. Gribakina 1992 to be published.
24. G. Miecznik, K. A. Berrington, P. G. Burke and A. Hibbert, *J. Phys. B: At. Mol. Opt. Phys.* **23** (1990) 3305.
25. V. K. Ivanov, in *Today and Tomorrow in Photoionisation. Proc. of the UK/USSR Seminar*, ed. M. Ya. Amusia and J. B. West (Daresbury Laboratory,1991).
26. M. Seman and L. M. Branscomb, *Phys. Rev.* **125** (1962) 1602.
27. V. K. Ivanov and L. P. Krukovskaya, in *13th International Conference in Atomic Physics* (Munich,1992).

IV. MOLECULES AND CLUSTERS

HIGH RESOLUTION CORE LEVEL PHOTOEMISSION OF MOLECULES

G.M. BANCROFT, J.D. BOZEK, J.N. CUTLER, Y.F. HU
Z.F. LIU, D.G.J. SUTHERLAND, K.H. TAN, AND J.S.TSE
Chemistry Department, University of Western Ontario
London, Ontario N6A 5B7 CANADA
and
Canadian Synchrotron Radiation Facility
University of Wisconsin-Madison
Stoughton, Wisconsin 53589 U.S.A.

ABSTRACT

High resolution core level Si 2p, I 4d and Xe 4d photoelectron spectra are reported for several Si, I and Xe molecules. In Si molecules, vibrational splittings from both symmetric and asymmetric vibrational modes have been observed and characterized on the Si 2p levels. For the molecules of heavier elements such as Ge, I or Xe, ligand field splitting, rather than vibrational splitting, dominates the 3d and 4d spectra. This ligand field splitting, as quantified by the C_2^0 crystal field term (which is proportional to the electric field gradient at the nucleus eq_n) is very sensitive to bonding and structure in the I and Xe molecules. For example, C_2^0 for IX molecules increases in the order X = CH_3 < I < CF_3 < H < Br < Cl. For the first time, a strong chemical dependence of the linewidth has been observed. The I 4d linewidth generally correlates well with both the I 4d chemical shift and the total I 5p electron density. As for eq_n, C_2^0 is structurally dependent. This dependence is illustrated by the determination of Xe-F bond angles in XeF_6 using the Xe 4d ligand field splitting.

1. Introduction

With recent advances in synchrotron radiation sources and beamlines, it has become possible to obtain core level photoelectron spectra of many gas phase molecules with unprecedented experimental resolution (photon plus electron resolution) of ≤ 0.1 eV.[1-5] Since most elements have one or more narrow core levels (usually the 1s, 2p, 3d or 4f level) with an inherent linewidth between 10 meV and 200 meV,[6,7] it has been possible for us to obtain total experimental linewidths close to the inherent linewidth for relatively low binding energy (< 150 eV) core levels.[1-5] The above resolution is substantially better than is currently being obtained (≥ 0.3 eV) using the best monochromatic Al Kα laboratory sources.[8,9] Moreover, it will soon be possible using synchrotron radiation to obtain even higher resolution spectra for the low binding energy levels, and to extend this resolution to all other narrow core levels.

High resolution is essential to resolve and characterize three effects in core level photoelectron spectra of molecules: vibrational splittings,[9,1-3] ligand field splittings,[4,5,7] and the chemical effect on core level lifetime widths.[4,10] An understanding of these three effects will be important for interpreting most high resolution photoelectron spectra of gases, solids and surfaces; and for bonding and structural studies in gases, solids and surfaces.

Many years have elapsed since the first observation and/or discussion of vibrational[11] and ligand field[12] splittings on the photoelectron spectra of core levels, and the chemical dependence of core level linewidths.[13] The rather slow progress has almost entirely been due to the lack of high resolution for the vast majority of core levels. Prior to the observation of the totally symmetric C-H vibrational frequency on the C 1s photoelectron spectrum of CH_4 in 1974, it was assumed that non-bonding core orbitals would show no vibrational structure.[13] The C-H vibrational structure on CH_4 was better resolved in 1984,[8] but it took until 1990 to resolve another vibrational splitting (the Si-H splitting on the Si 2p level in SiH_4 [1]) in another molecule using photoelectron spectroscopy. However, EELS[14] and later photoabsorption techniques[15,16] have resolved core level vibrational structures in many molecules.

Ligand field splitting was first observed using HeI radiation on the Zn 3d levels of Me_2 Zn,[12] but the splitting was not characterized. Ligand field splitting on the complex pre-edge Xe 4d photoabsorption spectra of XeF_2 and XeF_4 was reported in 1974[17]; and these spectra along with line broadening[18] on Sn 4d photoelectron spectra were interpreted using the common crystal field Hamiltonian. The splitting was observed and completely characterized between 1975 and 1982 using HeI and HeII photoelectron spectra on many compounds for low binding energy (< 25 eV) levels such as Zn 3d, Ga 3d, Cd 4d and In 4d.[7,19-22] It took until 1991 to extend these studies to higher binding energy levels such as I 4d and Xe 4d.[4,5,23]

In 1972, two papers appeared which discussed the chemical effect of core level lifetimes linewidths.[13] The resolution of these spectra was typical of the time (> 1 eV) for Mg Kα and Al Kα sources, and these studies did not consider the then unknown vibrational broadenings. There have been several theoretical papers which discuss the theoretical trends,[10] but only in 1991 was the existence of a strong chemical trend confirmed in the I 4d linewidths.[4]

In this paper, we review a number of recent high resolution molecular Si 2p, I 4d, and Xe 4d core level spectra to illustrate the three effects. We finish by mentioning possible applications of these results for future gas phase, polymer and surface photoelectron spectroscopy.

2. Experimental

All results reported here were recorded at the Canadian Synchrotron Facility (CSRF)[24] located on the 1 GeV storage ring Aladdin using a high resolution photoelectron spectrometer. The grazing incidence Grasshopper monochromator[25] was equipped with either an 1800 g/mm (Si 2p, I 4d and atomic Xe 4d spectra), or a 900 g/mm grating (molecular Xe 4d spectra) giving usual minimum photon linewidths of 80 meV, 50 meV and 110 meV, at photon energies of 130 eV for the Si 2p spectra[1-3] and ~ 90 eV for the I 4d[4,5] and Xe 4d spectra respectively.[23] The photoelectron spectrometer[26] is based on the 36 cm mean radius McPherson electron energy analyzer without any lens elements using either single channel (channeltron) or multichannel (Quantar #3395 position sensitive detector) counting. The theoretical electron resolution ($\Delta E/E$ = ~720) was ~ 40 meV at the ~ 30 eV kinetic energy used for most of these spectra. From the experimental Xe 5p spectra, this linewidth was 50 $\pm$ 10 meV. Overall resolution was ~ 60 meV for the I 4d and atomic Xe 4d spectra, usually ~ 100 meV spectra for the Si 2p spectra, and ~ 130 meV for the molecular Xe 4d spectra.

The I 4d and Xe 4d spectra were calibrated with the well characterized Xe $4d_{3/2}$ and Xe $4d_{5/2}$ lines of Xe gas at 69.525(10) and 67.541(9) eV binding energies respectively[27], while the Si 2p spectra were calibrated with the Kr $3d_{5/2}$ line of Kr gas at 93.798 eV. The spectra were deconvoluted with a symmetric Lorentzian-Gaussian lineshape using a non-linear least squares procedure.[28]

The Si and I compounds were obtained commercially in high purity. The Xenon fluorides XeF_2, XeF_4 and XeF_6 were prepared specially by G.J. Schrobilgen.[23]

3. Results and Discussion

a. *Vibrational Splittings*

Figure 1 shows the high resolution Si 2p photoelectron spectra of SiH_4, SiD_4 and Si_2H_6. These four spectra illustrate the different types of vibrational splitting that are seen in Si molecules.[1-3,29] In SiH_4, (Figure 1a), the two largest peaks correspond to the adiabatic $\nu = 0$ to $\nu' = 0$ transition for the Si $2p_{3/2}$ and Si $2p_{1/2}$ peaks separated by 0.613 eV. These two spin-orbit lines are split into a vibrational manifold of three peaks ($\nu' = 0,1,2$) with an energy spacing of 0.295 eV corresponding to the totally symmetric ν_1 vibrational mode of a_1 symmetry. At even higher resolution, (Figure 1b) the SiH_4 spectrum shows additional shoulders on all peaks indicating the presence of a second vibrational progression. The vibrational frequency of 0.11 eV corresponds to the asymmetric ν_2 bending frequency of SiH_4. The ν_1 frequency is much closer to that for the core equivalent species PH_4^+ or PH_3 than for the ground state SiH_4 molecule (Table 1).[1,2]

In SiD_4, there are four vibrational peaks corresponding to $\nu' = 0,1,2,3$ with an energy spacing of 0.212 eV. The vibrational frequency is again very close to that of the core-equivalent species PD_3 (Table 1).[3] The ratio of the two vibrational frequencies (0.212/0.295 = 0.719) is exactly that predicted from the square root of the ratio of the reduced masses (μ) of Si-H and Si-D.

The Si 2p spectrum of Si_2H_6 is very surprising (Figure 1d). Qualitatively, it is apparent that neither the symmetric Si-H or Si-Si vibrations (~ 0.3 eV and 0.06 eV respectively) are excited in this molecule. It is apparent (especially on the $Si2p_{3/2}$ peak at ~ 107.3 eV) that there exists distinct fine structure with an energy spacing of about 0.1 eV. By far the best fit results from a single vibrational progression corresponding to $\nu' = 0$-7 with an energy spacing of 0.102 eV (823 cm^{-1}). This is the same vibrational frequency observed for the small peaks in Figure 1b in SiH_4, and in the pre-edge photoabsorption spectrum of Si_2H_6.[30]

These are the first core level photoelectron spectra to show vibrational structure from non-totally symmetric vibrational modes. In molecules with degenerate or nearly degenerate core orbitals, ionization of core electrons may result in allowed excitations of non-totally symmetric vibrational modes[31] due to vibronic coupling between electronic and vibrational states. This theory has been tested on the C 1s photoabsorption spectra

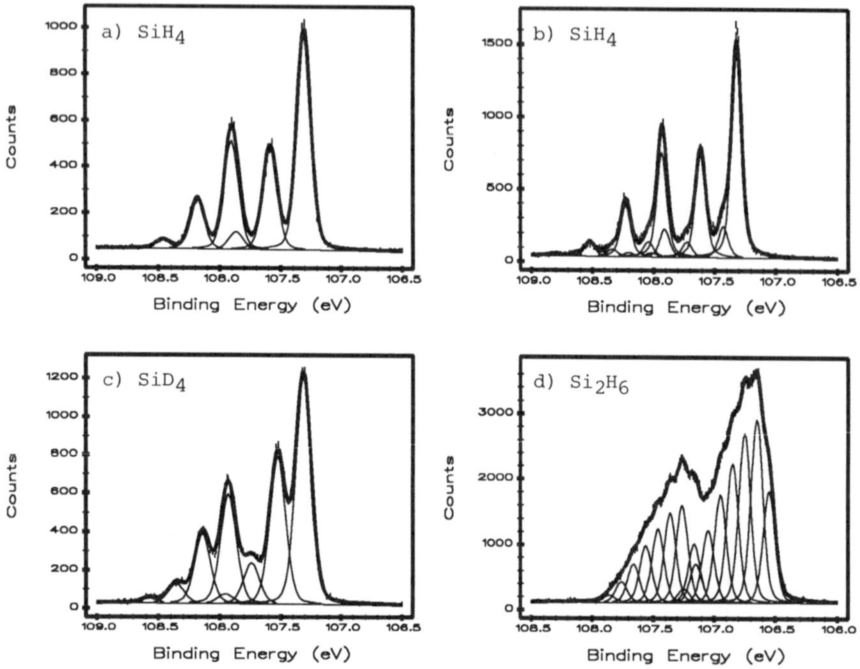

Figure 1: High resolution Si 2p photoelectron spectra of SiH_4, SiD_4 and Si_2H_6.[1-3,29] In a), c), d), the total linewidth of the lines are ≥ 130 meV, whereas in b) the total linewidths are 95 meV. The vibrational frequencies are given in Table 1.

TABLE 1

		Vibrational Frequency (cm^{-1})			
Molecule	Bond	Ground State	Core Equiv.	Experimental	Vibrational Mode
SiH_4	Si-H	2180	2327	2379	ν_1
	Si-H	970,910	-	887	$\nu_{2,4}$
SiD_4	Si-D	1545	1694	1710	ν_1
Si_2H_6	Si-H	844	-	823	ν_6
SiF_4	Si-F	801	893	847	ν_1
CH_4	C-H	2917	3335	3162	ν_1

of (several hydro-carbons such as ethylene[32] and acetylene[33].

In recent studies,[1,2,29] we have observed and characterized vibrational splitting and broadening on many Si molecules: e.g. SiF_4, SiD_xH_{4-x}, $Si(CH_3)_x(OCH_3)_{4-x}$, $Si(CH_3)_x\ [N(CH_3)_2]_{4-x}$, $SiH_x[Si(CH_3)_3]_{4-x}$ (x = 1-4). The totally symmetric vibrational frequency dominates these spectra, but even better resolution will be essential to completely characterize these splittings. Experimental and theoretical estimates show that the Si 2p inherent widths are 30 - 40 meV.[1,34]

We have also shown both experimentally and theoretically that core level vibrational structure decreases down a group (for example form SiH_4 to GeH_4), and across a row (for example from SiH_4 to PH_3 to HCl).[35] In general then, the heavy elements will show relatively little vibrational structure, and the core equivalent model also predicts this. However, our recent W 4f spectrum $W(CO)_6$ shows significantly vibrational splitting from the ligand ν_1(C-O) vibrational frequency.[36]

b. The Chemical Dependence of Ligand Field Splitting and Linewidths

High resolution I 4d core level spectra of ICl, IBr, HI and I_2 are shown in Figure 2.[4,5,37] Obviously the very different splitting patterns on the $d_{3/2}$ and $d_{5/2}$ levels shows that the dominant splitting in these spectra is not vibrational in origin. The spectra consist of five intense peaks due to the combination of spin-orbit splitting and ligand field splitting. The ligand field splitting splits the $d_{3/2}$ and $d_{5/2}$ peaks into a doublet and triplet respectively. We have also fit a small sixth peak at high binding energy which is vibrational structure on peak 1. Somewhat fortunately, then, ligand field splitting dominates for heavy elements such as I, while vibrational splitting dominates for lighter elements such as Si.* In other molecules of Ge and Sn, the two effects are of comparable magnitude, making the analysis very difficult (unpublished work).

The Hamiltonian used to describe the ligand field (neglecting the C_4^0 term) and spin-orbit splitting in the $4d^9$ hole state in the local I environment of $D_{\infty h}$ symmetry is:[20,22,38]

$$H = H_0 + C_2^0[3L_z^2 - L(L+1)] + \lambda[1/2\ (L_+S_- + L_-S_+) + L_zS_z] \qquad (1)$$

where H_0 is the Hamiltonian for the free ion (E_{4d}), C_2^0 and C_4^0 are the non-cubic and cubic crystal field parameters, respectively, λ is the spin-orbit coupling constant, and L and S are the orbital and spin angular momentum operators, respectively. Diagonalizing the Hamiltonian matrix in d electron subspace and treating d^9 and d^1 with the usual sign changes, the eigenvalues and eigenfunctions can be calculated.[20,22]

Several general trends in E_{4d}, C_2^0 (which dominates over C_4^0) and the linewidths for these four molecules (and several others) are shown in Figures 3 and 4.[37] First, an excellent linear correlation exists between C_2^0 and E_{4d} (Figure 3, left) for seven I molecules. Both parameters depend directly on the I 5p electron density. For example:

$$C_2^0 \propto F^2\,\Delta\rho \propto F^2\,[n_{p_z} - 1/2\,(n_{p_x} + n_{p_y})] \qquad (2)$$

where F^2 is the Slater-Condon integral between the valence 5p and core 4d orbitals,[7,22]

---

*Ligand field splitting would split the Si $2p_{3/2}$ level in the nontetrahedral molecules such as CH_3SiH_3 and Si_2H_6 but the splitting is < 50 meV in these molecules. However, for low symmetry surface Si atoms on the Si(100) 2×1 surface, Wertheim et al have shown a Si $2p_{3/2}$ splitting of 190 meV.[46]

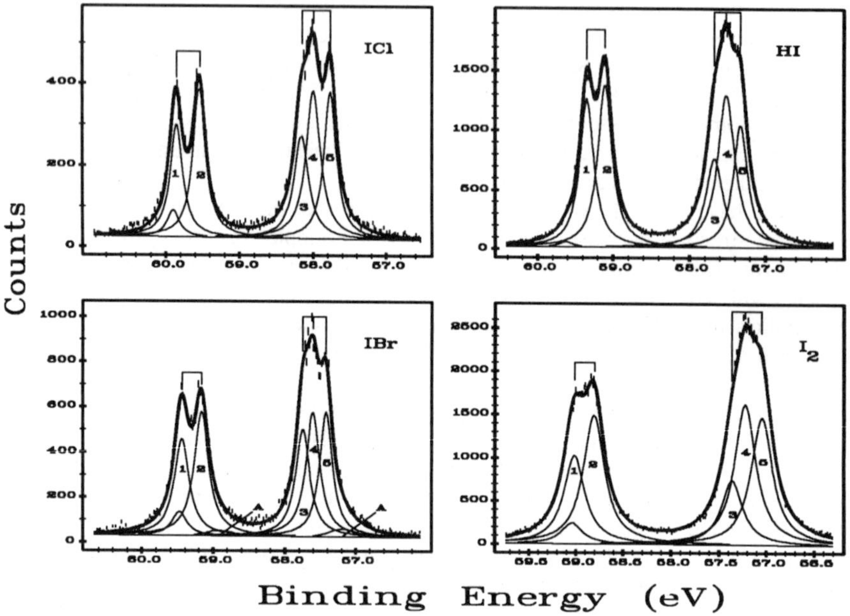

Figure 2: High resolution I 4d photoelectron spectra of ICl, IBr, HI and I_2 taken with 60 meV resolution at 90 eV photon energies.[4,5]

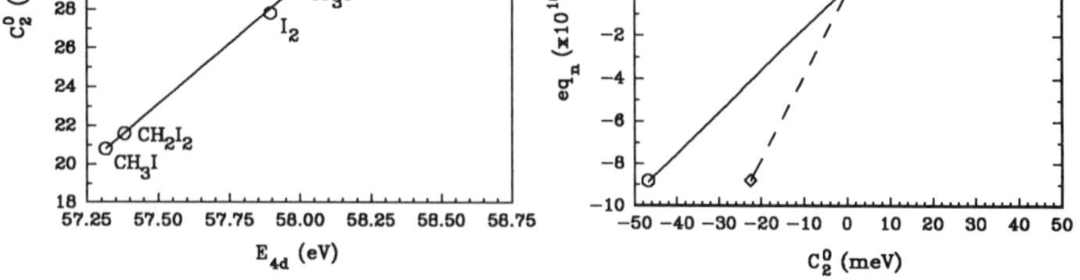

Figure 3: (left) A plot of C_2^0 versus E_{4d}. The line of best fit is shown, and R = 0.999; (right) A plot of the electric field gradient at the nucleus, eq_n, versus C_2^0 (solid line) and the $D_{3/2}$ splitting (dashed line) for (from left to right for the open ○ and ◇) XeF_4, CH_3I, CH_2I_2, I_2, CF_3I, IBr, ICl and XeF_2. The solid points ● and ◆ are for HI. The lines of best fit are shown and both have R = 0.998 excluding HI.

and $\Delta\rho$ measures the excess/deficiency of I 5p orbital density along the I-X bonding axis. An increase in electronegativity of the ligand bonded to I, then should increase both E_{4d} and C_2^0. However, HI is not in the expected order for both parameters. Since H is less electronegative than I, we would expect both E_{4d} and C_2^0 for HI to be <u>smaller</u> than I_2. Large relaxation effects apparently cause this one anomaly.[29,39]

Figure 3 (right) shows an excellent linear correlation (except for HI) between eq_n (the electric field gradient at the nucleus, as measured by ^{129}I and ^{129}Xe Mossbauer spectroscopy[40]) and C_2^0 (and the $D_{3/2}$ splitting). This correlation is expected because the C_2^0 crystal field term (equation 1) and the quadrupole Hamiltonian (H_Q) transform identically[22]:

$$H_Q = \frac{e^2 q_n Q}{4I(2I-I)} [3I_z^2 - I(I + 1)]. \quad (3)$$

C_2^0 (like e^2qQ) can now be a very useful photoelectron parameter for studying the bonding in many molecules. However, as for chemical shifts, one must be aware of relaxation effects (such as in HI).

Qualitatively in Figure 2, one can tell that the linewidths are in the order ICl < IBr ≈ HI < I_2. These lines are all Lorentzian showing that we are observing very close to the inherent lifetime width in all cases.[4] The change in linewidth is beautifully illustrated in Figure 4 by two correlations[37] of the FWHM versus, first, E_{4d} and second, the total I 5p valence electron density from ^{129}I Mossbauer studies.[40] These two correlations can be explained qualitatively. McGuire[41] showed that for the $N_{4,5}$ shells, the Auger process is the major relaxation channel to fill the core hole. The Auger rate depends on the square of the Coulomb matrix element:

$$< \psi_{4d}\chi \,|\, \frac{1}{r_{12}} \,|\, \psi_v \psi_{v'} > . \quad (4)$$

where χ and $\psi_v\psi_{v'}$ are, respectively, the continuum electron and the wave function for the valence electrons (I 5p) that participate in the Auger process. It is known from the Heisenberg uncertainty principle that the measured linewidths Γ are inversely proportional to the lifetime of the hole. Therefore, from measured linewidths, a quantitative understanding of the relationship between the Auger rate to fill the core hole and the electronegativity of the ligand is possible. Ligands with stronger electron withdrawing ability than I, such as Cl and Br, will decrease the valence I 5p electron density relative to I_2 (thus increasing E_{4d}), decrease the Auger rate, increase the lifetime of the core hole and give a smaller linewidth. Conversely, ligands such as CH_3 which are electron donating relative to I will increase the valence I 5p electron density (decreasing E_{4d}), increase the Auger rate, decrease the lifetime, and give a larger linewidth. Once again, HI does not fall on the line showing that relaxation effects are even more important for the linewidths than for E_{4d} or C_2^0.

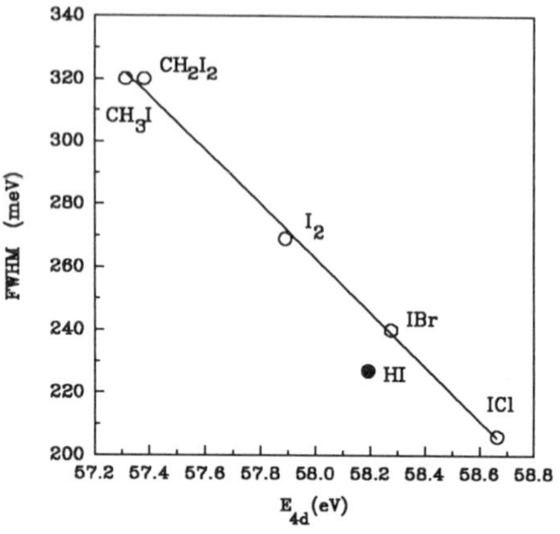

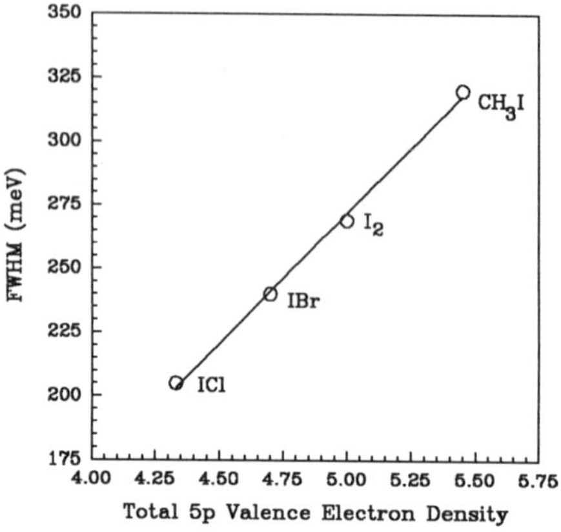

Figure 4: Plots of experimental full width at half maximum (FWHM) versus E_{4d} (left) and total I 5p electron density (right) as measured by ^{129}I Mössbauer spectroscopy The correlation coefficients are 0.998 (left) and 0.999 (right).

c. *Structural Dependence of Ligand Field Splittings*

The final example in this paper illustrates the use of ligand field splittings for structural determination.[23] The solid state structure of XeF_6 is well known. However, the gas phase structure of XeF_6 has long been a puzzle to both experimentalists and theoreticians. Various studies have suggested octahedral, and distorted C_{3v} and C_{2v} structures. Because $C_2^0 = 0$ for an octahedral XeF_6 structure, and C_2^0 increases with the distortion from octahedral symmetry, we felt that a high resolution study of the Xe 4d levels of the xenon fluorides (XeF_2, XeF_4 and XeF_6), would yield very useful information on the magnitude of the distortion from octahedral symmetry.

Figure 5 shows the Xe 4d spectra of XeF_2 and XeF_6. Linear XeF_2 gives a large C_2^0 (the largest positive value in Figure 3). Square planar XeF_4 gives a somewhat larger negative C_2^0. C_2^0 correlates well with eq_n for these Xe molecules as well as for the I molecules (Figure 3). The Xe 4d spectrum of XeF_6 is not resolved, but has very large linewidths (0.36 eV for $d_{3/2}$ and 0.40 eV for $d_{5/2}$) relative to Xe, XeF_2, or XeF_4. These large linewidths show that a signficiant ligand field splitting exists in XeF_6. The best fit to the spectrum (Figure 5, right) yields a C_2^0 of $\pm$ 18 $\pm$ 2 meV compared to + 39.1 meV and - 46.8 meV for XeF_2 and XeF_4 respectively. Using an additive model commonly used for e^2qQ studies,[40] and assuming a C_{3v} structure, we obtain two F-Xe-F bond angles of 50 $\pm$ 2° and 76 $\pm$ 4°. These angles are in good agreement with the latest theoretical values of 49° and 76° respectively,[42] but not in quite as good agreement with the electron diffraction results of 52° and 67°.[43]

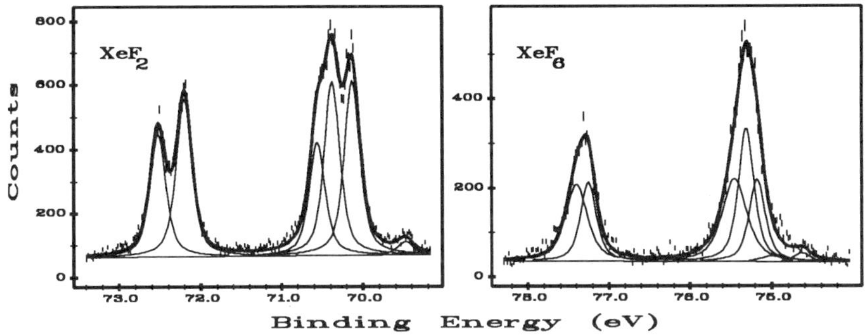

Figure 5: Xe 4d photoelectron spectra of XeF_2 and XeF_6 at experimental resolutions of 0.13 eV.[23]

4. Conclusions

Using instrumental resolutions of ≤ 0.1 eV, vibrational and ligand field splittings have been resolved and characterized on the Si 2p, I 4d and Xe 4d core levels of several molecules. These three effects will be important for the complete understanding of all high resolution core level spectra of gases, solids and surfaces. For example, we have already shown that vibrational splitting must be considered for the interpretation of the Si 2p spectra for H adsorbed on Si surfaces,[3,44] or CO adsorbed on metals.[36] Also the C-H vibrational splitting (from both symmetric ν_1 and asymmetric ν_2 modes) must be considered for the interpretation of C 1s polyethylene polymer spectra.[45] Ligand field splitting will be important for photoelectron spectra of surface atoms, as shown recently by the Si $2p_{3/2}$ splitting on Si surface atoms on the Si(100) 2× 1 surface.[46]

5. Acknowledgements

We are very grateful to NSERC and NRC for financial support, and to the staff at the Aladdin Synchrotron for their continued help and cooperation.

References

1. J.D. Bozek, G.M. Bancroft, J.N. Cutler, and K.H. Tan, *Phys. Rev. Lett.* **65** (1990) 2757.
2. J.D. Bozek, G.M. Bancroft, and K.H. Tan, *Phys. Rev. A.* **43** (1991) 3597.
3. D.G.J. Sutherland, G.M. Bancroft, and K.H. Tan, *Surf. Sci. Lett.* **262** (1992) L96.
4. J.N. Cutler, G.M. Bancroft, D.G. Sutherland, and K.H. Tan, *Phys. Rev. Lett.* **67** (1991) 1531.
5. J.N. Cutler, G.M. Bancroft, and K.H. Tan, *J. Phys. B: At. Mol. Opt. Phys.* **24** (1991) 4897.
6. M.O. Krause, *J. Phys. Chem. Ref. Data* **8**, (1979) 307.
7. R.P. Gupta, J.S. Tse, and G.M. Bancroft, *Philos. Transactions of the Royal Society* **293** (1980) 535 and references.
8. A. Nilsson, N. Martensson, S. Svensson, L. Karlsson, D. Nordfors, U. Gelius and H. Ågren, *J. Chem. Phys.* **96** (1992) 8770, and references therein.
9. U. Gelius, L. Asplund, E. Basilier, K. Helenelund and K. Siegbahn, *Nuc. Instr. Methods* **B229** (1984) 85.
10. M. Coville and T.D. Thomas, *Phys. Rev. A.* **43** (1991) 6053, and references.
11. U. Gelius, S. Svensson, H. Siegbahn, E. Basilier, A. Faxlav, and K. Siegbahn, *Chem. Phys. Lett.* **28** (1974) 1.

12. J.H.D. Eland, *Int. J. Mass. Spect. Ion. Physics* **4** (1970) 37.
13. R.W. Shaw, Jr., and T.D. Thomas, *Phys. Rev. Lett.* **29**, 689 (1972); R.M. Friedman, J. Hudis, and M.L. Perlman, *Phys. Rev. Lett.* **29** (1972) 692.
14. G.R. Wright, C.E. Brion, and M.J. Vander Weil, *J. Elect. Spect. Relat. Pheom.* **1** (1973) 457.
15. C.T. Chen, and F. Sette, *Phys. Scripta* **T31** (1990) 159, and references.
16. A.P Hitchcock, *Phys. Scripta* **T31** (1990) 159, and references.
17. F.J. Comes, R. Haensel, U. Nielsen and W.H.E. Schwarz, *J. Chem. Phys.* **61** (1974) 3581.
18. G.M. Bancroft, I Adams, H. Lampe, and T.K. Sham, *Chem. Phys. Lett.* **32** (1975) 173.
19. G.M. Bancroft, I. Adams, D.K. Creber, D.E. Eastman and W. Gudat, *Chem. Phys. Lett.* **38** (1976) 83.
20. G.M. Bancroft, D.K. Creber, and H.J. Basch, *J. Chem. Phys.* **67** (1977) 4891.
21. G.M. Bancroft, D.K. Creber, M.A. Ratner, J.W. Moskowitz, and S. Topiol, *Chem. Phys. Lett.* **50** (1977) 233.
22. G.M. Bancroft and J.S. Tse, *Comm. Inorg. Chem.* **5** (1986) 89, and references therein.
23. J.N. Cutler, G.M. Bancroft, J.D. Bozek, K.H. Tan, and G.J. Schrobilgen, *J. Am. Chem. Soc.* **113** (1991) 9125 and references.
24. G.M. Bancroft, J.B. Bozek, and K.H. Tan, *Phys. Can.* **43**, 113 (1987).
25. K.H. Tan, G.M. Bancroft, L.L. Coatsworth, and B.W. Yates, *Can. J. Phys.* **60** (1982) 131.
26. J.D. Bozek, J.N. Cutler, G.M. Bancroft, L.L. Coatsworth, K.H. Tan, D.S. Yang, and R.G. Cavell, *Chem. Phys. Lett.* **165** (1990) 1.
27. M.O. Krause, *Synchrotron Radiation Research*, ed. H. Winick and S. Doniach (Plenum Press, New York, 1980).
28. G.M. Bancroft, I. Adams, L.L. Coatsworth, C.D. Bennewitz, J.D. Brown, W.D. Westwood, *Anal. Chem.* **47** (1975) 586.
29. D.G.J. Sutherland, G.M. Bancroft and K.H. Tan, *J. Chem. Phys.*, accepted.
30. D.G.J. Sutherland, J.D. Bozek, G.M. Bancroft, and K.H. Tan, *Chem. Phys. Lett.* submitted.
31. L.S. Cederbaum and W. Domcke, *J. Chem. Phys.* **66** (1977) 5084.
32. F.X. Gadea, H. Koppel, J. Schirmer, L.S. Cederbaum, K.J. Randall, A.M. Bradshaw, Y. Ma, F. Sette, and C.T. Chen, *Phys. Rev. Lett.* **66** (1991) 883.
33. Y. Ma, C.T. Chen, G. Meigs, K.J. Randall, and F. Sette, *Phys. Rev. A.* **44(3)** (1991) 1848.
34. J. McColl and F.P. Larkins, *Chem. Phys. Lett.* to be published.
35. Z.F. Liu, G.M. Bancroft, J.N. Cutler, D.G.J. Sutherland, and K.H. Tan, *Phys. Rev. A.* in press.
36. Y.F. Hu, G.M. Bancroft, J.D. Bozek, Z.F. Liu, D.G.J. Sutherland, and K.H. Tan, *Chem. Commun.*, accepted.
37. J.N. Cutler, G.M. Bancroft, and K.H. Tan, *J. Chem. Phys.*, accepted.
38. C.J. Ballhausen, *Introduction to Ligand Field Theory* (McGraw-Hill, New York, 1962).
39. E.J. Aitken, M.K. Bahl, K.D. Bomben, J.K. Gimzewski, G.S. Nolan, and T.D. Thomas, *J. Am. Chem. Soc.* **102** (1980) 4873.
40. G.M. Bancroft and R.A. Platt, *Advances in Inorganic and Radiochemistry* **15** (1972) 59.
41. E.J. McGuire, *Phys. Rev. A.* **9** 1840 (1974).
42. M. Klobukowski, S. Huzinaga, L. Seijo, Z. Barandiaran, *Theor. Chim. Acta* **71** (1987) 237.
43. K.S. Pitzer, and L.S. Bernstein, *J. Chem. Phys.* **63** (1975) 3849.
44. L. Ley, J. Reichardt and R.L. Johnson, *Phys. Rev. Lett.* **49** (1982) 1664.
45. G. Beamson, D.T. Clark, J. Kendrick and D. Briggs, *J. Elect. Spect* **57** (1991) 79.
46. G.K. Wertheim, D.M. Riffe, J.E. Rowe and P.A. Citrin, *Phys. Rev. Lett.* **67** (1991) 120.

ANGLE RESOLVED PHOTOION SPECTROSCOPY OF CORE-EXCITED FREE-MOLECULES

A. YAGISHITA and E. SHIGEMASA
Photon Factory, National Laboratory for High Energy Physics, Tsukuba, Ibaraki 305, Japan

and

J. ADACHI and N. KOSUGI
Division of Molecular Engineering, Kyoto University, Yoshida, Kyoto 606, Japan

ABSTRACT

Angle-resolved photoion measurements have been performed in the inner-shell photoionization process. As a useful application of the angle-resolved measurements to spectroscopy, "symmetry-resolved" photoabsorption spectra of N_2, O_2, and NO have been observed. Not only inner-shell excited states with definite symmetries but also degenerate ionization channels in the continua have been clearly decomposed in the "symmetry-resolved" spectra.

1. Introduction

Theoretical treatments of molecular photodissociation were presented by Zare,[1] who derived the general expression for angular distribution of neutral products. After Zare's work, Dehmer and Dill obtained the formula for photoion angular distribution in dissociative photoionization.[2] The photoion angular distribution is determined by the total symmetry of molecular ion plus photoelectron, because the photoionization process also involves the ejection of photoelectrons. Below the ionization threshold, the formula by Dehmer and Dill[2] is in agreement with the results of Zare[1]. The formula is applicable to photodissociation processes after both inner-shell excitation and ionization,[3] if the molecular orientation is kept between the creation of inner-shell vacancy and its subsequent Auger decay.

As typical examples of the molecular orientation depending on photon energies, we have studied the K-shell photoexcitation processes of diatomic molecules containing first-row atoms.[4-10] According to the dipole selection rules, a K-shell electron is promoted to a σ or π excited orbital. The excited states have pure symmetries below the K-shell ionization threshold, and the photoion angular distribution would be $\cos^2\theta$ for $\Sigma \rightarrow \Sigma$ or $\Pi \rightarrow \Pi$ transitions ($\Delta\Lambda = 0$) and $\sin^2\theta$ for $\Sigma \rightarrow \Pi$ or $\Pi \rightarrow \Sigma$, Δ transitions ($\Delta\Lambda = \pm 1$) induced by linearly polarized photons, where θ is the angle between the electric polarization vector and the molecular axis. On the other hand, above the ionization threshold, the $\Delta\Lambda = 0$ and $= \pm 1$ ionization channels are degenerate, and the angular distribution would be $A\cos^2\theta$ +

$B\sin^2\theta$. However, angle resolved photoion measurements afford us the valuable opportunity to decompose the degenerate ionization channels.[6-10]

In this paper, we will review a significant new molecular photoabsorption spectroscopy based on the angle resolved photoion measurements.

2. Undulator Beamline and Experimental Setup

We are using the soft x-ray undulator beamline equipped with a 10-m grazing incidence monochromator,[11] because intense and highly polarized photons are demanded for the angle resolved photoion measurements. The first harmonic peak of undulator radiation ranges from 280 to 950eV. At the maximum of the first harmonic, the photon flux of about 10^{12} photons/sec is available with the resolving power of about 700.

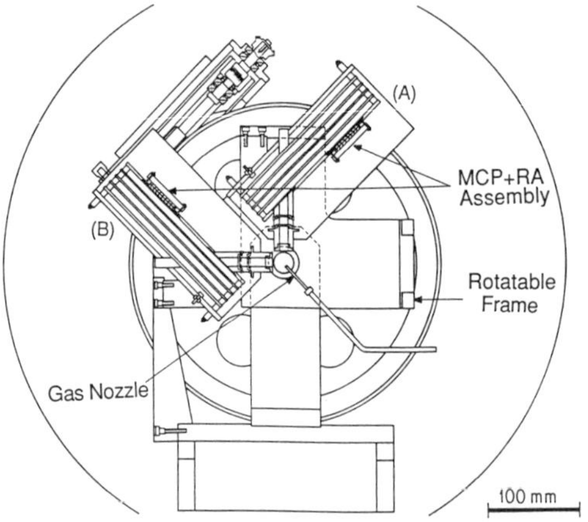

Fig. 1 Schematic drawing of the experimental apparatus. Two parallel-plate electrostatic analyzers labelled (A) and (B) are positioned at 90° and 0°, respectively, relative to the electric vector of the incident light. The analyzer (A) is rotatable around the incident light. MCP stands for a microchannel plate, and RA an one-dimensional resistive anode.

The experimental setup is shown schematically in Fig. 1. Two parallel-plate electrostatic energy-analyzers with position-sensitive detectors are installed in a vacuum chamber.[7] One of the analyzers is positioned in the parallel direction to the electric vector of the incident photons. Another one is rotatable around the incident photon beam. The photoions emerging from the interaction region are detected simultaneously and independently by the two analyzers. The whole kinetic-energy spectra of the photoions can be measured with a given deflection voltage of the analyzers. The spectra are composed of two major peaks; a peak near zero kinetic-energy corresponds to molecular ions with thermal energy, and a broad peak centered at around several eV is attributed to fragment ions. In order to detect only the fragment ions ($\geq$ 2eV) from the whole spectra, an appropriate deflection voltage is applied to each analyzer.

3. Angular Distribution of Fragment Photoions

The axial-recoil approximation[1] is valid under the condition that the photofragments dissociate with a kinetic energy much larger than the rotational spacing. In this case, the direction of emitted photoions can be equated to the molecular orientation at the time of photoexcitation. For randomly oriented molecules with cylindrical symmetry, the expression for the angular distribution $d\sigma/d\Omega$ of the photoions produced after the excitation with linearly polarized photons may be written in the form[2,3]

$$\frac{d\sigma}{d\Omega} = \frac{\sigma}{4\pi}[1 + \beta P_2(\cos\theta)] \quad , \tag{1}$$

where θ is the angle between the electric vector of the incident photons and the emitted photoion direction, β is the asymmetry parameter characterizing the photoion angular distribution, and the Legendre polynomial $P_2(\cos\theta) = 3/2 \cdot \cos^2\theta - 1/2$. The asymmetry parameter β is defined by the transition probabilities d_0^2 for the $\Delta\Lambda = 0$ transition and d_1^2 for the $\Delta\Lambda = \pm 1$:

$$\beta = \frac{2(d_0^2 - d_1^2)}{d_0^2 + 2d_1^2} \quad . \tag{2}$$

The total photoabsorption cross section σ is expressed as the sum of the transition probabilities :

$$\sigma = \frac{4\pi^2}{3}\alpha h\nu\,(d_0^2 + 2d_1^2) \equiv \sigma_0 + \sigma_1 \quad , \tag{3}$$

where α is the fine-structure constant, $h\nu$ is the photon energy, and σ_0 and σ_1 are the partial photoabsorption cross sections for the $\Delta\Lambda = 0$ and $\Delta\Lambda = \pm 1$ transitions. Because the asymmetry parameter β has no interference terms between the ionization channels, Eq. (1) can be rewritten using Eqs. (2) and (3) :

$$\frac{d\sigma}{d\Omega} = \frac{3}{4\pi}\left(\sigma_0 \cos^2\theta + \frac{\sigma_1}{2}\sin^2\theta\right) \quad . \tag{4}$$

As can be seen from Eq. (4), the pure $\Delta\Lambda = 0$ transition will lead to $d\sigma/d\Omega = 3/4\pi \cdot \sigma_0\cos^2\theta$, and the $\Delta\Lambda = \pm 1$ transition will lead to $d\sigma/d\Omega = 3/4\pi \cdot \sigma_1/2 \cdot \sin^2\theta$. Figure 2 shows the angular distribution of the photoions emitted after the $1s\sigma \rightarrow 2p\pi^*$ ($\Delta\Lambda = +1$) excitation of O_2. The experimental results behave in the predicted $\sin^2\theta$ distribution.

Although the angular distribution is composed of the $\cos^2\theta$ and $\sin^2\theta$ distribution when the two ionization channels exist, the angle resolved photoion measurements can decompose them. That is, the second term of Eq. (4) becomes zero at $\theta = 0°$, and the differential cross section $d\sigma/d\Omega|_{\theta=0°}$ for the direction parallel to the electric vector gives the partial cross section σ_0 :

$$\left.\frac{d\sigma}{d\Omega}\right|_{\theta=0^\circ} = \frac{3}{4\pi}\sigma_0 \quad . \tag{5}$$

On the other hand, the differential cross section $d\sigma/d\Omega|_{\theta=90^\circ}$ for the direction perpendicular to the electric vector gives the partial cross section σ_1:

$$\left.\frac{d\sigma}{d\Omega}\right|_{\theta=90^\circ} = \frac{3}{4\pi}\cdot\frac{\sigma_1}{2} \quad . \tag{6}$$

Then, one can obtain the "symmetry-resolved" photoabsorption spectra, if one measures the $d\sigma/d\Omega|_{\theta=0^\circ}$ and $d\sigma/d\Omega|_{\theta=90^\circ}$ as a function of photon energy.

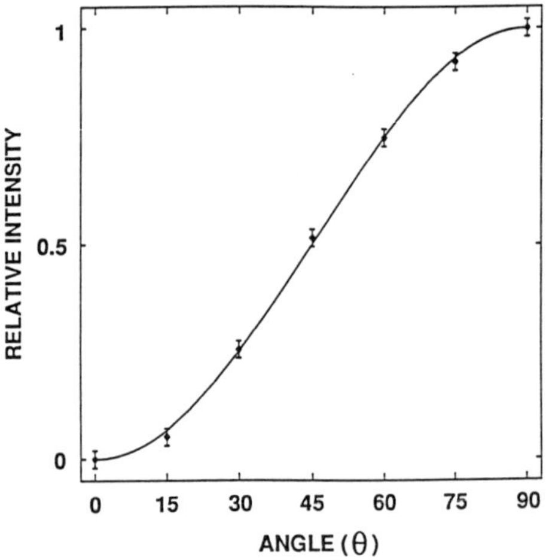

Fig. 2 The angular distribution of photoions emitted after the $1s\sigma \rightarrow 2p\pi^*$ excitation of O_2. Very weak signals at $\theta = 0°$ was set to the zero level, and the strongest signals at $\theta = 90°$ was normalized to unity. Full circles with error bars are date points, and a solid curve indicates the $\sin^2\theta$ distribution.

4. Symmetry-resolved Photoabsorption Spectra

As the powerful application of the angle resolved photoion measurements, we present the symmerty-resolved photoabsorption spectra of diatomic molecules in this section. The different detection efficiencies of the two analyzers have been corrected referring to asymmetry parameters for N 1s photoelectrons from N_2.[7] The constant valence contributions in the measured spectra have been subtracted from the spectra. Then, the $d\sigma/d\Omega|_{\theta=90^\circ}$ spectra have been multiplied by 2 to produce the σ_1 spectra (see Eq. (6)). In the following, the $d\sigma/d\Omega|_{\theta=0^\circ}$ and $2 \times d\sigma/d\Omega|_{\theta=90^\circ}$ spectra are displayed as the σ_0 and σ_1, respectively.

4.1. N_2 ($^1\Sigma_g^+$)

The symmetry-resolved nitrogen K-edge spectra σ_0 and σ_1 of N_2 were measured by the procedure described in the previous section.[7] They are shown in Fig. 3. The intense peak of $1s\sigma \rightarrow 2p\pi^*$ excitation at 401eV predominates in the σ_1 spectrum. The $1s\sigma \rightarrow 3p\pi$ Rydberg excitation at 407eV is clearly seen in the spectrum. The broad band around 415eV corresponding to doubly-excited states, $1s\sigma^{-1}2p\pi^{-1}(2p\pi^*)^2$, are superimposed on the continuous Π ionization channel. The σ^* shape resonance is observed as the very broad enhancement of photoabsorption in the σ_0 spectrum. The tiny peak at 406eV is due to the $1s\sigma \rightarrow 3s\sigma$ Rydberg excitation.

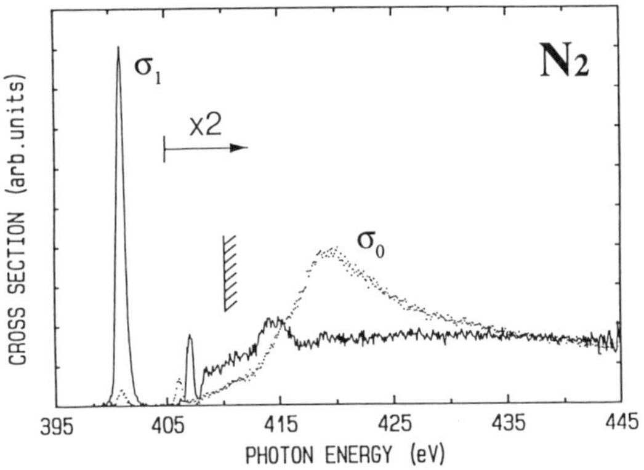

Fig. 3 σ_0 and σ_1 symmetry-resolved photoabsorption spectra of nitrogen K-edge of N_2. The dotted line represents the σ_0 spectrum, and the solid line the σ_1 spectrum.

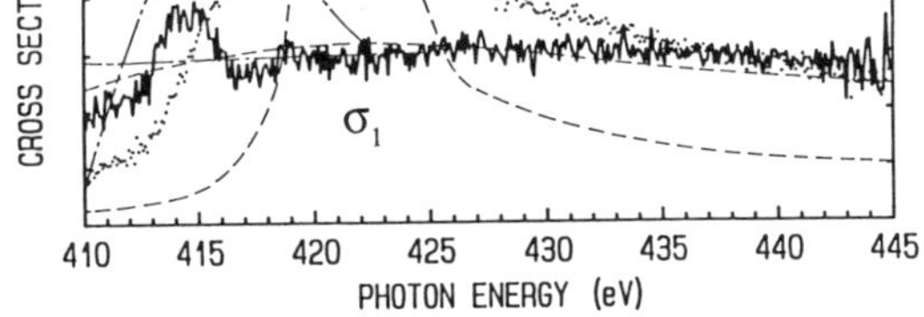

Fig. 4 The blowup spectra of σ_0 (dotted line) and σ_1 (solid line) above the K-shell threshold of N_2. The dashed curves are the Xα calculations by Dehmer and Dill[12], and the dot-dashed curves are the STMT calculations by Rescigno and Langhoff[13]. The theoretical σ_1's are normalized to the experimental at 445eV for the former calculations and at 425 for the latter.

The blowup spectra above the K-shell ionization limit are shown in Fig. 4, for comparison of them with *ab intio* calculations. The contribution of the Π ionization channel is 44% of that of the Σ ionization channel even at the maximum of the σ* shape resonance. The Π contribution in the continuum may introduce the deformation of the profile for the pure σ* resonance. Therefore, one should compare the experimental with theoretical results on each ionization channel. The cross section σ_0 of multiple scattering Xα calculations,[12] which is represented by the dashed curve, gives a poor agreement with the experimental. In our spectra, the σ_0 contribution is almost equal to the σ_1 even at 35eV above the threshold, differing from the statistical ratio of 1:2. The σ* shape resonance effect spreads over surprisingly wide energy ranges. However, the Xα calculations reach to the statistical ratio at only 25eV above the threshold. The results of Stieltjes-Tchebycheff moment theory (STMT) calculations[13], which are expressed by the dot-dashed curves, are somewhat better than the Xα results, but still unsatisfactory.

The discrepancies with the experimental results may be attributed to the neglect of the relaxation effects, the rearrangement of the other electrons to the presence of the core hole, in both the calculations. The importance of the relaxation effects in the K-shell photoionization cross section of CO is examined throughly in the work of Schirmer *et al.*[14] using the relaxed-core Hartree-Fock approximation.

4.2. O_2 ($^3\Sigma_g^-$)

The symmetry decomposition of an oxygen K-edge photoabsorption spectrum of O_2 is demonstrated in Fig.5. In the σ_1 spectrum, a π resonance is strongly observed at 530.5eV. In the σ_0 spectrum, two broad σ resonances are observed at about 539.5 and 542eV. It is reasonable that the π resonance is assigned to the 1sσ → 2pπ* transition[15] and the main features of the σ resonances are assigned to the 1sσ → 2pσ* transitions[16]. It has been revealed by our previous high-resolution spectrum and theoretical interpretation of it[8] that the 1sσ → 3sσ, 1sσ → 3pπ, and 1sσ → npσ (n = 3, 4, ···) Rydberg transitions are superimposed on the broad 2pσ* resonances. Although the present symmetry-resolved spectra have been measured with the resolving power of 1000, the 1sσ → 3pπ and 4pπ Rydberg transitions converging to the lower $^4\Sigma^-$ ionization limit are observed at 540.4 and 541.5eV in the σ_1 spectrum. The 1sσ → 3pπ and 4pπ Rydberg transitions converging to the higher $^2\Sigma^-$ ionization limit are also detected at 541.5 and 542.9eV in the spectrum. The 4pπ peak converging to the lower limit and the 3pπ peak converging to the higher limit are overlapped each other at 541.5 eV. In the σ_0 spectrum, the 1sσ → 4pσ and 1sσ → 3pσ Rydberg transitions converging to the $^4\Sigma^-$ and $^2\Sigma^-$ limits are overlapped at 542.0eV, and they are clearly seen on the upper 2pσ* resonance.

The σ_0 contribution in the continuum becomes about half of the σ_1 contribution at 11eV above the thresholds in contrast to the case of N_2 mentioned in the preceding section. This means that the convergence to the statistical ratio of 1:2 in the O K-edge of O_2 is much faster than in the N K-edge of N_2. This is the reflection of the fact that above the thresholds, there is no resonance structure in O_2 but the strong σ* shape resonance in N_2.

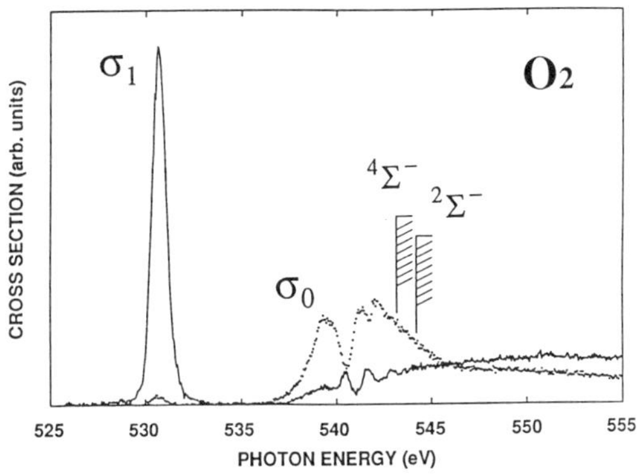

Fig. 5 σ_0 and σ_1 symmetry-resolved photoabsorption spectra of oxygen K-edge of O_2.

The theoretical study of *ab intio* SCF-CI and frozen-core calculations has been reported to give the full understanding of the oxygen K-edge spectra.[8] The core-ionized states allowed from the ground state $^3\Sigma_g^-$ correspond to the $^4\Sigma^-$ and $^2\Sigma^-$ states of O_2^+. The exchange splitting $\Delta E_{exch}(\infty)$ between the $^4\Sigma^-$ and $^2\Sigma^-$ states is given by

$$\Delta E_{exch}(\infty) = 3K_{c\pi} \simeq 0.80\text{eV (calc.)} \simeq 1.1\text{eV(exp.)} \quad , \qquad (7)$$

where c and π denote core (1sσ) and π^* (π_x^* or π_y^*) orbitals, and K_{ab} is the exchange integral between a and b. The calculated exchange splitting underestimates slightly the experimental value[17], but the agreement is improved in comparison with the previously reported result[18] (0.61eV). There are two 1s$\sigma \rightarrow$ 2pσ^* triplet excited states with the dipole-allowed $^3\Sigma^-$ symmetry. In one of them $^3\Phi_q(\sigma^*)$, the core and π^* electrons are coupled with the quartet ($^4\Sigma^-$) spin. In the other $^3\Phi_d(\sigma^*)$, they are coupled with the doublet ($^2\Sigma^-$) spin. The exchange splitting between the $^3\Phi_q(\sigma^*)$ and $^3\Phi_d(\sigma^*)$ states are given as follows :

$$\begin{aligned}\Delta E_{exch}(\sigma^*) &= E\{^3\Phi_q(\sigma^*)\} - E\{^3\Phi_d(\sigma^*)\} \\ &= -3K_{c\pi} + \frac{7}{3}K_{\sigma\pi} + \frac{2}{3}K_{c\sigma} \\ &\simeq 2.75\text{eV (calc.)} \\ &\simeq 2.5\text{eV (exp.)} \quad . \qquad (8)\end{aligned}$$

The theoretical splitting is in good agreement with the experimental. The exchange integrals between the valence orbitals and the core (1sσ) and σ^* orbitals, $K_{\sigma\pi}$ ($\simeq$ 1.2eV) and $K_{c\sigma}$ ($\simeq$ 1.1eV), are significantly larger than the integral between the core and π^* orbitals, $K_{c\pi}$ ($\simeq$ 0.3eV). Therefore, the $^3\Phi_d(\sigma^*)$ configuration state function is lower in energy than the $^3\Phi_q(\sigma^*)$. If σ^* is a Rydberg state, $K_{\sigma\pi}$ and $k_{c\sigma}$ are nearly zero and $^3\Phi_d$ becomes higher than $^3\Phi_q$. The transition probabilities to the $^3\Phi_d(\sigma^*)$ and $^3\Phi_q(\sigma^*)$ excited states have also been given. The theoretical ratio of 1:2 reproduces fairly well the observed spectrum (see

Fig. 5). Although there have been several interpretations proposed to explain the unusual structure of the σ* resonance, our theoretical consideration on the splitting and intensities has clarified that the σ* resonance is split by the exchange interaction. Wurth *et al.*[16] have recently calculated the 1sσ → 2pσ* transition energies and probabilities based on the Xα method, and reported that the exchange splitting is 1.3eV and that the lower state has larger intensity than the upper state. The quantitative disagreement is probably due to insufficient estimation for exchange interaction between the configurations state functions with four open shells.

4.3. NO ($^2\Pi$)

Figures 6 and 7 show symmetry-resolved nitrogen K-edge and oxygen K-edge spectra of NO. In the σ_1 spectra for both the N K-edge and O K-edge, intense and sharp π resonances are observed below the thresholds. In the σ_0 spectra, σ resonances appear prominently near, but above, the ionization thresholds for the N K-edge and O K-edge.

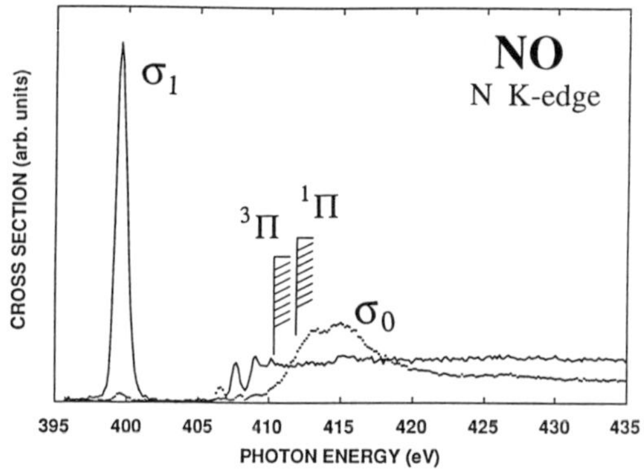

Fig. 6 σ_0 and σ_1 symmetry-resolved photoabsorption spectra of nitrogen K-edge of NO.

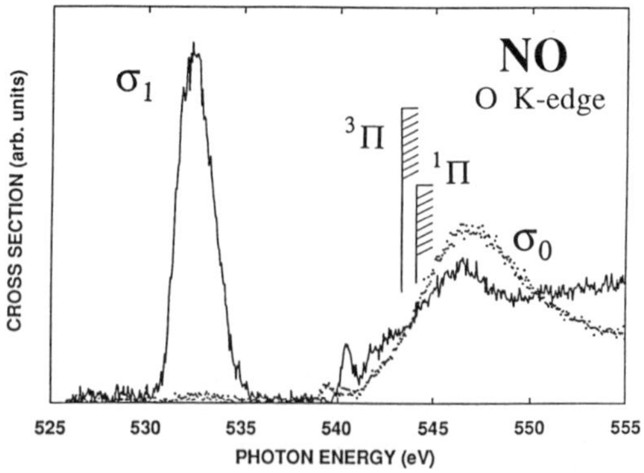

Fig. 7 σ_0 and σ_1 symmetry-resolved photoabsorption spectra of oxygen K-edge of NO.

These resonances are regarded as the $1s\sigma \rightarrow 2p\pi^*$ and $1s\sigma \rightarrow 2p\sigma^*$ transitions; the former is a bound-state resonance and the latter a shape resonance. The $1s\sigma \rightarrow 2p\pi^*$ transitions form the dipole-allowed excited states belonging to the $^2\Sigma^-$, $^2\Delta$, and $^2\Sigma^+$ symmetries ($\Delta\Lambda = \pm1$). Discussions on the ordering of these excited states, based on our high-resolution spectra and theoretical consideration, will be presented elsewhere.[9]

The peaks of the σ^* shape resonances are located close to the ionization thresholds, and the shape resonance enhancement spreads over rather narrow energy ranges. The σ_0 contribution is about half of the σ_1 at 23eV above the thresholds for the N K-edge. The convergence to the statistical ratio (1:2) in the N and O K-edges of NO is much faster than the N K-edge of N_2, and slower than the O K-edge of O_2. The double structure of the σ^* resonance in the N K-edge σ_0 spectrum of NO may be splitting caused by the exchange interaction. Namely, this splitting is due to the same origin for the exchange splitting in the $2p\sigma^*$ resonance of O_2, which has been described in the Section 4.2.

The weak peaks for the Rydberg transitions below the thresholds are very similar to the ones in the N_2 spectra: the lowest $1s\sigma \rightarrow 3s\sigma$ Rydberg transitions converging to the $^3\Pi$ N K-shell ionized state and the $^3\Pi$ O K-shell ionized state are seen in the σ_0 spectra of Figs. 6 and 7, and the second lowest $1s\sigma \rightarrow 3p\pi$ Rydberg transitions are seen in the σ_1 spectra. The peak of $1s\sigma \rightarrow 4p\pi$ Rydberg transition converging to the $^3\Pi$ ionization limit is located at 409eV in the N K-edge σ_1 spectrum, and the peak is overlapped with the one of the $1s\sigma \rightarrow 3p\pi$ transition converging to the $^1\Pi$ ionization limit. The $1s\sigma \rightarrow 4p\pi$ transition associated with the $^1\Pi$ limit is seen as a tiny peak at 410.1eV. Above the thresholds, broad structures are superimposed on the continuous Σ and Δ ionization channels in the σ_1 spectra, especially for the O K-edge. Some of them come from two-electron transitions, $1s\sigma^{-1}2p\pi^{-1}(2p\pi^*)^2$, as known in the K-shell excitation spectra of N_2 and CO. Because the contribution of the two-electron transitions for the O K-shell is fairly intense, and the energy positions of them are in the σ^* shape resonance region, the shape resonance enhancement must be overstimated when one evaluates the resonance effect from a conventional photoabsorption spectrum.

Acknowledgments

We are grateful to the staff of the Photon Factory for the stable operation of the storage ring during the course of the experiments. We would also like to thank Professors T. Sasaki and T. Matsushita for their extensive interest in our works.

References

1. R. N. Zare, *Mol. Photochem.* **4** (1972) 1.
2. J. L. Dehmer and D. Dill, *Phys. Rev.* **A18** (1978) 164.
3. D. Dill, J. R. Swanson, S. Wallence, and J. L. Dehmer, *Phys. Rev. Lett.* **45** (1980) 1393.
4. A. Yagishita, H. Maezawa, M. Ukai, and E. Shigemasa, *Phys. Rev. Lett.* **62** (1989) 36.
5. E. Shigemasa, K. Ueda, Y. Sato, T. Hayaishi, H. Maezawa. T. Sasaki, and A. Yagishita, *Phys. Scr.* **41** (1990) 63.

6. A. Yagishita and E. Shigemasa, *Rev. Sci. Instrum.* **63** (1992) 1383.
7. E. Shigemasa, K. Ueda, Y. Sato, T. Sasaki, and A. Yagishita, *Phys. Rev.* **A45** (1992) 2915.
8. N. Kosugi, E. Shigemasa, and A. Yagishita, *Chem. Phys. Lett.* **190** (1992) 481.
9. N. Kosugi, J. Adachi, E. Shigemasa, and A. Yagishita, to be submitted to *J. Chem. Phys.*
10. E. Shigemasa, T. Hayaishi, T. Sasaki, and A. Yagishita, to be submitted to *Phys. Rev.* **A**.
11. A. Yagishita, S. Masui, T. Toyoshima, H. Maezawa, and E. Shigemasa, *Rev. Sci. Instrum.* **63** (1992) 1351.
12. J. L. Dehmer and D. Dill, *Chem. Phys.* 65 (1976) 5327.
13. T. N. Rescigno and P. W. Langhoff, *Chem. Phys. Lett.* **51** (1977) 65.
14. J. Schirmer, M. Braunstein, and V. Mckoy, *Phys. Rev.* **A41** (1990) 283.
15. G. R. Wight and C. E. Brion, *J. Electron Spectrosc. Real. Phenom.* **4** (1974) 313.
16. W. Wurth, J. Stöhr, P. Feulner, X. Pan, K. R. Bauchspiess, Y. Baba, E. Hudel, G. Rocker, and D. Menzel, *Phys. Rev. Lett.* **65** (1990) 2426.
17. K. Siegbahn, C. Nording, G. Johasson, J. Hedman, P. F. Hedin, K. Hamrin, U. Gelius, T. Bergmark, L. O. Werme, R. Manne, and Y. Baer, *ESCA applied to free molecules* (North-Holland, Amsterdam, 1969).
18. P. S. Bagus and H. F. Schraefer III, *J. Chem Phys.* **56** (1972) 224.

CORE IONIZATION IN THE GAS PHASE, PROBED BY MULTICOINCIDENCE TECHNIQUES

P. Morin[1,2] , M. Lavollée[1] and M. Simon[1,2]

1 LURE, Université Paris-Sud, Bt. 209d, 91405 Orsay Cedex, France.
2 SPAM/CEA/CEN Saclay Bt522, 91191 Gif sur Yvette Cedex, France.

Abstract

Core ionization of simple triatomic molecules is examined using H_2S (S2p), N_2O (N1s) and ICN (I3d) prototype molecules, by means of multicoincidence (PEPIPICO) technique. Dissociation dynamics is shown to be markedly different in each case. The light mass of Hydrogen in H_2S allows observation of dissociation prior to Auger relaxation. In N_2O, the kinetic momenta distribution is shown to depend on which nitrogen (terminal or central) has been excited. Finally, in the case of ICN dissociation dynamics seems not to depend on photon excitation. In any case, fragmentation of multiply charged ions appears as a very fast process; an impulsive model, proposed to describe the dissociation dynamics, is shown to be more relevant than the Coulomb explosion model.

INTRODUCTION

Core ionization of free molecules is an interesting tool to produce very excited and/or multicharged ions. Indeed such a process is very weak at threshold appearance (because of dipole selection rule) but is the main decay process when core levels are excited, where resonant and normal Auger mechanisms take place on a very short time scale. High degree of ionization may be reached, depending on the depth of the excited level and on the occurrence of cascade effects in the Auger relaxation. The use of monochromatized synchrotron radiation allows the core-excitation of selected transitions (discrete or into ionization continua) and greatly simplifies the study of the electronic and nuclear relaxation.

The resulting multiply charged ions are likely to dissociate into several charged fragments, which requires sophisticated multicoincidence technique for their study. In the PEPIPICO technique (developed some years ago [1,2]) time of flight of two ionic fragments are measured using the (un-analyzed) electron signal as a start. This can be generalized to higher coincidence level. This technique is very powerful because it allows not only branching ratio between all the dissociation channels to be determined but also kinetic energies released and dynamics (sequentiality of bond breakage) to be studied.

In the case of simple triatomic molecules, we will show in the present paper that dissociation dynamics is not governed by a Coulomb explosion model, even when deep core ionization takes place: the examples of H_2S(3), N_2O(4-6) or ICN(7) will be discussed. The role of the mass ratio, geometry and chemical bond strength will be underlined.

PROCESSES AND RELATED EXPERIMENTAL TECHNIQUES

When a core hole is created after photon absorption, electronic relaxation takes place through the Auger effect. We deal with normal Auger process if the primary electron is directly ejected into the continuum or with resonant Auger process if the primary electron is trapped into an empty orbital. Filling the hole is not very different in both cases and is governed by the overlap integral between the core hole and the involved outer (valence) orbitals. The localisation of the hole on one particular atom, implies that the Auger relaxation mostly involves orbitals with some significant weight on that particular atom. The main difference between the two processes is in the final states populated after the decay: in the normal Auger process one ends up with a doubly charged ion, even though in resonant Auger the excited electron is likely to stay as a spectator, so that one ends up with very excited states of the singly charged ion. More complicated decay processes may occur also where for instance two Auger electrons are simultaneously ejected (double Auger process) which favours population of states of higher ionization degree. In case of deep core hole, the first Auger decay may be followed by subsequent Auger relaxation (cascade effect) which leads also to the observation of multiply charged ions.

The stability of such excited and multiply charged molecular ions is now questionned. In a simple picture one would expect the Coulomb repulsion of the charges to govern the dissociation dynamics: in the Coulomb explosion model, one assumes a very fast charge localisation on the atomic (or molecular) fragments with an instantaneous breaking of the bonds. The kinetic energies of the resulting particles are given by the mutual Coulomb repulsion between the various charges. The various questions one want to answer at this point are:

1) what is the validity of the Coulomb explosion model? In other words, can we observe stepwise dissociation processes, and what is the exact role of the ions charge?

2) what is the time scale of the various processes? For instance, can we always describe the Auger decay as taking place before ionic dissociation?

3) do we expect a selectivity in the dissociation dynamics, as revealed by photon energy dependence?

We now briefly review the various existing techniques suited for such ionic dissociation studies.

The time of flight mass spectrometry is extensively used because it allows not only mass to charge ratios to be measured but also kinetic energies and angular distributions of the fragments to be determined. In addition it provides a high collection efficiency which is of crucial importance in coincidence experiments. In PEPIPICO technique, initiated some years ago in England[1,8] and more recently developed in Orsay[2] for core ionization studies, one measures the time of flight between un-analyzed electron giving the start and the arrival of successive ions giving related stops. As detailed later, correlation between the various kinetic momenta of the particles generates various kinds of contour profile in the coincidence pattern which allows to distinguish in favourable cases between stepwise and concerted fragmentation processes. One of the interesting feature of this technique is that high coincidence rates may be achieved (for instance several hundred electron-ion-ion coincidence per second) which allows precise contour analysis.

Nevertheless, one information of importance is missing: the internal energy of the dissociating ion. This problem can be overcome by analyzing the electron energy or by detecting zero energy electron. In the latter technique[9] one selects the internal energy of the core hole which may be of high interest when two identical atoms are present but with a different chemical environment. In that case the induced chemical shift allows to distinguish the two atoms. For instance in CH_3CF_3[10] molecules significant changes in the fragmentation have been observed depending on which carbon has been excited. When selecting an Auger electron (of high kinetic energy), one really selects the internal energy of the ion. In a pioneering experiment Eberhardt et al[11] succeeded in such an experiment on CO where coincidences between electron and one ion were performed. Similar measurements were done recently by Ueda et al[12] on BF_3 . Recently Hanson et al[13] measured electron ion -ion coincidences on N_2O. These very promising experiments are nevertheless suffering of very low counting rates and the yet achieved statistics do not allow precise contour determination. No doubt that the situation will change in the very next future.

In the following, we will present results obtained using the PEPICO or PEPIPICO techniques. Schematics of the experimental arrangement is given below in figure1.

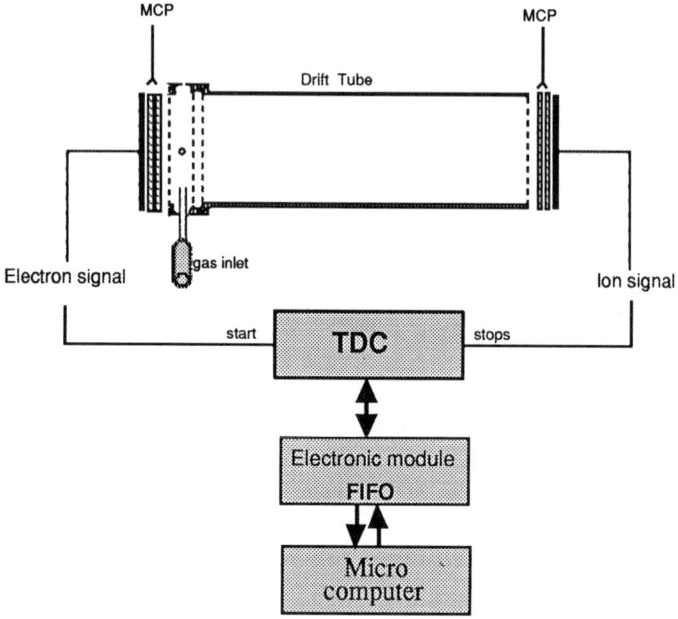

Figure 1: Schematic of the PEPIPICO apparatus

Synchrotron radiation from SuperAco storage ring is monochromatized using two grazing incidence monochromators, covering the energy range 10-250 eV (SA31 bending magnet beamline) and 150-800 eV (SU7 undulator beamline) respectively. The electrons and ions are extracted from the ionization region with a strong DC field (2000V/cm). The electrons are immediately detected by a channelplates detector which gives a start to the TDC. The ions drift into a field free tube before detection by channelplates giving the successive stops to the TDC. Reading and storage of the various times of flight is achieved in real time by a MacIntosh microcomputer.

H_2S: AN EXAMPLE OF VERY FAST DISSOCIATION

Hydrides molecules have been demonstrated recently [14-19] to have a specific relaxation behaviour after core excitation into the first empty orbital. Indeed, because of the antibonding character of this $\sigma*$ (X-H) orbital, and because of the light mass of the hydrogen atom, neutral dissociation has been shown to preceed or at least compete efficiently with Auger relaxation. Evidence for this effect has been first given by electron spectroscopy. For instance in HBr[14], observation of atomic Auger line originating from isolated bromine atom was a clear signature of the very fast dissociation. Mass spectrometry studies performed later(4) showed that a molecular decay channel had to be accounted for in competition with neutral dissociation, as evidenced by the observation of H^+/Br^+ ion pair formation along the σ* resonance.

The H_2S molecule is interesting because the neutral dissociation process in this case leads to the formation of the HS molecular fragment. Previous works (20-22) have demonstrated the adequacy of the fast neutral dissociation picture, by electron spectroscopy. Here we want to present the manifestation of such phenomenon on ionic fragmentation.

The total ion yield curve is presented in figure 2

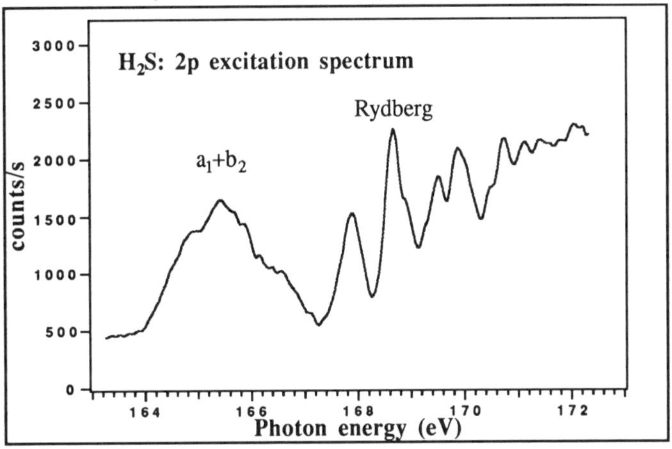

Figure 2: Total ion yield obtained on H_2S in the region of S2p

Transition from S2p to a_1 and b_2 empty orbitals gives rise to the first broad resonance at 165.4 eV. Four components are superimposed because each transition is split into two spin orbit components separated by 1.2eV. The unusually large width of the resonance is easily explained by the repulsive character of the a_1 and b_2 states. We have recorded mass spectra along the various resonances and above the S2p ionization limit. Our mass resolution was good enough to distinguish H_2S^+, HS^+, S^+, H_2S^{2+}, HS^{2+}, S^{2+}. The most striking observation is that the HS^{2+} doubly charged fragment is significantly observed only when the a_1+b_2 states are excited. This very selective behaviour is pointed out on figure 3 where the HS^{2+}/H_2S^{2+} ratio has been reported in the energy range of interest. An enhancement of about ten is observed for this ratio on the first resonance.

The interpretation for such a spectacular selectivity in the fragmentation is the following. When the a_1+b_2 resonances are excited, the H_2S molecule first undergoes neutral dissociation into H+HS*. The Auger relaxation then takes place on HS* fragment, leading to single but also double ionization (see previous section). Theoretical calculations by Parlant et al [23] have shown that the ground state of HS^{2+} is stable on the time scale of the experiment. Of course in addition to the ground state, other excited states are probably formed as well, that may dissociate into the charge separation $H^{+}+S^{+}$ channel but they do not contribute to the ratio determination.

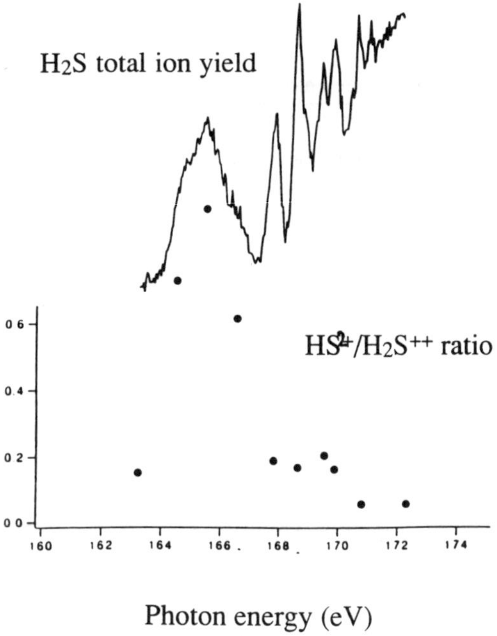

Figure 3: HS^{2+}/H_2S^{2+} branching ratio in the region of S2p excitation.

On the other hand the H_2S^{2+} formation is obviously signature of the Auger decay of undissociated molecules. In HBr, we have shown[4] that such a competitive decay was existing but not dominant. The same situation probably occurs in H_2S, leading to a weak abundance of H_2S^{2+} and explaining why theHS^{2+}/H_2S^{2+} ratio reaches a so high value on the a_1+b_2 resonances. To end up with our interpretation we have to explain why HS^{2+} is observed only on the first resonances. Indeed, for the Rydberg resonances and in the continuum, we expect double and triple ionization to increase significantly, due to double Auger and normal Auger processes. Our observations show that stable H_2S^{2+} is efficiently produced in that case (originating probably from ejection of the two electrons from the sulfur lone pair orbital). We also have to consider dissociative double ionization; the charge separation channel H^++HS^+ is the dominant channel but we also have to account for the H + HS^{++} channel that could contribute to the formation of HS^{2+}, but in a different manner than previously mentionned. Our results show that this last decay process is very unlikely. This is not surprising from energetics point of view: indeed the H + HS^{++} dissociation limit is located at much higher energy than the H^++HS^+ one. At this point one needs precise potential surface calculations of H_2S^{2+} to obtain a refine description of the (pre)dissociation processes.

In conclusion, we have with the H_2S molecule a clear example of selective dissociation: the HS^{++} production is possible only when starting from HS. This is the reason why it is observed only when the a_1+b_2 resonances are excited, which constitutes an a postriori confirmation of the fast neutral dissociation model.

DISSOCIATION DYNAMICS OF N_2O AFTER NITROGEN K SHELL EXCITATION

As a second exemple of core excitation study, the N_2O molecule is of special interest because of the presence of two unequivalent nitrogen atoms. Excitation from terminal (Nt) or central (Nc) nitrogen can be done selectively, taking advantage of the 4eV chemical shift of the two Nt1s and Nc1s binding energies. Previous works [4, 24, 25, 26] were dedicated to site specific fragmentation and we want here to focus on dissociation dynamics as a function of photon excitation. The three photon energies of interest are the following: 401eV (Nt1s-->Π^*), 405eV (Nc1s-->Π^*) and 430eV (N->σ^* in the continuum).

Dissociation dynamics can be obtained from PEPIPICO measurements using the momentum conservation law on one hand and the very simple formula giving the time of flight of a particle on the other hand: $t = t^0 - \alpha/q\ p\ cos(\theta)$ (I) where t^0 is the time of flight of the particle at rest, q its charge, α a constant of the apparatus, p the projection of the kinetic momentum and θ the ejection angle with respect to the detection axis. From this simple consideration, one easily deduces that in case of a two body dissociation for which $p_2=-p_1$, the time correlation between the two times of flight t_1 and t_2 is $t_2=-t_1$ (if we neglect t_1^0 and t_2^0). As a consequence the t_2 vs t_1 plot is a "cigar"-like picture of slope -1, as illustrated in figure 4a. More interesting is the situation where three particles are ejected, for instance two charged particles and a neutral. In case of stepwise dissociation, from momentum conservation applied at each step one deduces easely [27] that the t_2 vs t1 plot is now a parallelogramm, whose slope is given by the m_2/m_{12} mass ratio, assuming the following two step process:

(1) m^{2+}-->$m_{12}^+ + m_3^+$ (2) m_{12}^+-->$m_1^+ + m_2$

This is illustrated on figure 4b. Finally, in case of concerted three body dissociation, the angle and magnitude of the kinetic momenta are not simply related and the general picture[28] is an ellipse as shown in figure 4c. From the data, one can extract evidently the p_1 and p_2 momentum distribution, but also the kinetic momentum p_3 of the third undetected particle, which is of very high interest when discussing dissociation dynamics. Indeed, from momentum conservation one obtains $p_3=-(p_1+p_2)$ which implies, using formula (I) that the t_1+t_2 distribution reflects the one of p_3 .

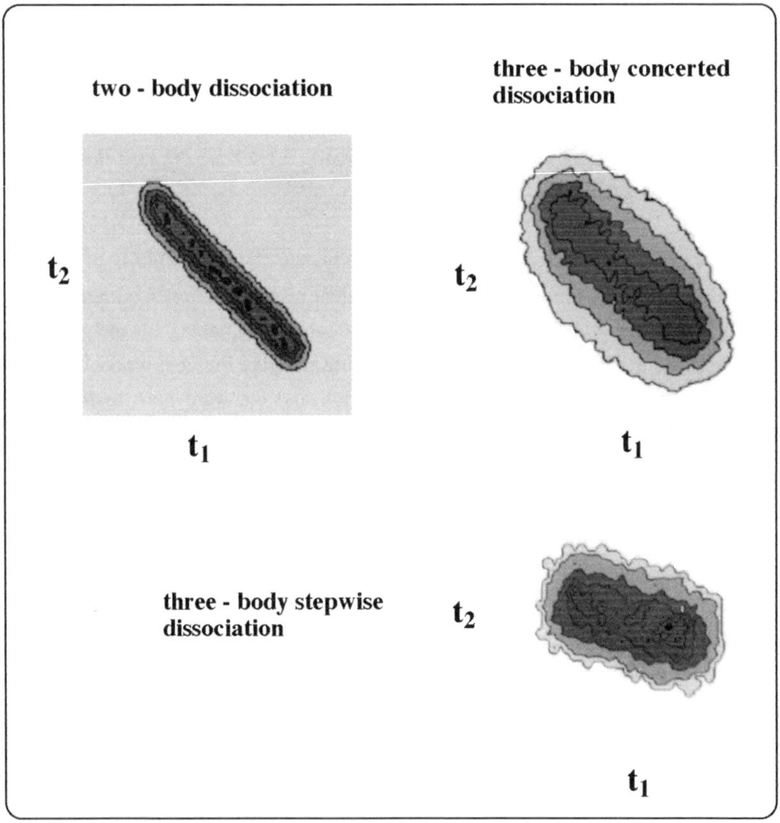

Figure 4: exemple of PEPIPICO spectra in case of two body dissociation (a), three body stepwise dissociation (b) and three body concerted dissociation (c).

We now focus attention on three body dissociation as resulting from nitrogen K shell excitation or ionization of nitrous oxide. For this study we have used isotopically labelled N_2O, namely $^{14}N^{15}NO$ in order to distinguish the central nitrogen from the terminal one in the atomization process. By atomization we mean formation of three atoms independent of their state of charge. PEPIPICO spectra recorded after Nt1s-->Π^* Nc1s-->Π^* and N->σ^* transitons are shown in figure 5a, 5b and 5c respectively. One clearly distinguishes on the right side of the figure the three Nt^+/Nc^+, Nt^+/O^+ and Nc^+/O^+ ion pairs, and on the left side six features corresponding to Nt^{2+}/Nc^+, Nt^{2+}/O^+, Nc^{2+}/Nt^+, Nc^{2+}/O^+, O^{2+}/Nt^+ and O^{2+}/Nc^+ ion pairs.

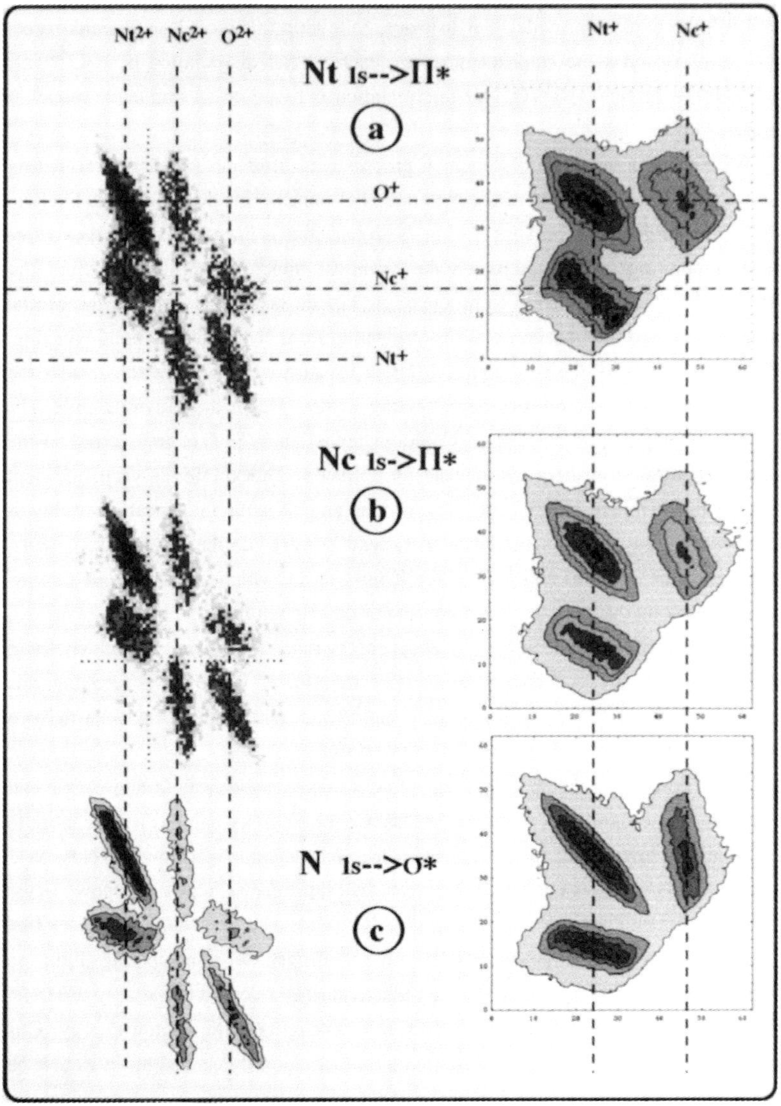

Figure 5: PEPIPICO spectra recorded at 401eV (a), 405eV (b) and 430eV (c).

The total number of coincidence events in each spectrum was typically of the order of 200 000, obtained in less than two hours in each case. Except for two pictures, statistics was good enough to make a contour level analysis of the peaks.

Above threshold (fig 5c), the central nitrogen is ejected with very little kinetic energy (about 1eV)[5] as compared to its other partners (about 10eV). The interesting finding is that this observation is valid whatever the charge is (0,1 or 2). This rules out the Coulomb explosion model. For instance, following this model one would expect Nc^{2+} to carry out some significant kinetic energy after dissociation of N_2O^{3+} into $Nc^{2+}+O^{+}+Nt$ or $Nc^{2+}+Nt^{+}+O$. Our interpretation is based on an impulsive model as described now.

We assume a very fast dissociation process (just after the Auger effect), with a quasi simultaneous breaking of the N-N and N-O bonds. This dissociation is likely to occur in linear geometry because N_2O in its ground state is linear and removal of a core electron does not affect too much the chemical bonds. Within this picture the $Nt^{+}+Nc+O^{+}$ reaction is likely to produce low kinetic energy Nc because the central atom is submitted to opposite impulses originating from N-N and N-O bond breaking. When Nc is charged, the situation is different because we now have to consider the Coulomb repulsion which was not affecting neutral Nc in the previous case. The additionnal impulse of Nc due to Coulomb repuslion is in fact transferred to the adjacent neutral atom because of head on collision. This tranfer is very efficient because the involved masses are very similar and because the linear geometry is preserved. Following this model, one expects a very similar behaviour when comparing $Nc^{n+}+Nt^{+}+O$ and $Nc^{n+}+O^{+}+Nt$ (n=1 or 2), which is exactly our finding. Note at this point that we cannot distinguish this description from the one where the two bonds are broken simultaneously, with equal energy released. We only tentatively try to explain that even if two unequivalent energies are released in the bonds breaking, this nevertheless result at the end in an equal momentum partitionning between the two external fragments.

When comparing the spectra obtained at the other resonances (a and b) one can clearly see some significant changes: for instance the Nc+/Nt+ pair has a slope close to 0 at 430eV (ie low kinetic energy Nc) and is then increasing to reach a value close to -1 at 401eV. This indicates that on discrete resonances, the central nitrogen is now ejected with a significant kinetic energy, especially after Nt1s-->Π* transition. Again we find this effect independent of the charge of the central nitrogen. When comparing all the ion pairs, the general finding is that the total kinetic energy released is not very different, but the partitionning between the particles is changing with photon energy. Our impulsive model is still valid to describe the dissociation dynamics, but it cannot be applied in the linear geometry. Indeed, when dealing with discrete resonances, we have to account for the extra electron temporarily trapped in the empty orbital. This additionnal electron may have a strong influence on the chemical bond because of its localisation close to the valence orbitals. Such an influence can be inferred from the so called equivalent core model: the structure of the excited molecule is given by replacing the excited atom by its Z+1 neighbour in the periodic table. In the present case, the molecule is equivalent to ONO after Nt1s-->Π* excitation and NOO after Nc1s-->Π* transition. The ONO molecule is known to be stable, with a bent angle of 134°. It is then likely to consider the impulsive model in a bent geometry instead of linear. Consequently, the collison of the central nitrogen with the adjacent atom is not a head on collision and the momentum transfer is not so efficient; the central atom may now escape perpendicularly to the molecular axis. This may explain why the observed peaks are significatively broad as expected from a concerted dissociation. For the Nc1s-->Π* transition, the equivalent core molecule is not known and theoretical calculations are needed. One expects an intermediate bent angle to explain our results.

In conclusion, we have shown that the dissociation of core excited N_2O was not governed by Coulomb explosion model, but rather well using an impulsive model. Dissociation dynamics has been shown to depend strongly upon the involved excited resonances.

THREE BODY DISSOCIATION IN ICN

The last example is ICN. This molecule is interesting because three different atoms are present (which allows selective core excitation as in N_2O) and because of very different chemical bonds (I-C simple and weak, C-N triple and strong). We have studied the fragmentation of the molecule over a very wide range of photon energies, starting at 40eV (close to the double ionization threshold) up to some 700eV (in the I3d ionization continuum). One of our finding is that the dissociation dynamics is not dependent upon photon excitation. For this reason, we only discuss here the PEPIPICO spectrum obtained at 700 eV, where a very large set of coincidence pairs are present.

Figure 6 is a PEPIPICO spectrum of ICN, obtained at 700eV, representing only the three body dissociation processes. Ionization of the iodine 3d shell allows to reach very high ionization degree. Auger processes involving two outer valence electrons lead to formation of doubly charged ions, that may dissociate into I^+/C^+ or I^+/N^+.This process is obviously weak, as compared to other channels, but is still easily measured. On the other side, the 3d hole may be filled in a first step by 4p or 4s electrons. Each of these vacancies may also be filled by 4d electrons and so on. Finally, because of such cascade effects, one may observe ICN^{8+} ions. In fact we observe up to ICN^{7+} as revealed by the

I^{5+}/C^{2+} and I^{5+}/N^{2+} ion pairs. All the peaks presented in figure 6 have a similar "cigar" like shape. At this point it can be interpreted either as a two step process with little kinetic energy released in one step, or as a dissociation process taking place in a linear geometry.

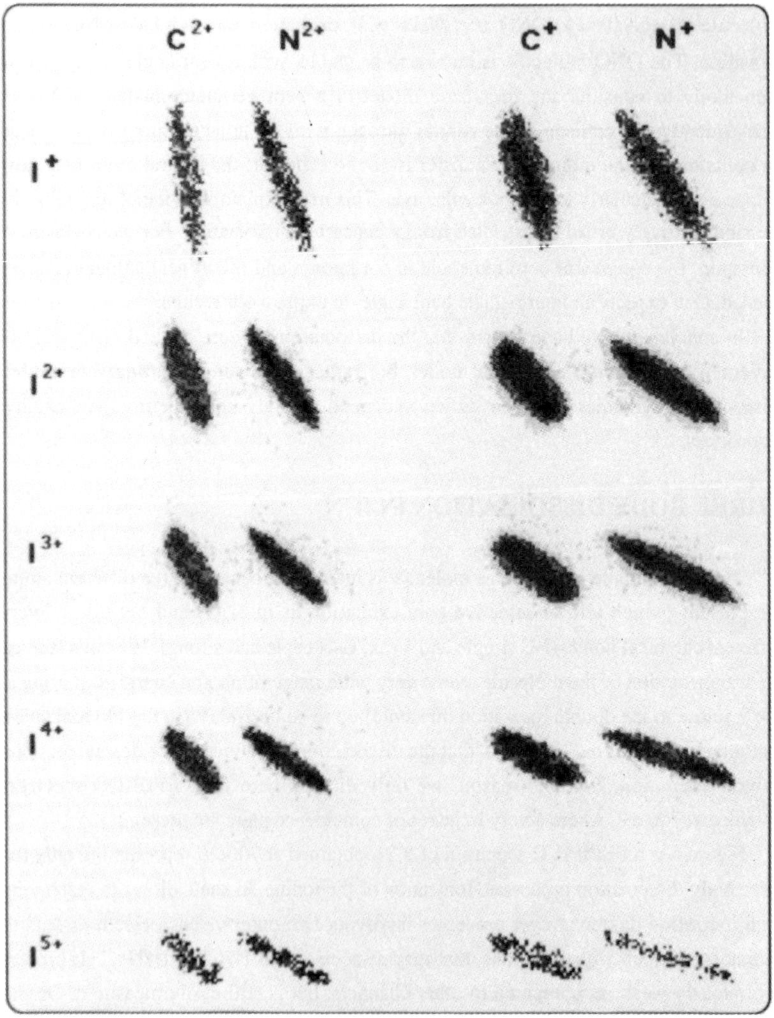

Figure 6: PEPIPICO spectrum of ICN recorded at 700eV photon energy. Only three body dissociation processes are reported.

One interesting point with our data, is the possibility to compare the different peaks where only the charge is changing. Because all the peaks are cigar like, they are characterized by their slope. Table 1 is a comparison of the various slopes. As shown by formula (I), the state of charge q of the fragments influences the time of flight as 1/q. Thus it is interesting when comparing the various slopes, to correct for this charge effect. Consequently, each measured slope corresponding for instance to I^{p+}/N^{q+} has to be multiplied by p/q. After correction for the charge effect, the slopes are found identical for any I^{p+}/C^{q+} or any I^{p+}/N^{q+} pairs and are independent of the photon energy.

pairs	mes.slope	calc.slope	pairs	mes.slope	calc.slope
I^{+}/C^{+}	-2.08	-2.16	I^{+}/N^{+}	-1.31	-1.33
I^{2+}/C^{+}	-1.15	-1.08	I^{2+}/N^{+}	-0.65	-0.66
I^{3+}/C^{+}	-0.71	-0.72	I^{3+}/N^{+}	-0.45	-0.44
I^{4+}/C^{+}	-0.50	-0.54	I^{4+}/N^{+}	-0.34	-0.33
I^{5+}/C^{+}	-0.44	-0.43	I^{5+}/N^{+}	-0.32	-0.27
I^{+}/C^{2+}	-6.67	-4.3	I^{+}/N^{2+}	-2.58	-2.66
I^{2+}/C^{2+}	-2.67	-2.16	I^{2+}/N^{2+}	-1.31	-1.33
I^{3+}/C^{2+}	-1.59	-1.44	I^{3+}/N^{2+}	-0.89	-0.88
I^{4+}/C^{2+}	-1.10	-1.08	I^{4+}/N^{2+}	-0.66	-0.66
I^{5+}/C^{2+}	-0.85	-0.87	I^{5+}/N^{2+}	-0.52	-0.54

Table1: measured and calculated slopes of the different I^{p+}/C^{q+} and I^{p+}/N^{q+} ion pairs measured at 700eV photon energy.

We conclude from the last point that dissociation dynamics is the same for all electronic states and occurs in a linear geometry. Moreover it is not governed by Coulomb repulsion but by the impulses transmitted to the particles immediately after chemical bond breakage (femtosecond time scale). We proposed an impulsive model to determine the kinetic momenta ratio: the C-I bond is first broken with an energy release E1. The C atom then collides the N atom (spectator in the first step). But CN is not a rigid system and behaves like an oscillator and we thus end up with an energy partitionning of E1 into translational energy kE1 and internal energy (1-k)E1 of CN. The CN system dissociates then rapidly, releasing its internal energy. We thus explain our results with k=.85 for I^{p+}/N^{q+} and k=1 for I^{p+}/C^{q+} as shown in table 1 where the slopes calculated within this model are reported.

With this simple model, already proposed to explain the dissociation of ICN in the UV regime[29], we can thus explain why the impulses are always found with the same ratio (i.e. same corrected slope), despite the very different kinetic energies released in each case.

CONCLUSIONS

Core ionization of H_2S, N_2O and ICN has revealed the breakdown of the Coulomb explosion model to describe the dissociation dynamics of these simple molecules. An impulsive model is proposed to interpret the dissociation dynamics and to account for fast dissociation processes. The role of the mass has been emphazised. In H_2S, the hydrogen is so light that dissociation can take place before Auger relaxation is achieved, resulting in a very selective production of HS^{2+} when the a_1+b_2 dissociative states are excite; dynamics in this case is governed by the resonant state. In ICN, dissociation mechanisms seem to be photon independent: dynamics is governed in that case by the final state (after electronic relaxation). N_2O corresponds to an intermediate situation: the life time of the resonant state seems long enough to influence significatively the dissociation dynamics.

ACKNOWLEGMENTS

We would like to warmly thank R.Thissen for the major improvements he brought to the experiment and M.Meyer for his active participation in obtaining the data, as well as T.LeBrun who started the experiment on N_2O. We are also grateful to A.Beswick, P.Millié and M.Miranda for helpful disscussions and comments.

REFERENCES

1: J.H.D.Eland, F.S.Wort and R.N.Royds, *J.Electr.Spectr.* 41(1986)297

2: M.Simon, T.LeBrun, P.Morin, M.Lavollée and J.L.Maréchal *Nucl.Instr.&Meth.*B62(1991) 167

3: P.Morin, M.Lavollée, M.Simon and R.Thissen *4th ECAMP conf., Riga, 1992*

4: T.LeBrun, M.Lavollée, P.Morin *AIP Conference Proceedings 215, X-Ray and inner shell processes, Knoxville, TN, 1990* T.A.Carlson, M.O.Krause, S.T.Manson eds, p846

5: T.LeBrun, M.Lavollée, M.Simon and P.Morin, submitted to *J.Chem.Phys.* (1992)

6: M.Lavollée, M.Simon, M.Meyer and P.Morin, *abstract VUV10, pMo97 Paris,* 1992

7: M.Simon, thèse d'Université, Orsay, 1992 and M.Simon, M.Lavollée, M.Meyer and P.Morin, *abstract VUV10, pTh79, Paris,* 1992

8: L.J.Frasinsky, M.Stankiewicz, K.J.Randall, P.A.Haterley, K.Codling *J.Phys.B* L819(1986)

9: K.Muller-Dethslef, M.Sander, L.A.Chewter, E.Schlag *J.Chem.Phys* .85(1986)5755

10: W.Habenicht, H.Barter, K.Muller-Dethlefs and E.W.Schlag *J.Chem.Phys.* 95(1991)6774

11: W.Eberhardt et al *Austr.J.Phys.* 39,633(1986)

12: K.Ueda, H.Chiba, Y.Sato, T.Hayaishi, E.Shigemasa and A.Yagishita Phys.Rev.A, 46(1992)R5

13: D.M.Hanson, C.I.Ma, K.Lee, D.Lapiano-Smith and D.YKim *J.Chem.Phys.* 93(1990)9200

14: P.Morin and I.Nenner *Phys.Rev.Lett.* 56(1986)1913

15: P.Morin and I.Nenner *Phys.Scripta* T17(1987)171

16: G.G.B.de Souza, P.Morin and I.Nenner *Phys.Rev.A* 34(1986)4770

17: S.Svensson, L.Karlsson,, N.Martensson,, P.Baltzer and B.Wannberg *J.Elec.Spec.* 50(1990)c1

18: H.Aksela, S.Aksela, O.P.Sairanen, M.Hotokka, G.M.Bancroft, K.H.Tan and J.Tulkki *Phys.Rev A* 41(1990)6000

19: H.Aksela, S.Aksela, M.Hotokka, A.Yagishita and E.Shigemasa *J.Phys.B* 25(1992)3357

20: H.Aksela, S.Aksela, A.Naves de Brito, G.M.Bancroft and K.H.Tan *Phys.Rev.A* 45(1992)7948

21: A.Naves de Brito and H.Agren *Phys.Rev.A* 45(1992)7953)

22: A.Naves de Brito, PhD thesis, Uppsala University, 1991

23: G.Parlant, J.Senekowitsch, S.Oneil and D.Yarkony *J.Chem.Phys.* 94(1991)7208

24: A.P.Hitchcock, C.E.Brion and M.J.Van der Wiel *Chem.Phys. Lett.* 66(1979)213

25: J.Murakami, M.C.Nelson, S.L.Anderson and D.M.Hanson *J.Chem.Phys.* 85(1986)5755

26: R.Murphy and W.Eberhardt *J.Chem.Phys.* 89(1988)4054

27: J.Eland *Accounts of chemical research* 22, 381, (1989)

28: T.LeBrun, *Thèse d'Université*, Orsay (1991)

29: K.Holdy, L.C.Klotz and K.R.Wilson *J.Chem.Phys.* 52(1970)4588

CORE EXCITATION IN ATOMIC AND MOLECULAR CLUSTERS

E. RÜHL, C. HEINZEL, A.P. HITCHCOCK* and H. BAUMGÄRTEL

Institut für Physikalische und Theoretische Chemie
Freie Universität Berlin, Takustr. 3, D-1000 Berlin 33, Germany

ABSTRACT

Core level excitation in atomic and molecular clusters is reported. Tunable synchrotron radiation is used for site specific excitation in the energy regimes of element K- and L-edges. Homogeneous clusters of argon, nitrous oxide and benzene are investigated. Two different aspects are discussed: (i) spectroscopic changes from the isolated gas phase to the condensed phase species and (ii) fragmentation of core-excited clusters. Spectroscopic properties of clusters are investigated by the use of total electron yield spectra as well as photoion yield curves of mass selected cluster cations. In the case of argon clusters, characteristic shifts of the energies of Rydberg states toward those of the corresponding exciton states in the solid are found. Evidence for extended x ray absorption fine structure (EXAFS) in free gas phase clusters is reported. The photoion-photoion coincidence technique (PIPICO) is applied to determine the fragmentation mechanisms of doubly charged clusters.

1. Introduction

Core-level excitation (element K- and L-edges) of isolated gas phase species, of the condensed phase and of adsorbates has been the subject of many studies.[1-5] Information on resonant states of valence and Rydberg character in the near edge (NEXAFS) regime has been obtained; the extended absorption fine structure (EXAFS) has been used for structural studies; the electronic relaxation dynamics and subsequent fragmentation processes of core excited states have been investigated.

Clusters can be regarded as an intermediate state of matter between the gas and the condensed phase. Properties of clusters, such as geometrical and electronic structure as well as dissociation dynamics have been studied in the last decade by the use of low energy photon light sources in the infrared, visible, UV and VUV regimes.[6,7]

We report in this paper recent progress in the field of core level excitation of weakly bonded atomic and molecular van der Waals clusters in the gas phase. The emphasis of this work is to investigate spectroscopic changes from isolated gas phase species to the condensed phase in the near-edge and EXAFS regimes. This is accomplished by studying clusters of atoms and molecules of different sizes with tunable soft x rays and mass spectral detection techniques.[8-12]

Fragmentation of singly and doubly charged clusters has been investigated using both photoionization and electron impact ionization techniques. One focus of these studies has been the occurrence of shell structures of singly charged cluster fragments (magic numbers),[13-15] as well as fragmentation properties of state or mass selected singly and doubly charged clusters.[16-18] It has been shown that clusters undergo massive fragmentation when they are ionized by photons, electrons, or neutral gas targets. Critical sizes of stable multiply charged clusters have been determined,[19,20] as well as indirect evidence for dissociative double ionization (Coulomb explosion).[21]

---

* permanent address: Department of Chemistry, McMaster University, Hamilton, Ontario, Canada

We report direct evidence for charge separation reactions within doubly charged clusters. Experimental techniques involving photoion-photoion coincidences (PIPICO)[9-11] and photoelectron-photoion-photoion-coincidences (PEPIPICO) are used.[22] The results are discussed in comparison with photon stimulated desorption (PSID) of cations from the condensed phase.

2. Experimental

The experimental setup consists of a continuous, skimmed, supersonic molecular beam expansion, a time of flight mass spectrometer (TOF-MS), and a threshold photoelectron spectrometer.[9-12] The latter device can be replaced by a total electron/cation yield detector consisting of a channeltron multiplier that is mounted approximately 1 cm from the ionization region. The TOF-MS is operated with pulsed and static cation extraction conditions fulfilling Wiley/McLaren space focussing.[23] Synchrotron radiation from the electron storage ring BESSY is used for the experiments (HE-TGM I and II beam lines). The compact design of the molecular beam machine allows the use of different beam lines and storage ring facilities, such as DESY and SUPER-ACO. The high photon flux (10^{12} photons/s) and medium spectral resolution ($E/\Delta E=500$) of the BESSY beam lines are well suited for cluster photoionization experiments. However, it should be noted that the experiment has already worked with considerably lower photon fluxes (approx. $5 \cdot 10^{10}$ photons/s) at the DORIS storage ring.[24]

Experiments that are performed with the setup are: (i) total yield curves of electrons and cations, (ii) photoion yield curves of mass selected clusters, (iii) coincidences between charged particles such as: photoion-photoion-coincidences (PIPICO), photoelectron-photoion-coincidences (PEPICO) as well as photoelectron-photoion-photoion-coincidences (PEPIPICO). These coincidence techniques are applied to study fragmentation of singly and multiply charged clusters.

Clusters are prepared by expanding the neat or diluted samples through a 50 μm nozzle. Argon and nitrous oxide are expanded as neat samples, whereas benzene clusters are formed by seeding benzene at room temperature in helium.

The size of the neutral clusters is estimated by using scaling parameters of the supersonic jet expansion and experimentally determined correlations with average cluster size $\overline{N}$.[25-27] Scaling parameters can be only derived for atomic species; this method fails for expansions of molecules.[28] Cluster size information can be obtained from the mass spectral distributions. However, there are inherent problems related to massive fragmentation of charged clusters, as well as mass discrimination effects for heavy cluster ions in the time of flight mass spectrometer.[12]

3. Results and Discussion

3.1 Argon Clusters

Figure 1 shows total electron yield curves of an argon cluster beam. The average cluster size $\overline{N}$ is adjusted by changing the stagnation temperature (T_0), while keeping the stagnation pressure (p_0) constant at 5 bar. The spectra recorded at $\overline{N}=1$ are obtained by expanding 1 bar argon at room temperature. The reduced scaling parameter Γ^* corresponds to 100. This is well below the threshold of cluster formation, which is verified to occur at $\Gamma^*>200$ in accordance with ref. 25.

There is a systematic change in the Ar $2p_{3/2}$->4s transition as $\overline{N}$ is increased. This transition occurs in the atom at 244.390 eV.[29] The signal can be satisfactorily fitted with only three components.[12] The energies of these components increase by approximately 0.35 eV over the sequence of decreasing nozzle

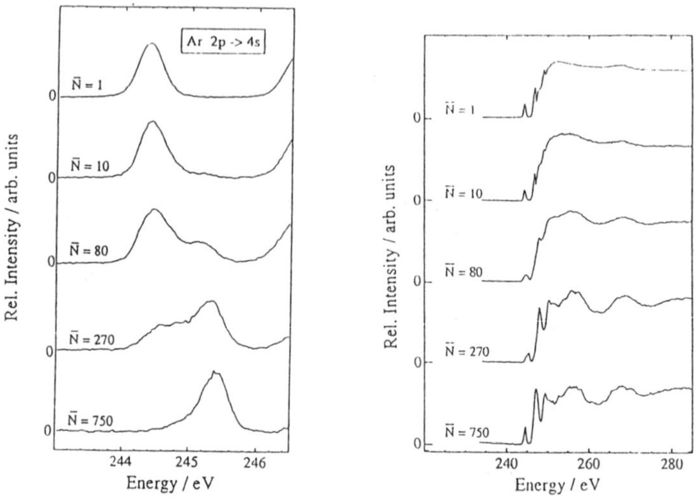

Figure 1: Total electron yield spectra of a argon cluster beam. $\overline{N}$ corresponds to the average cluster size.

temperature and thus increasing $\overline{N}$. We interpret these components as contributions from small, medium size, and large clusters. This is in accordance with earlier work on mass selected argon clusters ($\overline{N}$ =10), where a blue-shift of 0.35 eV was observed.[8] Clusters of this size have most of their atoms at the surface. It is therefore possible that the atoms on the surface of larger clusters have their 2p3/2->4s transition at 244.7 eV. The second component shifts from 244.7 eV to 245.0 eV as the cluster size increases. The energy of this component is very similar to the Ar 2p3/2->4s transition of argon physisorbed on Ru(001).[4] Component 3 has energies between 245.1 and 245.4 eV. The highest value is close to the 4s exciton in solid argon.[30,31] Similar spectral changes are also observed for the evolution of the 2p3/2->3d Rydberg state towards the corresponding exciton.[12] Differences occur between both exciton state with respect to their description: the 4s state corresponds to a localized Frenkel exciton, whereas the 3d exciton can be described with the Wannier-Mott model.[30] It is found that both exciton states develop similarly in the 2p core hole regime, whereas in valence shell excitation the 4s exciton develops faster than the 3d exciton.[27,32]

The energy regime above the 2p edges shows the development of undulations in the total electron yields as $\overline{N}$ is increased. These features have been interpreted as extended x-ray absorption fine structure (EXAFS) in argon clusters.[8,12,24] Additionally we observe fine structure superimposed on the broad EXAFS signal at $\overline{N}$ >100 which is interpreted as multiple scattering of the outgoing 2p electron by the surrounding argon neighbors (so-called XANES structure).[33] The XANES structure occurs at higher clusters sizes than the

EXAFS. A detailed study of the XANES structure as a function of average cluster size may provide complement information to electron diffraction of argon clusters.[34]

The EXAFS signal in the total electron yield curves has been analyzed using conventional techniques. Figure 2 compares the k-space EXAFS with that for Ar(2p) EXAFS calculated using the theoretical amplitude and phase functions of McKale[35] and Rehr et al. (Feff).[36] The agreement shown in Fig. 2 is obtained by using a k-scale origin (E_0) 5 eV below the experimental ionization threshold (246-248.6 eV)[4,31,32,37,38] Since inner potentials (E_0) can be as large as -15 eV, the -5 eV value is plausible.

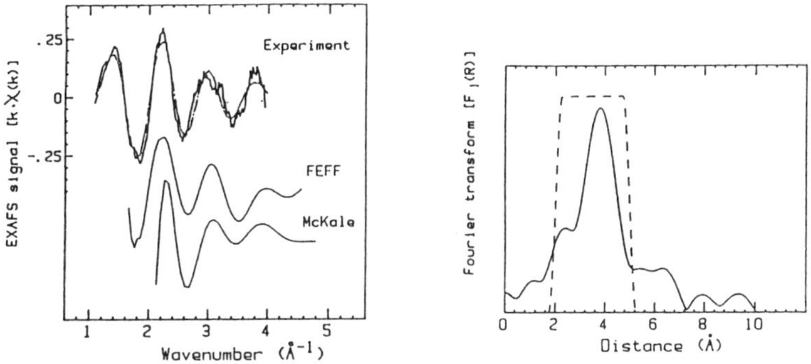

Figure 2: Analysis of the Ar 2p EXFAS from total electron yield spectra in comparison with the calculated EXAFS for the solid. See text for further details.

In addition the calculated EXAFS uses a coordination number of 12 and a Debye-Waller factor of 0.1 $Å^2$, which is much larger than expected simply from vibrational motion. This large value was needed to match the drop-off of the experimental EXAFS amplitude. It is therefore likely that clusters of different size and structure, which are present in the neutral cluster beam, contribute to a high degree of static disorder. An average Ar-Ar first shell distance of 3.78(5)Å was determined from the Ar 2p EXAFS of the clusters. This is in good agreement with the value for the solid (3.75 Å)[39] as well as results from mass selected argon clusters.[8] We have not found differences in nearest neighbor distance for different cluster sizes. This may be simply due to the limited energy range that could be analyzed in the case of Ar(2p) excitation. Measurements in the Ar(1s) regime should be more sensitive because of the larger accessible k-range. EXAFS experiments on free argon clusters above the Ar(1s) edge have been performed recently.[24] Inherent problems in this energy range are related to low cross sections as well as lower photon fluxes.

Core excitation of argon clusters leads to the formation of doubly charged clusters. Fragmentation pathways of argon cluster dications are determined by using the photoion-photoion-coincidence (PIPICO) technique.[9,40] This technique has been developed several years ago for studies on dissociative double ionization pathways of molecules.[41,42] Correlated pairs of cations are found in the case of argon clusters in the core excitation regimes[9, 24] as well as in the threshold region of dication formation.[40] The term 'Coulomb explosion' is frequently used for processes where high kinetic energy releases occur with electrostatically driven charge separation. A typical PIPICO spectrum of argon clusters is shown in Figure 3. The intense signal at $\Delta t=0$ corresponds to symmetric charge separations, where product ions of the same mass are formed. This assignment comes

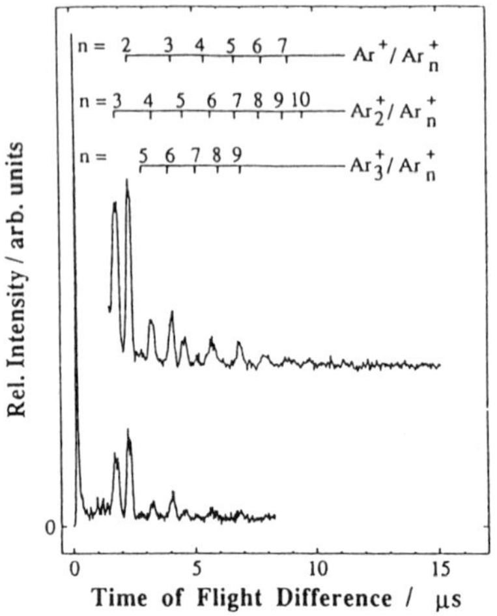

Figure 3: Photoion-photoion coincidence spectrum of argon clusters at 248 eV.

from recent PEPIPICO experiments, where the flight times of both singly charged fragments are measured.[22] In PIPICO spectra different symmetric charge separation routes are blended, because their time of flight difference is always zero. In the Ar(2p) PIPICO spectra one can identify three series of asymmetric charge separation pathways, where the lighter particle is either a singly charged monomer, dimer, or trimer. The other moiety in the charge separation process is a larger cluster cation. It is found that the series involving dimer cation formation is stronger in intensity than the other series. This is a result of the high stability of the dimer cation compared to the the neutral dimer.[43]

The kinetic energy release (KER) of charge separation in argon clusters can be estimated from the widths of the coincidence signals.[9] The widest part of the Ar^+/Ar_2^+ coincidence corresponds to a KER of 3.8(2) eV, if a two body dissociation of Ar_3^{++} is assumed to account for the formation mechanism. This corresponds to a charge separation distance of 3.7 Å which agrees with the EXAFS results. The maximum of the PIPICO distribution leads to a considerably lower KER. Possible reasons for this result are extensive evaporation of neutrals or internal excitation of the clusters. Recent PEPIPICO experiments indicate however, that the Ar^+/Ar_2^+ channel has contributions from sequential evaporation of neutrals after charge separation.[22] Low KER´s pose the question whether the simple picture of purely electrostatic charge separation applies for large clusters. This suggests that charge migration within the cluster is faster than spontaneous Coulomb explosion on the surface of the cluster. The internal energy that is left in the cluster after electronic relaxation is sufficient that singly charged cluster fragments are released with low KER. This assumption agrees with PEPIPICO results. The peak shapes indicate that the momenta of the singly charged fragments are not directed into opposite directions, as is typical for charge separation in small clusters and molecules.[22]

3.2 Nitrous Oxide Clusters

Figure 4 shows photoion yield curves of the nitrous oxide dimer cation in the energy regime of the nitrogen K-edge (390-460 eV) recorded at various stagnation conditions. The spectra are dominated by two distinct resonances below the nitrogen K-ionization energies. A considerable chemical shift of about 4 eV is found, where the 1s ionization limit of the terminal nitrogen N_t is observed at 408.75 eV and the central nitrogen N_c ionization is located at 412.62 eV.[44]

Therefore one expects two well separated N (1s)->π^* transitions which are found at 401.1 eV (N_t->π^*) and 404.7 eV (N_c->π^*), respectively.[45] A broad resonance centered at 422 eV has been assigned as the N(1s)->σ^* transition.

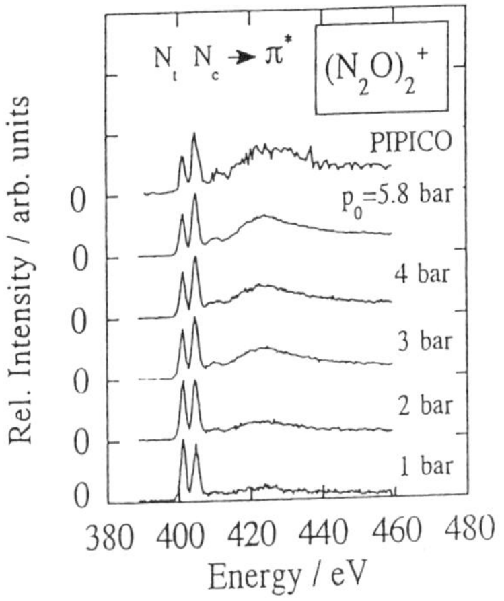

Photon stimulated desorption (PSID) experiments have shown that the energy position of the N(1s)->π* resonances is shifted relative to the isolated molecule.[46] We have therefore attempted to obtain similar results in cation yield curves of mass selected clusters. However, no visible shift is observed within the accuracy of the photon energy scale, which is found to be stable within 0.1 eV. Similar results are obtained from photoion yield curves of other cluster species[10] as well as for other molecular systems which were studied in the gas phase and the solid.[47] Figure 4 shows that there are considerable changes in relative intensity between both N(1s)->π* transitions in the dimer cation mass channel as the neutral cluster size is changed by increasing p_0. We observe an increase in the same sense in intensity above the nitrogen K-ionization potentials.

Figure 4: Photoion yield curves of the nitrous oxide dimer at different stagnation conditions and the $NO^+/(N2O)_2^+$ PIPICO yield (top spectrum)

These changes can be rationalized as follows: Small neutral clusters are present under mild expansion conditions, i.e. at low p_0 and high T_0. Under these conditions one observes the dimer (m/z=88) as the highest mass signal. The corresponding photoion yield of the dimer shows a more intense N_t->π* resonance and weak intensity in the N(1s) continuum (bottom curve in Fig. 4). It can be assumed that predominantly single ionization contributes to the dimer cation yield if only small amounts of larger neutral clusters are present in the beam. A similar behavior is observed for the isolated molecule: The parent cation N_2O^+ shows an intense N_t->π* resonance, whereas the N_c->π* intensity and the N(1s) continuum are weak. This finding is typical for all single ionization processes, because multiple ionization above the 1s ionization energies leading to multiparticle formation via dissociative double ionization. As the stagnation pressure is increased one observes that both N->π* resonances show the same height and the continuum intensity is increased. At the highest stagnation pressure (p_0=5.8 bar) the spectrum is remarkably similar to the PSID spectrum, where the N_c->π* is more intense than the N_t->π* resonance.[46] These changes in relative intensity are due to the increasing importance of dissociative double ionization via asymmetric charge separation of larger clusters, where small fragments such as dimer cations are formed (cf. Fig. 5). The dissociative double ionization yield of nitrous oxide clusters is also included in Fig. 4 (top spectrum). This spectrum resembles that of the dimer recorded at p_0=5.8 bar, which supports the assumption that charge separation dominates the photoion yields of mass selected small cluster fragments if large neutral precursors are present in the beam. PSID results have been interpreted in terms of the Knotek-Feibelman (KF) model.[48] In this model for electron stimulated desorption of cations from ionic solids it is assumed that a core hole is created in a primary step. Subsequent electronic relaxation via Auger decay leads to desorption of cations at a site which was initially negatively charged. From this one expects that PSID mimics photoabsorption spectra of the solid.[46] Electronic relaxation of nitrous oxide has been studied by de-excitation electron spectroscopy.[49] Different spectator autoionization decay routes are found. Interestingly, the ratio of single versus double ionization is almost unchanged at both N->π^* resonances. Differences in fragmentation of the isolated molecule have been explained by different de-excitation routes, where N_t->π^* populates the 2π cation ground state and N_c->π^* has contributions from the 1π excited cation state. This explains the more violent fragmentation into atomic and molecular species via single ionization at N_c->π^*. The Auger spectrum recorded at 418 eVexcitation energy is almost identical to the ordinary KLL-Auger spectrum, which is another indication of dominant double ionization processes in the N(1s) continuum.

Figure 5 shows the photoion-photoion-coincidence (PIPICO) spectrum of nitrous oxide clusters. The range of small time of flight differences is not shown, where charge separations of the bare molecule are found. We observe three series of cation pairs: (i) $NO^+/(N_2O)_n^+$, (ii) $N_2O^+/(N_2O)_n^+$, and (iii) $(N_2O)_2^+/(N_2O)_n^+$. Symmetric charge separation of clusters is identified via the PEPIPICO spectra.[22]

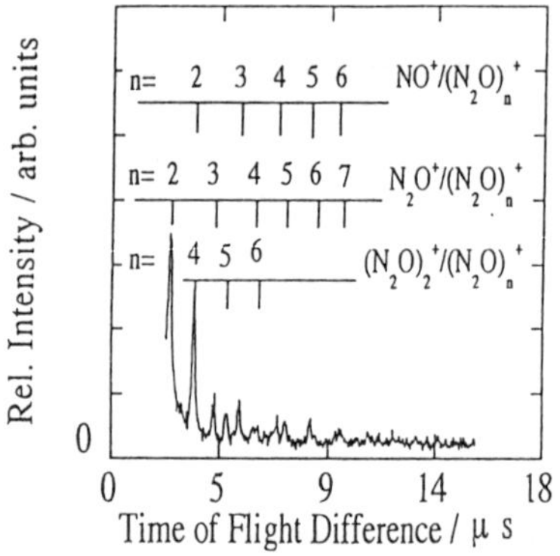

Figure 5: PIPICO spectrum of nitrous oxide clusters at 425 eV photon energy.

However, their relative strength compared to asymmetric charge separations is less intense than in the case of argon clusters. The KER´s indicate, as in the case of argon clusters, that they are lower than expected from the van der Waals distances.[10] This agrees with collision induced fragmentation of stable molecular clusters.[50]

3.3 Benzene Clusters

Figure 6 shows photoion yield curves of the benzene dimer (Bz_2^+) in the regime of the C(1s) excitation recorded at different stagnation conditions, as well as photoion yields for larger singly and doubly charged benzene cluster cations. The top spectrum shows the PIPICO yield of the Bz^+/Bz_2^+ charge separation channel. The spectral features in the 280-320 eV energy regime are well-known.[1] As in the case of nitrous oxide one observes all features in the gas and condensed phase,[51] although there are differences in relative intensities. The lowest energy feature is the C(1s)->π^* (e_{2u}) transitions at 285.2 eV. We find no shift of this resonance as a function of the cluster size. Three other maxima occur at higher energy: 288.9 eV (π^* (b_{2g})), 293.5 eV ($\sigma^*(e_{1u})$), and 300.2 eV ($\sigma^*(e_{2g}+a_{2g})$).[51] It should be noted that the feature occurring at 293.5 eV has been assigned as a doubly excited state decaying by autoionization.[52] The three photoion yield curves of Bz_2^+ show increased intensity in the C(1s) continuum as the cluster size is increased (see Fig. 6). The threshold of dimer formation is at about p_0=0.6 bar. The corresponding

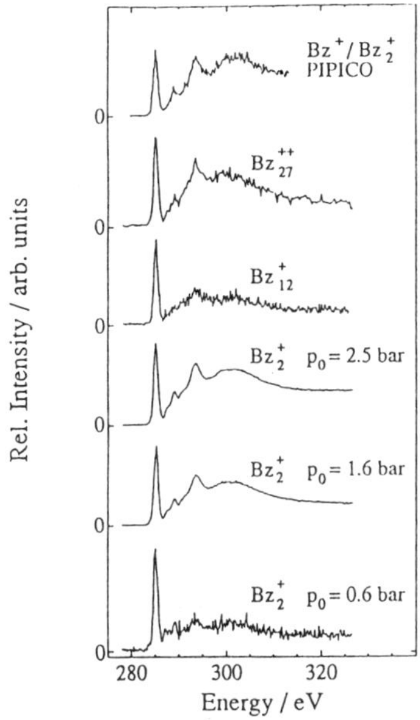

dimer spectrum consists mainly of a strong π* resonance and weak intensity at higher energy. A similar shape is observed for Bz^+ (m/z=78) under effusive jet conditions indicating that the dimer cation is formed as a result of single ionization, if no larger clusters are present in the beam. The photoion yields at p_0=1.6 bar and 2.5 bar are quite similar in shape but there is significantly greater intensity in the C(1s) continuum as a result of multiparticle fragmentation of C(1s) ionized large benzene clusters. The critical size for stable dications has been reported recently to be Bz_{23}^{++}.[50] The photoion yield of Bz_{27}^{++} is remarkably different from that of Bz_{12}^+ which is close in time of flight and both spectra are recorded at p_0=3.6 bar. The singly charged fragment shows small contributions from dissociative double ionization (as the dimer if no larger clusters are present), whereas the stable dication shows a spectrum of similar shape as the dimer cation yields and the PIPICO yield of the Bz^+/Bz_2^+ coincidences (top spectrum in Fig. 6).

Figure 6: Photoion yield curves of the benzene dimer (Bz_2^+) at different expansion conditions and the yield of the Bz^+/Bz_2^+ PIPICO channel.

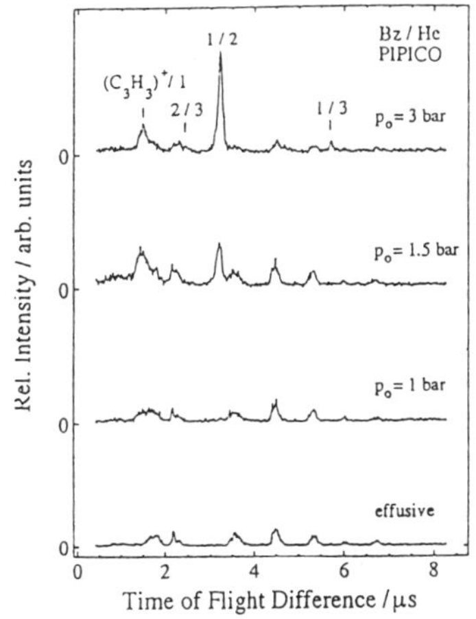

Figure 7 shows PIPICO spectra of benzene and its clusters recorded at different p_0. The spectra are normalized to the same number of starts in order to compare their relative intensities as the target density is increased with increasing p_0. The PIPICO spectrum of benzene has been studied by Richardson et al.[42] In agreement with this work we find the same PIPICO channels for the isolated molecule. The Bz^+/Bz_2^+ coincidence occurs at p_0>1 bar and is clearly visible at p_0=1.5 bar. Another cluster charge separation signal corresponds to $C_3H_3^+/Bz^+$, whereas the monomer/trimer (Bz^+/Bz_3^+) and dimer/trimer (Bz_2^+/Bz_3^+) fragmentations are considerably weaker in intensity even at the highest stagnation pressure (p_0=3.6 bar). Series of asymmetric charge separations are not observed, as in the case of

Figure 7: PIPICO spectra of benzene and benzene clusters recorded at 320 eV photon energy.

argon or nitrous oxide. This may be a result of the low critical size of stable doubly charged benzene clusters, which limits extensive evaporation of large neutral clusters.

4. Conclusions

Core level excitation of atomic and molecular clusters is a powerful technique to study spectroscopic changes from the isolated species to the condensed phase. Structural information is obtained from EXAFS in free clusters. Fragmentation of core excited clusters is investigated. Dissociative double ionization is the dominating process above the core ionization thresholds. These results are in agreement with photon stimulated desorption from the solids.

5. Acknowledgements

Financial support by the Bundesministerium für Forschung und Technologie (Grant No.: 05 5KEFXB 5 TP3) is gratefully acknowledged as is the support of a NATO travel grant.

6. References

1 A.P. Hitchcock, J. El. Spectrosc. Relat. Phenom. **25** (1982) 245, and update "Bibliography of Atomic and Molecular Inner-Shell Studies", Nov. 13, 1991.
2 D.M. Hanson, Advan. Chem. Phys. **77** (1990) 1
3 J. Stöhr, "NEXAFS Spectroscopy", Springer Ser. Surf. Sci., (Springer, Heidelberg, 1992)
4 D. Menzel, Appl. Phys. **A 51** (1990) 163
5 "X-Ray Absorption: Principles, Applications and Techniques", eds.: D.C. Koningsberger and R. Prins (Wiley, New York, 1988)
6 C.Y. Ng, Advan. Chem. Phys. **52** (1983) 263
7 "Structure and Dynamics of Weakly Bound Molecular Complexes", NATO-ASI Series **C 212**, ed. A. Weber (Reidel, Dordrecht, 1987)
8 E. Rühl, H.W. Jochims, C. Schmale, E. Biller, A.P. Hitchcock and H. Baumgärtel, Chem. Phys. Lett. **178** (1991) 558
9 E. Rühl, C. Schmale, H.W. Jochims, E. Biller, M. Simon, and H. Baumgärtel, J. Chem. Phys. **95** (1991) 6544
10 E. Rühl, Ber. Bunsenges. Phys. Chem., in press (1992)
11 E. Rühl, C. Schmale, H.W. Jochims, E. Biller, R. Locht, A.P. Hitchcock, and H. Baumgärtel, AIP Conference Proceedings, in press (1992)
12 E. Rühl, C. Heinzel, A.P. Hitchcock, and H. Baumgärtel, J. Chem. Phys., submitted (1992)
13 A. Ding and J. Hesslich, Chem. Phys. Lett. **94** (1983) 54
14 P. Scheier and T.D. Märk, Int. J. Mass Spectrom. Ion Proc. **76** (1987) R11
15 H.P. Birkhofer, H. Haberland, M. Winterer, and D.R. Worsnop, Ber. Bunsenges. Phys. Chem. **88** (1984) 207
16 W. Kamke, J. deVries, J. Krauss, E. Kaiser, B. Kamke, and I.V. Hertel, Z. Physik **D 14** (1989) 339; K. Mitsuke and K. Ohno, J. Phys. Chem. **93** (1989) 501
17 D. Kreisle, O. Echt, M. Knapp, E. Recknagel, K. Leiter, T.D. Märk, J.J. Saenz, and J.M. Soler, Phys. Rev. Lett. **56** (1986) 1551
18 A.J. Stace, P.J. Lethbridge, and J.E. Upham, J. Phys. Chem. **93** (1989) 333; N.G. Gotts and A.J. Stace, Phys. Rev. Lett. **66** (1991) 21
19 P. Scheier and T.D. Märk, J. Chem. Phys. **86** (1987) 3056
20 P. Scheier and T.D. Märk, Chem. Phys. Lett. **136** (1987) 423
21 M. Lezius and T.D. Märk, Chem. Phys. Lett. **155** (1989) 496
22 E. Rühl, C. Heinzel, H. Baumgärtel, M. Lavollée, and P. Morin, to be published

23 W.C. Wiley and I.H. McLaren, Rev. Sci. Instrum. **26** (1955) 1150
24 E. Rühl, C. Heinzel, H. Baumgärtel, H.C. Schmelz, C. Reynaud, W. Drube, R. Frahm, and A.P. Hitchcock, to be published
25 O.F. Hagena, Z. Physik **D 4** (1987) 291
26 J. Wörmer, Ph.D. thesis (University of Hamburg, Hamburg, 1990)
27 J. Wörmer, V. Guzielski, J. Stapelfeldt, G. Zimmerer, and T. Möller, Phys. Scr. **41** (1990) 490
28 O.F. Hagena and W. Obert, J. Chem. Phys. **56** (1972) 1793
29 G.C. King, M. Tronc, F.H. Read, and R.C. Bradford, J. Phys. **B 10** (1977) 2479
30 R. Haensel, G. Keitel, N. Kosuch, U. Nielsen, and P. Schreiber, J. Phys. (Paris) **C 4** (1971) 236
31 R. Haensel, N. Kosuch, U. Nielsen, U. Rössler, and B. Sonntag, Phys. Rev. **B 7** (1973) 1577
32 J. Wörmer, M. Joppien, G. Zimmerer, and T. Möller, Phys. Rev. Lett. **67** (1991) 2053
33 A. Bianconi in "X-Ray Absorption: Principles, Applications and Techniques", eds.: D.C. Koningsberger and R. Prins (Wiley, New York, 1988); D.D. Vvedensky, D.K. Saldin, and J.B. Pendry, Comp. Phys. Comm. **40** (1986) 421
34 J. Farges, M.F. deFeraudy, B. Raoult, and G. Torchet, J. Chem. Phys. **78** (1983) 5067; and J. Chem. Phys. **84** (1986) 3491
35 A.G. McKale, B.W. Veal, A.P. Paulikas, S.K. Chan, and G.S. Knapp, J. Am. Chem. Soc. **110** (1988) 3763
36 J. Mustre, Y. Yacoby, E.A. Stern, and J.J. Rehr, Phys. Rev. **B 42** (1990) 10843
37 W.L. Jolly, K.D. Bomben, and C.E. Eyermann, Atomic and Nuclear Data Tables **31** (1984) 433
38 O. Björneholm, Ph.D. thesis, (Uppsala University, Uppsala, 1992)
39 W. Niemann, W. Malzfeldt, P. Rabe, R. Haensel, and M. Lübcke, Phys. Rev. **B 35** (1987) 1099
40 E. Rühl, C. Schmale, H.C. Schmelz, and H. Baumgärtel, Chem. Phys. Lett. **191** (1992) 430
41 B. Brehm, and G. de Frenes, Int. J. Mass Spectrom Ion Phys. **26** (1978) 251; G. Dujardin, L. Hellner, D. Winkoun, and M.J. Besnard, Chem. Phys. **105** (1986) 291
42 P.J. Richardson, J.H.D. Eland, and P. Lablanquie, Org. Mass Spectrom. **21** (1986) 289
43 J. Jortner, Ber. Bunsenges. Phys. Chem. **88** (1984) 188
44 T.D. Thomas and R.W. Shaw, Jr., J. Electron. Spectrosc. Relat. Phenom. **5** (1974) 1081
45 G.R. Wight and C.E. Brion, J. Electron. Spectrosc. Relat. Phenom. **3** (1974) 191
46 R.A. Rosenberg, P.J. Love, P.R. LaRoe, V. Rehn, and C.C. Parks, Phys. Rev. **B 31** (1985) 2634
47 D. Menzel, G. Rocker, H.-P. Steinrück, D. Coulman, P.A. Heimann, W. Huber, P. Zebisch, and D.R. Lloyd, J. Chem. Phys. **96** (1992) 1724
48 M.L. Knotek and P.J. Feibelman, Phys. Rev. Lett. **40** (1978) 964; M.L. Knotek and P.J. Feibelman, Surf. Sci. **90** (1979) 78
49 F.P. Larkins, W. Eberhardt, I.-W. Lyo, R. Murphy, and E.W. Plummer, J. Chem. Phys. **88** (1988) 2948
50 N.G. Gotts, P.G. Lethbridge, and A.J. Stace, J. Chem. Phys. **96** (1992) 408
51 J.A. Horsley, J. Stöhr, A.P. Hitchcock, D.C. Newbury, A.L. Johnson, and F. Sette, J. Chem. Phys. **83** (1985) 6099
52 M.N. Piancastelli, T.A. Ferrett, D.W. Lindle, L.J. Medhurst, P. A. Heimann, S.H. Liu, and D.A. Shirley, J. Chem. Phys. **90** (1989) 3004

ELECTRONIC EXCITATIONS IN RARE GAS CLUSTERS STUDIED BY VUV FLUORESCENCE SPECTROSCOPY

M. Joppien, R. Müller, and T. Möller*

II. Inst. für Experimentalphysik, Universität Hamburg, FR Germany

**Hamburger Synchrotronstrahlungslabor HASYLAB at Deutsches Elekronensynchrotron DESY (Hamburg), FR Germany*

Abstract

The electronic structure of rare gas clusters has been probed with fluorescence methods. Some experimental aspects are discussed in connection with recent results on pure rare gas clusters. The discussion focuses on the character of electronically excited states (Rydberg states, bulk and surface excitons and cluster specific excitations) and the evolution with size.

Introduction

The development of the electronic structure as single atoms are put together to form a solid has been the subject of a great variety of research in the past decade [1]. In case of non-metallic clusters, the evolution of electronically excited energy levels is particularly appealing because a gradual transition from molecular valence and Rydberg states to the excitons of the solid is expected. Furthermore, the evolution of the ionization potential, of tightly bound valence and core levels is of considerable interest. Another interesting aspect concerns the fate of an electronic excitation, in other words, what happens after the electronic excitation has taken place. In this context, rare gas clusters play a key role because they are relatively easy to generate and due to their simple structure they may serve as a model system. The electronic excitations of rare gas clusters as well as other simple van der Waals bond species are located in the VUV spectral range. Therefore, most experiments in this field have been performed using synchrotron radiation for excitation. So far, measurements have been started to study ionization potentials [2-5], the level structure of electronic excitations [6,7] and of core levels [8] including also dynamic phenomena [9-11] These experiments have been performed using mass spectroscopic techniques in connection with coincidence methods [3,10] and fluorescence methods [11].

It has been shown recently, that the evolution of size of many basic features of clusters can be quantified by cluster size equations, which describe the gradual transition from a cluster to the infinite bulk system [12]. On the other hand, micro-clusters are not simply small solids but have unique properties [1,13], e.g.. the celebrated C_{60} is a beautiful example of this kind [14]. In the following, we present results recently obtained from time- and energy resolved fluorescence experiments. Fluorescence spectroscopy is suited very well for the examination of electronic excitations of van der Waals clusters because it allows a detailed investigation of the formation (excitation energy, oscillator strength) and the decay of the electronically excited states.

What makes rare gases especially interesting are tightly bound electronically excited states, namely excitons of the solid and Rydberg states of atoms and molecules which are to a certain extent equivalent. The radius r_e of these excitations (0.2 nm-10 nm, in principle up to infinity) covers the range of the radius of clusters containing up to 10^6 atoms. Therefore, strong variations of the observed spectral features with the cluster size are expected. In this article, we concentrate on the character of the primary excited states. Decay process are discussed in more detail elsewhere [11,15]. Recently, the experimental set-up has been considerably improved. As a result, the investigations could be extended to excitation of tightly bound valence levels of Ar_N and to excitations of Ne_N and He_N clusters.

Experimental

The measurements were performed at the experimental set-up Clulu (cluster luminescence) located at the high intensity VUV beamline Superlumi [16] at HASYLAB (Hamburg). Basically, state of the art experimental technique for VUV fluorescence measurement [17] is transferred to a molecular beam machine. However, some important changes have to be made to account for the specific constraints for doing experiments on clusters, e.g. the low particle density. The experimental arrangement is schematically outlined in figure 1. The set-up consists of a cluster beam, an atomic cross jet for doping the clusters and for mass separation, a time-of-flight mass spectrometer and several detection devices for the luminescence light.

Usually, large conical (convergent-divergent) nozzles (diameter between 80µm and 500µm, opening cone angle $2\Phi = 16^0$ - 30^0) are used for the preparation of the clusters in the size range from the dimer up to micro-crystals containing up to 10^6 atoms. After passing a skimmer, the cluster beam is crossed by the photon beam of monochromatized synchrotron radiation (band pass $\Delta\lambda$ =0.05nm - 0.25nm, $3*10^{10}$ - $2*10^{12}$ photons/s within the band pass in the spectral range 40nm - 160nm). The size distribution in the cluster beam is analysed with a time-of-flight mass spectrometer in a different set of measurements. Typically, the width ΔN (FWHM) of the size distribution corresponds to the average number N of atoms per cluster. The size of the clusters in the beam can be controlled by the choice of suitable condensation parameters (stagnation pressure p_0, nozzle temperature T_0). In order to prepare He_N clusters, the nozzle is mounted on a liquid Helium flow cryostat which

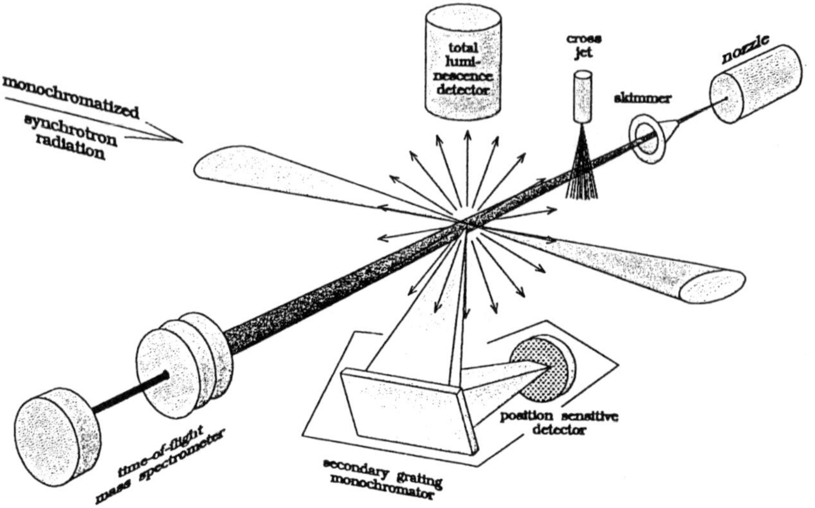

Fig. 1 Schematic illustration of the experimental set-up Clulu

can be operated between 4K and 400K.

The fluorescence light is detected either spectrally undispersed with different types of photomultipliers or analysed with a grating monochromator which is equipped with a position sensitive detector. Undispersed detection of the fluorescence light by scanning the excitation energy is favourable for the investigation of the primary excited states. Spectra recorded this way contain information which is in most cases equivalent to that of absorption spectra. However, the signal to noise ratio is considerably improved compared to the direct measurement of the absorption. Typically, counting rates in the scanning mode are between 10^2 and 10^6 photons per second.

The secondary monochromator takes the light spot of the synchrotron radiation as an entrance slit. The position sensitive detector is a multi-channel-plate detector with a resistive anode. In order to get an enhanced sensitivity at smaller photon energies (7 eV < hν <10eV), the first plate is coated with CsI. In case of pure rare gas clusters, total counting rates are in the order of 10 cps - 1000 cps, for doped clusters they are much smaller.

Results and Discussion

One of the main questions in cluster physics concerns the evolution of energy levels with size, e.g. the transition from atomic and molecular Rydberg states to the excitons of the solid. In both cases the excited electron moves around the positive charge. However, in the solid these bound electron hole states form bands classified according to their main quantum number n and the orientation of the polarization (transverse, longitudinal) relative to direction of propagation of the light [18]. As already pointed out in the introduction, there are three limiting situations according to the relationship between the radius r_e of the electronic excitation and the cluster radius R :

(I) $r_e < R$

(II) $r_e \cong R$

(III) $r_e > R$

Since the 'radius' of the excited orbital depends strongly on the main quantum number n, the different regimes might be present in clusters of a given size. In the following, we like to give an overview of the different kinds of excitations which are observed in pure rare gas clusters. As an example, absorption profiles of small Ar_N, Kr_N and Xe_N clusters are presented in figure 2. Energetically below the

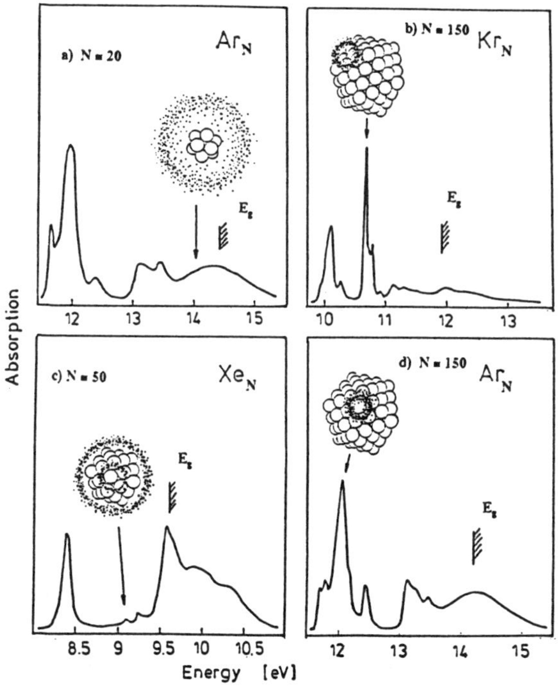

Fig. 2 Typical examples of absorption profiles of rare gas clusters and schematic illustration of the character of electronic excitations. The clouds represent the radial electron density.

ionization energy E_g strong absorption bands are observed. Some of them are rather sharp. The character of the electronically excited state is schematically visualized in the figure. The assignment of the different absorption bands comes from a comparison with the bands in rare gas solids and from that observed in gas phase experiments. In addition, model calculations have been performed which support the assignment [19,30].

Energetically close to the ionization limit a broad continuum is observed which is assigned to molecular Rydberg states (figure 2a)). Here, the excited electron is weakly bound in a state characterized by large main quantum numbers and an extended orbital with a radius much larger than the cluster radius R (regime III). Therefore, the excited electron is located mainly outside the cluster. The onset of absorption is dominated by several strong absorption bands which are assigned in accordance with corresponding bands in the solid either to tightly bound bulk (figure 2d)) or surface (figure 2b)) excitons. Depending on the 'radius' of the orbital they are classified as n=1 Frenkel or intermediate type or for larger main quantum numbers as Wannier type excitons. In case of surface excitons, the excitation takes place mainly in thd first and second monolayer of the cluster depending on the size of the orbital[6,20] . Excitonic absorption bands are only observed if the cluster radius R is considerable larger than the exciton radius [20].

In addition, small clusters ($N<500$) show cluster specific absorption bands, which are not observed in the solid. Some of them are rather sharp with width significantly smaller than that of excitonic absorption bands (figure 2c)). They are assigned to so called 'cluster excitons' [30] where the 'radius' of the excitation is comparable to the cluster radius R (regime II). While the hole is located in well defined shells of the cluster, the electron has a high probability to stay inside as well as outside the cluster. Consequently these excitations have a character in between molecular Rydberg states and excitons.

The evolution of excitonic energy levels with the cluster size is presented in figure 3 and 4. Figure 3 shows fluorescence excitation spectra of Kr_N clusters in the energy range of n =1 and n =1' excitons which correspond to the first atomic excitations ($4p \rightarrow 5s$, 5s', the prime denotes the spin-orbit partner). These states are split into transverse (t) and longitudinal (l) branches. Since the radius of these tightly bound excitations is small compared to the nearest-neighbour distance they are similar to atomic excitations (Frenkel exciton). Bulk and surface states are clearly separated. Corresponding bands are observed in the cluster. While the peak position changes only slightly with size the intensity of the absorption depends sensitively on the size. The contribution of bulk excitons drastically decreases with decreasing cluster size. In particular, small clusters ($N \cong 10$) only show surface excitations. More quantitatively, atoms located in the surface contribute to n=1 surface states whereas atoms located in the interior of the cluster contribute to n=1 bulk excitons [6]. Interestingly, in this energy range the shape of absorption profiles of small clusters ($N \cong 3$-10) is similar to that of surface excitons.

If weakly bound states characterized by extended orbitals are considered the situation becomes different. In this case, there is a minimum cluster size which is required for the formation of loosely bound Wannier excitons. Figure 4 shows fluorescence excitation spectra of Ar_N clusters covering the energy range from the first electronic excitation up to far above the ionization limit. Absorption bands in

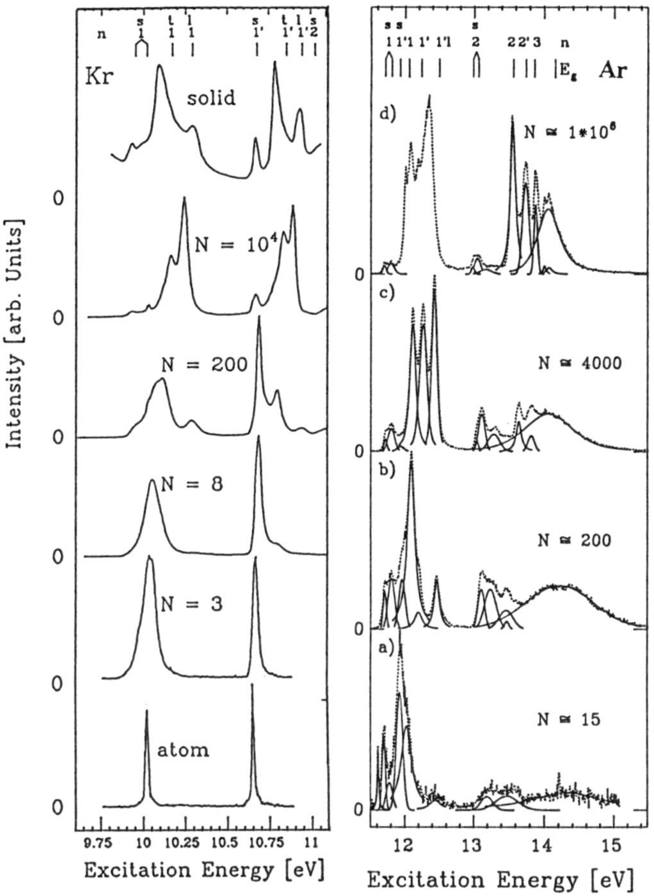

Fig. 3 and Fig. 4. Fluorescence excitation spectra of Kr_N (Fig.3, data taken from ref. 6 and ref. 31) and of Ar_N clusters (Fig.4, from ref.19). The peak position of excitonic absorption bands and the band gap energy E_g are given in top of the figures (from ref. 18).

large clusters have a 1:1 correspondence in the solid. In particular, with increasing cluster size the Wannier excitons ($2 \leq n$) show up . Again, the intensity of excitonic absorption bands sensitively depends on the cluster size and is directly related to the number of atoms in the interior of the cluster which are separated from the surface by a dead layer of the thickness d_n. The thickness of this layer is approximately equal to the radius of the corresponding exciton with the main quantum number n [20]. Therefore, weakly bound excitons (n=3 and n=4) are only observed in fairly large clusters. The energetic position of the excitonic absorption slightly changes with size. For the Wannier excitons the energetic shift relative to the bulk limit can be interpreted in terms of a confinement of the electronic

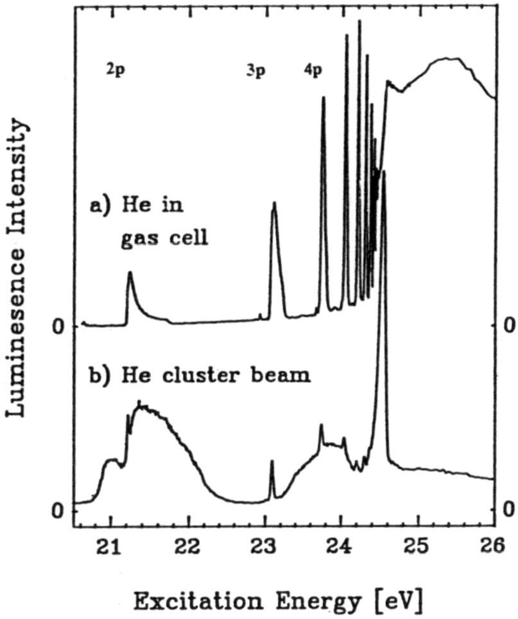

Fig. 5. Fluorescence excitation spectra of the total fluorescence of a) dense He gas (270 mbar, 300K) enclosed in a cell b) a heavily condensed He cluster beam (20μm nozzle, 7.5 bar, T=5K) [23] . The sharp lines are due transitions 1s→ np.

excitation in the cluster [19]. While bulk states are confined in three dimensions the confinement of surface states takes place in two dimensions. These differences can quantitatively describe the observed energetic shifts which are roughly proportional to R^{-2}.

While the heavy rare gases (Ne -Xe) have similar properties, Helium is in several respects a special case. Helium remains liquid even at 0 K and has many interesting thermodynamic properties, e. g. a suprafluid phase (^4He) below 2 K [21]. Furthermore, the electronic structure is very simple. Helium has the largest ionization energy of all materials. As a result of these unusual properties, the knowledge of the optical properties of condensed He is rather poor (there are no window material available which can withstand the large vapour pressure of liquid He). Up to now, the absorption spectrum of liquid or solid He has not been measured. Even the reflectivity is known only in small segments energetically in between a few atomic resonance lines [22]. Therefore, He_N clusters in a beam offer the unique possibility to do spectroscopy on liquid He. Figure 5 shows a comparison between the fluorescence yield of dense He gas enclosed in a gas cell and a heavily condensed cluster beam [23]. The clusters which are grown from the liquid phase have an estimated cluster size of $N \cong 10^5$. The absorption bands in the dense gas are due to molecular wings on the high energy side of the atomic resonance lines $1s\ ^1S_0 \rightarrow np\ ^1P_1$. These atomic excitations are also visible as sharp lines in the spectrum of the cluster beam sitting on top of the absorption bands of the cluster. The absorption bands of He_N clusters are in the same energy range, however, they have an unusual shape and are very broad. The bands close to the atomic 2p state state at 21.2 eV seem to be related to atomic 2s and 2p levels and can be interpreted in a molecular description. To a certain extent, they are equivalent to the n =1 exciton in solid Ar or Ar_N clusters. At higher energies (23.2 eV- 26 eV) broad absorptions bands are observed, which have no atomic parentage. On the other hand, these bands are completely different from Wannier excitons in large Ar_N clusters (see fig. 4d)). Moreover, the energetic position of exited states calculated in the Wannier model are far away from the observed maxima [23]. This means, that in contrary to the other rare gases (clusters, liquid or solid phase) excited states in He_N clusters can not be interpreted in the Wannier model. Presumably the radius of the excitons is smaller than the nearest neighbour distance even for excitons up to n=3, so that the Wannier model can not be applied.

In the following section some recent results concerning the decay of clusters are presented, in particular after excitation energetically far above the ionization threshold. In the preceding sections it was assumed that the measured fluorescence yield is roughly proportional to the number of absorbed photons. This should only be true as long as the excitation energy is smaller than the ionization energy. Above the ionization limit the fluorescence yield is expected to fall off very rapidly because the photo-ionization process is much faster than electronically relaxation into states below the ionization limit. Indeed, the fluorescence intensity decreases very rapidly above the ionization limit and approaches nearly zero. However, energetically far above the ionization limit the fluorescence yield again increases

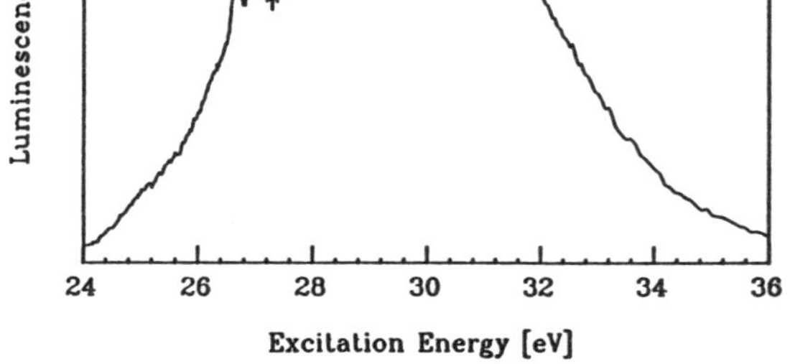

Fig.6. Fluorescence excitation spectrum of small Ar_N clusters energetically far above the ionization limit. The resonances are due to transitions $3s \rightarrow np$. [24]

[24]. The energy of the onset is approximately given by the sum of the ionization energy and of the energy which is needed to create an exciton. Consequently, it is assumed that the radiation is emitted from a center embedded in an ionized cluster. Interestingly, fluorescence excitation spectra recorded in this energy range show some fine structure (see figure 6).

This structure is due to excitation into discrete transitions ns → n'p (n=3 for Ar) from tightly bound valence levels into Rydberg-states. These resonances are well known for rare gas atoms as well as solid rare gases. While in the latter two cases these resonances are prototypes of window resonances with an asymmetry parameter q close to zero [25], fluorescence excitation spectra of Ar_N and Kr_N clusters can not be described with a small value of q. This means that the coupling between the two different excitation or relaxation channels changes with cluster size, but in the solid state limit it is again very similar to that in the free atom.

Detailed information on the relaxation and decay processes can be obtained from spectrally and/or time resolved fluorescence spectra. These technique has been successfully applied to $XeAr_N$ clusters [11, 26]. Important questions in this context concern the formation of atomic and molecular centers inside the cluster and desorption and fragmentation processes. Figure 7 shows fluorescence spectra of small Ar_N clusters excited at different energies. After primary excitation into

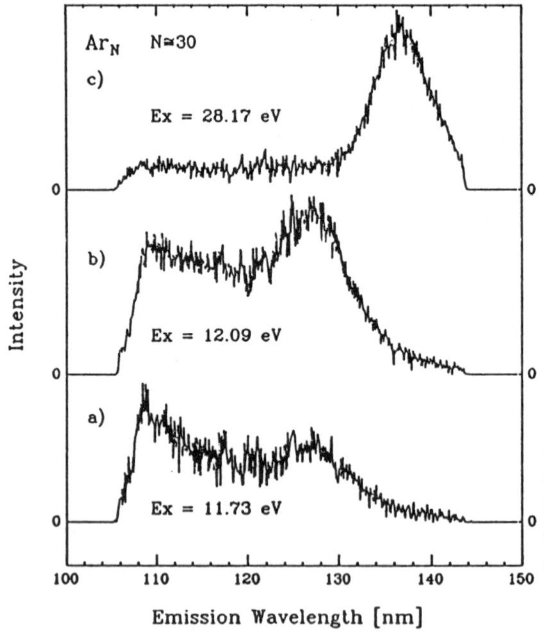

Fig. 7. Fluorescence of Ar_N clusters excited energetically below (a); b)) and above the ionization limit (c)) [24].

tightly bound exciton states broad bands corresponding to the well known emissions of solid Ar are observed [27]. They are assigned to a radiative decay of tightly bound excimer centers (M-band at 127nm) inside the cluster or excimers desorbing in high vibrational levels from the cluster surface (W-band at 109 nm). After excitation far above the ionization energy a different fluorescence band is observed which is not reported for solid Ar. It is assigned to the decay of excimer centers in an ionic cluster. Due to the polarization of the cluster by the ion inside the cluster the transition energy is shift compared to the M-band [28]. The assignment of the observed emission to the decay of excimers in ionic clusters is further supported by mass spectrometric measurements [29].

Conclusion

The examples presented in the preceding section have shown that fluorescence methods are a versatile tool for investigations of the electronic structure and the relaxation processes of rare gas clusters in a beam. Fluorescence spectroscopy offers the possibility to study systems of fundamental interest like liquid He. For future work it should be interesting to apply these methods to other interesting species e.g. doped He_N clusters. Furthermore, due to the large size range the results can substantially contribute to questions of current interest in the field of physics of clusters and small particles.

Acknowledgement

The authors like to thank colleagues from the Universität Hamburg which contribute substantially to the research in this field. In particular, helpful discussions and contributions of J. Wörmer, J. Stapelfeldt, V. Guzielski, M. Lengen, V. Heitmann, R. von Pietrowski, J. Szulkiewicz, S. Wolf and G. Zimmerer are kindly acknowledged. Furthermore, the authors like to thank P. Gürtler (HASYLAB) for critically reading the manuscript. Financial support from the Bundesministerium für Forschung und Technologie (BMFT) is kindly acknowledged.

References

[1] E. R. Bernstein, *Atomic and Molecular Clusters,* Elsevier Since Publishers, (Amsterdam 1990)
[2] P.M. Dehmer, S. T. Pratt , J.Chem.Phys.76, 843(1982)
[3] G. Ganteför, G. Bröker, E. Holub-Krappe, A. Ding, J.Chem.Phys.91, 7072 (1989)
[4] E. Holub-Krappe, G. Ganteför, G. Bröker, A. Ding, Z.Phys.D10, 319 (1988)
[5] W. Kamke, J. de Vries, J. Krauss, E. Kaiser, B. Kamke, I.V. Hertel, Z.Phys. D14, 33 (1989)
[6] J. Stapelfeldt, J. Wörmer, T. Möller, Phys.Rev.Lett. 62, 98 (1989)

[7] T. Möller, Z. Phys. D 20, 1 (1991)
[8] E. Rühl, H.W. Jochims, C. Schmale, E. Biller, A.P. Hitchcock, and H. Baumgärtel, Chem. Phys. Lett. 178, 558 (1991)
[9] M. Fieber, G. Bröker, and A. Ding, Z. Phys. D 20, 21 (1991)
[10] E. Rühl, C. Schmale, H.W. Jochims, E. Biller, A.P. Hitchcock, and H. Baumgärtel, J. Chem. Phys. 95, 6544 (1991)
[11] M. Lengen, M. Joppien, J. Wörmer, T. Möller , Phys. Rev. Lett. 68, 2362 (1992)
[12] J. Jortner, Z. Phys. D, in press
[13] O. Echt, E. Recknagel (eds) Small Particles and Inorganic Clusters ISSPIC 5, Z. Phys. D 19 & 20, (1991)
[14] H.W. Kroto, J.R. Heath, S.C. O'Brien, R.F. Curl, and R.E. Smalley, Nature 318, 162 (1985)
[15] M. Joppien, F. Grotelüschen, M. Lengen, R. Müller, J. Wörmer and T. Möller, Proceedings of the International Conference *Synchrotron Radiation and Dynamic Phenomena*, Grenoble 1991
[16] H. Wilcke, W. Böhmer, R. Haensel, N. Schwentner, Nucl.Instr.Meth. 208, 59 (1983)
[17] T. Möller, G. Zimmerer, Phys. Scr. T17, 177 (1987)
[18] N. Schwentner, E.E. Koch, J. Jortner, Electronic Excitations in Condensed Rare Gases, (Springer, Berlin 1985)
[19] J. Wörmer, M. Joppien, G. Zimmerer, and T. Möller , Phys. Rev. Lett. 67, 2053 (1991)
[20] J. Wörmer, T. Möller, Z.Phys.D 20, 39 (1991)
[21] M.L. Klein, J.A. Venables (eds.), *Rare Gas Solids*, Vol.1,2 (Academic Press, London, 1976, 1977)
[22] C.M. Surko, G.J. Dick, F. Reif and W.C. Walker, Phys. Rev. Lett. 23, 842 (1969)
[23] M. Joppien, R. Müller, T. Möller, to be published
[24] R. Müller, M. Joppien, T. Möller, to be published
[25] R.P. Madden, D.L. Ederer, and K. Codling, Phys. Rev. 177, 136 (1969)
[26] M. Lengen, Diplom-thesis, Hamburg 1990
[27] M. Joppien, F. Grotelüschen, T. Kloiber, M. Lengen, T. Möller, J. Wörmer, G. Zimmerer, J. Keto, M. Kytka, M.C. Castex , J. Luminesc. 48&49, 601 (1991)
[28] E.A. Bondarenko, E.T. Verkhovtseva, Yu.S. Doronin and A.M. Ratner, Chem. Phys. Lett. 182, 637 (1991)
[29] M. Foltin, G. Walder, A.W. Castleman Jr, T.D. Märk, J. Chem. Phys. 94, 810 (1991)
[30] M. Joppien, R. Müller, J. Wörmer, and T. Möller, Phys. Rev., submitted
[31] V. Saile, in ref. 18, page 44

ION ROTATIONAL DISTRIBUTIONS AT NEAR-THRESHOLD PHOTOELECTRON ENERGIES

KWANGSHI WANG, JEFFREY STEPHENS, AND VINCENT MCKOY
A. A. Noyes Laboratory of Chemical Physics
California Institute of Technology
Pasadena, CA 91125

Rotationally resolved photoelectron spectra can provide significant insight into the underlying dynamics of molecular photoionization. Recent advances in experimental techniques now make it possible to readily achieve rotational resolution in molecular photoelectron spectra. Here we present results of theoretical studies of rotationally resolved photoelectron spectra at low and very near-threshold energies for molecules such as HBr, NH, NO, and CO. These studies serve to reveal the rich dynamics of quantum-state specific studies of molecular photoionization and, where possible, to provide a robust description of key spectral features of interest in related experimental studies.

1. Introduction

Rotationally resolved photoelectron spectra and their associated photoelectron angular distributions can clearly provide much insight into the dynamics of molecular photoionization, one of the simplest of molecular fragmentation processes. These spectra are an obvious signature of the exchanges of energy and angular momentum between the photoelectron and molecular ion. Unfortunately, with the exception of hydrogen[1,2] and ionization from higher rotational levels of lighter systems such as NO,[3,4] OH,[5,6] and NH,[7] rotational resolution in photoelectron spectra has generally been beyond the reach of the techniques of conventional photoelectron spectroscopy. With the recent development of zero-kinetic-energy (ZEKE) photoelectron spectroscopy based on pulsed-field ionization (PFI) of very high Rydberg levels, however, it is now possible to exploit the narrow bandwidth of laser radiation to achieve rotational resolution in molecular ion spectra.[8] The unprecedented resolution of this technique has led to a surge of experimental activity in molecular photoelectron spectroscopy.[9] The underlying dynamics of such threshold photoelectron spectra can be expected to be rich in content and to present new theoretical challenges. Furthermore, laser-induced fluorescence (LIF) studies provide higher resolution than energy-resolved photoelectron measurements and may allow, for example, monitoring the populations of Λ components of ion rotational levels.[10]

In this paper we present some results of our studies of rotationally resolved photoelectron spectra at low and near-threshold photoelectron energies for HBr, NH, NO, and CO. These examples will include spectra at low photoelectron energies resulting from resonance enhanced multiphoton ionization (REMPI) of a single rotational level of HBr and NH and ZEKE photoelectron spectra for single-photon ionization of the ground states of NO and CO by coherent VUV and XUV radiation, respectively. These examples will serve to illustrate two significant objectives of our studies. These objectives are to provide insight into the rich underlying dynamics of highly resolved molecular photoelectron spectra and, where possible, a robust description of key spectral features of interest in related experimental studies.

2. Formulation and Calculations

The general theory of molecular photoionization processes used in the present studies has been described previously.[11,12] Here we present just a very brief outline of some essential features of our procedure as it is used to obtain ion rotational distributions and photoelectron angular distributions for (n+1) REMPI of diatomic molecules. We view (n+1) REMPI of a molecule as n-photon excitation from an initial unaligned (all M_J levels equally populated) ground state to an aligned intermediate state (levels with different $|M_J|$ values have different populations) which is subsequently ionized via absorption of another photon. For linearly polarized light, ionization originating from each of the $(2J_0 + 1)$ magnetic sublevels of the initial state forms an independent channel. The rotationally resolved differential cross section for ionization of a J level of the resonant intermediate state can be expressed in terms of Legendre polynomials as

$$\frac{d\sigma}{d\Omega} \propto \sum_{M_J, M_{J_+}} \rho_{M_J M_J} |\Gamma_{M_J M_{J_+}}|^2$$

$$= \frac{\sigma}{4\pi}\left[1 + \sum_{L=1}^{n+1} \beta_{2L}\, P_{2L}(\cos\theta)\right], \qquad (1)$$

where σ is the total cross section, β_{2L} the asymmetry parameters and $P_{2L}(\cos\,\theta)$ the Legendre polynomials. In Eq. (1) $\rho_{M_J M_J}$ is the population of a specific M_J level in the optically aligned resonant state and $|\Gamma_{M_J M_{J_+}}|^2$ is the probability for photoionization of the M_J level of the resonant state $|i\rangle$ leading to the M_{J_+} level of the ionic state $|f\rangle$. An expression for $\Gamma_{M_J M_{J_+}}$ which explicitly considers the spin coupling associated with multiplet-specific final-state wave functions and an intermediate coupling scheme between Hund's cases (a) and (b) for the resonant and ionic states has been given by Wang and McKoy.[12] This ionization probability is proportional to $|\langle f_{\beta,M_{J+}}|D_{\mu_0}|i_{\alpha,M_J}\rangle|^2$ where D_{μ_0} is the dipole operator in the laboratory frame and $\alpha, \beta = e/f$ designates parity.[12]

A central quantity in these studies is the matrix element for photoejection of an electron from a bound molecular orbital ϕ_i into a photoelectron continuum orbital $\Psi^{(-)}_{f,\mathbf{k}}(\mathbf{r})$. Here $\mathbf{k}$ is the momentum of the photoelectron and $(-)$ denotes incoming-wave boundary conditions. The partial wave components $\psi^{(-)}_{k\ell m}$ of $\Psi^{(-)}_{f,\mathbf{k}}(\mathbf{r})$ are defined by an expansion in spherical harmonics about $\hat{\mathbf{k}}$

$$\Psi^{(-)}_{f,\mathbf{k}}(\mathbf{r}) = \left(\frac{2}{\pi}\right)^{\frac{1}{2}} \sum_{\ell m} i^{\ell}\ \psi^{(-)}_{k\ell m}(\mathbf{r}) Y^*_{\ell m}(\hat{\mathbf{k}}). \tag{2}$$

Single-center expansions of $\psi^{(-)}_{k\ell m}(\mathbf{r})$ and $\phi_i(\mathbf{r}')$, e.g.,

$$\psi^{(-)}_{k\ell m} = \sum_{\ell'\lambda} g_{\ell m,\ell'\lambda}(k,r)\ Y_{\ell'\lambda}(\hat{\mathbf{r}}'), \tag{3}$$

define partial wave photoelectron matrix elements $r^{\ell\lambda\mu}_{fi}(R)$ in the molecular frame for ionization out of orbital $\phi_i(\mathbf{r}')$, i.e.,

$$r^{\ell\lambda\mu}_{fi}(R) = \sum_{\ell',\ell_o} \langle g_{\ell m,\ell'\lambda}(k,r,R) Y_{\ell'\lambda}(\hat{\mathbf{r}}')|r\ Y_{1\mu}(\hat{\mathbf{r}}')|\phi_{i\ell_0}(r,R) Y_{\ell_0,\delta_0}(\hat{\mathbf{r}}')\rangle, \tag{4}$$

where R denotes a dependence on internuclear distance. Eq. (4) reveals an important underlying dynamical aspect of molecular photoelectron wave functions. Whereas only $\ell = \ell'$ terms are allowed in Eq. (4) for the central fields of atomic systems, where the angular momentum of the photoelectron must be conserved, $\ell \neq \ell'$ terms arise in Eq. (4) due to the nonspherical potential fields of molecular ions. This angular momentum coupling between partial waves ℓ and ℓ' is brought about by the torques associated with the molecular ion potential and makes a molecular photoelectron orbital an admixture of angular momentum components. Viewed very simply, this means that whatever angular momentum ℓ a photoelectron may have at a given instant can be changed as the photoelecron "scatters" off the molecular ion. These angular-momentum changing collisions between the photoelectron and molecular ion play a crucial role in rotationally resolved molecular photoelectron spectra. The use of molecular photoelectron orbitals which correctly incorporate such angular momentum coupling is essential at the low photoelectron energies of interest here. For nonlinear molecules this angular momentum coupling becomes more pervasive and terms with $\ell m \neq \ell' \lambda$ arise in Eq. (4).[13]

In applications of our procedure to REMPI of Rydberg states we use the improved virtual orbital (IVO) method[14] to obtain the wave functions of the resonant state. The core orbitals of these Rydberg states are taken to be those of the fully relaxed ion. For the final state we assume a frozen-core Hartree-Fock model in which the core orbitals are taken to be those of the ion and the photoelectron orbital is obtained as a solution of a one-electron Schrödinger equation containing the Hartree-Fock potential of the molecular ion, $V_{ion}(\mathbf{r}, R)$, i.e.,

$$\left(-\frac{1}{2}\nabla^2 + V_{ion}(\mathbf{r}, R) - \frac{k^2}{2}\right)\psi^{(-)}_{k\ell m}(\mathbf{r}, R) = 0. \tag{5}$$

To obtain the partial wave photoelectron orbitals $\psi^{(-)}_{k\ell m}$, we use an iterative procedure, based on the Schwinger variational principle, to solve the Lippmann-Schwinger equation associated with Eq. (5).[15] A few iterations of this iterative variational method provide highly converged solutions for these molecular photoelectron orbitals. In studies of rotationally resolved photoelectron spectra at low photoelectron energies it is essential to use such converged solutions so as to faithfully represent the angular momentum coupling present in the molecular photoelectron wave function.

3. Results

As our first example we look at the rotational distributions of HBr^+ ions in their X $^2\Pi_{1/2}$ ground state resulting from (2 + 1) REMPI of HBr via the S(2) branch of the $F\ ^1\Delta_2(5p\pi)$ Rydberg state.[16] These ion distributions are of specific interest to us because they have been recently characterized by Xie and Zare using laser-induced fluorescence.[10] The high resolution provided by LIF made it possible to monitor the populations of the Λ components of opposite parity of the rotational levels of the ion.[10] These spectra show a strongly (−) parity-favored ion rotational distribution for photoionization of the (+) parity component of the J = 4 level of the $5p\pi$ Rydberg orbital of the $F\ ^1\Delta_2$ state which has about 97% p character. These (−) parity-favored ion distributions can be readily understood on the basis of the parity selection rule,[16–18] $(J_+ - J) + (S_+ - S) + (p_+ - p) + \ell = even$, which applies here, and the expected dominant even (ℓ = 0, 2) partial wave character of the photoelectron matrix element. p_+ and p are the parity indices for the ion and resonant states.

In Fig. 1 we compare our calculated ion rotational distribution for (2+1) REMPI of HBr via the S(2) branch of the $F\ ^1\Delta_2$ state with the experimental data of Xie and Zare.[10] The photoelectron energy here is about 2.33 eV. Calculated spectra are shown for photoionization of both the appropriately aligned J = 4 level and an unaligned level. The agreement between the calculated and measured ion distributions is very encouraging. Note that the 20% population seen in the (+) parity component of the Λ doublet is due to odd partial wave components of the photoelectron matrix elements which arise from strong ℓ mixing in molecular photoelectron orbitals. Such behavior is entirely nonatomic-like.[16,19] Comparison of the ion distributions for an unaligned J = 4 level with those for an aligned level and with the measured spectra serves to illustrate that, although not large, the effect of alignment is important.

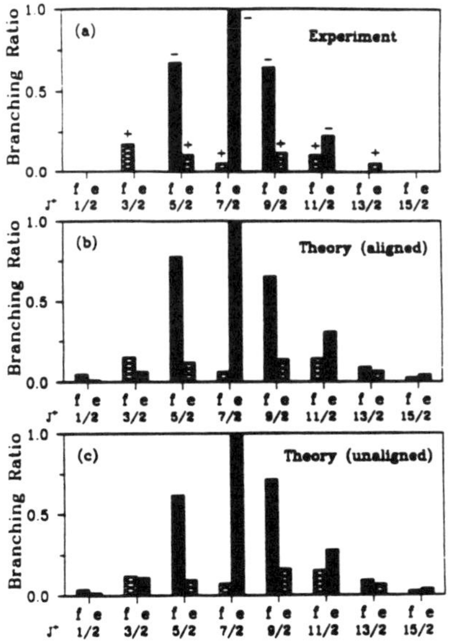

Fig. 1. Ion rotational distributions for HBr^+ resulting from (2+1) REMPI via the S(2) branch of the f $^1\Delta_2$ state: (a) measurements of Ref. 10; (b) and (c) calculated distributions for the aligned and unaligned resonant states, respectively, of Ref. 16.

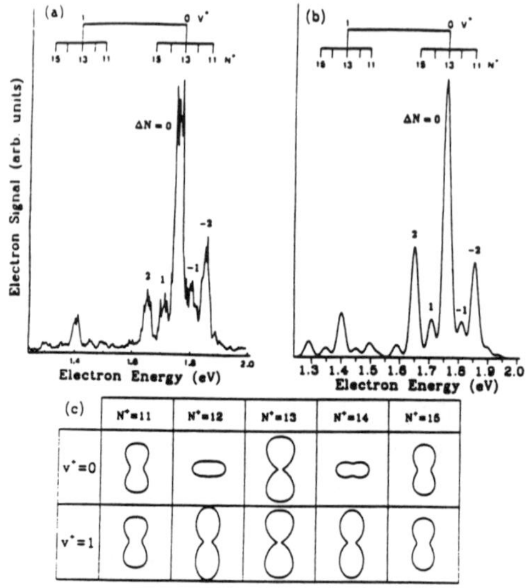

Fig. 2. (a) measured and (b) calculated photoelectron spectra along with the (c) calculated photoelectron angular distributions resulting from (2+1) REMPI via the R(12) branch of the f $^1\Pi(3p\sigma)$ state of NH for the 0−0 and 0−1 vibrational bands. Data taken from Ref. 7.

Fig. 2 shows (a) measured and (b) calculated rotationally resolved photoelectron spectra along with (c) calculated photoelectron angular distributions resulting from (2+1) REMPI via the R(12) branch of the f $^1\Pi(3p\sigma)$ state of NH for the 0−0 and 0−1 vibrational bands.[7] The calculated spectra are convoluted with a Gaussian detection function having an FWHM of 30 meV. In these spectra the spin-orbit components of the $X\ ^2\Pi$ state of NH^+ and the parity components of the resonant and ionic states are not resolved. The most striking feature of these spectra is the appearance of strong $\Delta N = N_+ - N' = even$ peaks, particularly for $\Delta N = 0$, for photoionization of this $3p\sigma$ Rydberg orbital of NH. On the basis of the simple selection rule $\Delta N + \ell = odd$ which applies here,[7] these even ΔN ($= 0, \pm 2$) peaks must be associated with odd ($\ell = 1, 3$) partial wave components of the photoelectron matrix element. Here ΔN is the change of angular momentum, exclusive of spin, between the resonant and ionic states. With its 67% p character one would expect significant contributions from $\ell = 0$ and 2 partial waves in photoionization of this resonant orbital. The strong even ΔN peaks in these spectra can, in fact, be shown to be due to the presence of low-energy Cooper minima in the $\ell = 2$ wave of the photoelectron channels.[7] Depletion of the d wave ($\ell = 2$) contribution to the photoelectron matrix element in the vicinity of these Cooper minima subsequently enhances the relative importance of the odd ℓ waves and, hence, that of the even ΔN rotational peaks. Similar effects of Cooper minima on rotationally resolved photoelectron spectra have also been observed for REMPI via the D $^2\Sigma^-(3p\sigma)$ state of OH[6] and the D $^2\Sigma^+(3p\sigma)$ state of NO.[20] The presence of Cooper minima in photoionization of excited states of diatomic molecules was first predicted for (3+1) REMPI of the D $^2\Sigma^-$ state of OH.[21] Finally, Fig. 2(c) shows the photoelectron angular distributions for the ion rotational levels in the spectra of Fig. 2(b). These angular distributions reflect the underlying angular momentum composition of the

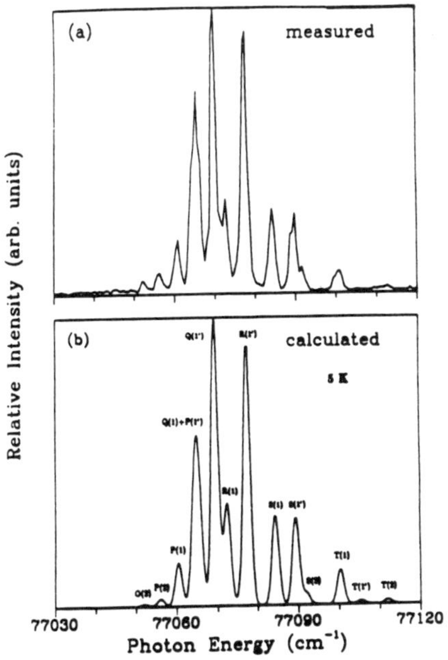

Fig.3. (a) measured and (b) calculated ZEKE photoelectron spectra for single-photon ionization of rotationally cold NO(X $^2\Pi_{1/2}$) by coherent VUV radiation. Calculated spectra are for 5 K. The measured spectra are from the work of M. G. White and collaborators. See text.

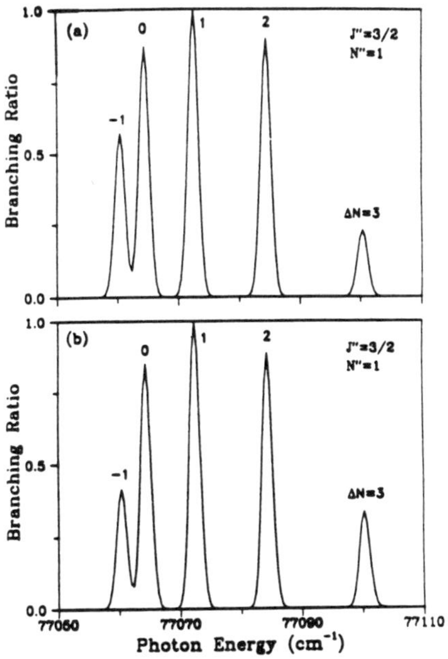

Fig. 4. (a) measured and (b) calculated ZEKE photoelectron spectra for single-photon ionization of the $N'' = 1$, $J'' = 3/2$ level of the X $^2\Pi_{1/2}$ state of NO by coherent VUV radiation. See text.

molecular photoelectron orbital. Note, furthermore, that the photoelectron angular distributions for the $\Delta N = \pm 1$ peaks are quite different for the $v_+ = 0$ and 1 ion levels, indicating an R-dependence of the photoelectron matrix element, even at this low vibrational excitation.

Next we look at the ZEKE photoelectron spectra resulting from single-photon ionization of rotationally cold NO(X $^2\Pi_{1/2}$) molecules by coherent VUV radiation.[22] Fig. 3 shows the jet-cooled PFI spectrum for the $v_+ = 1$ level of NO^+(X $^1\Sigma^+$) along with the calculated ion rotational distributions assuming a temperature of 5 K. These spectra were calculated for a photoelectron energy of 50 meV and convoluted with a Gaussian detection function with an FWHM of 2 cm^{-1}. In Fig 3(b) each branch is associated with a letter designation which refers only to the change in angular momentum apart from spin, i.e., $\Delta N = N_+ - N''$. The label $1'$ denotes the $N'' = 1$, $J'' = 1/2$ level. The agreement between these measured and calculated ZEKE spectra is quite encouraging.

To provide further insight into the underlying dynamics of these ZEKE photoelectron spectra, Fig. 4 shows calculated and measured ion rotational branching ratios for photoionization of the $N'' = 1$, $J'' = 3/2$ rotational level of the X $^2\Pi_{1/2}$ ground state of NO. The experimental results shown here are estimated from the actual measured spectra. These rotational branching ratios for ionization of a valence electronic state differ somewhat from those generally seen in ZEKE spectra of Rydberg states. This behavior reflects the broad range of angular momentum components ($\ell = 0$, 1, 2, and 3) which make significant contributions to the photoelectron matrix element.

Finally we report on the ZEKE photoelectron spectra for single-photon ionization of rotationally cold CO (X $^1\Sigma^+$) molecules by coherent XUV radiation.[23] Fig. 5 shows the jet-cooled PFI spectrum for the $v_+ = 0$ level of CO^+ (X $^2\Sigma^+$) and the calculated ion rotational distributions assuming a temperature of 8 K. These spectra were again calculated for a photoelectron energy of 50 meV and convoluted with a Gaussian detection function with an FWHM of 2 cm^{-1}. The letter designation on each branch also refers only to the change of angular momentum apart from spin. The agreement between these calculated and measured rotational branching ratios

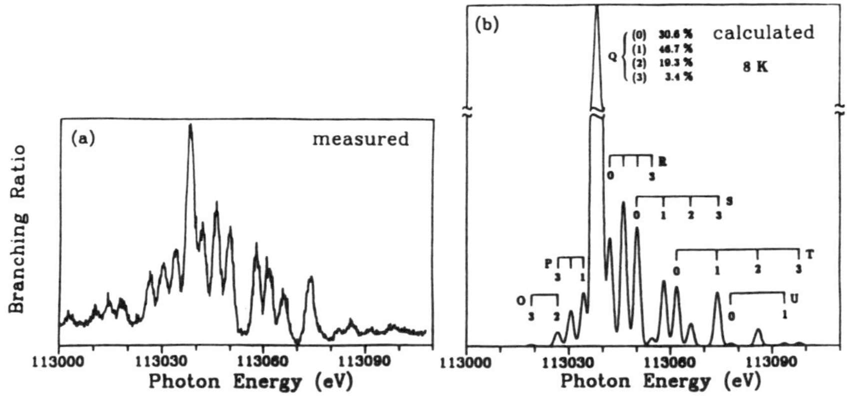

Fig. 5. (a) measured and (b) calculated ZEKE photoelectron spectra for single-photon ionization of rotationally cold CO ($X^1\Sigma^+$) by coherent XUV radiation. Calculated spectra are for 8 K. The measured spectra are from the work of J. Hepburn and collaborators. See text.

is excellent for all branches except the Q ($\Delta N = 0$) branches. Although the Q($N'' =$ 0, 1, 2) branches are the strongest in both spectra, the calculated spectral intensity is about 10 times larger than that of the R(1) branch. This behavior is indicated by the broken scale in Fig. 5(b). The strength of the Q branches in these spectra is consistent with the parity selection rule, $\Delta N + \ell = odd$, applicable here[16–18] and atomiclike behavior for photoionization of the 5σ orbital (45% s and 25% d character). The origin of this disagreement is not yet clear.

Acknowledgement

This work was supported by grants from the National Science Foundation, Air Force Office of Scientific Research, and the Office of Health and Environmental Research of the U. S. Department of Energy. We also acknowledge use of the resources of the Jet Propulsion Laboratory/California Institute of Technology CRAY Y-MP2E/116 Supercomputer.

References

1. L. Åsbrink, *Chem. Phys. Lett.* **7** (1970) 549.
2. S. T. Pratt, P. M. Dehmer, and J. L. Dehmer, *J. Chem. Phys.* **87** (1987) 3288.

3. W. G. Wilson, K. S. Wiswanathan, E. Sekreta, and J. P. Reilly, *J. Phys. Chem.* **88** (1984) 672.
4. S. W. Allendorf, D. J. Leahy, D. C. Jacobs, and R. N. Zare, *J. Chem. Phys.* **91** (1989) 2216.
5. E. de Beer, M. P. Koopmans, C. A. de Lange, Y. Wang, and W. A. Chupka, *J. Chem. Phys.* **94** (1991) 7634.
6. E. de Beer, C. A. de Lange, J. A. Stephens, K. Wang, and V. McKoy, *J. Chem. Phys.* **95** (1991) 714.
7. K. Wang, J. A. Stephens, V. McKoy, E. de Beer, C. A. de Lange, and N. P. C. Westwood, *J. Chem. Phys.* **97** (1992) 211.
8. K. Müller-Dethlefs and E. W. Schlag, *Ann. Rev. Phys. Chem.* **42** (1991) 109.
9. See, for example, E. R. Grant and M. G. White, *Nature* **354** (1991) 249.
10. J. Xie and R. N. Zare, *Chem. Phys. Lett.* **159** (1989) 399 and *Private Communication* (1991)
11. S. N. Dixit and V. McKoy, *J. Chem. Phys.* **82** (1985) 3546.
12. K. Wang and V. McKoy, *J. Chem. Phys.* **95** (1991) 4977.
13. M.-T. Lee, K. Wang, V. McKoy, R. G. Tonkyn, R. T. Wiedemann, E. R. Grant, and M. G. White, *J. Chem. Phys.* **96** (1992) 7848.
14. W. J. Hunt and W. A. Goddard. *Chem. Phys. Lett.* **3** (1969) 414.
15. R. R. Lucchese, G. Raseev, and V. McKoy, *Phys. Rev.* **A25** (1982) 2572.
16. K. Wang and V. McKoy, *J. Chem. Phys.* **95** (1991) 7872.
17. J. Xie and R. N. Zare, *J. Chem. Phys.* **93** (1990) 3033.
18. G. Raseev and N. Cherepkov, *Phys. Rev.* **A42** (1990) 3948.
19. H. Lefebvre-Brion, *Chem. Phys. Lett.* **171** (1990) 377.
20. K. Wang, J. A. Stephens, and V. McKoy, *J. Chem. Phys.* **95** (1991) 6456.
21. J. A. Stephens and V. McKoy, *Phys. Rev. Lett.* **62** (1989) 889.
22. Spectra measured by M. G. White and collaborators of Brookhaven National Laboratory.
23. Spectra measured by J. Hepburn and collaborators of the University of Waterloo.

THRESHOLD PHOTOELECTRON SPECTROSCOPY WITH COHERENT VACUUM ULTRAVIOLET: STUDIES OF IONIC EXCITED STATES.

W. Kong, D. Rodgers, and J.W. Hepburn
Department of Chemistry
University of Waterloo
Waterloo, Ont., N2L 3G1
CANADA

ABSTRACT

Coherent vacuum ultraviolet light generated by resonant four wave mixing in Xe or Kr has been used to photoionize molecules in a supersonic molecular beam. By using the technique of pulsed field threshold photoelectron spectroscopy, we have recorded rotationally resolved TPE spectra for a number of molecular ion excited states. As with previous studies using synchrotron radiation, we have observed TPE bands in the Franck-Condon gap region, for both O_2^+ and N_2O^+. In the case of N_2O^+, there are some significant differences between our results and the previous work.

1. Introduction

While the use of synchrotron radiation has greatly increased our understanding of photoionization dynamics, in the last decade lasers have played an increasingly important role in this work, largely through the use of various forms of multiphoton or multiple resonant spectroscopy and excitation. Although these forms of spectroscopy will continue to play a vital role in our continuing exploration of highly excited states, there are strong advantages to a complementary laser technique, in which short wavelength coherent light is used for single photon excitation.

The use of short wavelength coherent light for single photon spectroscopy experiments above and below the ionization threshold has a history almost as long as that of multiphoton work, beginning in the 1970's. Because of the high spectral brightness of coherent VUV sources, it is possible to use them in a wide range of experiments, including photofragment spectroscopy in supersonic molecular beams[1,2,3], and molecular spectroscopy[4]. The value of these coherent light sources for high resolution photoionization spectroscopy has been demonstrated in numerous different experiments, including work done by our group on the rotationally resolved photoionization spectroscopy of HI[5]. At this conference the work of several different groups using coherent short wavelength light will be reported, and it is clear that coherent sources are superior for photon energies up to 18eV. This 18eV limit is imposed by tunable laser technology, with 6.26eV being the current upper tuning limit for standard pulsed dye laser systems. One can anticipate the possibility of being able to generate intense coherent VUV up to and even beyond 20eV within a few years (less intense coherent sources in this energy range are already available[6]).

Two experimental techniques, particularly adapted to pulsed light sources, are time-of-flight photoelectron spectroscopy and threshold photoelectron spectroscopy. In fact, the same apparatus can be used for both types of experiments and although both have been extensively used in conjunction with synchrotron light, the pulsed coherent VUV allows for some improvements, particularly in the case of threshold photoelectron spectroscopy. This technique, developed more than 20 years ago[7,8], uses a tunable VUV source to ionize the molecule or atom of interest, and electrons with essentially zero kinetic energy are detected. By measuring the threshold electron signal as a function of ionizing photon energy, a high resolution photoelectron spectrum can be measured. A more recent modification, called ZEKE (for Zero Electron Kinetic Energy) spectroscopy by its creators[9], uses a small pulsed electric field delayed several hundred nanoseconds or even several microseconds after excitation to greatly improve the resolution for threshold detection. It is now widely accepted that under almost all circumstances, this technique relies on field ionization of metastable high principal quantum number Rydberg levels just below the ionization limit for threshold detection[10,11].

The PFI-ZEKE (for pulsed field ionization ZEKE) technique, originally developed in conjunction with multiphoton ionization, has been more recently used for single photon ionization, and to date several small molecules have been studied by this technique. The vast majority of the work has focused on the determination of rotational linestrengths for PFI-ZEKE spectra of ground states of molecular ions[11,12,13,14]. The only published exception to this is some work on Xe_2, in which the highly vibrationally excited Xe_2^+ favoured by Franck-Condon factors were probed by PFI-ZEKE[15]. If this technique is to be useful for ion state selection, or general high resolution spectroscopy of molecular ions, it must also be able to probe excited states of ions, or ionization channels that make up a very small fraction of the total ionization cross section. At the same time, the dynamics of these processes can be probed in more detail if excited states are studied, due to the presence of many open ionization and dissociation channels. Of particular interest is the study of the effects of open dissociation channels and ionic dissociation limits on the ionization process.

We have recently built a new apparatus for VUV spectroscopy of supersonic molecular beams. One of the features of this machine is a broad tuning range for the coherent VUV, a range that extends to 18 eV. This allows us to study excited states of molecular ions, along with ground states, even for ions with fairly high ionization potentials. In this paper, we will describe this new apparatus and provide a brief survey of some of our recent results on PFI-ZEKE spectroscopy of ionic excited states. In these experiments, we have been especially interested in demonstrating the general applicability of the PFI-ZEKE technique, and comparisons with previous TPES work using synchrotron radiation. More detailed discussions of the results presented here will be published elsewhere.

2. Experimental

A schematic diagram of the new photoionization apparatus is shown in figure 1. The coherent VUV is generated by resonant four wave sum mixing in pulsed jets of either krypton or xenon. Two Nd:YAG pumped dye lasers are used, one (ω_1) tuned to a variety of two photon resonances in Kr or Xe, the second (ω_2) tunable over the entire visible range and well into the ultraviolet. By overlapping these two beams and focusing them into the pulsed jet, radiation tunable over the 92 nm to 70 nm (13.5 eV to 17.7 eV) range is generated ($2\omega_1+\omega_2$). For wavelengths longer than 92 nm, the conversion efficiency in Xe varies strongly with frequency. Other techniques, such as resonant four wave sum mixing in Mg[4] or Hg[5] vapour, or resonant difference mixing in Kr[16], can be used to fill in the tuning range between 92 nm and 195 nm, at which point frequency mixing in crystals can be used. We need to use a monochromator to resolve the desired wavelength from other VUV wavelengths that are generated at the same time through other processes because of the two input frequencies necessary for four wave mixing. To preserve the controlled polarization of the coherent VUV, we have used a one metre normal incidence monochromator to refocus the coherent VUV into the experimental chamber. In the centre of this chamber, the coherent VUV crosses a supersonic molecular beam, formed by expansion of gas from a pulsed valve and collimated by a skimmer with a 0.5 mm diameter orifice. The pulsed supersonic beam is produced by a corrosion resistant pulsed valve with a 0.5 mm nozzle, which is housed in a separately pumped source chamber.

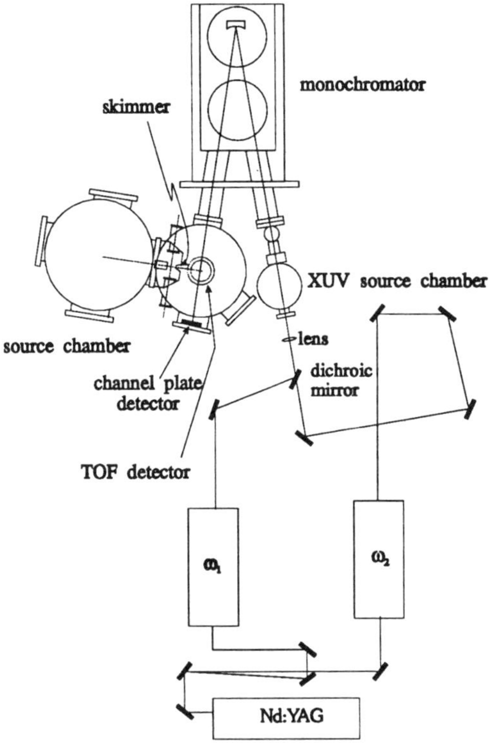

Figure 1 Schematic of the apparatus.

The experimental chamber, pumped by a 1000 l/s turbomolecular pump, has a pressure of 5×10^{-7} torr when the molecular beam is on. The crossing point of the molecular beam and the VUV laser beam is located at the centre of a double time of flight spectrometer, which can be configured for time of flight mass spectroscopy on the positive photoions and time of flight energy analysis on the photoelectrons. The spectrometer is surrounded by a double layer mu-metal magnetic shielding, and the flight path from the ionization volume to the channel plate detectors is 25 cm. For time of flight photoelectron spectroscopy, the amplified signal from the channel plates is digitized by a LeCroy 1.2 GHz transient recorder. A time of flight spectrum for photoionization of CO at 83.7 nm is shown in figure 2. This spectrum, recorded by averaging 4000 laser shots (400 seconds collection time) shows the sensitivity of our photoelectron spectrometer. The spectrum shown was produced with no extraction or focusing fields applied, meaning that it is an angle resolved photoelectron spectrum. Because the coherent VUV is linearly polarized, and the polarization is easily varied by varying the input laser polarization, we can also record photoelectron angular distributions.

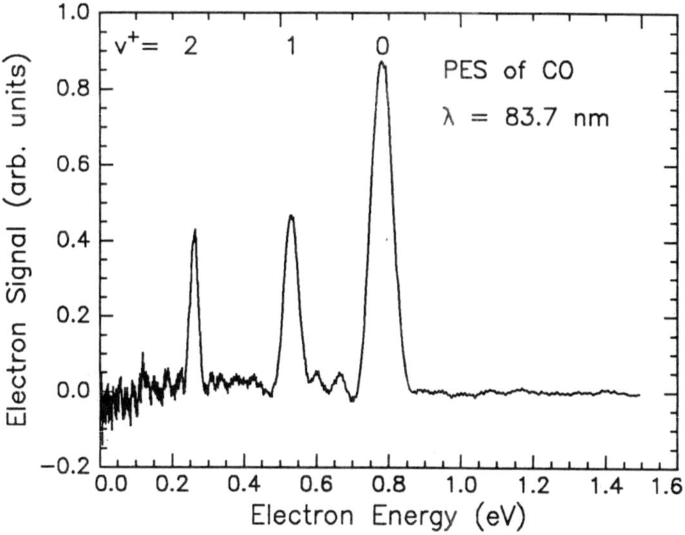

Figure 2 Time of flight photoelectron spectrum of CO, ionized at 83.7 nm. The relative intensities of the PES peaks are not corrected by the transmission function of the spectrometer.

In this paper we will focus on a different type of experiment, which is performed in the same apparatus. By applying a pulsed electric field on the grids surrounding the ionization volume, and detecting the delayed electrons formed as a result of this pulsed field, the apparatus can be used for PFI-ZEKE spectroscopy. This pulsed field is applied to the otherwise field free excitation volume typically 1 sec after the VUV pulse has excited the molecules in the beam. The resulting PFI-ZEKE electrons are then detected at the channel plates, well separated in time from the energetic electrons. In some cases, a weak DC field (on the order of 0.1 V/cm) is applied across the excitation region, and the stronger pulsed field is applied after a delay. In both cases, the pulsed field was kept as low as possible, between 0.4 and 1.6 V/cm.

The VUV wavelength was determined by calibration of the fundamental of the dye lasers using optogalvanic spectroscopy and laser-induced fluorescence in an iodine cell. The estimated absolute accuracy of our VUV photon energy is 1 cm^{-1}. The VUV linewidth was 1 cm^{-1} for the experiments reported here, but can be narrowed to less than 0.5 cm^{-1}. The resolution for the PFI-ZEKE spectra was between 1.5 cm^{-1} and 2.5 cm^{-1}, determined by the magnitude of the pulsed field. To illustrate the capabilities of the technique, the PFI-ZEKE spectrum of the ground state of HF^+ is given in figure 3.

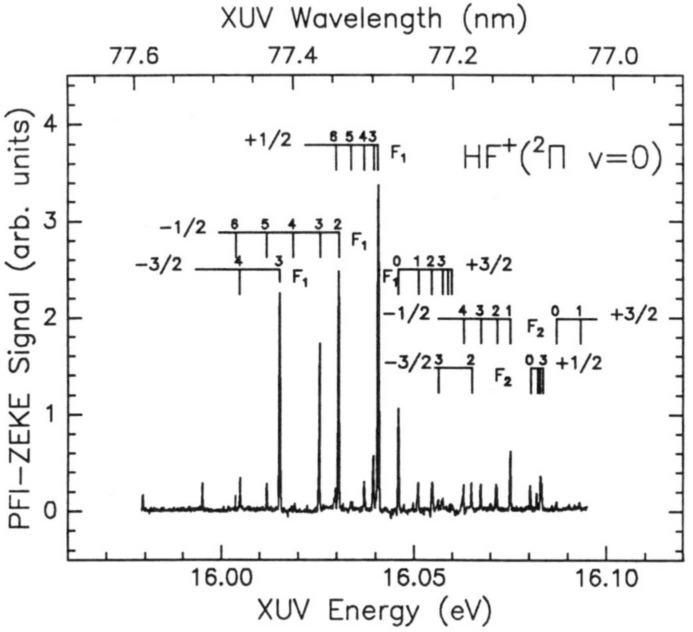

Figure 3 PFI-ZEKE spectrum of jet-cooled HF. Line assignments are labelled by ΔJ, J", and the spin orbit sublevel for HF^+ (F_1 or F_2)

To date we have carried out PFI-ZEKE studies on the following systems: $NO^+(a^3\Sigma^+)$, $CO^+(X\,^2\Sigma^+)$, $CO^+(A\,^2\Pi)$, $N_2O^+(X\,^2\Pi)$, $N_2O^+(A\,^2\Sigma^+)$, $O_2^+(X\,^2\Pi_g$, v^+=15-21), $O_2^+(a^4\Pi_u)$, $HF^+(X\,^2\Pi)$, and Kr^+ and Ar^+. In this paper, we will describe only our results on NO^+, O_2^+, and N_2O^+, which illustrate some of the features and capabilities of PFI-ZEKE spectroscopy.

3. $NO^+(a^3\Sigma^+)$ v^+=0,1,2

The spectroscopy of the $a^3\Sigma^+$ state of NO^+ has been studied by emission spectroscopy of the molecular ion[17]. However, as is often the case, the absolute energy of this state with respect to neutral NO, is not known to high accuracy. One obvious result from the PFI-ZEKE spectrum is an absolute term value for the observed state, with an accuracy of 1 cm^{-1}. We have recorded the spectra of the v^+ = 0, 1, and 2 levels of the $a\ ^3\Sigma^+$ state, and the spectrum for the v^+ = 0 level is shown in figure 4. From these data, we get an absolute energy of 126,389 cm^{-1} for the $a\ ^3\Sigma^+$ state, with respect to J=1/2 of the $^2\Pi_{1/2}$ state of NO. The lower half of the figure is the simulation result based on the Buckingham model.[18] Angular momentum transfer is calculated assuming a hydrogenic wavefunction for the outgoing electron, and modulation originated from the radial function is fitted using a least square fitting routine. More information can be obtained by analyzing the fitted coefficients.

The $^3\Sigma^+$ state of NO^+ is a metastable electronic state, as it is the lowest energy triplet state. This makes it suitable for state selection for reaction dynamics, since the state will survive long enough to allow it to collide with other species after it has been prepared. Our results show that PFI can be used to ro-vibrationally state select excited molecular ions for such dynamics studies.

The relative intensities of the three observed bands were not at all in accord with Franck-Condon factors. The strongest of the three was the v^+=1 band, while the v^+=2 band was the weakest, being ten times weaker than v^+=1. According to Franck-Condon factors, these two bands should have nearly the same intensity. The cause of this anomalous intensity is not clear, although loss of v^+=2 signal through autoionization is a possibility. The difficulty with this explanation is that the v^+=2 band is far from any obvious Rydberg resonances in the total ionization spectrum, while the v^+=1 band is essentially coincident with a very strong Rydberg resonance at 78.3 nm, corresponding to a state converging to the $b^3\Pi$ limit, which could provide oscillator strength.

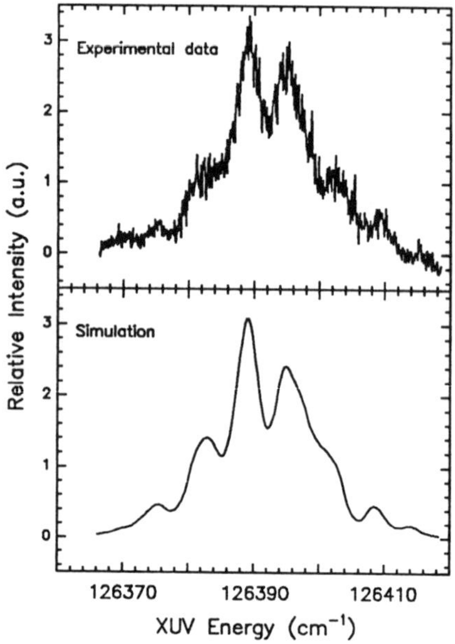

Figure 4 PFI-ZEKE spectrum for the v=0 level of the $a^3\Sigma^+$ state of NO^+.

4. O_2^+ ($X\ ^2\Pi_g$); v^+= 15 to 21

The photoionization of oxygen has been extensively studied, including a recent study by PFI-ZEKE[11]. One of the interesting features about oxygen photoionization is that the TPE spectrum shows significant intensity at energies corresponding to formation of highly vibrationally excited oxygen ions[19]. The formation of these high vibrational levels, which have essentially zero Franck-Condon integrals from the ground state of neutral oxygen, has been exploited for ion-molecule reaction dynamics studies, using the TPEPICO method[20]. Although the dynamics of the process that gives rise to formation of these Franck-Condon forbidden levels is not well understood, it has been ascribed to resonant autoionization, where excitation of degenerate Rydberg states coupled to dissociating states gives rise to formation of the non Franck-Condon levels.

If this were the case, one would not expect to observe these same levels in PFI-ZEKE spectra, as this process relies on excitation of long lived high Rydberg states converging on the ionic state in question, and any coupling to a dissociative state would remove these states before field ionization. In the absence of any such coupling, the Franck-Condon factors for excitation of highly vibrationally excited Rydberg states is vanishingly small. However, we have observed these levels in a PFI-ZEKE study on oxygen, and the spectrum of the v^+=15 level is shown in figure 5. To date, we have recorded spectra for v^+ = 15, 17, 20, and 21 (we have not yet investigated the other possible bands). The strongest signals were observed for the v^+ = 15 band, consistent with the synchrotron TPES spectrum, but we have not yet carried out a careful study of the relative band intensities. The identity of the observed bands is absolutely certain from the spectroscopy, both for band positions and rotational fine structure, and the simulation using the Buckingham model gives a very good fitting result. We still need to conduct more experiments on this system to measure the relative intensities of these observed bands with respect to other Franck-Condon allowed bands in O_2.

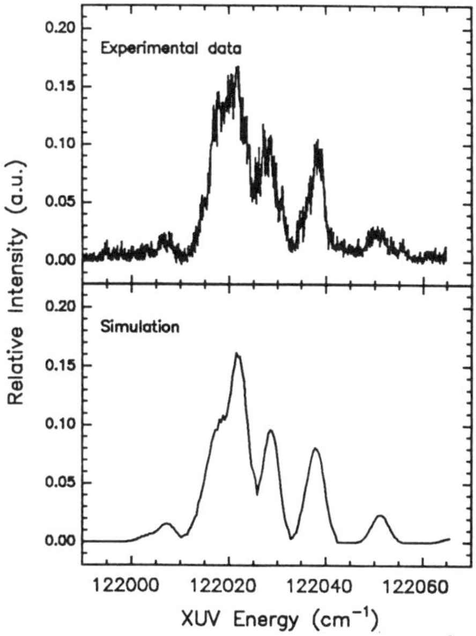

Figure 5 PFI-ZEKE spectrum of $O_2^+(X^2\Pi_{g1/2}\ v^+=15) \leftarrow O_2(X^3\Sigma_g^-)$.

Although our study of these high vibrational levels is not completed, several things are clear at this point. The first is that the PFI method can be used for state selection of highly excited ionic states, even in cases where those states are not accessible through Franck-Condon type transitions. Our results also indicate that the mechanism for resonant autoionization needs to be investigated in more detail. Although a coupling with dissociative neutral states may indeed be involved in this process, an explanation must be found for how such a coupling can give rise to metastable high Rydberg states.

5. N_2O^+ levels below the $A^2\Sigma^+$ state

The PFI-ZEKE spectra of ground state N_2O^+ has been studied previously[21]. We have collected and analyzed spectra for the first excited doublet state of N_2O^+, the $A\ ^2\Sigma^+$ state, for the (000), (100), and (001) vibrational bands. The analysis of these bands was relatively straight forward, and will be reported elsewhere[22]. Of special interest here is a series of experiments we carried out at photon energies just below the A state threshold.

Previous TPES work in this spectral range found that highly vibrationally excited states of N_2O^+ were formed through excitation of predissociating Rydberg states which converge on the $A\ ^2\Sigma^+$ and $B\ ^2\Pi$ limits[23]. This TPE signal in the Franck-Condon gap region was postulated to arise from an interaction between neutral dissociation and ionization continua, through a process of resonant autoionization[24]. One important feature of the TPES data was the coincidence between the peaks in the TPE spectra and the observed Rydberg resonances.

We have concentrated our attention on the 77 nm to 80 nm region where the previous workers had found several resonances in the TPE spectrum. Over this range, we observed only two bands in the PFI-ZEKE spectrum, one at 78.1 nm, the other at 79.7 nm. The other bands observed in previous work were not seen by us. One of the bands we observed is shown in figure 6, along with the total ionization spectrum. The PFI-ZEKE band is not coincident with the strong Rydberg resonance, but is just beside it, 200 cm^{-1} higher in energy. The two bands we observe come from metastable Rydberg states, since they persist even when a dc field is applied to the excitation region. If these two bands correspond to highly excited vibrational levels of the X state of N_2O^+, then we have shown that it is possible to excite high vibrational levels of metastable Rydberg states, possibly through a coupling with neutral dissociative states. This is exactly what was observed in O_2, where the identity of the PFI-ZEKE bands is unambiguous. What is not yet clear is why only two bands were observed in the wavelength range investigated, and why these bands were displaced from the autoionizing Rydberg states. Further work on this system and other similar systems is underway to clarify the mechanism that is giving rise to these highly excited levels.

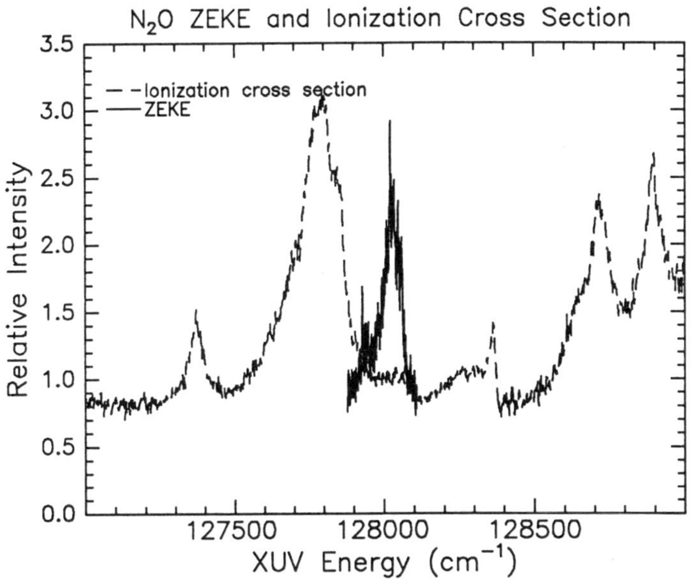

Figure 6 PFI-ZEKE spectrum of the 78.1 nm band in N_2O. The intensity axis is for the total photoionization signal, the ZEKE peak shown has been shifted upwards by 0.9 for clarity.

6. Acknowledgements

Support for this research was provided by NSERC(Canada), and the Network of Centres of Excellence program, administered by NSERC. We wish to acknowledge many useful discussions with Vince McKoy, and thank Denise Lacombe for her assistance with this research. WK thanks NSERC for a graduate scholarship, and JWH wishes to thank the Sloan foundation for a research fellowship. We also wish to acknowledge Andreas Mank who participated in the PFI-ZEKE experiments on HF, which will be published elsewhere.

7. References

1. I.M. Waller and J.W. Hepburn, *J. Chem Phys.* **88** 6658 (1988)
2. N. Sivakumar, G.E. Hall, P.L. Houston, J.W. Hepburn, and I. Burak, *J. Chem. Phys.* **88** 3692 (1988)

3. J.F. Schlenker, F. Bouchard, I.M. Waller, and J.W. Hepburn, *J. Chem. Phys.* **93** 7110 (1990)
4. D.J. Hart and J.W. Hepburn, *J. Chem. Phys.* **86** 1733 (1987)
5. D.J. Hart and J.W. Hepburn, *Chem. Phys.* **129** 51 (1989)
6. J.W. Hepburn, in *Vacuum Ultraviolet Photoionization and Photodissociation of Molecules and Clusters* ed. C.Y. Ng (World Scientific, 1991)
7. W.B. Peatman, T.B. Borne, and E.W. Schlag, *Chem. Phys. Lett.* **3** 492 (1969)
8. W.B. Peatman, *J. Chem. Phys.* **64** 4093 (1976)
9. K. Muller-Dethlefs and E.W. Schlag, *Ann. Rev. Phys. Chem.* **42** 109 (1991)
10. G. Reiser, W. Habenicht, K. Muller-Dethlefs, and E.W. Schlag, *Chem. Phys. Lett.* **152** 119 (1988)
11. R.G. Tonkyn, J.W. Winniczek, and M.G. White, *Chem. Phys. Lett.* **164** 137 (1989)
12. R.G. Tonkyn, R.T. Wiedmann, and M.G. White, *J. Chem. Phys.* **96** 3696 (1992)
13. M.-T. Lee, K. Wang, V. McKoy, R.G. Tonkyn, R.T. Wiedmann, E.R. Grant, and M.G. White, *J. Chem. Phys.* **96** 7848 (1992)
14. F. Merkt and T.P. Softley, *J. Chem. Phys.* **96** 4149 (1992)
15. R.G. Tonkyn and M.G. White, *J. Chem. Phys.* **95** 5582 (1991)
16. G. Hilber, A. Lago, and R. Wallenstein, *J. Opt. Soc. Am. B* **4** 1753 (1987)
17. K.P. Huber and M. Vervloet, *J. Mol. Spec.* **146** 188 (1991)
18. A.D. Buckingham, B.J. Orr and J.M. Sichel, *Phil. Trans. Roy. Soc. Lond.* A **268** 147 (1970)
19. J.L. Dehmer, A.C. Parr and S.H. Southworth, in *Handbook of Synchrotron Radiation*, edited by G.V. Marr (North-Holland, Amsterdam, 1986), Vol II.
20. P.M. Guyon and E.A. Gislason, *Topics Curr. Chem.* **151** 161 (1989)
21. R.T. Wiedemann, E.R. Grant, R.G. Tonkyn, and M.G. White, *J. Chem. Phys.* **95** 746 (1991)
22. W. Kong, D. Rodgers, and J.W. Hepburn, *to be published*
23. T. Baer, P.M. Guyon, I. Nenner, A. Tabouche-Fouhaille, R. Botter, L.P.A. Ferreira and T.R. Govers, *J. Chem. Phys.* **70** 1585 (1979)
24. P.M. Guyon, T. Baer and I. Nenner, *J. Chem. Phys.* **78** 3665 (1983)

Rotationally Resolved Photoionization with Coherent VUV Radiation

Ralph. T. Wiedmann, Russell G. Tonkyn,† Michael G. White

Chemistry Department, Brookhaven National Laboratory, Upton, NY 11973

Abstract

Pulsed field ionization (PFI) has been used in conjunction with coherent VUV radiation to investigate the rotational state distributions of molecular cations following single photon ionization. The rotational state distributions for several linear cations (O_2, NO, OH(OD), HCl and N_2O) can be interpreted predominately on the basis of the near-threshold, one-electron photoionization dynamics; however, field-induced autoionization is often the dominate ionization pathway for rotational branches involving negative changes in core angular momentum. For photoionization of the H_2X (X = O, S) non-linear triatomic molecules, transitions between asymmetric top levels involving the rotational angular momentum projections, K_a and K_c, permit resolution of the photoelectron continua according to symmetry. The observed spectra clearly demonstrate the importance of the non-spherical nature of the molecular ion potential which leads to photoelectron final states which are unexpected from atomic-like analogies.

1. Introduction

Photofragment spectroscopy provides information on the dynamics of the fragmentation process by revealing the internal state distribution of the usually unobserved molecular fragment through the measurement of the kinetic energy distribution of the "light" or atomic fragments. In the same way, photoelectron spectroscopy reveals the photoionization dynamics through detection and momentum resolution of the ejected photoelectron. The dynamics are considerably different than for neutral fragmentation, yet both involve similar concepts, especially as regards the evolution of a photoexcited "complex" into the observed asymptotic channels and the molecular forces which influence their branching ratios. With the development of laser-based, zero kinetic energy (ZEKE) and pulsed field ionization (PFI) techniques it is has now become possible to extend photoelectron spectroscopy to measurements

†Present address: Materials and Chemical Sciences Center, Pacific Northwest Laboratory, Richland WA

of cation *rotational* state distributions and hence, fully resolve quantum state distributions following photoionization.[1,2] We have used the PFI technique in conjunction with a high-resolution VUV laser source to investigate the spectroscopy and one-photon ionization dynamics of a number of small molecules in which we have explored the angular momentum constraints in rotational photoionization transitions.[3 - 9] Our concentration on one-photon processes arises from the extensive body of information which exists for molecular VUV photoionization using laboratory or synchrotron radiation sources, but for which only H_2 has been rotationally resolved. In this paper, we present a summary of results of joint theoretical and experimental studies on several diatomic and triatomic molecules which demonstrate the utility of rotationally-resolved threshold photoelectron spectroscopy for examing photoionization dynamics.

2. Experiment

The photoelectron spectrometer and VUV source used for threshold photionization measurements is described in detail elsewhere.[10] Briefly, we use a small, separately pumped frequency tripling chamber attached to a photoelectron/photoion time–of–flight (TOF) spectrometer. Tunable VUV radiation in the range 150 nm to 90 nm is produced by non-resonant third harmonic generation ($\omega_{vuv} = 3\omega_{uv}$) as well as resonant and non-resonant four-wave sum/difference frequency mixing ($\omega_{vuv} = 2\omega_{uv} \pm \omega_{vis}$) in free jet expansions of krypton, xenon and molecular nitrogen.[11] The collinear laser beams ($\omega_{uv}, \omega_{vis}$) are focused by a 100 mm focal length achromatic lens into the tripling chamber and very near the 1 mm dia. exit nozzle of a piezoelectric pulsed-valve.[12] The diverging fundamentals (visible and UV) and harmonically generated VUV radiation are captured by a pyrex capillary tube (30 cm long; 1 mm i.d.) and directed into the TOF spectrometer, where it passes between two parallel plates which define the extraction field for the TOF spectrometer. The capillary acts as an efficient differential pumping barrier separating the VUV generation chamber ($\sim 10^{-3}$ Torr) and the spectrometer chamber ($\leq 1 \times 10^{-7}$ Torr), while also acting as a light guide which provides a collimated VUV beam without refocussing. The capillary is also used as a gas filter which effectively separates the sum and difference VUV frequencies via total absorption of the high frequency radiation by a suitable rare gas injected into the center of the long capillary ($\leq$ 400 mTorr). The VUV intensity at the spectrometer is estimated to be $10^9 - 10^{10}$ photons per pulse, with an energy bandwidth of 0.7 cm^{-1}.

High-resolution threshold photoelectron spectra are obtained by the delayed, pulsed field ionization (PFI) method as first demonstrated by Reiser, *et al.*[13] PFI takes advantage of the well known Stark shift of an ionization threshold in an external electric field (given by $\sim 6\sqrt{F}$ cm^{-1}, where F is the field strength in volts/cm). Bound Rydberg levels very near the ionization threshold, which are stable in a field free environment, become open ionization channels when a Stark shift larger than their binding energy is applied. For fields used in these studies (0.3-0.5 V/cm) only Rydberg levels with $n \geq 150$ can be field ionized. Rydberg states with such high principal quantum numbers are very long lived due to rapid l mixing induced by a small DC electric field (~ 0.05 V/cm).[14] The small DC voltage on the repeller (lower) plate also sweeps out any slow or near ZEKE photoelectrons produced directly by the laser pulse. After a delay of 700 ns, a fast, negative pulse ($0.3 - .5$ V) is applied to the repeller plate which field ionizes the metastable, high-n Rydberg states. Electrons produced by this pulsed field are readily distinquished from photoelectrons arising from direct ionization as only the arrival time of the former depends on the pulse delay and voltage.

3. Threshold Photoionization of Linear Molecules

The ejection of a photoelectron with both spin and orbital angular momentum leads to selection rules for photoionization which look quite different compared to those of bound state spectroscopy. Photoabsorption takes place at short range where the excited electron strongly interacts with the ion core and where the general selection rule governing dipole transitions, i.e., $\Delta J = 0, \pm 1$, is valid. As the electron escapes the molecular ion potential, the interaction becomes weak and the coupling of angular momenta becomes more appropriate to a free photoelectron and well defined rovibronic level of the ion core. This uncoupling can be viewed as a scattering process in which the photoexcited electron exchanges energy and angular momentum with the ion core. The net result is that changes in core angular momentum well beyond $\Delta J = 0, \pm 1$ are possible. The conservation of angular momentum requires only that $\Delta J = J^+ - J'' = l + 3/2, l + 1/2, \ldots, -l - 3/2$, where l is the orbital angular momentum associated with s, p, d or higher outgoing waves.[15] For example, an outgoing d wave could result in eight rotational branches, with changes in the angular momentum of the core as large as $\pm 7/2$. The measurement of the relative branch intensities directly reflects the probability for the various partial waves of the outgoing electron. Such information is a sensitive probe of the ionization dynamics as the partial wave distribution can be strongly influenced by electron correlation and interactions of the photoexcited electron and the nuclear degrees of freedom, e.g., autoionization and shape resonances. Consequently, characterization of the quantum state distribution of the molecular ion and photoelectron pose challenges to theoretical descriptions of the photoionization process since both the electronic and nuclear wavefunctions must be accurately treated. Substantial progress has been made in theoretical approaches to photoionization and recent *ab initio* calculations using Hartree-Fock[4,16,17] and multichannel quantum defect theory (MQDT)[18,19] approaches have provided realistic cation rotational distributions for several molecules.

For the open shell diatomics, O_2[3] and NO, rotationally-resolved threshold photoelectron spectra obtained by the PFI technique were interpreted assuming that the observed branch intensities directly reflect the partial wave distribution of the ejected photoelectron at threshold. This approach is supported in both cases by one-electron, *ab initio* calculations which are in near quantitative agreement with the experimental spectra. For O_2, the calculations show that the observation of large changes in core rotation ($|\Delta J| \leq 9/2$) results from the presence of an $f\sigma(l = 3)$ shape resonance at the ionization threshold.[4] By contrast, threshold photoionization of NO results in a photoelectron continuum characterized by a broad partial wave distribution with $l = 0 - 3$ but with a propensity for small changes in core rotation ($|\Delta J| \leq 5/2$).

Unlike O_2 and NO, spectral simulations for N_2O^+ based on the Buckingham, Orr and Sichel photoionization line strength expressions[20] cannot reproduce a general feature of the data in which rotational branches with $\Delta J < 1/2$ have intensities uniformly larger than branches with $\Delta J \geq 1/2$.[5] Large differences in intensities between branches that only differ in the sign of ΔJ are not expected since only the magnitude of the angular momentum transfered to the core is relevant to determining line strengths (see equations (34)-(37), reference 20). Asymmetries in rotational branch intensities are also evident in the PFI spectra of the OH (OD) radical,[9] and particularly for HCl in which rotational branches with large negative changes in core angular momentum, $\Delta J = -5/2$ and $-7/2$, are relatively intense and, more importantly, they have no observable positive ΔJ counterparts. Such intensity asymetries have also been observed in the ZEKE-PES data of other laboratories as well, suggesting a mechanism which is general in origin.[2,21-24]

The most obvious source of complication in threshold measurements is from autoionization. Indeed, conventional threshold photoelectron spectroscopy using cw sources (e.g. synchrotron radiation) is often dominated by autoionizing states which yield low energy electrons. Such "resonant" autoionization processes have been well documented and lead to additional lines in non-Franck-Condon regions of the photoelectron spectrum.[25] The PFI technique used in our experiments discriminates against such electrons, but is susceptible to "forced" autoionization processes which are induced by the pulsed electric field.[26,27] Field-induced autoionzation involves the dipole coupling of high-n Rydberg states lying near the ionization threshold with near degenerate low-n Rydberg states converging to higher rotational limits.[28] Transitions to the low-n Rydberg states are strongly allowed and represent bound members of rotational branches with increasing ΔJ. Although the low-n Rydberg states lie too far below their respective thresholds to be field ionized, these levels can autoionize once the external field is applied and the high-n Rydberg levels become open ionization channels. For the most negative ΔJ branches, the number of posssible interactions with near-degenerate Rydberg levels is a maximum. Consequently, field induced autoionization can lead to substantial intensity enhancement

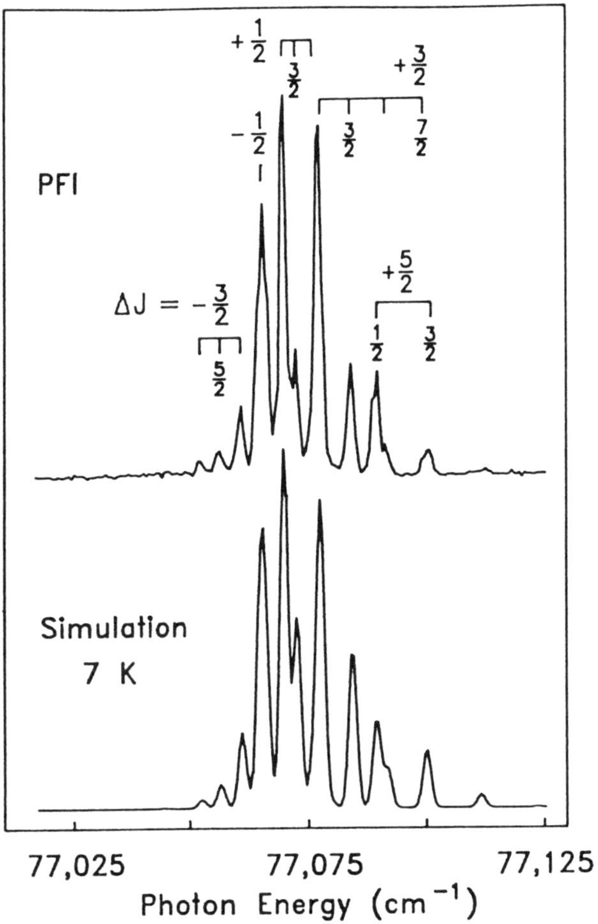

Figure 1: Upper trace: Pulsed field ionization spectrum of rotationally cold NO via $NO(\tilde{X}\,^2\Pi_{1/2},\, v'' = 0, J'') \rightarrow NO^+(X^1\Sigma^+,\, v^+ = 1, J^+) + e^-$ one-photon photoionization transitions. Lower trace: Simulated photoionization spectrum of NO at 7 K using best fit dynamical parameters in conjunction with the rotational photoionization line strength expressions of Buckingham, *et al* (see reference 20).

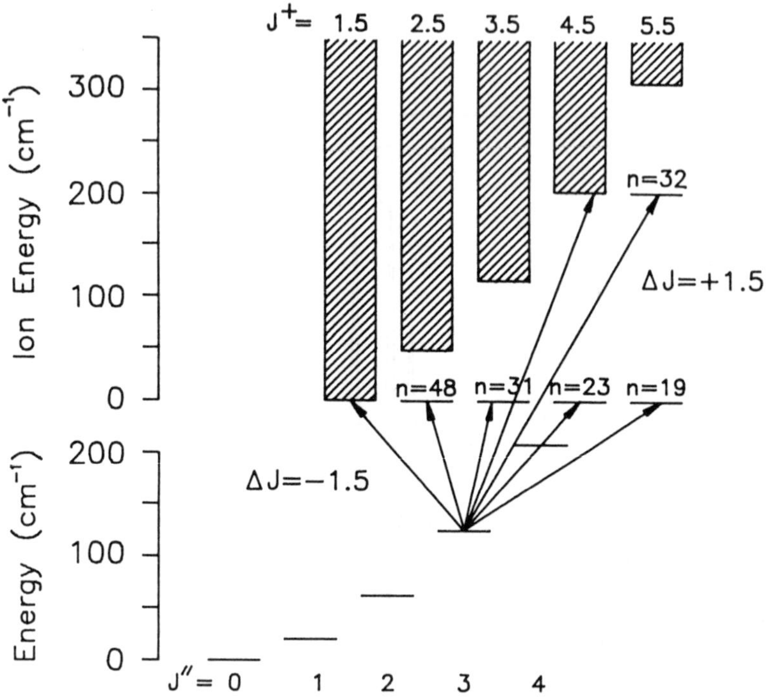

Figure 2: Upper trace: Scaled diagram illustrating field-induced autoionization in the PFI-ZEKE spectrum of HCl. The arrows mark transitions from $J'' = 3$ of the neutral ground state to Rydberg levels lying within 2 cm^{-1} of the ionization thresholds for $J^+ = 3/2$ ($\Delta J = -3/2$) and $J^+ = 9/2$ ($\Delta J = +3/2$). Principal quantum numbers for the low-n Rydberg states are estimated from accurate spectroscopic constants of HCl^+ and assuming zero quantum defect.

for branches with large negative ΔJ. By contrast, the number of possible Rydberg-Rydberg state interactions will be less for branches involving large positive ΔJ branches since there are fewer available low-n Rydberg states which can be optically connected to the ground rotational state. This mechanism is illustrated in Figure 2 for the specific case of HCl.

The implication of these results is that rotational line intensities obtained by ZEKE or PFI methods can be strongly influenced by Rydberg-Rydberg state interactions which ultimately result in autoionization. Since the experimentalist has some control over the sensitivity of PFI-ZEKE measurements to field-induced processes (delay, field strength, excitation bandwidth), this technique may give rise to new opportunities to study multichannel interactions in molecules with very high selectivity.

4. Threshold Photoionization of H_2X (X = O, S)

High resolution (~1 cm^{-1}) threshold photoelectron spectra for jet-cooled (15 K) samples of H_2O and H_2S were obtained for rotational transitions between the X^1A_1, (000) neutral ground state and the $\tilde{X}\,^2B_1$, $(\nu_1\nu_2\nu_3)$ vibrational levels of the ionic ground state.[6,8] For H_2O, rotationally-resolved data were obtained for ionization into the (000), (100) and (010) vibrational levels, while for H_2S^+ only the (000) and (010) levels could be probed due to inefficient VUV production at wavelengths near the (100) symmetric stretch. The H_2X^+ spectra could be readily assigned to two types of rotational photoionization transitions corresponding to specific changes in the asymmetric top angular momentum projection quantum numbers, K_a and K_c. Most of the stronger lines can be classified as type C rotational transitions ($\Delta K_c = 0, \Delta K_a = \pm 1$) but type A ($\Delta K_c = \pm 1, \Delta K_a = 0$) transitions are also clearly evident, particularly in the jet-cooled spectra. The utility of this classification stems from the fact that general symmetry arguments show that these rotational transitions are associated with only one photoelectron symmetry, i.e. ka_1 with type C and ka_2 with type A. The appearance of type A transitions with $\Delta K_a = 0$ is in variance with the predictions of a multichannel quantum defect theory (MQDT) analysis of photoionization of asymmetric top molecules by Child and Jungen.[19] These authors predict that photoionization will only involve a subset of type C transitions with $|\Delta K_a| = |K_a^+ - K_a''| = 1$ and $|\Delta N| = |N^+ - N''| \leq 1$, where N is the total angular momentum exclusive of spin. These limits on the changes of core angular momenta arise from the assumption that the $1b_1$ molecular orbital can be described exclusively in terms of an atomic p_x orbital. More recently, Gilbert and Child[28] have presented a model based on field-induced autoionization in an effort to explain the appearance of type A transitions in the threshold spectra of H_2O. Although such rotational autoionization processes as discussed above could give rise to the appearance of rotational final states which are nominally forbidden by optical selection rules, the specific application to H_2O also invokes an atomic-like description of photoexcitation from the $1b_1$ molecular orbital.

A more straightforward explanation for the appearance of type A transitions can be derived from a very recent *ab initio*, Schwinger variational photoionization calculation on H_2O.[29] In this calculation, no assumptions concerning the atomic character of the initial and final states are made and the continuum is calculated in the full anisotropic potential of the ion core. The calculations predict the appearance of type A photoionization transitions without recourse to final state interactions (field-induced autoionization) and the calculated H_2O^+ rotational spectrum is in

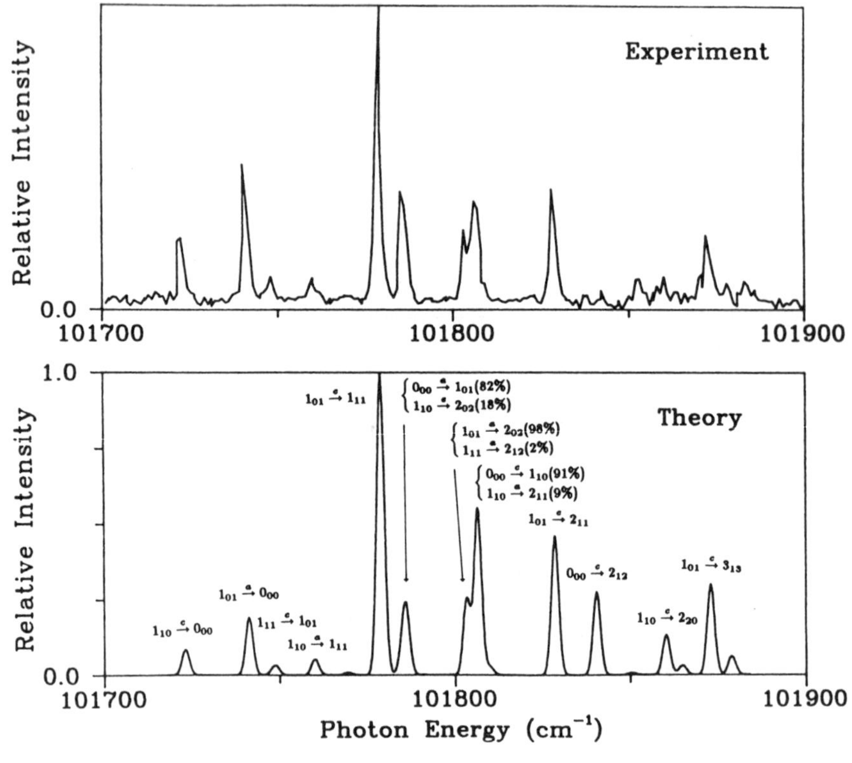

Figure 3: Experimental (top) and calculated (bottom) threshold photoelectron spectrum for photoionization from the $1b_1$ orbital of H_2O. The labels *a* and *c* refer to type A and type C rotational branch transitions.

nearly quantitative agreement with experiment. A comparison of the H_2O^+ $\tilde{X}\,^2B_1$, (000) PFI spectrum and the theoretical near threshold photoionization calculation is shown in Figure 3.

A partial wave analysis shows that type A transitions are accompanied by nearly pure *p* wave photoelectrons. Furthermore, the type C transitions are not exclusively *d* partial waves as expected from the atomic analogies, but include significant and for some lines (e.g. $0_{00} \leftarrow 1_{10}$) dominant *s* wave character. These calculations emphasize the importance of the non-spherical nature of the molecular ion core which can "scatter" or torque the escaping photoelectron into different partial waves. For H_2S, we expect that similar dynamics are applicable since the $1b_1$ molecular orbital is primarily localized on the sulfur atom. However, the participation of the nominally unoccupied $3d$ levels centered on the heavier sulfur atom could lead to additional high l components in the continuum and modify the rotational line strengths.

5. Acknowledgements

This work was performed at Brookhaven National Laboratory and supported by the US Department of Energy, Office of Basic Energy Sciences under contract No. DE-AC02-76CH00016.

References

[1] K. Müller-Dethlefs and E. W. Schlag, *Annu. Rev. Phys. Chem.* **42** (1991) 109.

[2] E. R. Grant and M. G. White, *Nature* **354** (1991) 249.

[3] R. G. Tonkyn, J. W. Winniczek and M. G. White, *Chem. Phys. Lett.* **164** (1989) 137.

[4] M. Braunstein, V. McKoy, S. N. Dixit, R. G. Tonkyn and M. G. White, *J. Chem. Phys.* **93** (1990) 5345.

[5] R. T. Wiedmann, E. R. Grant, R. G. Tonkyn and M. G. White, *J. Chem. Phys.* **95** (1991) 746.

[6] R. G. Tonkyn, R. T. Wiedmann, E. R. Grant and M. G. White, *J. Chem. Phys.* **95** (1991) 7033.

[7] R. G. Tonkyn, R. T. Wiedmann and M. G. White, *J. Chem. Phys.* **96** (1992) 3696.

[8] R. T. Wiedmann and M. G. White, *Proceedings of the SPIE: Optical Methods for State- and Time-Resolved Chemistry* **1638** (1992) 273.

[9] R. T. Wiedmann, R. G. Tonkyn, M. G. White, K. Wang and V. McKoy, *J. Chem. Phys.*, in press.

[10] R. G. Tonkyn and M. G. White, *Rev. Sci. Instrum.* **60** (1989) 125.

[11] R. H. Page, R. J. Larkin, A. H. Kung, Y. R. Shen and Y. T. Lee, *Rev. Sci. Instrum.* **58** (1987) 1616.

[12] D. Proch and T. Trickl, *Rev. Sci. Instrum.* **60** (1989) 713.

[13] G. Reiser, W. Habenicht, K. Müller-Dethlefs and E. W. Schlag, *Chem. Phys. Lett.* **152** (119) 1988.

[14] W. A. Chupka, *private communication.*

[15] J. Xie and R. N. Zare, *J. Chem. Phys.* **93** (1990) 3033.

[16] X. Song, E. Sekreta, J. P. Reilly, H. Rudolph and V. McKoy, *J. Chem. Phys.* **91** (1989) 6062; H. Rudolph, J. A. Stephens, V. McKoy and M. T. Lee, *J. Chem. Phys.* **91** (1989) 1374.

[17] E. de Beer, C. A. de Lange, J. A. Stephens, K. Wang and V. McKoy, *J. Chem. Phys.* **95** (1991) 714.

[18] H. Lefebvre-Brion, *Chem. Phys. Lett.* **171** (1990) 377.

[19] M. S. Child and Ch. Jungen, *J. Chem. Phys.* **93** (1990) 7756.

[20] A. D. Buckingham, B. J. Orr and J. M. Sichel, *Phil. Trans. Roy. Soc. Lond. A* **268** (1970) 147.

[21] M. Takahashi, H. Ozeki and K. Kimura, *Chem. Phys. Lett.* **181** (1991) 255.

[22] F. Merkt and T. P. Softley, *J. Chem. Phys.* **96** (1992) 4149.

[23] G. Reiser and K. Müller-Dethlefs, *J. Phys. Chem.* **96** (1992) 9.

[24] G. P. Bryant, Y. Jiang, M. Martin and E. R. Grant, *J. Phys. Chem.*, to be published.

[25] I. Nenner and J. A. Beswick, in *Handbook on Synchrotron Radiation*, edited by G. V. Marr (North Holland, Amsterdam, 1987) Vol. 2.

[26] Chr. Bordas, P. Brevet, M. Broyer, J. Chevaleyre and P. Labastie, *Europhys. Lett.* **3** (1987) 789.

[27] C. R. Mahon, G. R. Janik and T. F. Gallagher, *Phys. Rev.* **A41** (1990) 3746.

[28] R. D. Gilbert and M. S. Child, *Chem. Phys. Lett.* **187** (1991) 153.

[29] M. T. Lee, K. Wang, V. McKoy, R. G. Tonkyn, R. T. Wiedmann, E. R. Grant and M. G. White, *J. Chem. Phys.* **96** (1992) 7848.

MULTIPHOTON IONIZATION AND ION-PAIR FORMATION IN MOLECULAR HYDROGEN

S. T. PRATT, E. F. McCORMACK, J. L. DEHMER, and P. M. DEHMER
Argonne National Laboratory
Argonne, IL 60439 U.S.A.

ABSTRACT

We have investigated both photoionization and ion-pair formation in molecular hydrogen by using double-resonance excitation via the E,F $^1\Sigma_g^+$, v=6 level. The energetic threshold for ion-pair formation occurs just below the H_2^+ X $^2\Sigma_g^+$, v^+=9 ionization threshold. The spectrum in this region was studied by using both conventional and constant-ionic-state photoelectron spectroscopy, as well as by monitoring the H^- production. The decay dynamics in this region are extremely rich, because excited levels may decay by rotational and vibrational autoionization, by predissociation to neutral H + H^* (n=2,3,4), by predissociation to the ion pair $H^+ + H^-$, and by fluorescence. In addition, the dissociative potential curve of the $2p\sigma_u 3s\sigma_g$ $^1\Sigma_u^+$ doubly excited electronic state crosses the H_2^+ X $^2\Sigma_g^+$ potential curve in the same energy region, and the electronic autoionization of this state is found to significantly influence these decay processes.

1. Introduction

The process of photon-induced ion-pair formation in molecular hydrogen, that is, $H_2 + h\nu \rightarrow H^+ + H^-$, was first studied with conventional single-photon techniques by McCulloh and Walker[1] and subsequently at much higher resolution by Chupka et al.[2] The energetic threshold for ion-pair formation occurs just below the H_2^+ X $^2\Sigma_g^+$, v^+=9 ionization threshold, and, because the cross section for direct excitation into the ion-pair continuum is extremely small, the dominant mechanism for ion-pair formation involves predissociation of Rydberg states with $v' \geq 9$.[1,2] In addition, predissociation of doubly excited electronic states may be important in ion-pair formation. In the present work, we have used double-resonance excitation via the E,F $^1\Sigma_g^+$, v=6 level to reexamine the H_2 spectrum in this region. This overall three-photon process used in the present study accesses the same ungerade manifold of states as the single-photon process. Previous two-photon studies by Kung et al.[3] focused on the gerade manifold.

Figure 1 shows the potential energy curves relevant to the present study.[4-7] The dominant configurations for the $^1\Sigma_u^+(1)$ and $^1\Sigma_u^+(2)$ doubly excited states are $(2p\sigma_u 2s\sigma_g)$ and $(2p\sigma_u 3s\sigma_g)$, respectively.[6,8] The potential curve of a third doubly excited state, which corresponds to $^1\Pi_u(1)$ $(2p\sigma_u\ 3d\pi_g)$, is nearly coincident with that

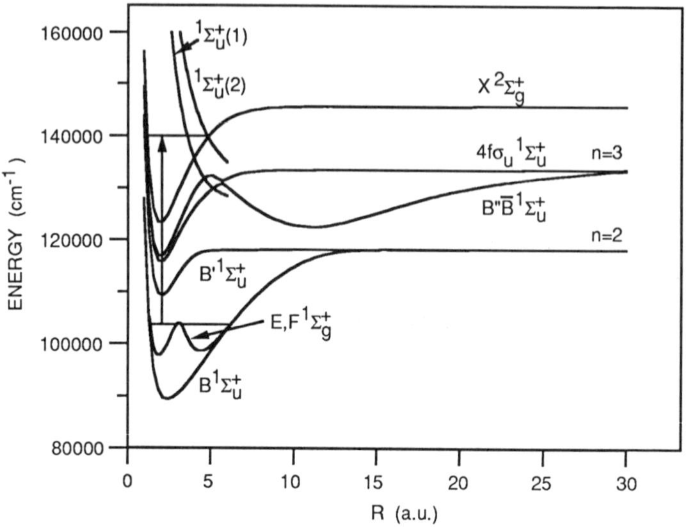

Figure 1. Relevant potential energy curves in the energy region of interest.

of the $^1\Sigma_u^+(2)$ state.[6] The autoionization widths of the $^1\Sigma_u^+(1)$ and $^1\Sigma_u^+(2)$ states increase with internuclear distance and are 0.4 and 0.2 eV, respectively, at R=1.4 a.u.[8] The width of the $^1\Pi_u(1)$ state is considerably smaller (0.004 eV at 1.4 a.u.) and has a more complex R-dependence.[8]

Because the energy range of interest is well above the threshold for both ionization and neutral dissociation, ion-pair formation via resonantly excited levels occurs in competition with autoionization and predissociation. In general, ion-pair formation is an extremely weak process, corresponding to only a fraction of a percent of the total ion yield.[2] To better understand the interplay among the different decay channels, we have used photoelectron spectroscopy to study the autoionization of the excited levels[9] and mass spectrometry to detect the H^- product of ion-pair formation.[10] We are currently examining the neutral products of predissociation as well.

2. Photoelectron Studies

Two different types of photoelectron measurements have been performed. In the first, constant-ionic-state (CIS) spectroscopy was used to determine the wavelength dependence of photoelectron intensity corresponding to different values of v^+. Monitoring photoelectrons with high values of v^+ reduces the contribution of direct ionization and thus enhances the resonant structure relative to the continuum background. For example, when photoelectrons with $v^+=8$ are monitored, the contribution to the spectrum from autoionization of $v' \geq 9$ Rydberg states is strong (because of the $\Delta v=-1$ propensity rule),[11-13] while the contribution to the spectrum from direct ionization is weak (since direct ionization of the E,F $^1\Sigma_g^+$, v=6 intermediate state produces very little $v^+=8$).[14] The comparison of CIS spectra obtained by monitoring photoelectrons with different values of v^+ also provides information on the autoionization mechanisms. In the second type of photoelectron measurement, these mechanisms are explored in even greater detail by recording the complete photoelectron spectra at selected wavelengths of interest.

Figure 2 shows the CIS spectrum obtained by pumping the Q(0) E,F $^1\Sigma_g^+$, v=6 $\leftarrow$ X $^1\Sigma_g^+$, v"=0 transition and probing the transitions to levels in the region of the $v^+=9$ threshold while monitoring the $v^+=8$ photoelectrons. As expected, this spectrum is dominated by resonant transitions to Rydberg series converging to the X $^2\Sigma_g^+$, $v^+\geq 9$ limits. Because the inner well of the E,F $^1\Sigma_g^+$ state is best described by the $1s\sigma_g 2s\sigma_g$

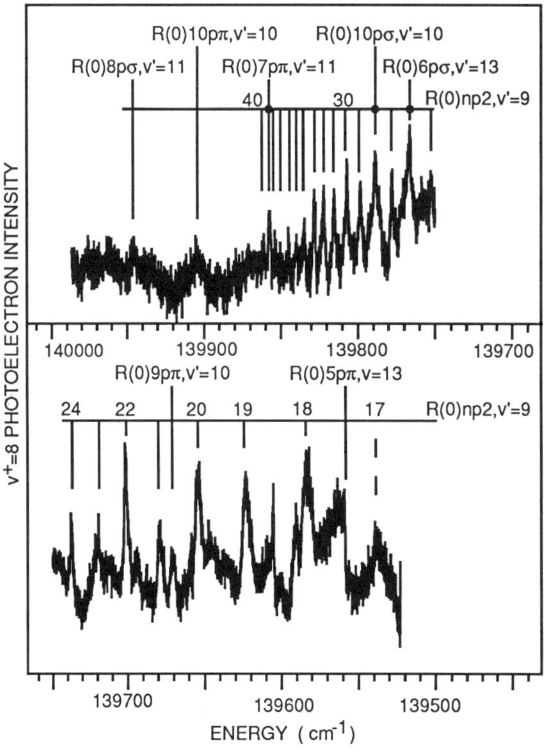

Figure 2. The CIS spectrum obtained by pumping the two-photon Q(0) E,F $^1\Sigma_g^+$, v=6 $\leftarrow$ X $^1\Sigma_g^+$, v"=0, transition, scanning the probe laser, and monitoring the H_2^+ X $^2\Sigma_g^+$, $v^+=8$ photoelectron band. The energy is with respect to the X $^1\Sigma_g^+$, v"=0, J"=0 level.

configuration, the dominant probe transitions are to (X $^2\Sigma_g^+$)np Rydberg series. For the Q(0) pump transition, two Rydberg series that converge to the X $^2\Sigma_g^+$, v+=9 ionization threshold are allowed. These correspond to the (X $^2\Sigma_g^+$)np, v'=9, N'=0 and (X $^2\Sigma_g^+$)np, v'=9, N'=2 series, which converge to the X $^2\Sigma_g^+$, v+=9, N+=0 and 2 limits, respectively. These two series may be denoted R(0)np0, v'=9 and R(0)np2, v'=9. Low-lying members of the analogous series with v'≥10 can also be observed in the wavelength region of Fig. 2.

Between 139750 and 139850 cm^{-1}, Fig. 2 displays a single, intense series, which converges to the N+=2 threshold and clearly corresponds to the R(0)np2, v'=9 series. This series is also quite strong in the ion-pair spectrum discussed below. The X $^2\Sigma_g^+$, v+=9, N+=0 threshold is at 139831.135 cm^{-1}.[15-17] In the energy region above 139750 cm^{-1}, the R(0)np0, v'=9 series produces a quasi-continuum that does not appear to affect the regularity of the R(0)np2, v'=9 series; however, below 139750 cm^{-1}, the R(0)np2, v'=9 series is more irregular because of perturbations by members of the R(0)np0, v'=9 series and by interloping members of other series. It is still possible to extrapolate the R(0)np2, v'=9 series to lower energies, and most of the major features fall reasonably well into this series. However, the assignments at lower energies must be considered more tentative.

The photoelectron spectrum of H_2 obtained after single-color, two-photon-resonant, three-photon ionization via the E,F $^1\Sigma_g^+$, v=6 ← X $^1\Sigma_g^+$, v"=0 transition has been reported in several previous studies.[14,18,19] Because of the v=2 Rydberg character of the inner-well portion of the E,F $^1\Sigma_g^+$, v=6 level, the photoelectron spectrum is strongly peaked at v+=2, with the v+=1, 2, 3, and 4 bands containing approximately 13%, 42%, 12%, and 12% of the total intensity.[14,19] None of the remaining bands contains more than 5% of the total intensity, except for the v+=13 band, which is enhanced by ionization from the outer well of the E,F $^1\Sigma_g^+$ state and contains approximately 7% of the total intensity.[14,19] If the branching ratios are renormalized to include only the v+=0-9 ionization continua, the v+=7, 8, and 9 bands all contain less than 3% of the total intensity.

Figure 3 shows typical two-color photoelectron spectra for the Q(1) E,F $^1\Sigma_g^+$, v=6 ← X $^1\Sigma_g^+$, v"=0 pump transition with the probe laser tuned in the region just below the H_2^+ $^2\Sigma_g^+$, v+=9 ionization threshold. Figure 3a corresponds to a total energy of 139796.3 cm^{-1}, which is between two resonances in the wavelength spectrum; Fig. 3b corresponds to the peak of the R(1)22p3, v'=9 resonance at a total energy of 139799.5 cm^{-1}. In both spectra the v+=7 and 8 bands are significantly more intense than the corresponding bands in the single-color photoelectron spectrum. In general, the summed branching ratio for the v+=7 and 8 bands in spectra obtained throughout this region of the spectrum is ~0.15; this is ~3 times greater than that of the single-color photoelectron spectrum. In contrast, the v+=5 and 6 photoelectron bands are comparable in intensity in the single-color and two-color spectra. In addition to this overall difference in the appearances of the one- and two-color photoelectron spectra,

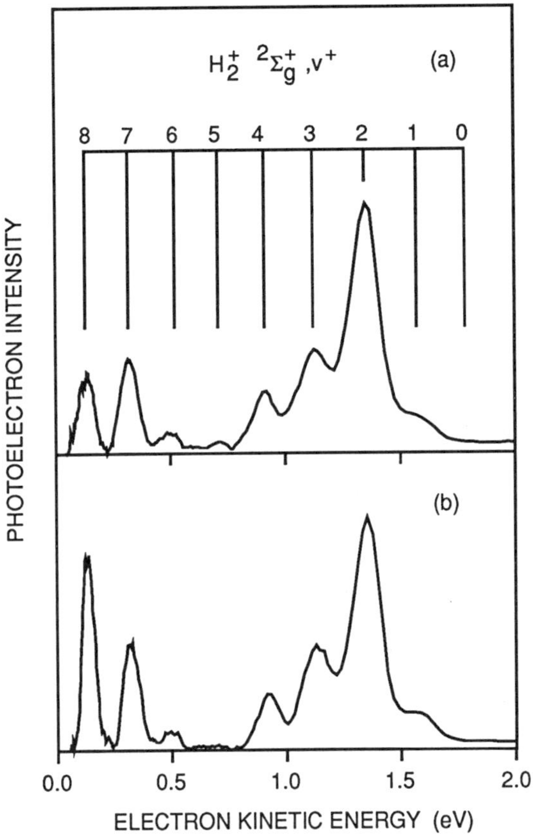

Figure 3. Photoelectron spectra obtained after two-color ionization via the E,F $^1\Sigma_g^+$, v=6, J=1 level. (a) Off-resonance spectrum at a total energy of 139976.3 cm^{-1}. (b) On-resonance spectrum at 139799.5 cm^{-1}, corresponding to the R(1)22p3, v'=9 transition.

Fig. 3 clearly indicates that the two-color photoelectron spectra display a significant wavelength dependence, reflecting the resonant structure in the excitation spectrum.

Because the one- and two-color photoelectron spectra were recorded by photoionization of the same intermediate level, the observed differences in the vibrational branching ratios must be due to the final state. Two-color photoionization via the E,F $^1\Sigma_g^+$, v=6 level in the energy region just below the v^+=9 threshold is expected to be affected by the resonant structure produced by Rydberg states converging to the $v^+\geq9$ ionization thresholds. These resonances decay by vibrational autoionization and by predissociation, and they are expected to be relatively sharp. Thus, as discussed in more detail below, these resonances will influence the detailed wavelength dependence of the vibrational branching ratios; however, they probably are not responsible for the relatively high intensity of the $v^+=7$ and 8 photoelectron bands that is generally observed throughout the region of the present study.

Figure 1 shows that the $^1\Sigma_u^+(2)$ potential curve crosses the X $^2\Sigma_g^+$ potential curve at an energy just above the $v^+=9$ level. Although the vibrational overlap between the inner well of the E,F $^1\Sigma_g^+$, v=6 level and the doubly excited states is negligible near the X $^2\Sigma_g^+$, $v^+=9$ ionization threshold, excitation from the outer well of the E,F $^1\Sigma_g^+$, v=6 level is expected to have a more favorable vibrational overlap and thus produce observable consequences in the photoelectron spectrum. The large widths of the $^1\Sigma_u^+$ doubly excited states are also expected to influence the observed branching ratios over a broad energy range. This assumption is consistent with the observation of the unexpectedly high intensity of the X $^2\Sigma_g^+$, $v^+=7$, 8 photoelectron bands over a relatively large energy region near the $v^+=9$ limit. The electronic transition from the $(1s\sigma_g 2s\sigma_g)$ inner well of the E,F $^1\Sigma_g^+$ state to the $(2p\sigma_u 2s\sigma_g)$ $^1\Sigma_g^+(1)$ state is fully allowed, and the transition moment is expected to be large. The corresponding transitions to the $(2p\sigma_u 3s\sigma_g)$ $^1\Sigma_u^+(2)$ and $(2p\sigma_u 3d\pi_g)$ $^1\Pi_u(1)$ states are two-electron transitions, which are nominally forbidden. However, the $^1\Sigma_u^+(2)$ state is expected to contain some $(2p\sigma_u 2s\sigma_g)$ character, and the transition may gain intensity in that manner. The outer well of the E,F $^1\Sigma_g^+$ state has the nominal configuration $(2p\sigma_u)^2$, from which transitions to all three doubly excited states are allowed.

Using the δ-function approximation for the Franck-Condon overlap, an examination of the region of R sampled by the E,F $^1\Sigma_g^+$, v=6 wavefunction indicates that autoionization of the $^1\Sigma_u^+(1)$ state is more likely to populate lower lying vibrational levels (e.g., $v^+=4$, 5, 6) than the $v^+=7$ and 8 levels. However, because the $^1\Sigma_u^+(2)$ state crosses the ionic curve at the energy of the $v^+=9$ level, autoionization of this state is expected to populate the $v^+=7$ and 8 levels. The same is true for the $^1\Pi_u$ state, although its smaller autoionization width may reduce its contribution. Thus, it appears that the $^1\Sigma_u^+(2)$ doubly excited state is the most likely cause of the observed effects, although the $^1\Pi_u(1)$ state may contribute to some extent. Note that just below the X $^2\Sigma_g^+$, $v^+=9$, $N^+=0$ threshold, autoionization of the quasicontinuum of the R(0)np0, v'=9 series of autoionizing levels may also enhance the $v^+=8$ continuum.

The dramatic increase of the relative intensity of the $v^+=8$ band in Fig. 3b (on resonance) with respect to Fig. 3a (off resonance) is consistent with the propensity rule for vibrational autoionization.[11-13] This behavior, i.e., the increased intensity of the $v^+=8$ band on resonance, appears to be general throughout the region to the X $^2\Sigma_g^+$, $v^+=9$, $N^+=J^+$ threshold. As indicated by the CIS spectrum of Fig. 2, the resonant structure is most clearly reflected in the $v^+=8$ bands. Although the corresponding $v^+=7$ CIS spectra do show resonant structure analogous to that observed in the $v^+=8$ CIS spectra, the intensities of the resonant features are considerably weaker, and the background continuum is larger than in the former spectra. The intensity of the $v^+=5$ and 6 photoelectron bands is small throughout this region. In summary, both electronic and vibrational autoionization appear to play a strong role in determining the photoionization dynamics in this region of the spectrum.

3. H⁻ Studies

Figure 4 shows the H^- spectrum obtained by pumping the Q(0) E,F $^1\Sigma_g^+$, v=6 ← X $^1\Sigma_g^+$, v"=0 transition and scanning in the region of the threshold for ion-pair formation. The spectra were recorded by using a delayed, pulsed electric field to extract the H^-, so that the spectra display the field-free structure. Although the same series are observed as in the CIS spectrum of Fig. 2, the intensities are quite different. Figure 5 displays a small portion of the v^+=8 CIS and H^- spectra together. The overall

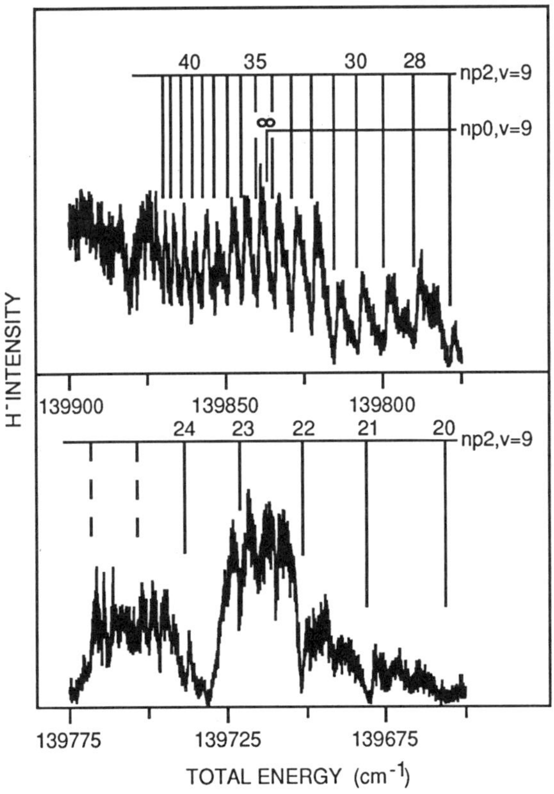

Figure 4. A portion of the field-free excitation spectrum for ion-pair formation obtained by pumping the Q(0) E,F $^1\Sigma_g^+$, v'=6 ← X $^1\Sigma_g^+$, v"=0 transition. The total energy is with respect to the X $^1\Sigma_g^+$, v"=0, J"=0 level.

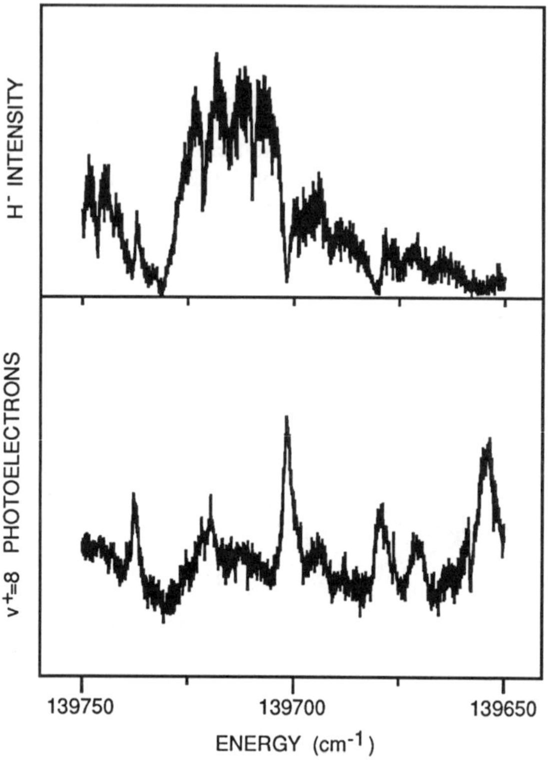

Figure 5. A portion of the $v^+=8$ CIS spectrum of Fig. 2 plotted with the corresponding portion of the H^- spectrum of Fig. 4.

intensity of the H^- signal is a small fraction of a percent of that of the $v^+=8$ CIS signal, reflecting the weak nature of the ion-pair process. Figure 5 clearly shows strong variations in the relative intensities and profiles of the two decay pathways, as some features appear as peaks in both spectra whereas others appear as peaks in the CIS spectrum and dips in the H^- spectrum. This variation must result from the complex interactions among the initially excited Rydberg states and the singly and doubly excited predissociative states at small-R as well as from the crossings and avoided crossings in the intermediate-R region (10-270 a.u.). A detailed understanding of the dynamics of ion-pair formation must await detailed theoretical calculations. These might best be performed within the framework of multichannel quantum defect theory (MQDT). Experimentally, we are now attempting to monitor the neutral H(1s) + $H(n\ell)$ decay pathways as a function of wavelength in an effort to understand more fully the competing decay pathways for this complex region of the spectrum.

Beyond the crossing with the n=4 asymptotic curves at ~270 a.u., the ion-pair curve should be unperturbed by crossings with any other potential curves. In this region, the ion-pair system experiences a true Coulomb potential, and it is possible to draw analogies between the $H^+ + H^-$ system and the $H^+ + e^-$ system. In particular, the phenomenon of field ionization,[20-22] that is, the depression of the observed ionization threshold as a function of applied electric field, is well known in atoms and molecules. For m=0 levels in a pure Coulomb field, the classical shift of the ionization threshold is given by $\Delta = \alpha F^{1/2}$, where $\alpha = -6.11\ \mathrm{cm}^{-1}/\sqrt{\mathrm{V/cm}}$ and F is the electric field strength.[21,22] (The situation for m≠0 is complicated by a centrifugal term that has been discussed by Cooke and Gallagher.[22] The effect of this term is typically a few percent.) For applied fields of 100-500 v/cm, the saddle point in the effective potential of the Coulomb plus external field occurs at several thousand a.u. At this separation, the ion-pair system is expected to follow the pure Coulomb potential, and the ideas of field ionization should be directly transferable to the phenomenon of an external field-induced reduction in the threshold for ion-pair formation.

Figure 6 shows a plot of the experimental threshold for H^- formation as a function of the square root of the applied DC electric field.[10] The linear dependence of

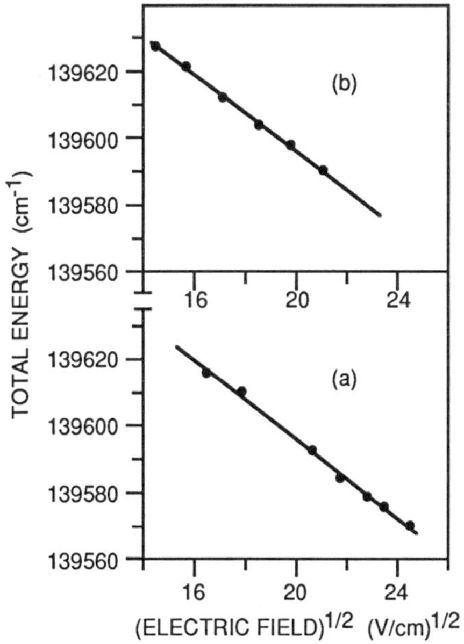

Figure 6. The threshold for H^- formation vs. the square root of the applied electric field, obtained by using the E,F $^1\Sigma_g^+$, v=6, J=0 intermediate state. The energy is with respect to the X $^1\Sigma_g^+$, v"=0, J"=0 level. (b) Same as (a) but for the E,F $^1\Sigma_g^+$, v=6, J=1 intermediate state.

the threshold on the square root of the applied electric field is clearly evident. The slopes of the two curves are -5.9 ± 0.1 and -5.7 ± 0.2 $cm^{-1} / \sqrt{V / cm}$ in Figs. 6(a) and (b), respectively. Thus, the analogy between the $H^+ + H^-$ system and the $H_2^+ + e^-$ system appears to be quite appropriate at large R.

Taking the analogy with a Coulombic system one step further, at very large R the vibrational levels of the ion-pair system are expected to fall into a Rydberg series converging to the ion-pair threshold. The energy levels would be described by $E_{v,J} = D_0(H^+ + H^-) - R_{(H^+ + H^-)}/(n-\mu_{v,J})^2$, where D_0 is the ion-pair dissociation energy, $R_{H^+ - H^-}$ is the Rydberg constant scaled by the reduced mass of the system, and $\mu_{v,J}$ is the quantum defect, which depends on v and J. At small R, the ion-pair curve deviates strongly from the 1/R potential, as molecular effects resulting from the multiparticle character of H^- dominate the potential. Unlike the electron-ion potential, which is dominated by the 1/r term at small r as well, the resulting ion-pair potential is much more weakly bound than expected on the basis of the 1/R potential alone. The effect is to introduce a much stronger energy dependence to the quantum defect. Pan and Mies[23] have calculated the quantum defects for the vibrational energy levels of the lowest ion-pair state of LiI and have clearly demonstrated the Rydberg character of the high vibrational levels. As expected, the energy dependence of μ is much stronger than in the corresponding electron-ion system; however, for $v \gtrsim 1000$, the vibrational quantum defect is approximately constant as a function of v. The experimental study of Rydberg vibrational levels in the ion-pair potential of a simple molecule such as H_2 may provide new insight into the long-range dynamics of simple molecular systems.

4. Acknowledgements

This work was supported by the U.S. Department of Energy, Office of Energy Research, Office of Health and Environmental Research, under Contract W-31-109-Eng-38. EFM was also supported by the Hollaender Distinguished Postdoctoral Fellowship Program of DOE-OHER.

5. References

1. K. E. McCulloh and J. A. Walker, *Chem Phys. Lett.* **23** (1974) 439.
2. W. A. Chupka, P. M. Dehmer, and W. T. Jivery, *J. Chem. Phys.* **63** (1975) 3929.
3. A. H. Kung, R. H. Page, R. J. Larkin, Y. R. Shen, and Y. T. Lee, *Phys. Rev. Lett* **56** (1986) 328.
4. L. Wolniewicz and K. Dressler, *J. Chem. Phys.* **82** (1985) 3292.
5. L. Wolniewicz and K. Dressler, *J. Chem. Phys.* **88** (1988) 3861.
6. S. L. Guberman, *J. Chem. Phys.* **78** (1983) 1404.
7. H. Wind, *J. Chem. Phys.* **42** (1965) 2371; D. M. Bishop and R. W. Wetmore, Mol. Phys. **26** (1973) 145.
8. H. Takagi and H. Nakamura, *Phys. Rev. A* **27** (1983) 691.
9. S. T. Pratt, P. M. Dehmer, and J. L. Dehmer, *J. Chem. Phys.* in press.

10. S. T. Pratt, E. F. McCormack, J. L. Dehmer, and P. M. Dehmer, *Phys. Rev. Lett.* **68** (1992) 584.
11. R. S. Berry, *J. Chem. Phys.* **45** (1966) 1228.
12. J. N. Bardsley, *Chem. Phys. Lett.* **1** (1967) 229.
13. G. Herzberg and Ch. Jungen, *J. Mol. Spectrosc.* **41** (1972) 425.
14. S. L. Anderson, G. D. Kubiak, and R. N. Zare, *Chem. Phys. Lett.* **105** (1984) 22.
15. Calculated from the ionization potential of Ref. 16 and the dissociation energy of Ref. 17.
16. E. F. McCormack, J. M. Gilligan, C. Cornaggia, and E. E. Eyler, *Phys. Rev. A* **39** (1989) 2260.
17. L. Wolniewicz and T. Orlikowski, *Mol. Phys.* **74** (1991) 103.
18. M. A. O'Halloran, P. M. Dehmer, F. S. Tomkins, S. T. Pratt, and J. L. Dehmer, *J. Chem. Phys.* **89** (1988) 75.
19. E. Y. Xu, A. P. Hickman, R. Kachru, T. Tsuboi, and H. Helm, *Phys. Rev. A* **40** (1989) 7031.
20. See, for example, D. Kleppner, M. G. Littman, and M. L. Zimmerman, in *Rydberg States of Atoms and Molecules*, eds. R. F. Stebbings and F. B. Dunning (Cambridge University Press, New York, 1983) p. 73.
21. M. G. Littman, M. M. Kash, and D. Kleppner, *Phys. Rev. Lett.* **41** (1978) 401.
22. W. E. Cooke and T. F. Gallagher, *Phys. Rev. A* **17** (1978) 1226.
23. S. H. Pan and F. H. Mies, *J. Chem. Phys.* **89** (1988) 3096.

V. ELECTRONIC STRUCTURES IN SOLIDS

Photoelectron Spectroscopy in Heavy fermions: Inconsistencies With the Kondo Model*

A.J. Arko, J.J. Joyce, R.R. Blyth, P.C. Canfield, J.D. Thompson, R.J. Bartlett and Z. Fisk
Los Alamos National Laboratory, Los Alamos, NM 87544
and
J. Lawrence and J. Tang
University of California, Irvine Campus, Irvine CA
and
P. Riseborough
Polytechnic University, Brooklyn, NY

ABSTRACT

We have investigated a number of Ce and Yb heavy fermion compounds via photoelectron spectroscopy and compared the results to the predictions of the Imurity Anderson Hamiltonian within the Gunnarson-Schonhammer approach. For the low T_K materials investigated we find little or no correlation with T_K, the only parameter that can be determined independent of photoemission.

1. Introduction

Photoelectron spectra (PES) of Ce and Yb heavy fermions have been the subject of intense study[1-4] for a number of years. The valence band 4f excitations exhibit a complex structure[4] near the Fermi energy which cannot be reconciled with a band structure density of states[5,6] (DOS). Indeed, the heavy electron masses imply an extremely nasrrow structure at E_F which is not obtained from ordinary band theory. In UPt_3, a 5f heavy electron material for which calculations exist, for example, enhancements of the band mass (using the local density approximation, LDA) of order of 20 are needed[7] to compare with measurement. Clearly, the 4f electron interacts with a broad conduction band in a complex fashion which poses a challenge to the theorist.

While a number of models have been proposed[8] to explain the PES data, there appears to be a growing concensus that the Gunnarson-Schonhammer (GS) approach[9-11], as well as the similar non-crossing approximation[12] (NCA), give a comprehensive explanation of valence and conduction band PES for Ce and Yb systems. This elegant and quantitative theory, based on the Anderson single impurity Hamiltonian[13], assumes that the 4f electrons in heavy electron systems interact very weakly with the conduction band so that a description in terms of this localized model is appropriate. The single impurity Hamiltonian has been extremely successful in describing various bulk properties of heavy electron compounds[8,14], despite the fact that a 4f "impurity" exists at every site, and thus the concept of a "Kondo lattice" has gained acceptance. The greatest impetus toward accepting the GS and NCA approaches was given by PES results[1-4] which for Ce compunds usually show a multi-peaked structure even though the 4f state is occupied by only one 4f electron. A semblance of systematics has been obtained in the past on Ce systems[2] but with instrument resolution far too large to look at fine details. Resolution has steadily improved until now we are ready to re-examine the systematics based on the GS model.

There are several key predictions[9-12] of the GS model. The first is the existence of the so-called Kondo Resonance (KR) which is predicted to be: (i) of width kT_K (where

T_K is the Kondo temperature), (ii) to be of integrated weight kT_K/Γ (where Γ is the hybridization width), and (iii) to lie a distance kT_K above (below) the Fermi energy E_F in Ce (Yb) compounds. For Ce compounds the KR is most appropriately viewed in inverse photoemission (BIS) since only the tail of the resonance lies below E_F ; for Yb compounds the KR can be directly observed in photoemission spectra (PES). An important feature of the KR is the prediction (iv) that the amplitude of the KR will dramatically decrease with temperature, while the total spectral weight decreases only in the amount that the f-occupancy (n_f) changes. This latter will be only a small effect for small T_K materials.

In the presence of spin orbit and crystal field splitting the predicted spectra are considerably more complicated[2]; in addition to the KR of the ground state multiplet, crystal field (CF) and spin orbit (SO) sidebands are predicted above and below E_F, with characteristic temperature scales for renormalization of the spectral weight which can be substantially different[9] than T_K. Within the NCA approach it is these sidebands which yield the bulk of the intensity in the near-E_F region, with the KR itself contributing only a small portion. CF states have been observed as quasielastic peaks in neutron diffraction data[15-17]. If we assume that to first order the neutron quasielastic linewidth is a measure of the temperature scale (T_{Kcf}) of the CF state[18], then these states would be expected to decay with a Kondo temperature of T_{Kcf}. The effective Kondo temperature of the SO sideband (T_{Kso}) can be substantially larger than T_K and renormalization is not likely to be observed[2]. The proponents of the GS model further claim that an additional complication in the near-E_F PES intensity is the effect of finite coulomb correlation[2,9-12] ($U_{ff} << \infty$) which allows a finite amount of double f-occupancy. This presumably adds substantial f-spectral weight in the KR region which will be temperature independent.

We thus see that the predictions of the theory are complex and difficult to separate out from simple mechanisms which also must be part of any metallic compound. Part of the difficulty can be overcome by working with Yb compounds where the KR is fully occupied so that one has the benefit of the full T-dependence[12].We will show in this paper that despite the subtleties of the effects there are predictions which can be tested and which seem to be inconsistent with results. In the discussion we will look at each prediction in turn and see how it agrees (or disagrees) with experiment.

2. Experimental

Specimens of heavy fermions were either prepared in single crystal form (by flux growth[19]) or melted with an electric arc or in an enclosed capsule as an on-stoichiometry melt. In each case they were characterized using X-ray diffraction as well as susceptibility to ensure that the compounds were single phase and displayed the proper T_K. Three materials ($CeSi_2$, $CeAl_3$ and $YbAgCu_4$) were poly-crystalline. Measurements were carried out at the Los Alamos U3C beamline at NSLS with specimens cleaved in-situ and at temperature to expose a clean surface. Typical base pressures in our UHV chamber were $1x10^{-10}$ Torr. The instrument resolution varied from ≈ 60 meV at 60 eV photon energy (primarily used for Yb compounds) to 90 mev at the Ce 4f resonance of ≈ 120 eV. Tempearture control was effected by having the specimens in contact with a cryostat cold finger with either liquid nitrogen or liquid helium as the cryogen. Typical spectra were collected in less than one hour.

3. Results and Discussion

In our recent publications[18,20,21] we call into question the systematics expected on the basis of the Kondo picture. In Fig. 1 we show low-resolution valence band PES spectra taken at the 4f resonance (hν = 120 eV) for a series of heavy fermion compounds having T_K's ranging from ≈ 3° K to ≈ 300° K. At this photon energy the 4f and 5d features are expected to dominate the spectrum. At first glance it would appear that the predictions of the GS theory are borne out since three 4f-related features are seen in the spectra corresponding to: i) the KR and the cluster of CF states nearest E_F, ii) the spin-orbit split sideband at -0.3 eV, and iii) the d-screened f^0 state at ≈ -2 eV sometimes called the "main peak" (actually, only the γ-Ce data of Ref. 22 have sufficient resolution in Fig. 1 to clearly resolve the two near-E_F features.) On closer inspection, however, problems develop. Let us look at each property in detail below.

3.1. *Spectral Weight and Systematics with T_K*

One is immediately struck in Fig. 1 by the relatively constant ratios between the spectral weights of the main peak and the two features near E_F. Indeed, the near-E_F peaks are smaller in $CeBe_{13}$ (T_K = 300K) than in materials with smaller T_K's (recall that the weight of the KR, and to a smaller degree the sidebands, should vary[12] as kT_K/T). High resolution data in Fig. 2 for the two near-E_F features show also that the weight ratio of the so-called KR to the SO-sideband seems to likewise remain moreor less constant with T_K (in contrast to Ref. 23 where systematics with T_K are presumably observed). This is especially striking for $CeSb_2$, a material with a ferromagnetic transition[24] at 10° K and possibly a vanishingly small T_K.

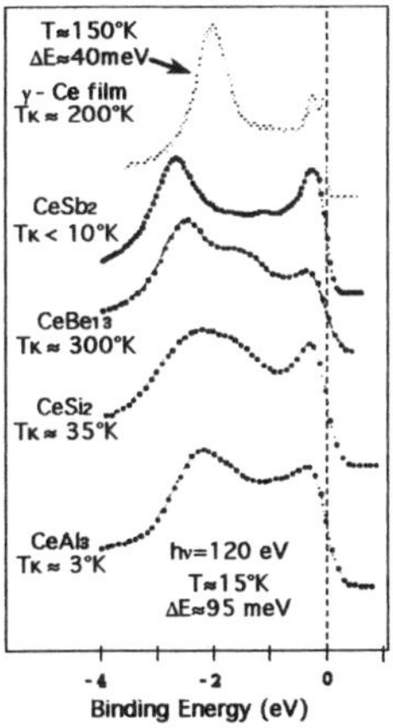

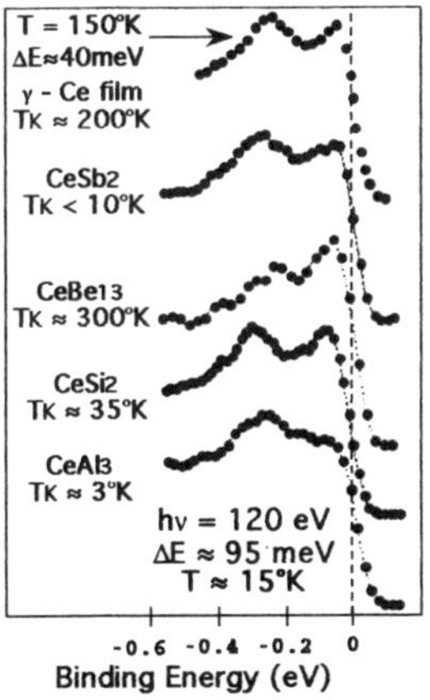

Fig. 1. (Left) Low resolution (ΔE ≈ 150 meV) valence band spectra for several Ce heavy fermions taken at ≈20 K. The high resolution spectrum for Γ-Ce is from Ref. 22. The T_K's vary by two orders of magnitude.

Fig. 2. (Right) High resolution spectra for the compounds in Fig. 1 within 600 meV of the Fermi energy. Note the similarity for all materials.

Two distinct mechanisms have been proposed to resolve the nearly constant spectral weight of the near-E_F features on the basis of the GS or NCA models. The first involves the introduction of the already mentioned CF and SO split states combined with lifetime broadening[23,25]. The difficulties with this approach are that i) the spectral weight (and width) of even these states is to a considerable degree a function of T_K, and 2) one would anticipate *a priory* that the lineshape and amplitude would be a strong function of crystal structure. This does not seem to be the case. While the lineshape will be considered in greater detail in the next sub-section, one can see from Fig. 3, where we show our GS calculation superimposed on the data of Ref. 25, that the mere introduction of CF and SO states results in a very poor fit to the data. Moreover, it would yield far too small a spectral weight for very small T_K materials (e.g., CeAl or $CeSb_2$) as discussed below.

A second proposed mechanism[2,9-11] stresses the importance of the effects of a finite U_{ff} such that doubly occupied 4f states are allowed. It is claimed that these f^2 states yield a dramatic amplitude increase at E_F which accounts for the nearly constant spectral weight in low-T_K materials. This effect is particularly important for very small-T_K materials where the model would otherwise yield only minimal spectreal weight at E_F (see inset of Fig. 4). We have tested this hypothesis with our own calculations for infinite and finite U_{ff} within the GS model. We find that if T_K is constrained to remain invariant for different values of U_{ff}, the net effect of double occupancy is to increase the 4f spectral weight throughout the valence band region without any significant change in relative intensities, and especially no change in widths of the near-E_F features (some narrowing occurs for the main peak since the hybridization must be decreased for small U_{ff} in order to maintain T_K). This is shown in Fig. 3 where we display the two calculated spectra (for $U_{ff} = \infty$ and $U_{ff} = 7$ eV) superimposed on the data of Ref. 25 which represent the highest resolution data to date (18 meV) on any heavy fermion system. Here T_K was constrained to 37°K (≈ thermodynamic value[14] for $CeSi_2$) while the calculated spectra have been broadened by 18 meV to simulate instrument resolution[25] and have been normalized to each other at the main peak at -2 eV. In Fig.4 we also reproduce the spectra displayed in Ref. 2 using their published parameters. Curves A and B are identical to those in Ref. 2 and represent $U_{ff} = \infty$ and 6.5 eV respectively where T_K is allowed to vary from 0.004°K to 0.6°K. Curve C represents U_{ff}=6.5 eV but with T_K = 0.004°K. The unbroadened version of curve A is shown in the inset of Fig.4 superimposed on the broadened curve. Clearly a finite U_{ff} will not yield the

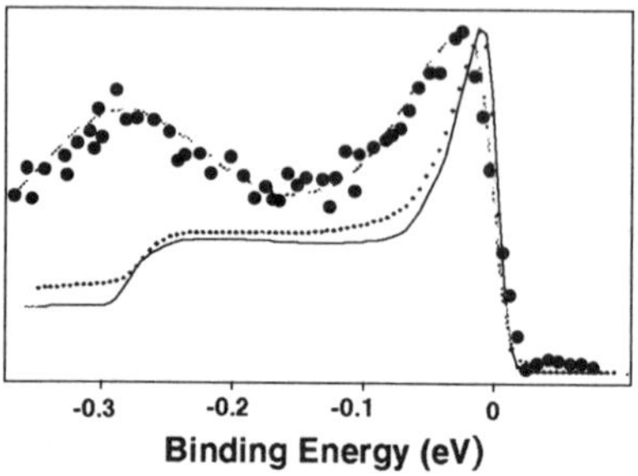

Fig. 3. Near E_F data from Ref. 25 for $CeSi_2$ taken at He I and He II energies with 18 meV resolution. The line through the data is our Voigt function fit. The solid and the dashed curves represent our GS calculations for infinite and finite U_{ff} (7 eV) respectively. T_K was held at the thermodynamic value of 37 K.

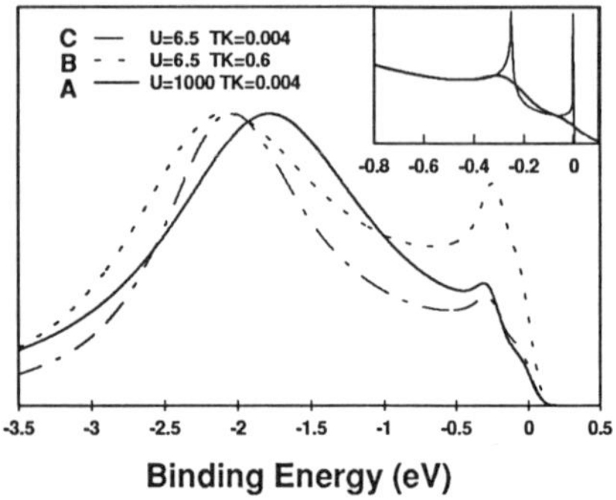

Fig. 4. Spectra obtained from GS calculations to show ineffectiveness of U_{ff} (see text). Parameters are from Ref. 2 for curves A and B. For curve C, U_{ff} = 6.5 eV but T_K is kept invariant with curve B yielding the same spectral weight at E_F. The inset shows the broadened and unbroadened versions of curve A to emphasise the spiked nature of the KR and SO sidebeand.

desired results and we must look for other mechanisms to explain the nearly constant spectral weight with T_K in the near-E_F region. Moreover, from the inset it is obvious that the weight and particularly the width of the SO sideband is strongly dependent on T_K.

3.2. Spectral Width and Lineshape

The width of the apparent KR and the SO sideband likewise is nearly constant with T_K in the data, and far broader than our resolution as well as expectations[12]. It is evident from Fig. 3 that the calculated spectrum, even after including CF (determined from neutron diffraction[15-17]) and SO states while at the same time working with U_{ff} = 7 eV, is still much narrower than the measured spectrum. The authors of Ref. 25, whose data we show, recognized that there was a problem with the width and assumed that lifetime broadening was responsible for the extra width, while in Ref. 2 the extra width is attributed to f^2 occupancy. We have already shown the inadequacy of the latter argument above. As for lifetime broadening, while it is included within the NCA model, this broadening cannot exceed kT_K, while the hole lifetime broadening should be insignificant at E_F. If we allow hole lifetime effects, such a large effect at E_F would translate into a 1eV wide feature[26] at 0.3 eV which is not the case. Thus the inordinately large width of the features at E_F relative to calculations remains unexplained. This is especially true for small-T_K materials where from the inset in Fig.4 one sees that the SO width should likewise be very small. Instrument resolution is not a limiting factor in seeing these effects.

The lineshape of these features does not agree with the calculated spectrum. This is particularly evident for the SO feature at -0.3 eV. We, on the other hand, have fit the data of Ref. 25 with Gaussian-broadened Lorentzians (Voigt functions) and obtained the excellent fit shown by the dashed lines through the data. Since Voigt functions are the conventional lineshapes used for fitting to core level spectra, the excellent fit obtained strongly suggests (but does not prove) a core level nature of these states. Indeed, we find that the lineshapes of all the Ce heavy fermions we investigated can be approximated by Voigt functions and vary between 100 meV and 200 meV in width, with the larger value generally applying to the SO peak at -0.3 eV.

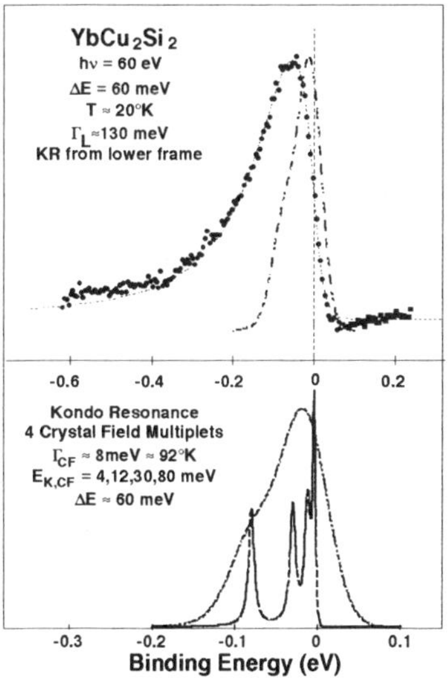

Fig. 5. The peak nearest to E_F in $YbCu_2Si_2$ with the remaining valence band spectrum subtracted off (upper frame). The dashed curve is a simulated Kondo resonance and its sidebands obtained in lower frame by broadening the measured (neutron diff.) crystal field states with a 60 meV Gaussian.

The nearly constant shape and width of the near-E_F features (≈ 100 meV to ≈ 200 meV) extends to Yb heavy fermions as well[27]. In Fig. 5 we show the peak nearest E_F (presumably the KR) for $YbCu_2Si_2$ where all the remaining parts of the spectrum have been subtracted. Here we approximate the crystal field states with equal amplitude doublets having a width determined from neutron diffraction. We have

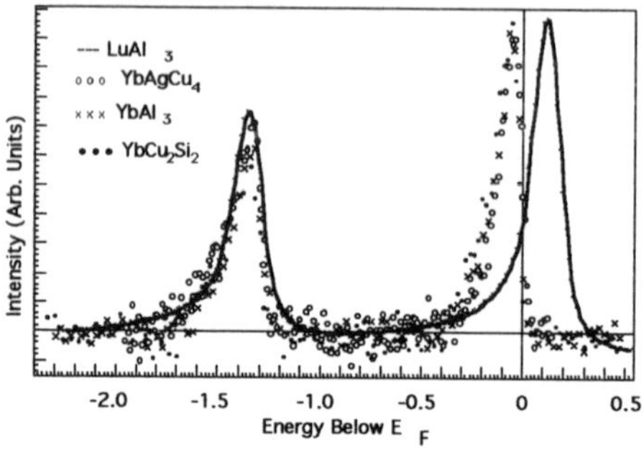

Fig. 6. The two near-E_F spin-orbit split features in three Yb heavy fermions obtained from fitting to raw data (taken at 20K) and stripping away the surface and underlying DOS features (see text). Note that they exactly fall on top of each other, including the branching ratio. The solid line is the same feature in $LuAl_3$ but shifted by 6 eV in order to fall on top of the Yb features. The spin-orbit splitting is somewhat larger, but the similarity to Yb 4f features is striking.

recently shown[18] that this actually overestimates the width of the near-E_F feature compared to a GS calculation. The bottom frame shows these states at the measured energies, with the envelope representing broadening by a 60 meV Gaussian (our resolution). This broadened line is then compared in the top frame to the data. One can see that the same problems are encountered in Yb systems that are encountered in Ce heavy fermions, namely, the width of the features is too large compared to the model and is entirely similar to the widths encountered in Ce systems.

In Fig.6 we show the KR and the SO sideband for three Yb heavy fermions in which the underlying DOS, the ever-present surface features, as well as the secondaries have been subtracted away (more on these subtractions below). The data were taken at ≈ 20°K. The T_K's for these materials vary by more than an order of magnitude[14] (35°K for $YbCu_2Si_2$, 100°K for $YbAgCu_4$, and 400°K for $YbAl_3$) and yet it is possible to almost exactly overlay all the spectra after normalizing at peak amplitude. For effect we also show the 4f spectrum for $LuAl_3$ in which the 4f electrons are situated 6 eV below E_F and cannot possibly be interpreted as the KR and its sideband. We note that they have nearly the same width and branching ratio as the heavy fermion materials, though the spin-orbit splitting is somewhat larger, as expected. This nearly constant width becomes especially difficult to understand for YbAgCu4 where neutron diffraction and bulk property data are consistent[28] with a lack of CF states. If lifetime broadening were responsible for the large widths, then the lack of CF states should result in a much narrower line for this material. On the other hand, the near constancy of the intensity, width, and lineshape is much easier understood if we assume that we are dealing with screened 4f core levels[29-32] at E_F.

3.3. Temperature Dependence

Despite the fact that the Kondo resonance is expected to renormalize dramatically with temperature[12], the observation of this temperature dependence can be subtle and difficult owing to the existence of CF and SO states. Moreover, most of the temperature may only manifest itself as a broadening of the KR and its associated sidebands[25,34]. The situation is not entirely clear. Within NCA it appears[12] that the actual loss of spectral weight is predicted to vary with n_f which is a slow function of temperature. The exact Monte Carlo approach[33], on the other hand, suggests a rapid loss of spectral weight in the KR, much faster than the variation of n_f, with the main peak gaining the lost weight. Within NCA, the broad SO sideband may show no temperature dependence up to room temperature, while the CF sidebands may vary on a scale which is slower than T_K (we call it T_{Kcf}).

Bickers et al.[12] show the unoccupied states in Ce as having a different temperature dependence than the occupied states. Here both the KR and the SO sideband display equivalent variation with temperature unlike the SO state below E_F which is nearly temperature independent. This is very significant since a loss of intensity away from E_F is easier to distinguish from normal processes than a loss at E_F (recall that the Fermi function is temperature dependent). But now, because of particle-hole symmetry, the KR for Yb compounds occurs on the occupied side of E_F so that one obtains the full KR intensity rather than just a tail as in Ce materials, and the SO sideband should display a significant temperature dependence. Thus one anticipates a much larger temperature dependence in Yb compounds vs Ce compounds.The bottom line is that a search for temperature dependence over and above simple mechanisms is best done with high resolution PES in Yb compounds. Unfortunately, as we will see below, there are serious (but solvable) complications with surface-related features in Yb materials.

In Fig.7 we show the raw spectra in the near-E_F region for $CeSi_2$ and $YbCu_2Si_2$ for three temperatures (20°K, 80°K, and 300°K). $T_K \approx 35$°K for both materials. The spectra were normalized using the Si-2p core levels for $CeSi_2$ (Fig.7a) and Cu-3d valence bands for $YbCu_2Si_2$ (Fig.7b). Very small, and nearly identical, temperature dependences are seen in both materials, suggesting that spectral wight losses (if any) are nearly the same in both materials. One would have anticipated far more weight loss (perhaps an order of magnitude) in the Yb compound owing to the full occupancy of the KR. We have previously shown[21] that any temperature dependence in $CeSi_2$ is entirely explainable from simple mechanisms (phonon broadening, surface effects, etc.). Thus it is likely that in Yb compounds simple mechanisms are the also the only source of the very small temperature dependence[18,27]. Fig.7c shows an apparent temperature effect seen in an arc-melted button of $YbCu_2Si_2$. We believe that the use of poly-crystalline materials vs single crystals with their stable surfaces can yield an apparent temperature effect (see below for discussion).

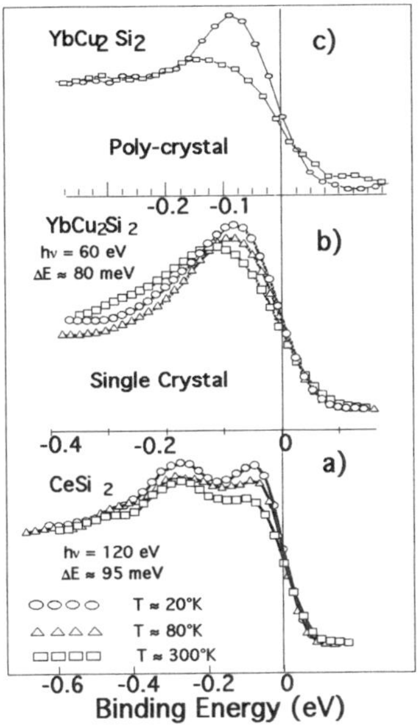

Fig. 7. Raw data at the indicated temperatures (see frame a) for the near-E_F peaks in $CeSi_2$ and $YbCu_2Si_2$. Very little temperature dependence is seen. Frame c is for poly-crystalline $YbCu_2Si_2$. Note that it has a large temperature dependence most likely due to changes in surface structure. For Yb compounds it appears that single crystals are mandatory.

3.4. Problems With Extracting a Temperature Dependence in Yb Materials

Yb compounds tend to be far more complex in their surface structure than Ce compounds. Indeed, various surface features can be a strong function of temperature and/or surface adsorbates. In our example above we showed the temperature dependence (or lack thereof) for a very stable single crystal of a two dimensional material, $YbCu_2Si_2$, which has been cleaved in vacuum to expose the most stable plane. Even here we discovered that adsorbates (with changing temperature) had a strong effect on the surface related features, although happily the bulk features remained unaffected. Three dimensional materials, and especially poly-crystalline materials, have a less stable surface and are thus much more difficult to analyze.

In any attempt to extract a temperature dependence one must account for at least the following effects as one raises the temperature from 20°K to 300°K: i) changes in amplitudes of surface-related features since Yb is known to be divalent[1,2] at the surface; ii) appearance of additional components due perhaps to oxide formation; iii) changes in underlying DOS due to chemical changes beyond the first crystallographic layer; iv) increase in secondary electron emission due to adsorbates; v) increases in surface core-level shifts. If the bulk feature constitutes only a small percentage of the total spectral weight near E_F (≈20% for $YbAl_3$), then even small changes in the surface features can yield an apparent temperature effect. Some, or all, of these effects have been observed by us in the study of single crystal $YbAl_3$ and poly-crystalline $YbAgCu_4$, as well as in poly-crystalline $YbCu_2Si_2$ (Fig.7c). Indeed, it was the apparent large temperature dependence of the near-E_F peak of the latter which launched our attempts to observe a KR in heavy fermions with our improved resolution. Use of the highly stable single crystals eliminated all interesting effects (Fig.7b).

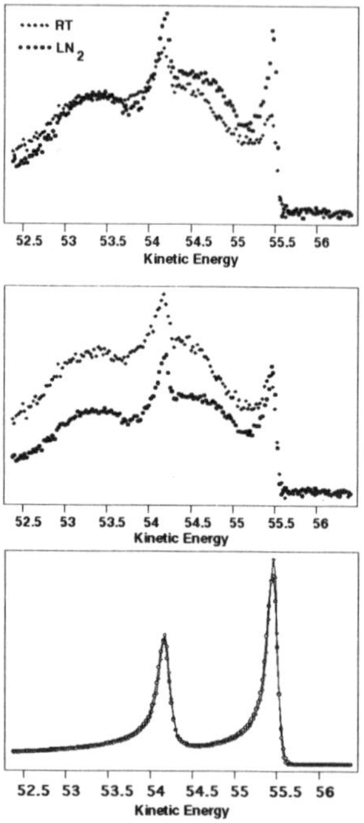

Fig. 8. Spectra at 20 K and 300 K for $YbAl_3$. In top frame the normalization is to surface states while in the middle frame the spectra are correctly normalized to Al-2p core levels. Fitting to the data to extract the bulk 4f's yields no temperature dependence (bottom frame).

In $YbAl_3$ even the use of single crystals fails to suppress the surface driven temperature effects which yield an apparent temperature dependence for the bulk peaks with improper analysis. We show these effects in Fig.8 where in frame A the 20°K and 300°K valence band spectra have been normalized to the surface state at -1 eV with a resulting large apparent temperature effect for the bulk features. In frame B we show the same spectra, but now more appropriately normalized to the Al-2p core levels. One immediately sees that the feature at E_F has not changed much with temperature, most of the changes being evident in the surface features. Thus normalization to surface features, no matter how reproducible the lineshape, can yield misleading results. One can only extract the 4f components from a fit to the data using the best estimates for the behavior of elements i) through v) above. We have performed such fits, which however, because of the complexity of effects are not entirely unique (the differences are mostly in the choice of the particular surface features). Nevertheless, nearly all fitting attempts resulted in a temperature-independence for the bulk 4f peaks. Frame C represents our most reasonable attempt to extract the bulk 4f states (at 20°K and 300°K) from the surface states and the underlying DOS, using a least squares minimization technique with the 4f's represented by Voigt functions. The fitting technique will be described in detail in a later publication[35]. There appears to be no temperature effect at all, which from our point of view is understandible since the first 4f peak occurs at -60 meV where it would be only slightly affected by the Fermi function.

4. Conclusions

We have shown that many of the trends and systematics anticipated on the basis of the GS model are not realized in spectroscopic heavy fermion data. While the GS model predicts the correct number of 4f-related fetaures and their energy position, it fails to correctly predict the spectral weight, the width of both the KR (predicts much too narrow) and the main peak (predicts much too broad a feature, not discussed), temperature dependence (from NCA), as well as lineshape of the KR, the SO sideband and the main peak. The typical 4f-spectrum of a Ce (or Yb) heavy fermion seems to first order to be unrelated to T_K, crystal structure, or the s-p-d DOS of the conduction band. Indeed, the Yb KR and SO sideband very much resemble the 4f core levels of Lu and W. While we are not prepared to say that the GS model is to be abandoned, we point out that any successful model must be able to satisfactorily explain the near constancy of the data, as well as the magnitude of the spectral widths. Bulk property data are very successfully explained by the Anderson Hamiltonian so that we believe that it is only a matter of some invalid approximation which fails this otherwise elegant GS theory.

*Work supported by the U.S. Department of Energy.

5. References

1. J.M. Lawrence and D.M. Mills, *Comm. Cond. Mat. Phys.* **15**, 163 (1991).
2. J.W. Allen, S.-J. Oh, O. Gunnarson, K. Schonhammer, M.B. Maple, M.S. Torikachvili and I. Lindau, *Advances in Physics* **35**, 275 (1986).
3. J. W. Allen, in *Synchrotron Radiation Research: Advances in Surface and Interface Science, Volume I: Techniques,* edited by R.Z. Bachrach (Plenum Press, New York, 1992) pp. 253-323.
4. D.W. Lynch and J.H. Weaver, in *Handbook on the Physics and Chemistry of Rare Earths, Vol. 10,* edited by K.A. Gschneider, Jr. L. Eyring and S. Huffner, (Elsevier Publishers, Amsterdam) pp. 231-300, 1987.
5. R.C. Albers, *Phys. Rev.* B **32**, 7646 (1986); R.C. Albers, A.M. Boring and N.E. christensen, *Phys. Rev. B* **33**, 8116 (1986).
6. O. Eriksson, R.C. Albers, A.M. Boring, G.W. Fernando, Y.G. Hao, and B.R. Cooper, *Phys. Rev. B* 43, 3137 (1991).
7. T. Oguchi and A.J. Freeman, *J. Magn. Magn. Mater.* **52**, 174 (1985).
8. P. A. Lee, T. M. Rice, J. W. Serene, L. J. Sham, J. W. Wilkins, *Comm. Cond. Matt. Phys.* **12**, 99 (1986).
9. O. Gunnarson and K. Schonhammer, *Phys. Rev. B* **31**, 4815 (1985).
10. O. Gunnarson and K.Schonhammer in *Theory of Heavy Fermions and Valence Fluctuations*, T. Kasuya and T. Saso, editors (Springer-Verlag, Berlin, 1986) p. 110 Gunnarson-Schonhammer

11. O. Gunnarson and K. Schonhammer in *Handbook on the Physics and Chemistry of Rare Earths, Vol. 10,* edited by K.A. Gschneider, Jr. L. Eyring and S. Huffner, (Elsevier Publishers, Amsterdam), pp. 103-163, 1987.
12. N. E. Bickers, D.L. Cox and J.W. Wilkins, *Phys. Rev. B* **36**, 2036 (1987)
13. P.W. Anderson, *Phys. Rev.* 124, 41 (1961).
14. G.R. Stewart, *Rev. Mod. Phys.* **56**, 755 (1984).
15. R. Currat, R.G. Lloyd, P.W. Mitchell, A.P. Murani and J.W. Ross, *Physica B* **156&157**, 812 (1989)
16. H. Yashima and T. Satoh, *Solid State Comm.* **41**, 723 (1982)
17. R.M. Galera, A.P. Murani and J. Pierre, *Physica B* 156&157, 801 (1989)
18. J.J. Joyce, A.J. Arko, J.M. Lawrence, P.C. Canfield, R.J. Bartlett, Z. Fisk, and J.D. Thompson, *Solid State Comm.*, in press.
19. P.C. Canfield and Z. Fisk, *Phil. Mag. B* **65**, 1117 (1992).
20. J. J. Joyce, A. J. Arko, J. M. Lawrence, P. C. Canfield, Z. Fisk, R. J. Bartlett, J. D. Thompson, and J. L. Smith, *J. Less Comm. Met.***181**, 161 (1992).
21. J. J. Joyce, A. J. Arko, J. M. Lawrence, P. C. Canfield, Z. Fisk, R. J. Bartlett, and J. D. Thompson, *Phys. Rev. Lett.* **68**, 236 (1992).
22. C. Laubschat, E. Weschke, C. Holtz, M. Domke, O. Strebel,and G. Kaindl, *Phys. Rev. Lett.* **65**, 1639 (1990).
23. F. Patthey, J.-M. Imer, W.-D. Schneider, H. Beck and Y. Baer, *Phys. Rev. B* 42, 8864 (1990).
24. P.C. Canfield, J.D. Thompson, and Z. Fisk, *J. Appl. Phys.* **70**, 5992 (1991); T. Suzuki, A. Isobe, Y. Haga, Y.S. Kwon, and T Kasuya, Int. Conf. on Magnetism, Edinburgh, 2 - 6 Sept., 1991, Abstract Book, pg. 155.
25. F. Patthey, W.-D. Schneider, Y. Baer, B. Delley, *Phys. Rev. Lett.* **58**, 2810 (1987).
26. J.B. Pendry and J.F.L. Hopkinson, *J. Phys. F* **8**, 1009 (1978).
27 J.M. Lawrence, A.J. Arko, J.J. Joyce, P.C. Canfield, Z. Fisk, J.D. Thompson, and R.J. Bartlett, *J. Magn. Magn. Mater.* **108**, 215 (1992).
28. A. Severing, A.P. Murani,J.D. Thompson, Z. Fisk, and C.-K. Loong, *Phys. Rev. B* **41**, 1739 (1990); C. Rossel, K.N. Yang, M.B. Maple, Z. Fisk, E. Zirngiebel and J.D. Thompson, *Phys. Rev. B* **35**, 1914 (1987).
29. P.S. Riseborough, *Solid State Comm.*, 57, 721 (1986).
30. S. Huffner, *J. Phys. F: Met. Phys.* **16**, L31 (1986).
31. S. Huffner, *Z. Phys. B - Cond. Mat.* 86, 241 (1992).
32. M.R. Norman, *Phys. Rev. B* 31, 6261 (1985); M.R. Norman, D.D. Koelling, and A.J. Freeman, *Phys. Rev. B* **31**, 6251 (1985).
33. R.N. Silver, J.E. Gubernaitis, D.S. Silva, M. Jarre, *Phys. Rev. Lett.* **65**, 496 (1990).
34. D. Maltere, M. Grioni, P. Weibel, and Y. Baer, *Phys. Rev. Lett.* **68**, 2656 (1992).
35. J.M. Lawrence, A.J. Arko, J.J. Joyce, P.C. Canfield, Z. Fisk, J.D. Thompson, and R.J. Bartlett, (unpublished).

Angle-Resolved Photoemission of Single Crystal C_{60}

Z.-X. Shen

Department of Applied Physics and the Stanford Synchrotron Radiation Laboratory
Stanford University, Stanford, CA 94305

We report angle-resolved photoemission (ARPES) data from single crystals of C_{60} cleaved in UHV. Our ARPES spectra show only minimal energy dispersion of the C_{60} valence band features. In addition, we found that the valence band spectra of C_{60} consist of many sharp features that can be essentially accounted for by the quantum chemical calculations describing individual molecules. These results suggest that the electronic structure of solid C_{60} is mainly determined by the bonding interactions within the individual molecules, and the intermolecular band width of solid C_{60} is very small.

I. Introduction

C_{60} is a molecule with 60 carbon atoms that occupy the 60 vertices formed by the intersections of 20 hexagonal and 12 pentagonal faces to form a hollow cage, as depicted in Fig.1.[1, 2] Large quantities of C_{60} molecules become available only recently,[3] and this makes it possible to study C_{60} in solid state. Unlike graphite and diamond, solid C_{60} is a molecular crystal with an fcc crystal structure. The C-C bonds within an individual C_{60} molecule are 1.4 Å and 1.46 Å long. This length is significantly shorter than the smallest C-C nearest neighbor distance between carbon atoms in different molecules in solid C_{60}, which is about 3 Å. Although the C-C bond within the C_{60} molecule is shorter than the bond in graphite (1.42 Å and 3.4 Å) and diamond (1.54 Å), its density (1.7 g/cm^3) is smaller than both graphite (2.2 g/cm^3) and diamond (3.5 g/cm^3). This leads us to a picture of the solid C_{60} that is composed of well-separated C_{60} molecules.

In this shortened version of our long paper,[4] we focus on two primary issues: (i) electronic structure as mainly determined by the bonding interactions within individual molecules; (ii) the intermolecular band width as induced by electron hopping between the C_{60} molecules. These issues are important for the understanding of many interesting properties of the fullerenes, in particular the mechanism of superconductivity in alkali doped C_{60}.[5]

II. Experimental

The details of the growth of C_{60} single crystal is described in our previous publication.[4] The quality of the single crystal samples has been verified by transmission Laue pattern, as shown in Fig.2. The Laue pattern was obtained with the X-ray beam perpendicular to the cleaved surface plane. We typically obtain two types of diffraction patterns, showing clear diffraction spots in three or six-fold symmetry, respectively. These diffraction patterns do not have any stripes, implying that the samples indeed are high quality single crystals. We believe that three-fold Laue pattern is taken along the (111) direction of an fcc crystal, while the six-fold Laue pattern is taken either along the (0001) direction from a hcp crystal, or along the (111) direction of an fcc structure with ordered stacking faults.

The photoemission study of C_{60} was performed in a VG ADES 400 system with a hemispherical energy analyzer at the TGM beam line I-2 of the Stanford Synchrotron Radiation Laboratory. The base pressure of the chamber during the experiment was 2×10^{-10} torr, and the overall energy resolution was about 200 meV. The samples on the holder were introduced into the UHV chamber through a fast entrance load-lock and cleaved in situ. The data reported here were collected after the charging had stabilized.

III. Results & Discussions

Fig.3 shows two photoemission spectra recorded at normal emission with photon energies of 60 and 100 eV, respectively. For comparison, we have also included in Fig.4 angle-integrated spectra from graphite and diamond. The valence band spectra of graphite and diamond consist of broad features from dispersive bands. In contrast, the photoemission data from C_{60} crystals exhibit very distinct valence band features as marked. These valence band features agree remarkably well with a cluster calculation represented by vertical bars in Fig.3,[6] indicating that the electronic structure of the C_{60} crystal is mainly determined by the molecular bonding of the individual C_{60} molecules, and the effect of the energy dispersion is relatively small. This assertion also receives very strong support from the fact that photoemission spectra from solid film and from the gas phase are very similar.[7-11]

The difference between the spectra of C_{60} and that of graphite and diamond can be understood from their crystal structures. Because of the relatively short atomic distances involved in graphite(1.42Å) and diamond(1.54Å), there is large wave function overlap and these bands are very dispersive, resulting in broad valence band features. However, the energy dispersion of C_{60} is determined by the ~ 3Å bond length in adjacent molecules. Because of this significantly larger bond length, the energy dispersion is much smaller.

The above discussion brings us to the issue of the conduction band width that is important to understand the mechanism of superconductivity in doped C_{60}. Due to the repulsive nature of the electrons, a narrow band does not favor the traditional electron-phonon coupling mechanism because the retarded screening of the electron-electron repulsion is very weak. As depicted in Fig.4, there are two kinds of band width discussed in the literature. The first is the intramolecular band width, W_{intra}, that includes all the carbon 2s and 2p states. This band width is mainly determined by the intramolecular interactions, and is of the order of 20 eV. The second is the intermolecular band width, W_{inter}, that is mainly caused by the intermolecular interactions in solid C_{60}. This band width is more relevant to the transport properties of the Alkali doped C_{60}. As shown in the left part of the figure, the highest occupied state is the 5 fold degenerate h_u orbital. The three electrons donated by Alkali doping in A_3C_{60} will occupy the t_{1u} level. As C_{60} molecules form solid, the t_{1u} level is broadened into a band with a width of W_{inter}. As depicted in the right side of Fig.4, this band is well separated from the rests of the carbon 2s and 2p bands, and is partially occupied. Therefore, this band is mainly responsible for the transport properties of A_3C_{60}.

The experimental results of the intermolecular band width are presently controversial. Measurements such as IR reflectivity, NMR, critical field, and magnetic susceptibility give intermolecular band widths of about 80-200 me.[12-16] This range of values was obtained indirectly from the density of states measured in these experiments. Contradicting these results, the peak widths for both the highest valence band state and the lowest conduction band state range from 800 meV to 1200 meV.[7-11] If these values from photoemission data are taken as the intermolecular band width,[14] then the photoemission results and the results obtained by other measurements differ by about an order of magnitude.

In the following, we use angle-resolved photoemission as a tool to address this issue. What we will discuss is actually the width of h_u band instead of t_{1u} band. However, one would expect these two bands to have similar widths. Fig.5 presents two complementary sets of angle-resolved photoemission data from a single crystal of C_{60} cleaved in ultra-high vacuum. Panel (a) gives off-normal emission data with a photon energy of 25 eV, while panel (c) gives normal-emission data at various photon energies. In the normal-emission case, we fixed the direction of the **k** vector but changed its amplitude. In the off-normal emission case, we fixed the amplitude of the **k** vector but changed its direction. In both cases, we change the **k** vector by relatively small amounts.

However, this change is very sizable as compared to the small Brillouin zone size of the C_{60} crystal (approximately 60% and 20% for the off- and normal emission cases, respectively).[17] To first order, the most obvious conclusion from Fig.5 is that, although the photoemission feature of C60 has a sizable width, there is minimal energy dispersion of the features. The upper bound for the energy dispersion of the first and second photoemission features is 50 meV that is shown more clearly in panel (b). Although we can not rule out the possibility that this small shift is caused by a change in the spectral background, or that each photoemission feature contains several states and their relative intensity may change with emission angle and photon energy, we are confident to say that the energy dispersions of the C_{60} photoemission features are very small.

The lack of energy dispersion in the photoemission features of this material does not necessarily mean that there are no dispersions of individual bands. All the photoemission features resolved contain several bands (e.g., the first feature is five fold degenerate), and the individual bands may disperse differently. We might have a situation that the individual bands have some dispersion while the centroid of these bands is basically non-dispersive. This scenario leads to a change of the width of the photoemission features; however, since we didn't see any sizable changes of the widths of photoemission features, we think the dispersion of individual bands is also small. Another possible explanation for the lack of energy dispersion in our data is the rotational disorder. It has been found that the C_{60} molecules spin with high frequencies in the solid. [18] The hopping matrix elements and therefore the band width will be affected by the relative orientations of the C_{60} molecules. Nevertheless, the rotational disorder alone can not explain why the photoemission features observed are much broader than the instrumental resolution. A completely disordered system naturally gives narrow bands.

The above discussion makes it very clear that the width of the photoemission feature in C_{60} does not reflect the intermolecular band width of solid C_{60}, in contrast to an earlier paper.[9] The band width of C_{60} is likely to be much narrower than the width of the photoemission feature, even though it is difficult to determine exactly how narrow the real band width is. Now the question that must be answered is how does the broadening of the photoemission peak arise.

We gave a detailed discussion of this issue in our previous paper.[4] An important point to remember in interpreting photoemission data is that photoemission is an excited state measurement. The final state effect in the photoemission process could introduce additional complications. We suggested that the photoemission peak is broadened by the non 0-0 transitions via Franck-Condon principle. Although one might not necessarily expect such a broadening mechanism in C_{60}, this suggestion finds its support in a gas phase experiment by Baltzer et al..[19] With an exceptionally good energy resolution of 22 meV, Baltzer et al. can resolve three features for the highest molecular orbital state h_u (HOMO). Since HOMO is a mono-energetic state, the three resolved features have to be the vibrational states and the non 0-0 transitions have to be strong. Similarly, we expect vibrational final state effects to be important to the photoemission data from the solid.

IV. Summary

Our photoemission spectra from single crystal C_{60} show sharp molecular features, indicating the molecular orbitals are relatively undisturbed in solid C_{60}. Our angle-resolved photoemission data reveals that the C_{60} single crystal has narrow bands that exhibit minimal energy dispersion. This result reconciles the controversy about the intermolecular band width of the C_{60} as measured by different techniques.

Acknowledgment

This work is done in collaboration with following colleagues: J. Wu, D.S. Dessau, R. Cao, D.S. Marshall, P. Pianetta, X. Yang, J. Terry, D.M. King, B.O. Wells, D. Elloway, H.R. Wendt, C.A. Brown, Heinrich Hunziker, and M.S. de Vries. This research is performed at the Stanford Synchrotron Radiation Laboratory (SSRL) which is operated by the Department of Energy, Office of Basic Energy Sciences, Divisions of Chemical Sciences. The Office's Division of Materials Sciences has provided support for this research. This work is also supported by the NSF grant through the Center of Materials Research at the Stanford University and the NSF grant #9121288.

References:

1. E. Osawa, Kagaku, 25, 854 (in Japanese), (1970).
2. H. W. Kroto, J. Heath, S. C. O'Brien, R. F. Curl and R. E. Smalley, Nature 318 (1985) 162.
3. W. Krätschmer, L. D. Lamb, K. Fostiropoulos, D. R. Huffman, Nature 347 (1990) 354.
4. J. Wu, Z.-X. Shen, D.S. Dessau, R. Cao, D.S. Marshall, P. Pianetta, X. Yang, J. Terry, D.M. King, B.O. Wells, D. Elloway, H.R. Wendt, C.A. Brown, Heinrich Hunziker, and M.S. de Vries, To be published in Physica C
5. A. F. Hebard, M. J. Rosseinsky, R. C. Haddon, D. W. Murphy, S. H. Glarum, T. T. M. Palstra, A. P. Ramirez, A. R. Kortan, Nature 350 (1991) 600.
6. Susumu Saito and Atsushi Oshiyama, Phys. Rev. Lett. 66 (1991) 2637.
7. D. L. Lichtenberger, K.W. Nebesny, C. D. Ray, D. R. Huffman and L. D. Lamb, Chem. Phys. Lett. 176 (1991) 203.
8. J. H. Weaver, J. L. Martins, T. Komeda, Y. Chen, T. R. Ohno, G.H. Kroll, N. Troullier, R. E. Haufler, and R. E. Smalley, Phys. Rev. Lett. 66 (1991) 1741.
9. C. T. Chen, L. H. Tijeng, P. Rudolf, G. Meigs, J. R. Rowe, J. Chen, T. P. McCauley, Jr, A. B. Smith III, A. R. McGhie, W. J. Romanow, and E. W. Plummer, Nature 352 (1991) 603.
10. G. K. Wertheim, J. E. Rowe, D. N. E. Buchanan, E. E. Chaban, A. F. Hebard, A. R. Kortan, A. V. Makhija, R. C. Haddon, Science 252 (1991) 1420.
11. T. Takahashi, T. Morikawa, S. Sato, H. K. Yoshida, A. Yuyama, K. Seki, H. Fujimoto, S. Hino, S. Hasegawa, K. Kamiya, H. Inokuchi, K. Kikuchi, Physica C, 185-189 (1991) 417.
12. C. S. Yannoni, R. D. Johnson, G. Meijer, D. S. Bethune, and J. R. Salem, J. Phys. Chem. 95 (1991) 9.
13. L. D. Rotter et al., preprint.
14. R. Tycko, G. Dabbagh, M. J. Rosseinsky, D. W. Murphy, R. M. Fleming, A. P. Ramirez, J. C. Tully, Science 253 (1991) 884.
15. A. P. Ramirez, private communication. Susceptibility and critical field measurements also yield a narrow band width (about 7 or 8 times narrower than the photoemission data).
16. D.S. Bethune, G. Meijer, W.C. Tang, H.J. Rosen, W.G. Golden, H. Seki, C.A. Brown, and M.S. de Vries, Chem. Phys. Lett. 179 (1991) 181.

17. In our estimation, we compared **k** to half of the fcc Brillouin zone since it is symmetric with respect to the zone center. For the normal emission case, an inner potential of 8 eV is used. This approximation will not affect our message that we have covered a sizable portion of the zone. We have additional normal emission data which give the same result.
18. C.S. Yannoni, R.D. Johnson, G. Meijer, D.S. Bethune, and J.R. Salem, J. Phys. Chem. 95 (1991) 9.
19. P. Baltzer, W. J. Griffiths, A. J. Maxwell, P. A. Brühwiler, L. Karlsson and N. Martensson, to be published.

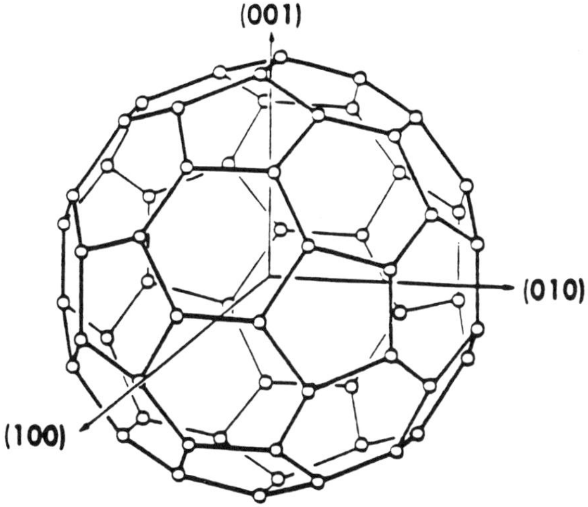

Fig.1. Illustration of C_{60} molecular structure.

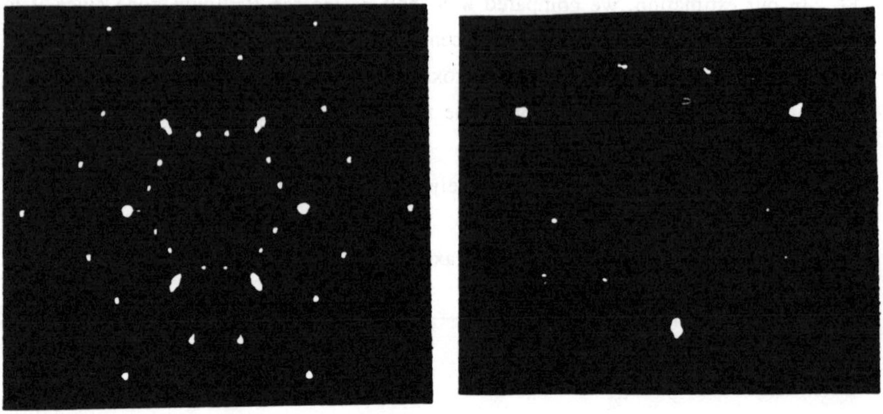

Fig.2 Transmission Laue pattern along a cleaved direction of C_{60} single crystals.

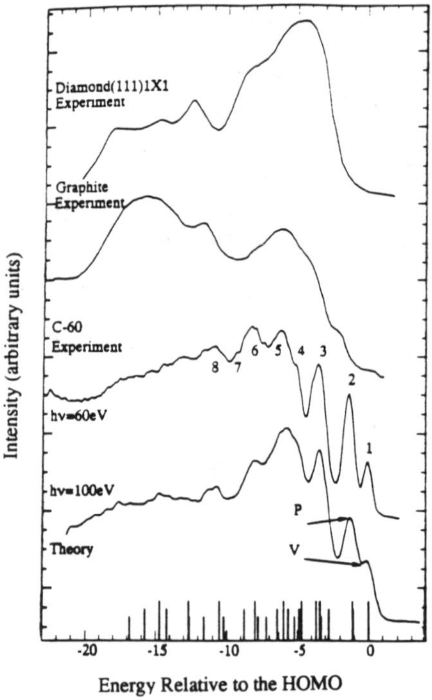

Fig.3. Valence band spectra of C_{60} together with that of graphite and diamond.

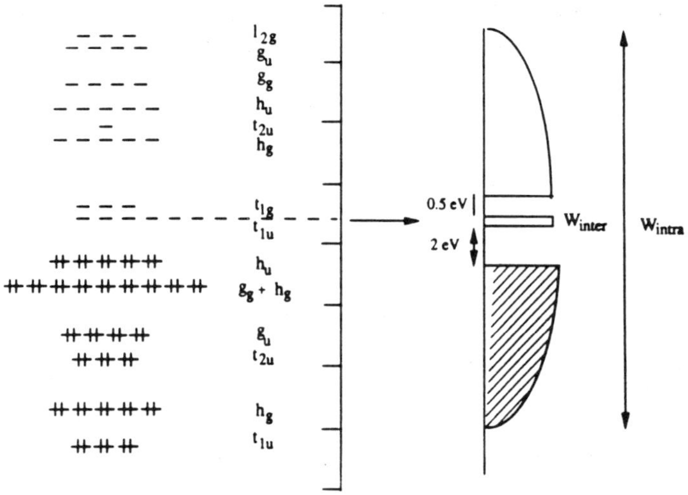

Fig.4. Illustration of the electronic structure of C_{60} and the definations of W_{intra} and W_{inter}.

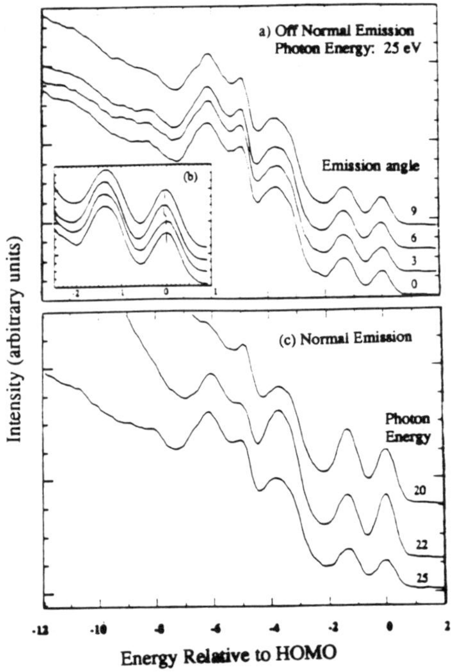

Fig.5 (a) Off-normal emission data from the C_{60} single crystal, the inset (b) shows details of the first two photoemission peaks. (c) Normal-emission data from the C_{60} single crystal.

FULL MULTIPLE SCATTERING INTERPRETATION OF THE Cu L_3 EDGE XANES IN HIGH T_c SUPERCONDUCTORS

M. Pompa [1,2], A. Bianconi[1,3], S. Della Longa[3], P. Lagarde[2]

[1] *INFM Research Unit, Dipartimento di Fisica, Università di Roma, Italy*
[2] *LURE, Université Paris Sud, 91405 Orsay, France*
[3] *Università di L'Aquila, Via S. Sisto 20, L'Aquila, Italy*

ABSTRACT

The linearly polarized Cu L_3 XANES (X-ray Absorption Near Edge Structure) spectra of several high T_c superconductor single crystals and of their parent insulating compounds have been measured. The spectra have been interpreted in the frame of the full multiple scattering (MS) theory in the real space. Over a range of 20 eV, the polarized Cu L_3 XANES spectra have been predicted in term of a one-electron dipole ($\Delta \ell$=+1) transitions Cu 2p→ εd, probing the partial density of states of the conduction bands projected on the Cu site with the selected orbital angular momentum ℓ=2 and its projections m_ℓ. The oscillator strenght for the dipole allowed transitions Cu 2p→ εs has been found to be a factor 100 weaker than the Cu 2p→ 3d transitions. The physical origin of features in the continuum XANES spectra and the effect of the electron-hole Coulomb interaction in the final states have been identified.

The difference spectra between the metal and the insulating phase allows to identify the symmetry of the itinerant $3d^9\underline{L}$ states induced by doping, that is deduced to be a mixture of Cu 3d holes with m_ℓ=0, and 2 orbital angular momentum and oxygen 2p holes on planar oxygen in the molecular orbital combination of local a_1 and b_1 symmetry. Evidence for the more delocalized character of $3d^9{}_{x^2-y^2}\underline{L}$ states versus the more localized character of $3d_{3z^2-r^2}\underline{L}(a_1)$ states has been found. The presence of $3d_{3z^2-r^2}\underline{L}(a_1)$ states is associated with distorted Cu site configurations, and the presence of a second band with a_1 character below but close to the Fermi level.

1. INTRODUCTION

X-Ray Absorption Spectroscopy (XAS) at the L-edges of copper in high Tc superconductors has proved to be a powerful tool in order to determine the new electronic configuration related to the superconducting behavior. The polarization of the synchrotron light has then be used to give the symmetries of these new states : for instance, in La and Bi compounds, they have been found polarized in the (ab) plane in agreement with the average symmetry of the CuO_2 planes. Nevertheless a spectroscopic explanation is still missing for some of the features of the experimental spectra, for instance the spectroscopic weight present mainly on the z-polarized spectrum between the two L_3 and L_2 white lines as well as the presence of a doping-dependent z-component of these white lines.

This paper has therefore two goals :

- first, an attempt by means of the multiple scattering approach, of a quantitative interpretation of the overall L_3 edge of copper in La_2CuO_4 and $Bi_2Sr_2CaCu_2O_8$; it is worth to point out that this approach has never been applied in such a low energy domain.

- second, a comparison between a doped system and its insulating parent which shows the z-character of part of the new states induced by doping.

2. EXPERIMENTAL

We have investigated the following set of samples: a) a $Bi_2Sr_2CaCu_2O_8$ single superconducting crystal with T_c=85K, b) an insulating $La_2CuO_{4-\delta}$ ($\delta < 0.03$) single crystal, c) the same La_2CuO_4 oxygen depleted single crystal, c) a superconducting $La_{1.92}Sr_{0.08}CuO_4$, d) a $Bi_2Sr_2CuO_6$ single superconducting crystal.

The Cu-L_3 x-ray absorption experiments have been carried out on the super-ACO storage ring of the synchrotron radiation facility LURE at Orsay. The details of the experimental set up are reported elsewhere [1].

In Fig.1, we report the experimental spectrum of a La_2CuO_4 powder normalized to the calculated atomic like absorption coefficient at the high energy value $\alpha_{0\infty}$ (50 eV above V_{0i}) [1] which can be considered as the classical atomic jump.

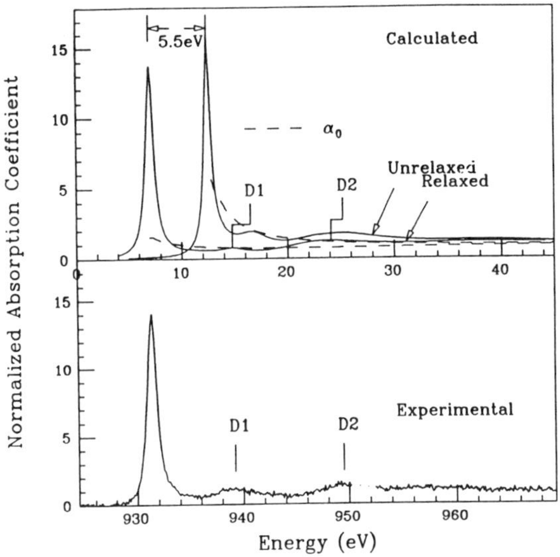

Fig.1 Experimental unpolarized Cu L_3-edge XANES spectra of La_2CuO_4 powder (lower panel). The corresponding calculated spectra in the relaxed and unrelaxed potentials are shown (upper panel).

3. XANES SPECTRA: NUMERICAL CALCULATIONS

The Cu L_3 XANES spectra of $Bi_2Sr_2CaCu_2O_8$, La_2CuO_4 and $La_{1.875}Sr_{0.125}CuO_4$ have been calculated, in the real space, using the Multiple Scattering approach [1].

The calculations start with the construction of clusters of 38, 27 and 45 atoms respectively for $Bi_2Sr_2CaCu_2O_8$, La_2CuO_4 and $La_{1.875}Sr_{0.125}CuO_4$ using the atomic coordinates given in literature [2]. The obtained clusters are divided into 5 concentric shells each one with a different mean distances from the central absorbing Cu ion.

The final state wave function of the excited electron is calculated in a potential of a non overlapping muffin tin type using Mattheis prescription. It is generally regarded as an approximation to the LDA (Local Density Approximation) where exchange and correlation effects are included in a one electron Hamiltonian by means of a local potential function. It consists of the following steps:

i. Calculation of charge densities for neutral atoms on each site from a self-consistent field atomic calculations of the relativistic Hartee-Fock-Slater kind.
ii. Superimposition and spherical averaging of the atomic charge density about the atoms;
iii. Resolution of the Poisson's equation for the Coulomb part of the potential. The local exchange correlation contribution is found using an approximation of the $X\alpha$ type (α=1).

The muffin tin approximation has been tested for close packed metals and it has been found good enough, while it has been demonstrated that the more open the structure, the worse the approximation becomes because the interstitial potential undergoes large variations around its mean value. We then have found a set of muffin tin radii giving an interstitial potential V_{0i} nearly constant (the condition for nearly touching and not overlapping spheres), leaving a discontinuity on the La (in La_2CuO_4 and $La_{1.875}Sr_{0.125}CuO_4$) and Ca (in $Bi_2Sr_2CaCu_2O_8$) atoms and choosing the same potentials for Cu and O. The average potential V_0 of the crystal has been taken as the mean value of V_{0i}. The muffin tin radii of Cu and O atoms have been found to be nearly the same showing the large covalency of the Cu-O bonds in the systems. The obtained values of the muffin tin radii and energy constants are reported elsewhere [1].

In order to compare the calculated XANES spectra with the corresponding experimental ones, it has been necessary to introduce a broadening due to the experimental resolution, the core-hole life time and the energy-dependent photoelectron lifetime due to the inelastic scattering with valence electrons.

The final factor entering the calculation of XANES spectra concerns the effect of the core-hole. The description of the final state is a many body problem, as the passive electrons experiment the attractive interaction of the less screened nucleus. In order to reduce the problem to a one electron problem, we have applied the so-called final state rule: the relevant dipole matrix elements are calculated from a valence wave function obtained in the potential of the final state of the x-ray process. The spectra have then been recalculated in the fully relaxed potential obtained by substituting the central absorbing atom by the one with atomic number increased by one according to the Z+1 approximation [3].

4. PHYSICAL ORIGIN OF THE FEATURES IN THE CONTINUUM XANES SPECTRA

The Cu $2p_{3/2} \rightarrow d$ channel to final states with ℓ=2 is given by the sum of the partial absorption cross sections to final states which can be classified as (ℓ, m_ℓ) = (2, 0), (2, 1) and (2, 2) final states. We have ignored the second photoabsorpion channel allowed by the dipole selection rule $\Delta\ell$=-1 because we have found it negligible.

The first step is the calculation, starting from the photoionization cross sections, of the polarized $I_{//}(\omega)$ and $I_{\perp}(\omega)$ and of the unpolarized XANES spectra for an increasing number of shells.

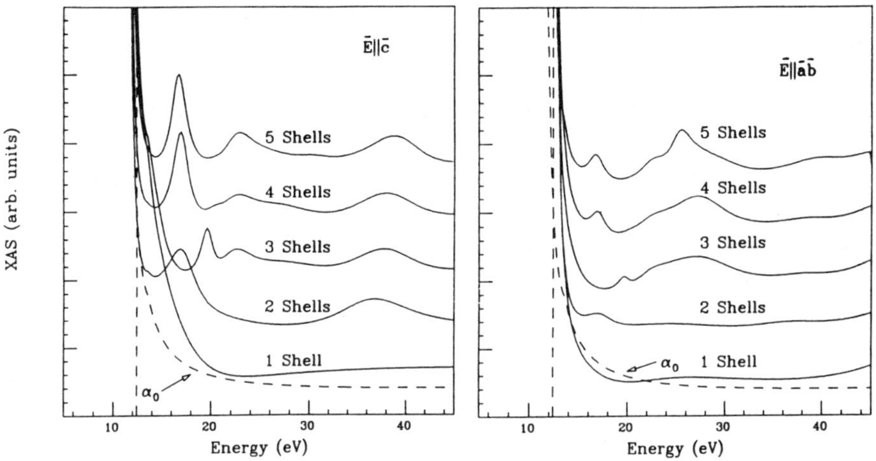

Fig.2 Calculated polarized E//c and E⊥c Cu L_3-edge XANES spectra of La_2CuO_4 as a function of the cluster size. The zero of the energy scale has been fixed at the average interstitial potential V_0.

For sake of simplicity we report here only the results obtained for La_2CuO_4. Analougous results have been obtained for $La_{1.875}Sr_{0.125}CuO_4$ and $Bi_2Sr_2CaCu_2O_8$ with slightly different values in energy positions.

In Fig.2, we plot the calculated polarized E⊥c (panel a) and the E//c (panel b) Cu L_3 XANES spectra for different sizes of the La_2CuO_4 cluster. The zero of the energy scale has been fixed at the interstitial potential V_{0i}. The Fermi level E_0 has been determined above V_{0i} and it has been found, in La_2CuO_4, to be at about 12.5 eV.

It is clear that the experimental XANES spectra can be reproduced in the calculation only by considering the largest cluster because, for a lower number of shells, features are smoothed and don't show the fine structures of the experimental spectra.

For a single cluster CuO_4 (the first shell) no structure in the continuum has been obtained in both polarizations. Including the apex oxygen (the second shell) a new absorption feature appears at about 17 eV, i.e. about 4.5 eV above the Fermi energy mainly in the E//c XANES spectrum. This feature is due to the peak in the m_ℓ=0 partial cross section. Therefore, the calculation shows that the apex oxygen gives the main contribution to the formation of the conduction band identified by the peak of the partial density of states at about 4 eV above the Fermi level. It is usually associated to the LaO layers in band structure calculation and so this band has also a relevant Cu ℓ=2, m_ℓ=0 character in the CuO_2 plane. The same result has been found for $Bi_2Sr_2CaCu_2O_8$, but with a position of the corresponding BiO band closer to the Fermi level. As far as the Fermi energy is concerned, what we have here called E_0 is not expected to be the Fermi level E_F of the system as it will be pinned to the states induced by doping in the gap. A second peak in the E⊥c spectrum (at 26 eV in Fig.2., i.e. at about 13.5 eV above E_0) grows up by including in the cluster five CuO_4 square plane units and it arises from the m_ℓ =2 partial cross section [1].

5. THE ELECTRON-HOLE COULOMB INTERACTION IN THE FINAL STATE

Concerning the effect of the core hole final state on the XANES spectra we can separate the spectra into two energy regions:

a) the Cu $2p_{3/2} \rightarrow 3d$ region giving the white line at the Fermi level

b) the continuum XANES region due to the d-like partial density of states of the high energy conduction bands.

In Fig.1, the calculated unpolarized spectra of La_2CuO_4 are compared with the experimental spectrum of a powder sample. The threshold of the Cu $2p_{1/2} \rightarrow$ 3d transition, giving the spin orbit split ($\Delta E \cong 20$ eV) L_2 white line at 951 eV [4], overlaps the L_3 XANES continuum. We have extracted the unpolarized XANES spectrum in the high energy continuum by subtracting the renormalized L_3-edge XANES in order to suppress the L_2-edge contribution and to better resolve the experimental XANES spectrum in the range 937-970 eV.

As we switch on the Z+1 approximation we obtain a 5.5 eV rigid red shift of the calculated white lines. The same energy shift has been obtained in the case of $Bi_2Sr_2CaCu_2O_8$ and $La_{2-x}Sr_xCuO_4$. On the contrary, the XANES peaks in the continuum show only a rigid red shift of about 1.7 eV in the fully relaxed potential (1 eV in the $Bi_2Sr_2CaCu_2O_8$). Because of the different core hole final state effects, we find that the energy separation between the white line and the first XANES feature in the continuum spectra which is 7.8 eV (5.2 eV in $Bi_2Sr_2CaCu_2O_8$) in the experimental spectra, turns out to be 8 eV (6.3 eV) and 4 eV (2.1 eV) in the fully relaxed and unrelaxed potentials, respectively. This is the most direct experimental evidence of the validity of the fully relaxed potential.

The different effects of the core-hole-induced relaxations on the Cu 3d and εd final states in the continuum are due to the much more localized character of the first ones. The Coulomb interaction between the core hole and the Cu 3d electrons is as large as the Coulomb repulsion between the 3d holes in the valence band U_{dd}=6 eV.

6. THE POLARIZED XANES SPECTRA

Starting from the assumption of the validity of the Z+1 approximation, the polarized XANES spectra of different systems have been analysed and the differences between them, in going from the insulating to the metallic phase, interpreted.

In Fig.3, we report the blow up of the continuum XANES normalized experimental spectra (upper panel) and the corresponding calculated one (lower panel) for the insulating La_2CuO_4. Both of them show an isosbestic point at 22 eV, i.e. about 15 eV above the white line which is polarization independent.

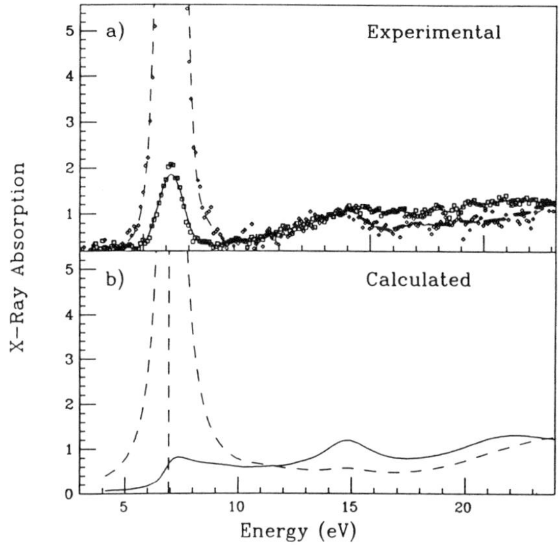

Fig.3 (a) Blow up of the experimental polarized E⊥c and E//c Cu L_3 edge XANES spectra in the continuum of La_2CuO_4. (b) Calculated polarized E⊥c and E//c Cu L_3-edge XANES spectra of La_2CuO_4 in the unrelaxed and fully relaxed potential for a cluster of 5 shells including the broadening.

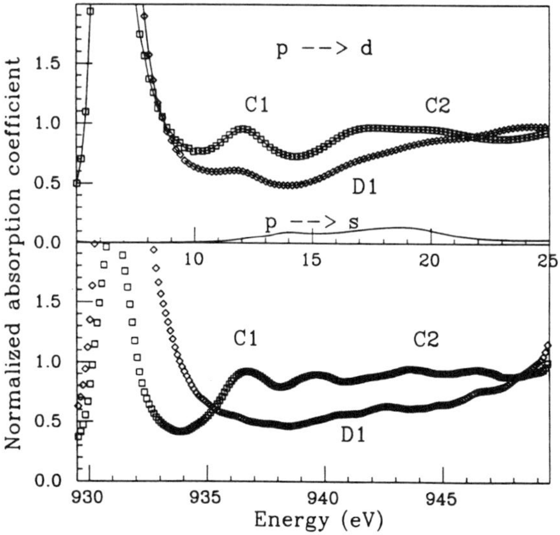

Fig.4 (a) Calculated polarized E⊥c and E//c Cu L_3-edge XANES spectra of $Bi_2Sr_2CaCu_2O_8$ in the unrelaxed and fully relaxed potential for a cluster of 5 shells including the broadening. (b) Blow up of the experimental polarized E⊥c and E//c Cu L_3 edge XANES spectra in the continuum of $Bi_2Sr_2CaCu_2O_8$.

Some important differences are present :

- First, there is a minimum in the experimental spectrum between the excitonic Cu $2p^5 3d^{10}$ final state and the continuum threshold estimated at 937 eV not predicted by the theoretical calculations.

- Second, the calculated intensity of the E//c white line is negligible while it is observed in the experimental spectrum. The experimental E//c component of the white line comes from $3d_{xz,\ yz,\ 3z^2-r^2}$ symmetries due to distortions of the local Cu site, and to the tilting of the CuO_6 elongated octahedra in the orthorombic phase [5]. However, it could also be associated to an incomplete degree of polarization of the synchrotron radiation beam and/or to a not perfect crystal surface. Therefore, we can conclude that the probability for the $3d_{x^2-y^2}$ symmetry of the 3d holes in the La_2CuO_4 crystal is in the range 95-100 %.

Going from the insulating to the metallic phases of $La_{2-x}Sr_xCuO_4$ and $Bi_2Sr_2CaCu_2O_8$, we find, both in calculated and experimental spectra, an increase of the E//c component of the white line.

According to our experimental and calculated Cu L_3 X-ray absorption data [6], the symmetry of the Cu 3d holes is partially converted from the Cu $3d_{x^2-y^2}$ symmetry in the insulating antiferromagnetic phase to the $3d_{3z^2-r^2}$ symmetry upon doping.

In the case of $Bi_2Sr_2CaCu_2O_8$ (Fig.4), the agreement between the experimental and the calculated spectra is not as good as in La_2CuO_4. The C1 feature in the continuum at 936.5 eV is well predicted by the calculation ; a second peak at 939.5 eV is not predicted by the calculation ; the third feature, C2, which appears as a broad band extending from 943 to 946 eV is predicted but at lower energy. The peak D1 of the in-plane polarized spectrum is weak but well predicted by the calculated E//ab spectrum as well as the minimum at 939 eV. The main disagreement between the calculation and the experiment concerns the peak at 939.5 eV measured in the out-of-plane polarization. It is at the same energy position of the $3d^8$ states, but it could depend on doping or it could be due to a multielectron final-state feature.

The calculation predicts exactly the dichroism of the XANES spectra and the main features and so, we can consider the agreement as satisfactory, taking into account that the coordinates of the atoms of the cluster are only averaged values over the five different Cu sites present in the real structure which exhibits a 5x1 modulation superstructure.

7. THE SYMMETRY OF $3d^9\underline{L}^*$ STATES INDUCED BY DOPING

The experimental investigation of the $YBa_2Cu_3O_7$ and $Bi_2Sr_2CaCu_2O_8$ by polarized XAS spectroscopy had already pointed out that the orbital angular momentum m_ℓ of the Cu 3d holes associated with the ligand holes could be modified giving the $3d^9{}_{3z^2-r^2}\underline{L}^*$ states beyond the expected Zhang-Rice singlets [7]. Here, we report the percentage of the $3d^9{}_{3z^2-r^2}\underline{L}^*$ and $3d^9{}_{x^2-y^2}\underline{L}^*$ states to the total number of $3d^9\underline{L}$ states in the gap by the analysis of changes in the polarized Cu 2p→ 3d transitions going from the insulating to the metallic phase. This difference analysis is not affected by systematic errors due to the particular experimental tecnique and it is then possible to determine the actual relative variation.

The difference between the polarized Cu L_3 x-ray absorption spectra between the superconductors $La_{2-x}Sr_xCuO_4$ [10], $Bi_2Sr_2CuO_6$ [8] and $Bi_2Sr_2CaCu_2O_8$ [1, 9] and their parent insulating compounds have been analysed. The total number of new states induced in the metallic phase are represented by the integral of the difference between the two unpolarized white lines. The relative contribution to this total increase comes from the differences of polarized E⊥c and E//c spectra.

In Fig.5, we report the polarized E//c and E//ab white lines of $Bi_2Sr_2CaCu_2O_8$ and their differences.

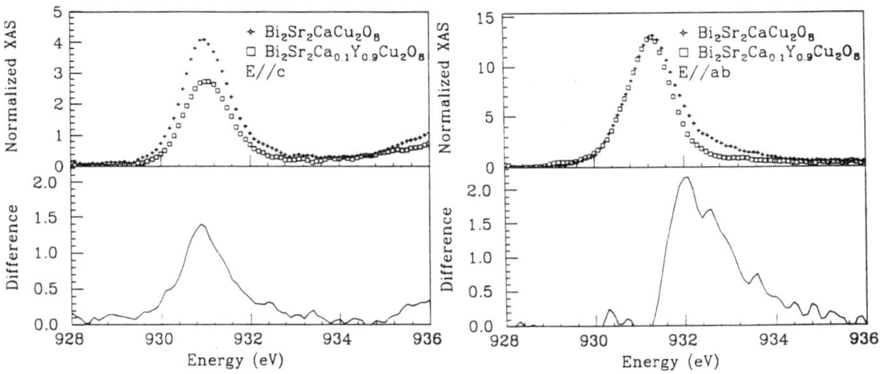

Fig.5 Polarized E⊥c (a) and E//c (b) L_3-XAS white line spectra of metallic $Bi_2Sr_2CaCu_2O_8$ compared with the parent insulating compound (upper panel) and their difference (lower panel).

The final states symmetry in the E⊥c and E//c polarizations can be expressed in terms of the b_1 and a_1 molecular orbitals symmetry. The b_1 molecular orbital is obtained as the linear combination of the Cu $3d_{x^2-y^2}$ with the p_{xy} planar oxygen orbitals with b1 symmetry, while the a1 molecular orbitals are formed by the Cu $3d_{3z^2-r^2}$, p_{xy} planar oxygen orbitals with a1 symmetry and p_z orbitals of apical oxygen.

The difference spectra of the polarized E//ab spectra probe mainly the transitions from Cu $2p^63d^9{}_{x^2-y^2}\underline{L}(b_1)$ to Cu $2p^53d^{10}\underline{L}(b_1)$ states with a small contribution of the Cu $2p^63d^9{}_{3z^2-r^2}\underline{L}(a_1)$.

The difference between the E//c spectra probes mainly the transitions from $3d^9{}_{3z^2-r^2}\underline{L}(a_1)$ to $2p^53d^{10}\underline{L}(a_1)$ with may be some extra contributions of a different symmetry due to the non perfect polarization of the beam. Because the energy position of the $2p^53d^{10}\underline{L}(b_1)$ has been determined at 932 eV, the additional states at 931 eV can be assigned to $2p^53d^{10}\underline{L}(a_1)$ final states. Therefore the energy splitting of ~ 1eV between the $2p^53d^{10}\underline{L}(b_1)$ and $2p^53d^{10}\underline{L}(a_1)$ probes the energy splitting between $\underline{L}(a_1)$ and $\underline{L}(b_1)$ in agreement with theoretical calculations by Otha [10]. Because the absorption white line is a bound excitonic state, as demonstated previously, and because the Cu $2p^53d^{10}\underline{L}(a_1)$ final states are at the same energy, we can deduce their localized character. On the contrary, the higher energy position of the Cu $2p^53d^{10}\underline{L}(b_1)$ final states indicates mobile carriers with b_1 symmetry screening the core hole.

In Fig.6 the integral of the difference spectra in the two polarisations as a function of doping, for the different samples, together with the percentage of new states induced by doping are shown. As it is clear the percentage of a_1 states remains constant with doping within the experimental errors while that of b_1 states follows the total doping level.

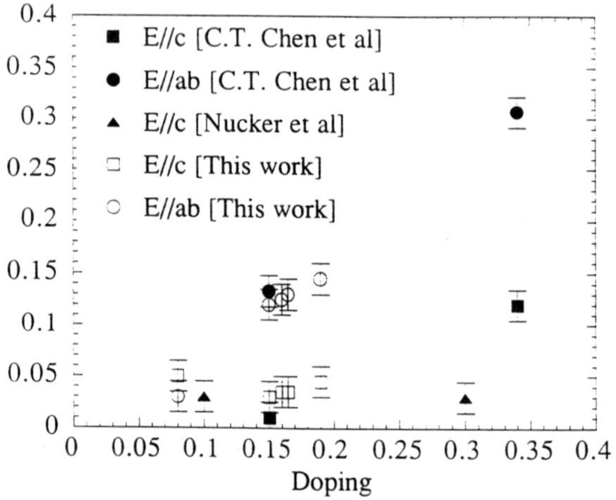

Fig.6 The integral of the difference spectra in the two polarisations for several samples as a function of doping together with the percentage of new states induced by doping.

8. CONCLUSIONS

In conclusion we have interpreted the experimental features of the Cu L_3-edge polarized XANES spectrum of $La_{2-x}Sr_xCuO_4$ and $Bi_2Sr_2CaCu_2O_{8+\delta}$. We have identified the physical origin of the peaks in the continuum and related some of them to the local structure around the copper atom.As it has been pointed out in other L-edges experiments, the oscillator strength for the 2p→ s channel is shown to be negligible. The electron-hole Coulomb interaction in the final states which is modellized by the Z+1 approximation has a strong effect on the localized white line which suffer an excitonic energy shift as large as 8 eV, while the delocalized d-states are slightly affected.

The change of the polarized Cu L_3 spectra has allowed the identification of the symmetry of the $3d^9\underline{L}^*$ states induced by doping. The weight of the $3d^9{}_{x^2-y^2}\underline{L}(b_1)$ and $3d^9{}_{3z^2-r^2}\underline{L}(a_1)$ states in the metallic phase has been obtained by the difference spectra between the superconductor and insulating systems of polarized E//c and E⊥c spectra respectively. The states $3d^9{}_{x^2-y^2}\underline{L}(b_1)$ is clearly found in the variation of the E⊥c spectra from the insulating to the metallic phase while the polarized E//c spectra show the presence of $3d^9{}_{3z^2-r^2}\underline{L}(a_1)$ states. Evidence for the localized character of $3d_{3z^2-r^2}$ $\underline{L}(a_1)$ states has been found. Therefore, Cu L_3 edge XANES spectra show that the cuprate insulating systems do not remain rigid upon doping concerning the Cu 3d hole symmetry.

ACKNOWLEDGMENTS

The present research has been supported by the Consorzio Interuniversitario Nazionale per la Fisica della Materia (INFM) and Consiglio Nazionale delle Ricerche, in the frame of the "Progetto Finalizzato per le Tecnologie Superconduttive". We thank the staff of Super-ACO for running the machine and also the support of LURE in the frame of the CEE program for the Large European Facilities.

REFERENCES

[1] M. Pompa, C. Li, A. Bianconi, A. Congiu Castellano, S. Della Longa, A.-M. Flank, P. Lagarde, D. Udron, *Physica C* **184**, 51-64 (1991); A. Bianconi, S. Della Longa, C. Li, M. Pompa, A. Congiu Castellano, D. Udron, A.-M. Flank, P. Lagarde, *Phys. Rev B* **44**, 10126 (1991)

[2] K. Yamada, E. Kudo, Y. Endoh, K. Tsuda, M. Tanaka, K. Kokusho, H. Asano, F. Izumi, M. Oda, Y. Hikada, M. Suzuki, and T. Murakami, *Jpn. J. Appl. Phys.* **27**, 1132 (1988); P. Bordet, J.J. Capponi, C. Chaillout, A.W. Hewat, E.A. Hewat, J.L. Houdeau, M. Marezio, J.L. Tholence and D. Tranqui, *Physica C* **156**, 189 (1988); R.J. Cava, A. Santoro, D.W. Jhohnson Jr., and W.W. Rhodes*Phys. Rev.* **35,** 6716 (1987); M. Francois, K.Yvon, P. Fisher, and M. Decroux *Sol. State Commun.* **63**, 35 (1987)

[3] A. Bianconi in *"X-ray absorption: principle, applications, techniques of EXAFS, SEXAFS, XANES"* ed. R. Prinz and D. Koningsberger, J. Wiley and Sons, New York, (1988); P.J. Durham in *"X-ray absorption: principle, applications, techniques of EXAFS, SEXAFS, XANES"* ed. R. Prinz and D. Koningsberger, J. Wiley and Sons, New York, (1988)

[4] A. Bianconi in *International Conference on Superconductivity -ICSC*, S.K. Joshi, C.N.R. Rao, and S.V. Subranyam editors, (Proc. Int. Conf. on Superconductivity, Bangalore, Jan. 10-14, 1990) World Scientific, Singapore, pag. 448 (1990)

[5] M. Pompa, P. Castrucci, C. Li, D. Udron, A.-M. Flank, P. Lagarde, H. Katayama-Yoshida, S. Della Longa, A. Bianconi, *Physica C* **184**, 102-112 (1991) and references therein

[6] C. Li, M. Pompa, A. Congiu Castellano, S. Della Longa, A. Bianconi, *Physica C* **175**, 369 (1991) and references therein

[7] F.C. Zhang and T. M. Rice *Phys. Rev.* B**37**, 3759 (1988); H. Eskes and G.A. Sawatzky *Phys. Rev. Lett.* **61**, 1415 (1988)

[8] M. Pompa, S. Turtù, A. Bianconi, F. Campanella, A.-M. Flank, P. Lagarde, C. Li, I. Pettiti, D. Udron, *Physica C* **185-189**, 1061-1062 (1991)

[9] A. Bianconi, S. Della Longa, M. Missori, I. Pettiti, M. Pompa, to be published in "Lattice Effects in High Tc Superconductors" edited by Y. Bar-Yam, World Scientific Pub., Singapore, (1992)

[10] Y. Ohta, T. Tohyama and S. Maekawa *Physica C* **166**, 385 (1990) and Y. Ohta, T. Tohyama and S. Maekawa *Phys. Rev. B* **43**, 2968 (1991)

HIGH-RESOLUTION PHOTOEMISSION STUDIES OF CORRELATIONS AND PHASE TRANSITIONS IN ONE AND TWO DIMENSIONAL SYSTEMS

D. MALTERRE, B. DARDEL, M. GRIONI, P. WEIBEL and Y. BAER,
Institut de physique de Neuchâtel, rue L. Breguet 1
Neuchâtel Switzerland

ABSTRACT

High resolution photoemission spectroscopy was exploited for the study of phase transitions in low-dimensional systems. It is shown that 1D-metals do not exhibit a finite spectral density at the Fermi level. Different mechanisms are invoked to explain this striking behaviour. In the 2-dimensional 1T-TaS_2, photoemission spectra clearly reflect the complex phase diagram. They indicate the Fermi surface collapse at the quasicommensurate-commensurate transition.

1. Introduction

Physical properties of solids, such as magnetic susceptibility, specific heat, etc., reflect the low-energy electronic excitation spectrum. Nevertheless, this low-energy behaviour usually results from interactions at a larger energy scale. For instance, in strongly correlated systems, the coulomb interactions of several eV strength are the determinant parameter for the heavy-fermion properties characterized by an energy scale of a few meV[1]. High-energy spectroscopies are considered as tools for the investigation of the high energy scale in such systems. The recent development of high-resolution photoemission allows a direct determination of the low-energy part of the excitation spectrum. We will illustrate the potentiality of the photoemission spectroscopy for the study of low-energy excitations in low-dimensional materials.

Low-dimensional ($D<3$) materials usually exhibit transitions, like the charge-density-wave transition (CDW), associated with significant modifications of the electronic structure. In particular, the one-dimensionality is characterized by several features which qualitatively differ from higher dimensionalities. As demonstrated by Peierls[2], the metallic state is unstable with respect to a lattice distortion yielding the formation of an insulating state. Nevertheless, with short range interactions, a 1-D system cannot present a thermodynamic transition at finite temperature owing to the very large fluctuations associated with the 1-D character[3]. The electronic correlations are also known to be very unusual[4]; the simple model of Landau Fermi liquid which generally accounts for the excitations in 2 and 3-D materials is no longer valid in 1-D systems where an infrared catastrophe leads to the Luttinger liquid pathology[5]. In real 1-D materials, electronic transitions are actually observed suggesting that transverse interactions play a significant role. However, in systems where these interactions are weak, the main characteristic features of one-dimensionality remain.

The paper is organized as follows. In part 2, we discuss the spectroscopic consequence of the Peierls transition in quasi-1D systems. In part 3, the high-resolution photoemission is exploited to study the electronic modifications associated with the phase transitions in the two-dimensional 1T-TaS_2 compound. Finally, part 4 is devoted to a few concluding remarks.

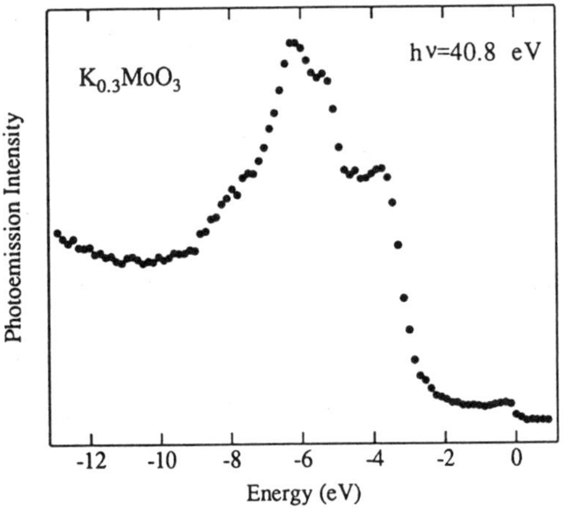

Figure 1: Photoemission spectrum of $K_{0.3}MoO_3$ with hν =40.8 eV excitation energy.

2. Peierls transition in 1D materials

Metallic quasi-one-dimensional materials usually exhibit an electronic transition toward a semiconducting phase at low temperature. This transition, called the Peierls transition, which is accompanied by the appearance of charge-density-waves and a distortion of the lattice, results from the instability of the electron gas at a wave vector $\mathbf{q} = 2\mathbf{k}_F$ ($\mathbf{k}_F$ is the Fermi wave vector). A mean field treatment predicts a modification of the spectral function similar to that encountered in the BCS theory of superconductors[6]. In going from the metallic to the semiconducting phases, a transfer of spectral weight from the vicinity of E_F to lower energy is expected, reflecting the opening of the Peierls gap.

The molybdenum blue bronze ($K_{0.3}MoO_3$) belongs to this class of materials. It is a metallic oxide with a crystalline structure formed by slabs of MoO_6 clusters separated by alkali atoms. This atomic distribution clearly shows a preferential direction, and leads to highly anisotropic physical properties, such as the electrical conductivity[7]. A metal-insulator transition is exhibited at 180 K as shown in many experimental studies[7-9]. In particular, the Peierls gap has been estimated between 100-150 meV from optical measurements[9].

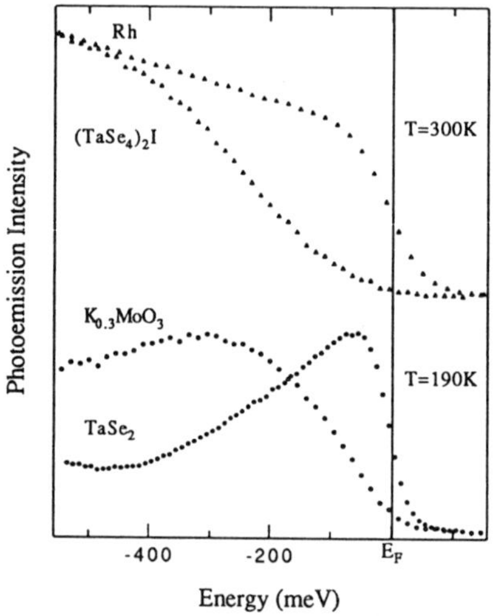

Figure 2: Photoemission spectra of $K_{0.3}MoO_3$ and $(TaSe_4)_2I$.

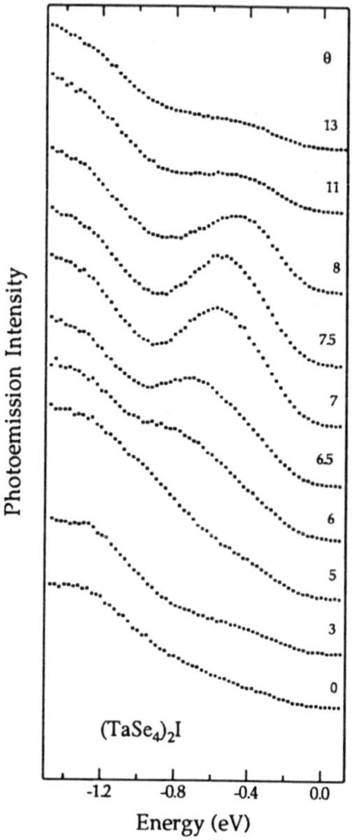

Figure 3: Angular dependence in $(TaSe_4)_2I$ for different polar angles (see also ref. 12).

2.1 The metallic phase

First let us consider the metallic phase above the Peierls temperature. As shown in figure 1, the UV-photoemission spectrum of this compounds is dominated by the valence band structure between -2 and -10 eV. An additional small feature appears near the Fermi level and reflects the conduction band which, in these oxides, results from the formation of hybridized Mo 4d and O 2p states[10]. This band, empty in the MoO_3 oxide, is partially filled by the alkali s electrons in $K_{0.3}MoO_3$. In figure 2, we focus on the spectral function in the vicinity of the Fermi level. Surprisingly, this spectrum of the metallic phase does not exhibit a step at the Fermi energy which is always characteristic of a metal[11]. A similar result is obtained with $(TaSe_4)_2I$. This quasi-1D compounds exhibit a Peierls transition at 263 K. This behaviour strongly contrasts with the situation observed for higher dimensionality as illustrated on the spectra of $TaSe_2$, a 2-D metal, and of rhodium metal. In these metals, the Fermi discontinuity is clearly observed and as the energy resolution is better than 20 meV, the width at E_F is essentially due to the thermal broadening. Although the spectral function of $K_{0.3}MoO_3$ do not exhibit an angular dependence, a dispersive feature in the chain direction is clearly observed in $(TaSe_4)_2I$ as shown in figure 3. This feature moves toward the Fermi level with increasing momentum but vanishes before reaching E_F near E=0.5 eV.

2.2 Discussion of the possible mechanisms

As the formation of a Peierls distortion is favoured by a high density of states at E_F ($N(E_F)$) and a strong electron-phonon coupling, a low $N(E_F)$ is unlikely. Thus the density of excitations (spectral function) differs from the quasi-particle density of states as a consequence of electron-electron and electron-phonon interactions. To explain this behaviour, we will investigate the influence of the 1-D specificities (fluctuations, correlations) on the spectral function.

Let us first consider the influence of fluctuations on the spectral density. Mean field theory shows that, if fluctuations are ignored, a second-order transition takes place at T_P^{MF} (mean field Peierls temperature). In strictly 1-D materials, fluctuations suppress the transition. The order parameter (the energy gap) fluctuates in time and space below T_P^{MF}. This behaviour leads to a pseudo-gap in the density of excitations near the Fermi level and then to a strong reduction of the spectral intensity at E_F[6]. To account for a transition at a finite temperature in real materials, the transverse coupling should be considered. This transverse coupling yields a reduction of the fluctuations and the system undergoes a transition at T_P ($0<T_P \ll T_P^{MF}$). Rice et al have investigated the combined influence of fluctuations and transverse coupling on the density of excitations[13]. Their analysis shows that the fluctuations induce the formation above T_P of a pseudo-gap reminiscent of the actual gap in the low-temperature insulating phase. Such a pseudo-gap was observed at room temperature in

$K_{0.3}MoO_3$ and $(TaSe_4)_2I$ by optical measurements[9,14]. A reduction of the density of excitations near E_F is then predicted and depends on the transverse coupling: for strong coupling, the density of excitations is weakly modified (strong 3-D effects) whereas, in the weak coupling range, the pseudo-gap is more pronounced and the metallic signature is strongly reduced. With this formalism a temperature dependent evolution of the spectral intensity at E_F is expected above T_P, and a normal metallic behaviour should be observed near T_P^{MF} (T_P^{MF}=330 K for $K_{0.3}MoO_3$). Moreover this pseudo-gap is accompanied by a peak of quasi-particle states centered at Δ (half-gap). As such a peak is not actually observed and as measurements up to room temperature do not restore the metallic signature, one has to look for additional mechanisms contributing to the reduction of the spectral function near E_F.

As mentionned in the introduction, correlations in 1-D are highly singular. In contrast to metals in higher dimension, correlations cannot be treated in perturbation and the excitation spectrum is completely different: while Fermi liquids are dominated at low-energy by quasi-particle type excitations, only collective excitations are present in the 1-D correlated electron systems (Luttinger liquid). In strongly correlated Fermi liquids, the spectral density significantly differs from the density of quasi-particle states. For example, in heavy fermion cerium based compounds, the quasi-particle density of states exhibits a narrow peak near E_F as revealed by the large value of the specific heat coefficient[1]. Although this high density of states is deduced from specific heat or magnetism measurements, a very low intensity near E_F is found in the photoemission spectra. The spectral function $\rho(E_F)$ and the quasi-particle density of states $N(E_F)$ at the Fermi level are related by the renormalization factor Z through the relation $\rho(E_F)= Z \cdot N(E_F)$. Therefore, the lower the renormalization factor, the lower the spectral function at E_F. This factor Z is also the discontinuity amplitude in the momentum distribution function of *4f* states at Fermi momentum in such systems. In 1-D metals the discontinuity disappears[5] and the singular limit Z—>0 must be considered. In this limit, the representation of excitations in terms of quasi-particles is no longer valid: the state of electrons on the Fermi surface becomes unstable with respect to the emission of electron-hole pairs. A classical example of such an infrared catastrophe is encountered in the edge singularity of X-ray absorption spectroscopy and results in a vanishing spectral weight at the edge[15]. Then, we expect the infrared catastrophe in 1-D system to induce a strong reduction of intensity near E_F. This conjecture is corroborated by a calculation in the Luttinger model which predicts a vanishing spectral density at E_F[16].

Final state effects resulting from the photoemission technique itself can also be invoked. In the photon absorption, high energy phonons corresponding to quasi-molecular modes can be excited. As a consequence, satellites corresponding to final states with a few phonons lead to a transfer of weight from the Fermi energy to higher binding energies. A similar mechanism has been previously proposed to account for the low density of excitations near E_F in 1-D organic metals like TTF-TCNQ[17,18]. Due to the narrowness of the conduction band in this material, the hole left behind can be considered as localized in the characteristic excitation time (~ 10^{-16} s) and various vibration modes can be excited according to Franck-Condon principle. In such approaches, the low intensity near E_F is only a consequence of a localized character of the hole left behind and could also be encountered in higher dimensions.

2.3 The Peierls transition in the blue bronze

In figure 4, we have reported the He I photoemission spectra of $K_{0.3}MoO_3$ for different temperatures. These spectra, normalized to their area, are all characterized by a pronounced structure centered at about -300 meV. With decreasing temperature, a deformation of the spectrum is shown, with an additional shift for temperatures below the transition temperature (T_c=180 K)[19]. This behaviour can be interpreted by a transfer of spectral weight from the vicinity of the Fermi level to higher binding energy. Moreover, the spectral function is modified over a much wider energy range (500 meV) than expected from the opening of a Peierls gap (50-75 eV). In the inset, the photoemission intensity at constant binding energy (E=-70 meV) is reported as a function of temperature. The spectral intensity slowly decreases with decreasing temperature in the high temperature range, and exhibits a sudden change of slope at T=167 K which is unambiguously correlated with the electronic transition in this compound. The characteristic temperature (167 K) is slightly reduced with respect to the nominal transition temperature (T_c=180K). Surface effects could be responsible for this observation. In order to investigate the modification of the electronic structure below T_c, we define a phenomenological parameter $\Delta(T)$, which is the shift of the inflection point of the photoemission spectrum with respect to its position at T=183 K (E=-80 meV). Above 167K, Δ does not significantly vary with temperature whereas below 167 K a rapid evolution is

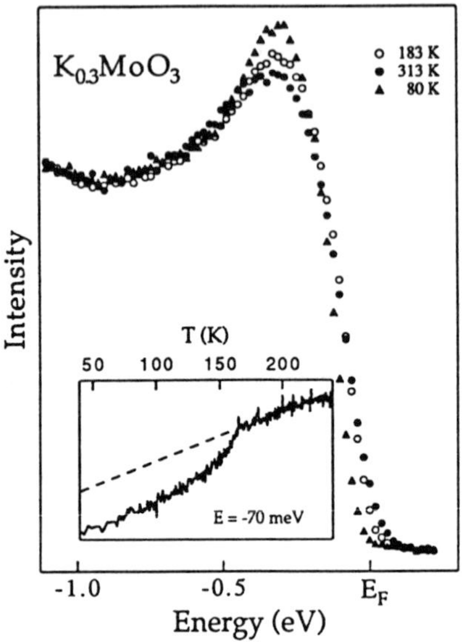

Figure 4: UPS spectrum at several temperatures. inset: T-dependence of the intensity at E=-70 meV.

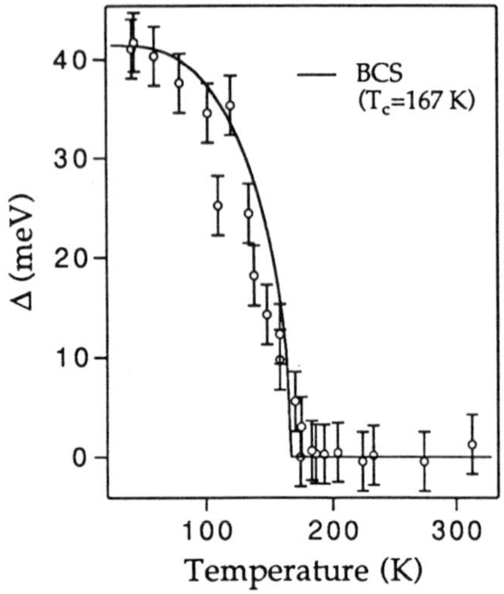

Figure 5: T-dependence of Δ. The solid line is the gap in BCS model.

observed as shown in figure 5. Moreover, the solid line shows that Δ(T) roughly follows the temperature dependence of the gap in a BCS model. This dependence suggests that this parameter is in some way correlated with the Peierls gap but cannot be identified with it since the evolution of the spectra through the phase transition does not evidence the low-energy feature corresponding to the Peierls gap. Then we can conclude that the photoemission spectra indirectly reveal the Peierls gap opening in this CDW system.

3. Charge-density-wave transition in 1T-TaS_2

1T-TaS_2 exhibits unique physical properties that are the consequence of its quasi-two-dimensional (2D) character, and of an unusually complex CDW phase diagram[20]. The most striking observations concern the resistivity which undergoes a sudden tenfold increase in coincidence with the first order phase transition (with a temperature hysteresis) from the *quasi-commensurate* (QC) to the commensurate (C) CDW structure. This anomalous behaviour indicates that important modifications of the electronic structure, with the disappearance of a large portion of the Fermi surface, take place at the QC-C transition.

Near-normal emission photoelectron spectra measured at increasing temperatures are shown in figure 6. The bottom spectrum (356 K) is representative of the incommensurate phase. New structures appear in the QC phase (230K<T<350K) initially at binding energies of 100 meV, 300 meV and 800 meV. These structures sharpen and shift towards higher binding energy with decreasing temperature, and at 230 K the metallic Fermi edge is clearly separated from the first peak. Below 230 K, in the commensurate CDW phase, a substantial growth of the shallowest peak is accompanied by a rapid decrease of intensity at E_F. At 20 K the valence band emission is dominated by a sharp peak at 180 meV, and the emission at the Fermi level is vanishingly small. The whole low-temperature spectrum exhibits a strong angular dependence, but the intensity at E_F remains very small at all angles. The evolution of the spectral function reflects the splitting of the Ta d_{z^2} band into three manifolds under the effect of the CDW[21] and the occurrence of a Mott localization in the C phase. The prominent peak centered at 180 meV is interpreted as the lower Hubbard subband and yields an estimate of the on-site Coulomb correlation (U≈200 meV).

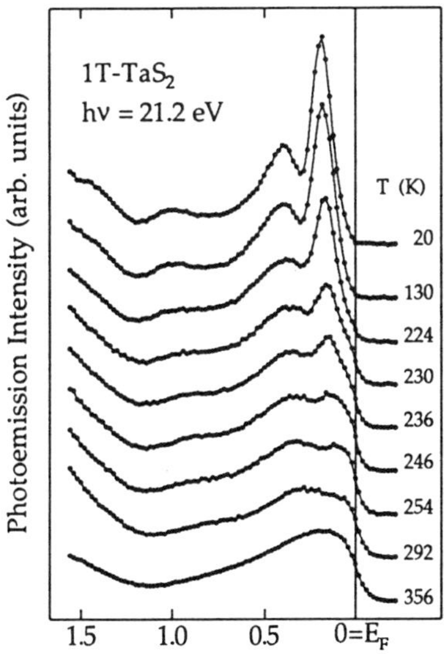

Figure 6: UPS spectra collected at increasing temperature.

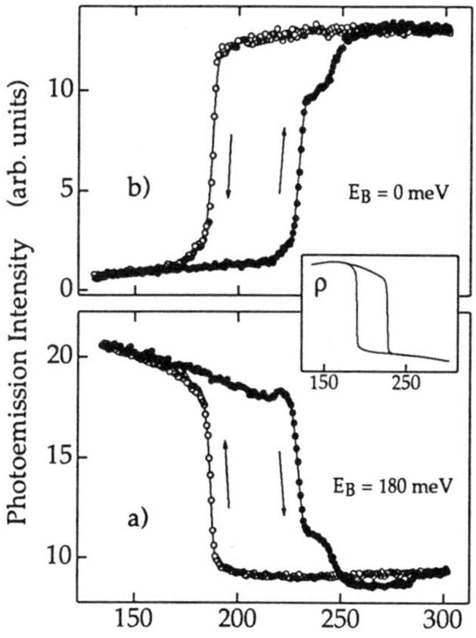

Figure 7: TCE curves during cooling and warming a) at E=-180 meV, b) at E_F.

The large anomaly of the electrical resistivity (inset fig. 7) at the QC-C transition suggests important modifications of the DOS, which should manifest themselves in the photoemission spectra. In figure 7a, we present a temperature-dependent constant energy curve (TCE) collected at a binding energy of 180 meV (corresponding to the main peak in the low-T spectra) during a complete cooling and warming cycle 300 K / 20 K /300K. We observe a 100 % rise of intensity at 187 K upon cooling, and a corresponding intensity drop at 230 K upon warming . The transition at 230 K is not as sharp as the one at 187 K, and the main step is followed by a second one around 245 K. A similar temperature dependence is shown on the photoemission signal at the Fermi level (fig. 7b). The sharp 90% intensity drop at 187 K (similar behaviours are shown at different emission angles) marks a collapse of the Fermi surface explaining the related tenfold increase in resistivity as shown in the inset. This Fermi surface collapse reflects the Mott transition at the QC-C critical temperature and corresponds to the localization of electrons in the distorted structure. Nevertheless, below 185 K the photoemission intensity at E_F decreases linearly with decreasing temperature, suggesting that the density of states at the Fermi level remains finite[22]. Thus, a real gap never opens and the QC-C transition must be regarded as a gapless Mott transition as proposed by Fazekas and Tosatti[23]. This result is consistent with the resistivity measurements which lead to a activation energy of 5 meV.

In conclusion, our photoemission data confirm that the QC-C transition is accompanied by an abrupt decrease of the density of states at the Fermi energy. In contrast to previous photoemission study which suggested a 200 meV gap[24], we provide a direct demonstration of the existence of a finite density of states at E_F in the commensurate phase.

4. Conclusion

In this study, we have used high-resolution photoemission for the study of CDW transitions in 1- and 2-D materials. The unusual absence of a Fermi edge in 1-D metals seems to result from the singularities of one-dimension. Moreover, the modification of the electronic structure accompanying the phase transitions are directly reflected on the photoemission spectra near the Fermi level. Such measurements show that owing to the high-resolution, the low-energy excitations which govern the physical properties of solids, are now accessible to the field of electron spectroscopies.

5. Acknowledgements

We would like to thank Prof. Y. Pétroff, C. Schlenker and F. Lévy for many useful interactions concerning this line of research.

7. References

1. P. Fulde, J. Keller and G. Zwicknagl, *Solid State Phys.* **41** (1988) 1.
2. R. F. Peierls, *Quantum theory of solids* (Clarendon Press, Oxford 1955).
3. L.D. Landau and E.M. Lifchitz, *Statistical Physics*, (Pergamon Press, 1959).
4. J. Solyom, *Adv. Phys.* **28** (1979) 201.
5. J.M. Luttinger, *J. Math. Phys.* **4** (1963) 1154.
6. P.A. Lee, M. Rice and P.W. Anderson, *Phys. Rev. Lett.* **31** (1973) 462.
7. *Low dimensional electronic properties of molybdenum bronzes and oxides*, edited by C. Schlenker (Kluwer Academic Publishers, Dordrecht, 1989).
8. J.P. Pouget, C. Schlenker and J. Marcus, *J. Phys. Lett.* **44** (1984) L113.
9. G. Travaglini, P. Wachter, J. Marcus and C. Schlenker, *Solid State Comm.* **37** (1981) 599.
10. M.H. Wanboo and L.F. Schneemeyer, *Inorg. Chem.* **25** (1986) 2424.
11. B. Dardel, D. Malterre, M. Grioni, P. Weibel, Y. Baer and F. Lévy, *Phys. Rev. Lett.* **67** (1991) 3144.
12. Y. Hwu et al, to be published.
13. M.J. Rice and S. Strässler, *Solid State Commun.* **13** (1973) 1389.
14. H.P. Geserich, G. Scheiber, M. Dürrler, F. Lévy and P. Monceau, *Physica* **143B** (1986) 198.
15. P.W. Anderson, *Phys. Rev. Lett.* **18** (1967) 1049.
16. H.J. Schulz, *Int. J. Mod. Phys.* **5** (1991) 57.
17. W.D. Grobman, R.A. Pollak, D.E. Eastman, E.T. Maas and B.A. Scott, *Phys. Rev. Lett.* **32** (1974) 534.
18. P. Nielsen, A.J. Epstein and D.J. Sandman, *Solid State Comm.* **15** (1974) 53.
19. B. Dardel et al, Europhys. Lett. (1992) in press.
20. J.A. Wilson, F.J. Di Salvo and S. Mahajan, *Adv. Phys.* **24** (1975) 117.
21. N.V. Smith, S.D. Kevan, and F.J. Di Salvo, *J. Phys. C* **18** (1985) 3175.
22. B. Dardel, M. Grioni, D. Malterre, P. Weibel, Y. Baer and F. Lévy, *Phys. Rev. B* **45** (1992) 1462; Phys. Rev. B (1992) in press.
23. P. Fazekas and E. Tosatti, *Phil. Mag. B* **39** (1979) 229.
24. R. Manzke, T. Buslaps, B. Pfalzgraf, M. Skibowski, and O. Anderson, *Europhys. Lett.* **8** (1989) 195.

COLLAPSE OF MOTT-HUBBARD FRAMEWORK BY HOLE DOPING IN $Bi_2Sr_2CaCu_2O_8$

T. TAKAHASHI, SATORU SUZUKI, T. KUSUNOKI,
and H. KATAYAMA-YOSHIDA
Department of Physics, Tohoku University, Sendai 980, Japan

ABSTRACT

A combined spectroscopic study with photoemission, inverse photoemission and polarized oxygen K absorption was performed on superconductive and nonsuperconductive $Bi_2Sr_2Ca_{1-x}Y_xCu_2O_8$ (x=0.0-0.6) single crystals and ceramics. It was found that the hole-doping into a nonsuperconductor causes a transfer of weight of the electronic states from the upper Hubbard band to the additional states in the vicinity of the Fermi level produced by the hole-doping itself. The combination of photoemission and inverse photoemission shows that the hole-doping does not cause a shift of the Fermi level comparable to a half of the gap (about 1 eV) nor produce a pseudo-gap between the upper Hubbard band and the additional states in the unoccupied states. This suggests that the hole-doping causes a gradual change of the electronic structure from the Mott-Hubbard picture to the Fermi-liquid (band) picture.

1. Introduction

The existence of a large Fermi surface as predicted by the band structure calculation in the high-Tc superconductor has been established by angle-resolved photoemission spectroscopy[1-3]. The next step for understanding the high-Tc mechanism is to elucidate the process how the Fermi-liquid-like states giving such a large Fermi surface are formed by the carrier-doping. In this paper, we report a combined spectrocopic study with photoemission, inverse photoemission, and polarized oxygen K absorption spectroscopies on the superconductive and nonsuperconductive $Bi_2Sr_2Ca_{1-x}Y_xCu_2O_8$ (x=0.0-0.6) single crystals and ceramics. Combination of photoemission and inverse photoemission gives a detail information on the electronic structure in the vicinity of the Fermi level while polarized x-ray absorption spectroscopy provides a direct information on the nature and the symmetry of the unoccupied electronic states above the Fermi level. The photoemission and inverse photoemission spectroscopies show that the hole-doping creates additional states in the charge-transfer gap of the insulator, but does not cause a shift of the Fermi level comparable to a half of the gap nor produce a pseudo-gap between the upper Hubbard band and the additional states in the unoccupied states. It is found from the polarized x-ray absorption spectroscopy that the lowest unoccupied electronic band in the nonsuperconductor is located about 1 eV above the Fermi level and has a strong in-plane (parallel to the CuO_2 plane) symmetry. This band is ascribed to the upper Hubbard band. Direct comparison of the spectra has revealed that the hole-doping diminishes the intensity of the upper Hubbard band and simultaneously produces additional electronic states near the Fermi level having also the in-plane symmetry. The present results indicate that the hole-doping causes a gradual collapse of the Mott-Hubbard framework of the electronic structure in the insulator by transferring the electronic states from the upper Hubbard band to the additional electronic states produced in the charge-transfer gap.

2. Experimental

Single crystals of $Bi_2Sr_2Ca_{1-x}Y_xCu_2O_8$ were grown with a floating-zone method and a self-flux method. The resistivity and magnetic susceptibility measurements showed that the crystal of x=0.0 is a superconductor with the T_c =81 K while that of x=0.6 does not show superconductivity even at 4 K. Ceramic samples were prepared by sintering a mixture of high purity Bi_2O_3, $SrCO_3$, $CaCO_3$, Y_2O_3, and CuO powders with respective molar ratios. The T_c's determined by resistivity and magnetic susceptibility measurements were 76, 88, and 47 K for x=0.0, 0.3, and 0.5, respectively.

The x-ray absorption measurement was performed at Photon Factory of the National Laboratory for High Energy Physics of Japan. The oxygen K absorption spectrum was recorded at room temperature with the total-photo-yield method at the energy resolution of about 70 meV. Single crystals were cleaved along the a-b plane in the spectrometer to obtain a clean surface for the measurement. The experimental procedure of photoemision and inverse photoemission has been already described elsewhere[4,5].

3. Results and Discussion

Figure 1 shows polarized oxygen K absorption spectra of superconductive $Bi_2Sr_2CaCu_2O_8$ (lower panel) and nonsuperconductive $Bi_2Sr_2Ca_{0.4}Y_{0.6}Cu_2O_8$ (upper panel) single crystals measured at two incidence-angles (θ) of 0° and 75°. Since the incidence-angle is referred to the surface normal, the spectrum of θ=0° represents the O 2p states parallel to the CuO_2 plane (O $2p_{x,y}$) while that of 90° is for the perpendicular component (O $2p_z$). The spectrum measured at θ=75° represents a major weight of the z component (about 93%) with a small contribution from the x,y component (about 7%). As shown in Fig. 1, the parallel component (θ=0°) of the spectrum for the superconductor shows a prominent peak at the absorption-threshold at 528.5 eV which is almost equal to the binding energy of the oxygen 1s core level, while the intensity of the peak is remarkably reduced in the spectrum of θ =75°. This is in good agreement with the previous reports[5-7], indicating a dominant O $2p_{x,y}$ nature of the electronic states at the Fermi level in the superconductor. As for the nonsuperconductor, such a peak is not observed in the spectra. Instead, a large shoulder emerges at about 529.5 eV in the parallel

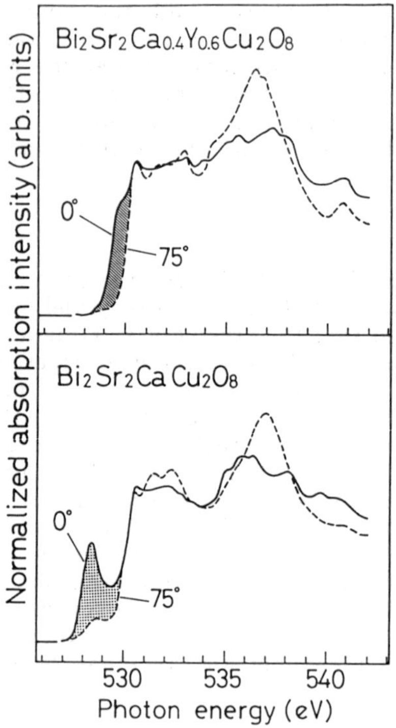

Fig. 1

Polarized oxygen K absorption spectra of superconductive $Bi_2Sr_2CaCu_2O_8$ (lower panel) and nonsuperconductive $Bi_2Sr_2Ca_{0.4}Y_{0.6}Cu_2O_8$ (upper panel) single crystals measured at two incidence-angles (θ= 0° and 75°) with linearly-polarized synchrotron-radiation light. The spectral intensity is normalized with respect to the integrated spectral intensity from 531 eV to 542 eV. Note that a prominent peak in the superconductor and a large shoulder in the nonsuperconductor at the absorption-threshold.

component (θ=0°). It is found in Fig. 1 that this shoulder has a strong in-plane symmetry. We ascribe this large shoulder observed only in the nonsuperconductor to the upper Hubbard band projected into the O 2p states through the p-d hybridization, since the shoulder is positioned at the energy about 1 eV above the Fermi level and has a strong in-plane symmetry. The direct observation of the upper Hubbard band with a strong in-plane symmetry indicates that the Mott-Hubbard framework of the electronic structure certainly stands in the nonsuperconductor (insulator).

Figure 2 shows comparison of the oxygen K absorption spectrum with the same incidence-angle of light (θ). While the spectrum of θ=75° shows little difference between the superconductor and the nonsuperconductor,

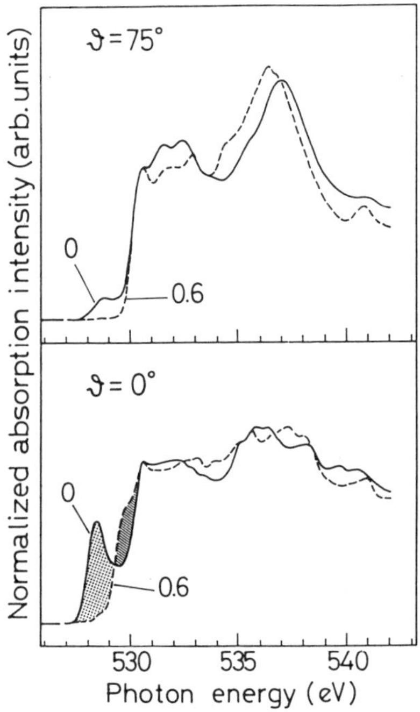

Fig. 2

Polarized oxygen K absorption spectra of $Bi_2Sr_2Ca_{1-x}Y_xCu_2O_8$ (x=0.0 and 0.6) single crystals measured at two incidence-angles (θ) with linearly-polarized synchrotron-radiation light. Note that there is a transfer of weight of the spectral intensity near the absorption-threshold in the parallel component (θ=0°, lower panel) between the superconductor and the nonsuperconductor.

the parallel component (θ=0°) exhibits a drastic change between the two. It is clearly seen in the lower panel of Fig. 2 that there is a transfer of weight of the electronic states between the shoulder at 529.5 eV (upper Hubbard band) in the nonsuperconductor and the peak at 528.5 eV in the superconductor. Although a similar transfer of weight of the electronic states has been reported in the polycrystalline samples of $La_{2-x}Sr_xCuO_4$[8,9], $YBa_2Cu_3O_{7-\delta}$[10], and $Bi_2Sr_2Ca_{1-x}Y_xCu_2O_8$[5], the present result shows more clearly the reduction of the intensity of the upper Hubbard band and the growth of the additional states with the hole-doping by measuring the polarization dependence of the spectrum of single crystals.

It is noted here that there is a different interpretation of the x-ray absorption result. Chen et al.[9] have analyzed their x-ray absorption spectra of $La_{2-x}Sr_xCuO_4$ in terms of the doped Mott-Hubbard picture, where the reduction of the spectral intensity of the upper Hubbard band is explained

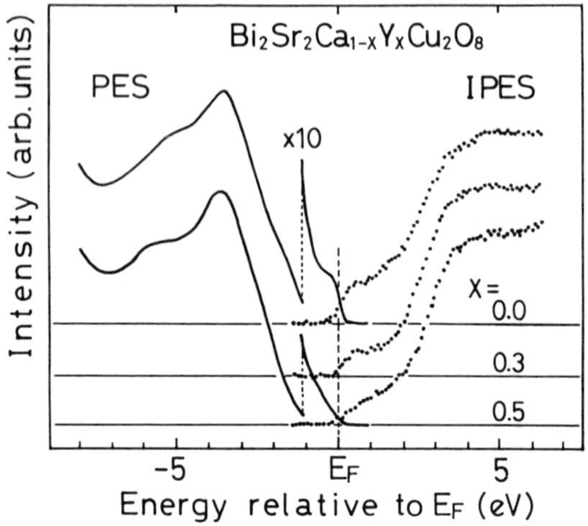

Fig. 3

Photoemission (PES) and inverse photoemission (IPES) spectra of $Bi_2Sr_2Ca_{1-x}Y_xCu_2O_8$ (x=0.0, 0.3, and 0.5). Note that hole-doping (namely by decreasing the Y content) produces an almost uniform distribution of additional states in the vicinity of the Fermi level.

by the increase of the hybridization in the final state[11]. If the Mott-Hubbard framework is retained even after the hole-doping as assumed by Chen et al., a substantial shift of the Fermi level comparable to a half of the charge-transfer gap (about 1 eV) and as a result a pseudo-gap between the upper Hubbard band and the doped O2p band should be observed in the photoemission and inverse photoemission spectra. Figure 3 shows comparison of photoemission and inverse photoemission spectra of $Bi_2Sr_2Ca_{1-x}Y_xCu_2O_8$ (x=0.0, 0.3, and 0.5) ceramics. It is found in Fig. 3 that the hole-doping creates additional states in the vicinity of the Fermi level, but does not cause a rigid shift of the Fermi level by about 1 eV nor produce a pseudo-gap. If the pseudo-gap would be formed by doping in the doped Mott-Hubbard picture, it should appear around 1 eV above the Fermi level since the charge-transfer gap is about 2 eV. However, the hole-doping produces an almost uniform distribution of additional states in the charge-transfer gap. Thus, the photoemission and inverse photoemission result obviously contradicts to the prediction from the doped Mott-Hubbard model where a distinct energy gap is retained between the upper Hubbard band and the doped O2p states.[12] This raises a serious question against the interpretation of x-ray absorption spectra based on the doped Mott-Hubbard model.

4. Conclusion

The present combined spectroscopic study with photoemission, inverse photoemission, and polarized oxygen K absorption spectroscopies has revealed the process of the formation of the Fermi-liquid-like states in the high-Tc superconductor. The experimental result of the polarized x-ray absorption spectroscopy has directly shown the existence of the upper Hubbard band with a perfect in-plane symmetry about 1 eV above the Fermi level in the insulator. This indicates that the Mott-Hubbard picture certainly stands in the case of insulator. On the other hand, it was found that the hole-doping into the insulator causes a transfer of weight of the electronic states from the upper Hubbard band to the additional states produced in the charge-transfer gap. This suggests that the Mott-Hubbard framework in the insulator gradually collapses by the hole-doping and finally the Fermi-liquid band picture is recovered in the metallic superconductor.

Acknowledgment

The authors thank A. Yamanaka, F. Minami, and S. Takekawa for supplying single crystals. This work was supported by the grants from the Ministry of Education, Culture and Science of Japan and from the Foundation for Promotion of Material Science and Technology of Japan.

References

1. T. Takahashi, H. Matsuyama, H. Katayama-Yoshida, Y. Okabe, S. Hosoya, K. Seki, H. Fujimoto, M. Sato, and H. Inokuchi, Phys. Rev. B **39** (1989) 6636.
2. R. Manzke, T. Buslaps, R. Claessen, M. Skibowski, and J. Fink, Physica C **162-164**(1989) 1381.
3. C.G. Olson, R. Liu, A.-B. Yang, D.W. Lynch, A.J. Arko, R.S. List, B.W. Veal, Y.C. Chang, P.X. Jiang, and A.P. Paulikas, Science **245** (1989) 731.
4. T. Watanabe, T. Takahashi, S. Suzuki, S. Sato, and H. Katayama-Yoshida, Phys. Rev. B **44** (1991) 5316.
5. H. Matsuyama, T. Takahashi, H. Katayama-Yoshida, T. Kashiwakura, Y. Okabe, S. Sato, N. Kosugi, A. Yagishita, K. Tanaka, H. Fujimoto, and H. Inokuchi, Physica C **160** (1989) 567.
6. F.J. Himpsel, G.V. Chandrashekhar, A.B. McLean, and M.W. Shafer, Phys. Rev. B **38** (1988) 11946.
7. N. Nücker, H. Romberg, X.X. Xi, J. Fink, B. Gegenheimer, and Z. Zhao, Phys. Rev. B **39** (1989) 6619.
8. H. Romberg, M. Alexander, N. Nücker, P. Adelmann, and J. Fink, Phys. Rev. B **42** (1990) 8768.
9. C.T. Chen, F. Sette, Y. Ma, M.S. Hybertsen, E.B. Stechel, W.M.C. Foulkes, M. Schluter, S.-W. Cheong, A.S. Cooper, L.W. Rupp, Jr., B. Batlogg, Y.L. Soo, Z.H. Ming, A. Krol, and Y.H. Kao, Phys. Rev. Let. **66** (1991) 104.
10. P. Kuiper, G. Kruizinga, J. Ghijsen, M. Grioni, P.J.W. Weijs, F.M.F. de

Groot, G.A. Sawatzky, H. Verweij, L.F. Feiner, and H. Petersen, Phys. Rev. B **38** (1988) 6483.
11. H. Eskes and G.A. Sawatzky, Phys. Rev. B **43** (1991) 119.
12. E. Dagotto, A. Moreo, F. Ortolani, J. Riera, and D.J. Scalapino, Phys. Rev. Lett., **67** (1991) 1918.

High Resolution Photoemission Study of $Bi_2Sr_2CaCu_2O_{8+x}$

Marshall Onellion*
Physics Department
University of Wisconsin
Madison, WI 53706
U.S.A.

Angle-resolved photoemission studies of $Bi_2Sr_2CaCu_2O_{8+x}$ have been performed, with total energy resolution as good as 15 meV. We find that the photoemission spectra exhibit a change between above and below the superconducting transition temperature (T_c). We observe a condensate forming [1] at a binding energy of 35-45 meV below the Fermi energy. The condensate has an intrinsic width of 20 meV, independent of location in the Brillouin zone at which we observe it. The lineshape of the photoemission spectrum is not in accord with weak-coupling s-wave BCS theory. Specifically, there is a reproducible dip in the density of states at a binding energy of 75-85 meV below the Fermi energy. The photoemission lineshape is consistent with d-wave pairing or s+id wave pairing. The spectral intensity is conserved between above and below T_c at the six places in the Brillouin zone at which we took measurements. This conservation is in contrast to a weak-coupling approximation of the resonating valence bond model,[2] which predicts a change in the spectral intensity between above and below T_c. We have used a BCS-like lineshape,[3,4] in lieu of a better model, to infer a value of the superconducting gap from our data. We obtain a value of the gap, delta, of 21-27 meV in the Gamma-to-Xbar direction, and of 15-21 meV in the Gamma-to-Mbar direction. Consequently, we are unable to state conclusively that there is an anisotropy in the order parameter. We observe a condensate forming for a range of angles in the Brillouin zone; this result, and our data in the normal state, are consistent with cluster calculations [5] of the dispersion of the electronic states for temperatures above T_c. Our data is consistent with tunneling results, both point contact [6,7,8,9] and break junction,[10] which yield results in quantitative agreement with our photoemission results. In particular, recent point contact tunneling results [6] suggest that d-wave or s+id wave pairing is the symmetry of the order parameter, as has been suggested by theoretical calculations.[11,12] Infrared spectroscopy studies of our samples, from the same batch, establish that at 20° K there is a difference in the optical absorption, at ~1% level, between the a-axis and the b-axis in the ab-plane, for photon energies well below twice the gap value.[13] This implies either that there are carriers outside the condensate at this temperature, or that the order parameter is not of s-wave symmetry. Studying the normal state electronic properties, we have established that the Bi-O planes contain mobile carriers, which would be candidates for providing interlayer coupling.[14] Further, we have demonstrated that oxygen annealing can directly affect the carrier density of the Cu-O planes, calling into question that picture that the Bi-O planes are the charge reservoirs and the Cu-O planes inviolable.[15]

Beginning in early 1989,[3,4,16-19] several investigators in France, Germany, Japan, Switzerland and the U.S.A. have achieved what many thought would never happen: the direct observation of the opening and details of the superconducting gap. Dr. Clifford Olson and colleagues [15] deserve the credit for the first unambiguous observation of the opening of the gap. Since then, one of the major thrusts has been to obtain as much detailed information on the normal state and superconducting state properties as possible. As the above abstract notes, my colleagues and I have concentrated on the superconducting state and the manner in which the various parts of the unit cell couple electronically.

In published work,[1] we have reported what is, to date, the highest energy resolution study, with the best signal-to-noise, of a superconducting system. In figure one of [1], we report raw data that demonstrates the formation of a superconducting condensate. There are several important aspects about the data. The binding energy of the condensate is between 35-45 meV. Recent results from tunneling,[6-10] having energy resolution of ~1 meV, indicate that the binding energy we have observed is intrinsic to the feature, not a limitation due to our energy resolution of 15 meV. In addition, although we observe a shift of the leading edge of the spectrum toward higher binding energy for temperatures below T_c,[1] we observe spectral intensities for all binding energies below the Fermi energy. Such a result is not consistent with s-wave pairing, for which there should be no excitations (states) in the gap. Tunneling results,[6-10] with better energy resolution, also report such excitations in the gap. Taking account of the different energy resolutions, the two spectroscopies report a condensate with binding energy of the same value.

In addition to the binding energy, the raw data also provide the intrinsic energy width of the condensate: 20 meV.[1] This is confirmed by tunneling results as well.[6-10] This is, to our knowledge, the most narrow feature ever reported in angle-resolved photoemission. By using the binding energy and width of the feature, and a simple, BCS-like lineshape model,[3] we can obtain an estimate of the value of the superconducting gap. Our best estimate is that the gap is between 21-27 mev in the Gamma-to-Xbar direction, and 15-21 meV in the Gamma-to-Mbar direction. Consequently, we are not as yet able to say that the order parameter is definitely anisotropic; we return to this point below. Using a BCS-like lineshape model is not ideal; it is merely the best currently available.

The condensate is not observed for all portions of the Brillouin. For example, it is not observed near the center of the Brillouin zone. It is, however, observed for a range of momenta larger than one would expect, based on L.D.A. calculations of the dispersion in the normal state. Cluster calculations,[5] however, predict a smaller dispersion and an electronic state very close to the Fermi energy for a larger portion of the Brillouin zone. These calculations are consistent with our experimental results, but the reader might do well to regard the issue of where the condensate is observed as an unsettled, interesting, issue.

The lineshape of the photoemission spectrum is not in accord with weak-coupling s-wave BCS theory. Specifically, there is a reproducible dip in the density of states at a binding energy of 75-85 meV below the Fermi energy. By contrast,[20] weak-coupling, s-wave BCS theory predicts no dip at all. There are two types of calculations that have reproduced the dip. Mueller, Arnold and Swihart [21] obtained a dip in the spectral density by considering strong-coupling. Without having to specify the exchange boson, they obtain quantitative agreement with experiment. However, the coupling constant they, and Wermbter and Tewordt [22] find necessary to obtain agreement is quite large (~10), and difficult to reconcile with other experimental work. More recently, D. Coffey and L. Coffey [12] have argued that d-wave pairing leads naturally to the dip observed in both photoemission and tunneling.[1,3,4,6-10,16-19,23] The argument is based on two points: whether the spectroscopy studies a S-I-S or S-N-S structure, and how pair breaking excitations manifest themselves in the two spectroscopies. It is interesting to note that such an analysis is able to predict, in agreement with experiment,[1,6] that the dip occurs at a different multiple of the gap energy for the two spectroscopies. What the photoemission data has clearly established is that s-wave pairing is not consistent with the data, and that some type of d-wave or s+id wave pairing is consistent with the data. The issue of p-wave pairing is difficult to determine from the photoemission data alone.

For angle-integrated photoemission, the Luttinger sum rule applies, indicating that one can test whether the system is a Fermi liquid or a Luttinger liquid by determining whether the total density of electronic states appears to be conserved.[2] The reason is that in a Luttinger liquid,[24,25] the "particle" is a composite, a holon and spinon, with different distribution functions. As Huber has shown in a weak-coupling approximation,[2] this should lead to observable differences in angle-integrated photoemission. In angle-resolved photoemission, however, there is no Luttinger sum guarantee that the spectral intensity should be conserved. For this reason, observing a change in the spectral intensity between $T<T_c$ and $T>_c$ would not be significant. However, for the six data sets we have of sufficient quality to compare spectral intensity between $T<T_c$ and $T>_c$, all six exhibit the same spectral intensity with temperature (the error bars are 2%). Since this was done at an energy resolution sufficient to observe a spinon energy well below the gap value,[1,3,4] and since Huber's calculations imply that we should see more than a 2% change, we conclude that a simple version of the RVB model is not consistent with our data.

Among several ways to test the symmetry of the order parameter, infrared absorption is a useful method. In absorption measurements, unlike obtaining the optical conductivity, there is no Kramer-Kronig analysis, no inference. Our colleagues and we [13] have performed infrared measurements on samples from the same batch as samples for which photoemission measurements were performed. We find that at a temperature of 20° K, there is a difference in the optical absorption, at ~1% level, between having the photon electric field vector in the a-axis versus the b-axis in the ab-plane, for photon energies well below twice the gap value.[13] Both a-axis and b-axis absorptions are nonzero. So the results are quite clear: having the photon electric field vector in the ab-plane results in a nonzero optical absorption, and there is a nonzero difference in the absorption for the electric field vector parallel to the a-axis versus parallel to the b-axis. There are two natural explanations for these data. Either there are carriers outside the condensate at this temperature, or the order parameter is anisotropic.[13]

In addition to observing the onset of superconductivity, we have performed measurements of the normal state electronic properties. Among the most interesting and important questions is whether there are mobile carriers anywhere in the unit cell other than the Cu-O planes. Earlier reports on this issue were inconclusive.[26,27] Using resonant photoemission in both the constant initial state and constant final state modes, we have established two important points.[14,15] We have observed a resonant response of the Fermi level electrons across the Bi5d core levels. The resonant response exhibits a threshold that is 1.5-2.0 eV below the core level binding energy. This response is characteristic of a correlate plasmon excitation,[28] which means there are mobile carriers on the Bi-O planes.[14] Consequently, we believe that there are mobile electron carriers on the Bi-O planes; such planes are not pristine insulators. This is significant due to the various models analyzing interlayer versus intralayer coupling in the cuprates.[28] The other point of importance is that oxygen annealing affects the electron carrier density on the Cu-O planes. We have determined this by measuring the Fermi edge strength, normalized to the mostly Cu3d peak in the valence band. As the oxygen annealing increases the hole carrier density, and lowers the electron carrier density, the Fermi edge loses strength.[15] At the same time, the fractional resonant enhancement of the Fermi edge electron electrons across the O2s threshold increases. Using a recently developed method for calculating the carrier density from the resonant enhancement,[30] we find that the change in carrier density determined by the two methods agree to within 10%. Consequently, the resonant enhancement of the Fermi energy electrons across the O2s threshold is a measure of carrier density. However, there is no corresponding change in the strength of the resonant enhancement across the Bi5d core levels. We conclude that the change across the O2s threshold is due to a change in the Cu-O planes, not the Bi-O planes.[15]

The data in [1] exhibit a characteristic difference between the superconducting and normal states. The electronic states for $T>T_c$ are much broader (100-150 meV) than for $T<T_c$. There is no ready explanation for this behavior, since the temperature difference is small (~50° K). A related matter is whether we can, currently, use the photoemission linewidths as a test of whether the materials are

Fermi liquids.[31] In contrast to some published work, we would argue that at present one cannot use photoemission linewidths in this manner. We recently published results [32] indicating that the linewidths of dispersing electronic states can increase as the states disperse through the Fermi energy. Earlier work by S. Kevan and colleagues [33] on surface states note the same behavior. Such behavior is not consistent with the argument that the linewidth of a Fermi liquid state should decrease to zero with the binding energy as $(E_b)^2$. However, such an argument assumes implicitly that the linewidth observed is entirely due to the photohole. As P. Thiry noted several years ago,[34] and N. Smith [35] and Smith, Thiry and Petroff have discussed in detail recently,[36] this assumption is not warranted, even for simple sp-band electronic states. Consequently, until the details of the electronic structure are worked out, we view using photoemission linewidths of high-temperature superconductors as a test of Fermi liquid behavior as premature. There is also a most intriguing possibility to the data in [1]. The simple picture of superconductivity says that the occupied electronic density of state is affected for binding energies up to ~x3 the gap value. Such a view does not explain why the electronic state becomes so much broader for $T>T_c$. An intriguing possibility, currently under investigation, is that the opening of a superconductivity gap has the effect of removing a broadening channel- an inelastic loss channel. At present, however, this possibility remains unproven.

(*) Various parts of this work done in collaboration with colleagues at Ecole Polytechnique Federale de Lausanne, Switzerland, the University of Rome and other institutions in Italy, and in the U.S.A., including: Helmut Berger, William Bonner, Yeh Chang, Paul David, Favia Gozzo, Yeukuang Hwu, David Huber, Robert Joynt, Ronald Kelley, Salvo La Rosa, Francis Levy, Luca Lozzi, Giorgio Margaritondo, Pat Morris, Marino Marsi, Manuel Quijada, Ned Stoffel, Ming Tang, David Tanner, Michael Winokur, and Roberto Zanoni.

References

1. Y. Hwu, L. Lozzi, M. Marsi, S. La Rosa, M. Winokur, P. Davis, M. Onellion, H. Berger, F. Gozzo, F. Levy and G. Margaritondo, Phys. Rev. Lett. 67, 2573 (1991).

2. D.L. Huber, Solid State Commun. 68, 459 (1988).

3. Y. Chang, Ming Tang, R. Zanoni, M. Onellion, Robert Joynt, D.L. Huber, G. Margaritondo, P.A. Morris, W.A. Bonner, J.M. Tarascon, and N.G. Stoffel, Phys. Rev. B39, 4740 (1989).

4. Y. Chang, Ming Tang, Y. Hwu, M. Onellion, D.L. Huber, G. Margaritondo, P.A. Morris, W.A. Bonner, J.M. Tarascon, and N.G. Stoffel, Phys. Rev. B39, 7313 (1989).

5. G. Dopf, J. Wagner, P. Dieterich, A. Muramatsu and W. Hanke, Phys. Rev. Lett. 68, 2082 (1992).

6. J.F. Zasadzinski et.al., Bull. Am. Phys. Soc. 37, 668 (1992) and unpublished.

7. A. Goldman, private communication and Jin-Xiang Liu, Ji-Chun Wan, A.M. Goldman, Y.C. Chang and P.Z. Jiang, Phys. Rev. Lett. 67, 2195 (1991).

8. A. Chang, Zhao Y. Rong, Yu. M. Ivanchenko, Faren Lu and E.L. Wolf, unpublished; E.L. Wolf, H.J. Tao and B. Susla, Solid State Commun. 77, 519 (1991).

9. Qun Chen and K.-W. Ng, Phys. Rev. B45, 2569 (1992).

10. L. Mihaly, private communication and D. Mandrus, L. Forro, D. Koller and L. Mihaly, Nature 351, 460 (1991).

11. Q.P. Li and R. Joynt, "Nuclear relaxation in the superconducting state of $YBa_2Cu_3O_{6+x}$: evidence for s+id state," Phys. Rev. B, in press; Q.P. Li, B. Koltenbah and Robert Joynt, unpublished.

12. D. Coffey and L. Coffey, private communication and unpublished.

13. M. Quijada, D.B. Tanner, R.J. Kelley and M. Onellion, unpublished.

14. R.J. Kelley, L. Dottl, H. Edelman, G. Margaritondo and M. Onellion, unpublished.

15. R.J. Kelley, G. Margaritondo, P.A. Dowben and M. Onellion, unpublished.

16. C.G. Olson et.al., Science 245, 731 (1989).

17. J.-M. Imer, F. Patthey, B. Dardel, W.-D. Schneider, Y. Baer, Y. Petroff and A. Zettl, Phys. Rev. Lett. 62, 336 (1989).

18. R. Manzke, T. Buslaps, R. Claessen and J. Fink, Europhys. Lett. 9, 477 (1989).

19. T. Takahashi, JJAP Series 7, "Mechanisms of Superconductivity," pp. 141-146, 1992 and references therein.

20. "Introduction to Superconductivity," Michael Tinkham, McGraw-Hill Book Co., New York, 1975, pp. 42f.

21. G.B. Arnold, F.M. Mueller and J.C. Swihart, Phys. Rev. Lett. 67, 2569 (1991).

22. S. Wermbter and L. Tewordt, Phys. Rev. B, in press (Nov., 1992) and references therein.

23. D.S. Dessau, B.O. Wells, Z.-X. Shen, W.E. Spicer, A.J. Arko, R.S. List, D.B. Mitzi and A. Kapitulnik, Phys. Rev. Lett. 66, 2160 (1991).

24. P.W. Anderson and Yong Ren, "Theory of the Normal State of Cuprate Superconductors," Proc. MRS Symp., Nov. 17, 1989.

25. P.W. Anderson, Phys. Rev. Lett. 64, 1839 (1990).

26. B.O. Wells, Z.-X. Shen, D.S. Dessau et.al., Phys. Rev. Lett. 65, 3056 (1990).

27. T. Takahashi and H. Katayama-Yoshida, Phys. Rev. Lett. 68, 416 (1992).

28. G. Wendin, "Breakdown of the One-Electron Pictures in Photoemission," Springer-Verlag, Berlin, 1981.

29. S.H. Liu and R.A. Klemm, Phys. Rev. B 45, 415 (1992) and references therein.

30. P.A. Dowben, D. LaGraffe, Dongqi Li, G. Vidali, L. Zhang, L. Dottl and M. Onellion, Phys. Rev. B43, 10677 (1991).

31. C.G. Olson, R. Liu, D.W. Lynch, R.S. List, A.J. Arko, B.W. Veal, Y.C. Chang, P.Z. Jiang and A.P. Paulikas, Phys. Rev. B42, 381 (1990).

32. Y.Hwu, L. Lozzi, S. La Rosa, M. Onellion, P. Aalmeras, F. Gozzo, H.Berger, F. Levy and G. Margaritondo, Phys. Rev. B. 45, 5438 (1992).

33. S.D. Kevan, Phys. Rev. Lett. 50, 526 (1983).

34. Pierre Thiry, Ph.D. thesis, "La photoemission angulaire dans les solides structure electronique de volume et de surface du cuivre," L'Universite Pierre et Marie Curie, 1981.

35. N.V. Smith, Comments Cond. Mat. Phys. 15, 263 (1992).

36. N.V. Smith, P. Thiry and Y. Petroff, unpublished.

VI. SURFACES

HIGH RESOLUTION ABSORPTION SPECTROSCOPY OF MOLECULAR ADSORBATES AFTER CORE LEVEL EXCITATION

D. Arvanitis, H. Rabus*, T. Lederer, and K. Baberschke
Inst. für Experimentalphysik, FU Berlin, Arnimallee 14,
D–1000 Berlin 33, Germany

Abstract

The photoabsorption of C_2H_4 molecules after core level excitation is measured at the Carbon K–edge with an energy resolution of the incident X–rays of $\lesssim$ 200 meV. The fine structure of the ($1s^{-1}$, π^*) state is studied for physisorbed and weakly chemisorbed species in submonolayer coverage on Cu(100) and Ag(100) surfaces. The fine structure is interpreted as due to intramolecular vibrations and analyzed in the framework of the linear coupling model. For the physisorbed species Rydberg excitations are also identified. By varying the X–ray incidence angle, the symmetry of these states is experimentally assigned.

1. Introduction

During the last few years, the quality of soft X–ray (≈0.1–1 keV) photoabsorption data after core level excitation has been greatly improved for excitations to bound final states of condensed molecular multilayers or gas phase molecules[1-5]. Using monochromators with a resolving power of up to 10^4, intramolecular excitations of vibrational character in the ($1s^{-1}$, π^*) state of N_2, C_2H_4 and C_6H_6 molecules were identified as well as in the region below the gas phase IP[2-4]. High resolution measurements have been shown to be possible also in the case of adsorbed molecules[5,6] at BESSY. Here we investigate the fine structure of adsorbed C_2H_4 molecules after core level excitation in a weakly chemisorbed and physisorbed states. We will focus therefore in moderate photon energy resolutions between 100 and 200 meV at the C K–edge and photon fluxes of the order of 10^8 photons/s, that allow still the study of adsorbed molecular submonolayers. Also the data of the correspondent gas and condensed phases will therefore be presented using this energy resolution. In the present Communication, the changes of the fine structure of the ($1s^{-1}$, π^*) state are analysed, as the strength of the molecular interaction to the substrate increases. A quantitative analysis in terms of intramolecular vibrations of C–C and C–H character allow to determine the evolution of the energy and natural width of the electronic ground state transition as the strength of chemisorption increases. It is also demonstrated that high resolution Near Edge X–ray Absorption Fine Structure (NEXAFS) measurements can be performed for submonolayer molecular coverages over long energy ranges of the order of 40 eV with an energy resolution of ~200 meV. This allows the determination of the symmetry of Rydberg excitations for physisorbed species. A systematic lowering of the photoabsorption continuum is found as one moves from the gas phase, to condensed multilayers and to physisorbed species.

2. Experimental

The measurements were performed at the SX–700 II beam line of the Freie Universität at BESSY[2,5]. Gas phase measurements were performed using a gas cell equipped with a AlSi window, at a pressure of ≈0.1 mbar[7]. Submonolayer experiments for weakly chemisorbed C_2H_4 molecules were performed at a temperature of ~50K on Cu(100) and Ag(100) surfaces. Dosages of single layer coverages (~0.1 monolayer) were used[7]. For condensed multilayers a dosis of 150 L was used corresponding to a coverage of more than 70 layers[7]. By direct dosing at 20K on the Cu(100) surface, the C_2H_4 molecules were found to adsorb in a physisorbed state[9]. Spectra were taken in this case for both 0.2 monolayer (ML) and 0.4 ML coverage data[8]. The spectra of the adsorbed molecules were taken in the partial electron yield mode using an incident photon energy resolution of 200 meV ($\lesssim 3\ 10^8$ photons/s) or better. An energy calibration resulting in an accuracy better than 50 meV was achieved by dosing multilayers after each submonolayer experiment[7]. The energy resolution was checked in situ using the multilayer or gas phase measurements and measurements of the LaM_5 edge (~ 835 eV) in LaAl_2[2].

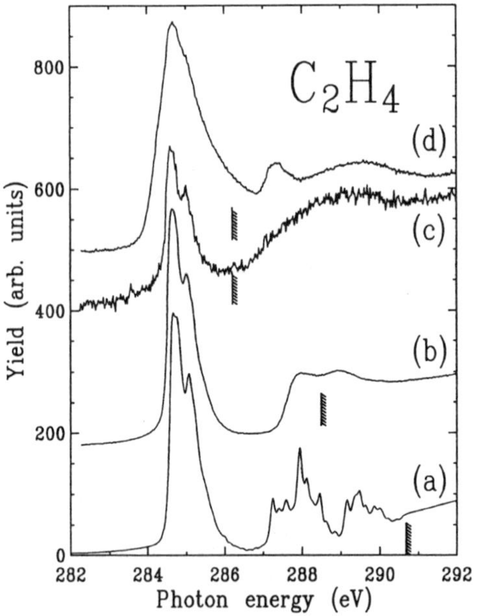

Figure 1: Photoabsorption measurements at the C K–edge of: (a) gas phase C_2H_4 with $\Delta(h\nu)$=90 meV, (b) Condensed multilayers, (c) 0.4 ML physisorbed molecules on Cu(100) at normal X–ray incidence, (d) as in (c) but for a grazing X–ray incidence angle of 15°. For (b), (c) and (d) $\Delta(h\nu)$=140 meV. The hatched marks indicate the onset of a continuum threshold in photoabsorption (see text).

3. Changes in the spectra upon condensation and adsorption

The evolution of photoabsorption spectra taken with a resolution better than 150 meV can be seen in Figure 1 for C_2H_4 molecules in the gas phase (a), condensed multilayers (b) and physisorbed on Cu(100) ((c),(d)). A comparison of curves (a) and (b) indicates that upon condensation: 1) Much of the fine structure in the range of 287–291 eV is quenched. 2) The continuum threshold is lowered in energy. For the gas phase data of Fig. 1 the onset of a continuum of final states is well defined as the Ionization Potential which can be clearly seen around 290.7 eV as the point of convergence of the Rydberg series, clearly resolved below that value. This onset, indicated by a hatched mark, is lowered in the case of the condensed multilayer due to the polarization of the medium surrounding the photoexcited ion. A standard approach of continous electrocynamics in media yields a threshold shift of ~ –2.6 eV with a dielectric constant for solid ethylene of ~2.1[7], close to the value obtained by a more detailed analysis of the data[7]. 3) The Rydberg excitations[4] around 287.25 eV also appear to be quenched, yielding a longer flat region in the multilayer spectrum up to 287.5 eV. No changes in the fine structure of the intensive ($1s^{-1}$, π^*) state resonance have been found[7] going from Fig. 1(a) to Fig. 1(b), the data of Fig. 1(a) being taken with a better energy resolution. This situation appears to be modified in the case of physisorbed C_2H_4 molecules (Fig. 1 (c), (d)). For normal X–ray incidence (Fig. 1 (c)), the fine structure of the ($1s^{-1}$, π^*) state appear to be conserved but no clear zero line above the π^* resonance can be found. For the grazing X–ray incidence data (Fig. 1(d)) the π^* resonance is broadened and appears to overlap with the still increased continuum intensity in the region between 286 and 287.5 eV. The XPS binding energy for the physisorbed species was found to be ~286 eV (hatched marks).

4. The evolution of the ($1s^{-1}$, π^*) state, upon increasing chemisorption strength

The fine structure of the ($1s^{-1}$, π^*) state is now well established to be of vibrational origin. In the case of the C_2H_4 molecule, the gas phase and condensed phase data could be analyzed assuming the excitation of the symmetric CH_2 and C–C stretch modes and: 1) The Born–Oppenheimer approximation, 2) No coupling of normal modes, 3) The same normal coordinates in the ground and excited states, for data taken with a photon energy resolution down to ~90 meV[7]. However, after a detailed analysis of even higher energy resolution data[10], these assumptions are found to be not fully justified and therefore natural linewidths of the order of 100 meV or below should be interpreted taking core–hole localization effects into account. Here we will apply assumptions 1)–3) and use the linear coupling model as we focus on the study of molecular adsorbates with an energy resolution of ≈ 200 meV.

Under these conditions one obtains for the vibrational excitation energies

$$E(n_1, \dots n_q) = \sum_{r=1}^{q} n_r \cdot \hbar\omega_r \qquad (1)$$

ω_r being the vibrational frequency of the rth mode. Using low temperatures (50 K or below) the Franck–Condon factors are given by

$$I(n_1, \ldots, n_q) = I(0, \ldots, 0) \cdot \prod_{r=1}^{q} \frac{S_r^{n_r}}{n_r!} \tag{2}$$

Taking into account the fact that the ground and excited vibrational states correspond to different frequencies, ($\omega_{r,gr}$ and $\omega_{r,ex}$ respectively) the bond length change in the excited ($1s^{-1}$, π^*) state is given by[11,7]:

$$|\Delta Q_r| = \sqrt{2 S_r} \cdot \frac{\omega_{r,gr}}{\omega_{r,ex}} . \tag{3}$$

For $\omega_{r,gr} > \omega_{r,ex}$ the bond has become weaker, $\Delta Q_r < 0$ and vice versa.

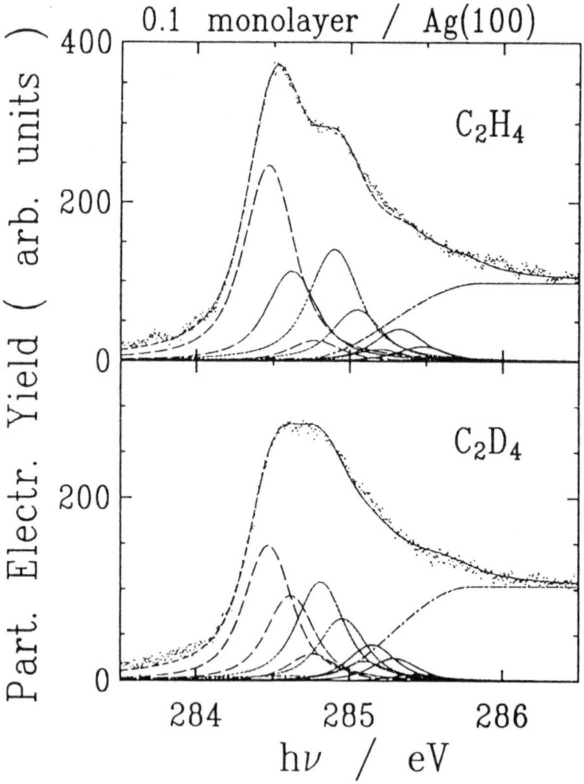

Figure 2: 0.1 ML data for the (1s,π^*) state of weakly chemisorbed C_2H_4 and C_2D_4 on Ag(100). The linear coupling model has been used with a CH_2 and C—C stretch modes. Here Δ(hv)=200 meV.

Fig. 2 shows a fit using Eqs. (1) and (2) for C_2H_4 molecules weakly chemisorbed on Ag(100) at 50K. In this case flat lying species have been found, with a desorption temperature of 143 K, weakly perturbed with respect to the gas phase[7]. From the ($1s^{-1}$, σ^*) resonance position a slightly weakened C–C bond was determined, the elongation with respect to the gas phase being $\Delta R_{cc}=0.03$ Å for the ground state[7]. From Eqs. (1) and (2) the $\omega_{r,ex}$ and S_r are determined. Voigt profiles were used, with the monochromator resolution being 200 meV. A deuteration effect is clearly visible even by eye inspection, the shoulder about 284.8 eV for C_2H_4 (Fig. 2, upper part) clearly shifts to lower energies. Both the CH_2 and C–C stretch modes are observed as in the gas phase. Using the linear coupling model the six fit parameters necessary to generate a fit with two vibrational series as in Fig. 2, are checked for self consistency through the effect of deuteration[7,8]. A natural linewidth γ of 280(15) meV for C_2H_4 and 250(15) meV for C_2D_4 was used for the fits of Fig. 2. For condensed multilayers, a value of 160(5) meV was determined for C_2H_4 (150(5) meV for C_2D_4). These values indicate that even in the case of a weak interaction with the surface the ($1s^{-1}$,π^*) state decays faster. The small systematic deviations of γ upon deuteration eventually indicate the limitations of the assumptions 1)–3) made in the beginning of this Section. These deviations are however almost within the error and therefore a more systematic interpretation cannot be attempted.

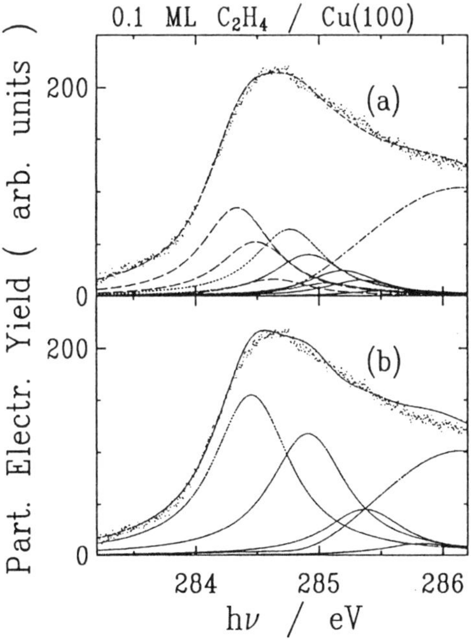

Figure 3: 0.1 ML data for the ($1s^{-1}$, π^*) state of chemisorbed C_2H_4 molecules on Cu(100). In (a) both CH_2 and C–C stretch modes has been used in the fit. In (b) only the CH_2 mode is used (see text). The data were taken with Δ(hv)=200 meV.

The interaction of the C_2H_4 molecules with a Cu(100) surface at 50K is even stronger. The desorption temperature is now 190 K, and the C–C bond is elongated by ΔR_{CC}=0.10Å[7]. Even by eye inspection a further broadening of the low energy side of the π^* resonance can be observed. A deuteration effect is still observable in this case[7,8]. Fig. 3 illustrates two possible fits. In Fig. 3(a) both CH_2 and C–C stretch modes are included. For the fit of Fig. 3(b) only the CH_2 mode is allowed. It is observed that even in this case, of a drastic variation in the input of the fit parameters, the position and width of the ground state electronic transition are not strongly affected. A natural width of 620(20) meV is found in this case[7] (Fig. 3(a)). Also an energy shift towards lower energies is observed for the ground state transition. As it is observed by comparing Fig. 3(a) to (b) the parameters that vary the most are the line intensities (S factors). The assumptions underlying the two different fits of Fig. 3 differ concerning the time evolution of the final state.

Introducing the same number of series as in the fit of Fig. 3(a), to the fine structure of the π^* resonance in the data of Figs. 2 and 3, corresponds to the sudden approximation: The absorption of the photon is an instantaneous process on the time scale of the molecular vibrations. The nuclei cannot rearrange during the excitation process. Therefore if the molecule has been initially in the vibrational ground state, this is in general not the case after the absorption process. The transition probabilities to the different vibrational states are given by the overlap integrals of the time independent nuclear wave functions of the electronic ground and excited states. The second approach shown in Fig. 3(b), assumes that the time evolution of the excited state has to be taken into account for the calculation of the Franck–Condon factors. In the presence of the substrate additional decay channels are available and the natural line width γ increases. Due to the reduced natural lifetime of the core excited state the vibrational wave functions are different from those for the free molecule. Now in the low energetic modes the nuclei do not have enough time to move before the core excited state decays. Therefore the probability P to excite these vibrations is reduced by a factor in the order of[5]:

$$P \propto \exp(-\gamma/\hbar\omega_r) \qquad (4)$$

The fits of Figs. 3(a) and 3(b) indicate that the broadening of the lines (620 meV) is such that it should influence the intensity distribution of the C–C stretch mode ($\hbar\omega_r$=150 meV). The increased natural width indicates that the final state decays within a time of the same order of magnitude as the intramolecular vibrations. A supplementary relaxation channel is now available through the chemisorption to the substrate, the extra electron in the π^* final state can escape to the substrate. However if Eq. (4) is used a factor P of 0.02 is obtained (Fig.3(b)), which does not yield a fit as satisfactory as in Fig. 3 (a). It is possible that time effects indeed exist but Eq. (4) is too simplistic to describe them quantitatively. In this context it should be mentioned that time dependent effects have also been seen in photo–emission experiments[12,13] and interpreted in a similar way, using the same type of analysis.

The fits shown in Figs. 2 and 3 depend also on the position of the continuum threshold. The data are fitted by introducing the ground state transition line in the low energy side of the resonance. The subsequent lines are added to reproduce the shape of the π^* resonance yielding the Franck–Condon envelope. This is matched to higher energies to a continuum step modelling the continuum of final states due to the metal bands. A fit starting with the assumption of a threshold directly at the low energetic side of the resonance does not yield consistent results. As it has been observed before[14,15] the height of the π^* resonance is related to the intramolecular bonding strength and the C–C bond length. Fits where the continuum threshold is positioned at the low energy side of the resonance do not yield consistent results concerning the C–C bond length[14,7]. For example in the data of Fig. 3 such an input would yield a very weak π^* intensity, compatible only to an almost fully saturated C–C bond. This would imply a σ^*(C–C) resonance position in clear contradiction to the experiment[14].

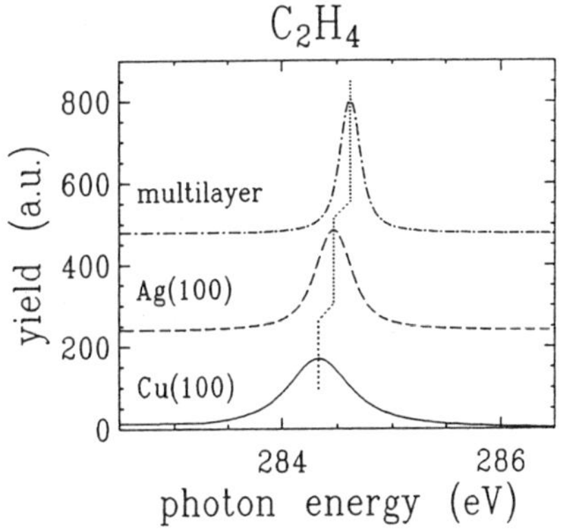

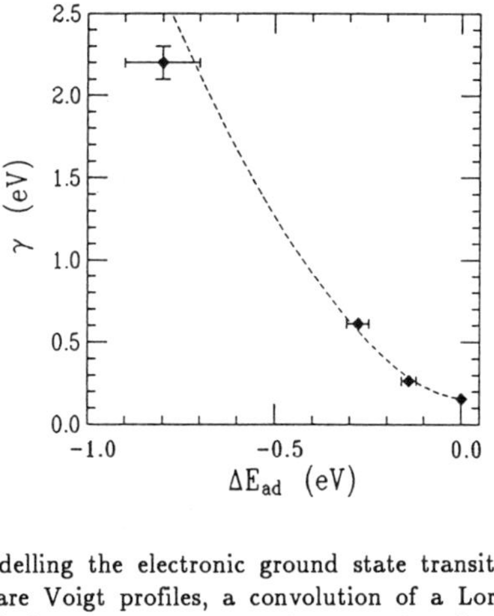

Figure 4: The line modelling the electronic ground state transition is shown. The lines used throughout this work are Voigt profiles, a convolution of a Lorentz to a Gaussian line. The experimental width Δ(hv)=200 meV here, is the same for all the data shown. A systematic broadening due to life–time effects (γ) and shift ΔE_{ad} of the ground state transition can be seen. The Lorentzian part of the lines γ is also shown as a function of ΔE_{ad}.

The results of the data analysis are summarized in Fig. 4 where only the ground state transition components of the fits are shown. A systematic energy shift ΔE_{ad} and broadening are observed. This is also seen in the correlation diagram of the natural width γ versus ΔE_{ad} for this transition. Together with the C_2H_4 data of the present work a point corresponding the the system of even stronger chemisorption CO/Ni(111)[16] is shown. Trends in the same direction are observed as for the data of the present work, which indicates that the phenomena described here are not specific of the particular systems chosen and depend only on the chemisorption strength between a molecule and a metal surface.

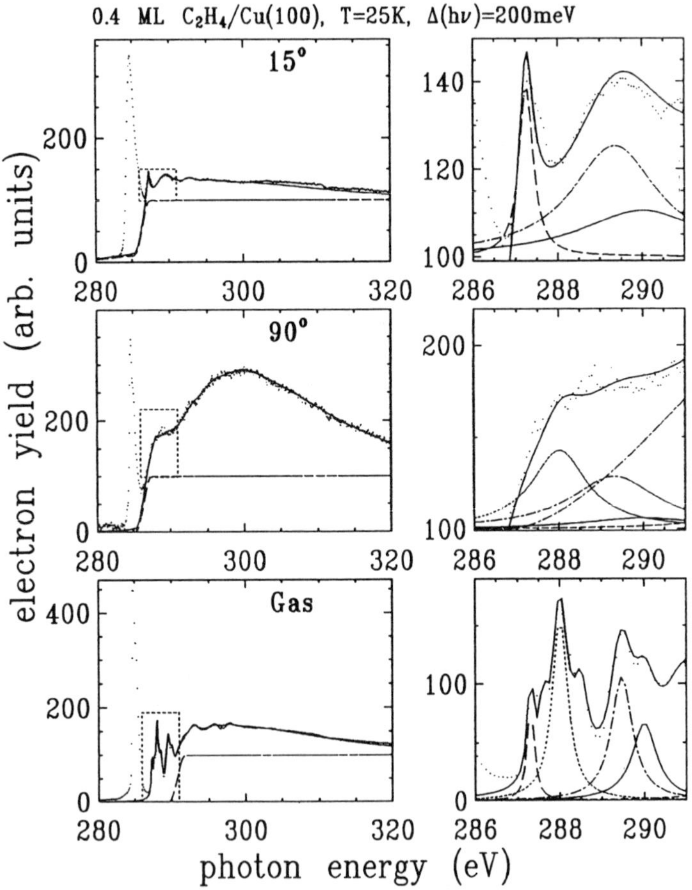

Figure 5: Data of physisorbed C_2H_4 on Cu(100) are shown, together with gas phase data using the same monochromator resolution. Rydberg states can be identified for the physisorbed species, and their symmetry character experimentally determined (see text).

5. High resolution NEXAFS on physisorbed molecules

High photon energy resolution data for C_2H_4 molecules on Cu(100) at a temperature of 20K are shown in Fig. 5. The σ^* resonance intensity at 15° grazing X–ray incidence is very weak, indicating that the molecules lie with the C–C axis parallel to the surface. The observation of π^* intensity at normal X–ray incidence indicates that there exist molecules adsorbed with the π orbital parallel to the surface. A quantitative analysis indicates that 58% of the molecules are found in this geometry, the rest being adsorbed with their π orbital perpendicular to the surface[9]. For these species the data of Fig. 1 indicate that the fine structure of the ($1s^{-1}$,π^*) state is exactly the one observed for gas phase molecules[8]. This is an indication for the abscence of a chemical bond to the surface, in view of the results of Section 4. Another one is given from the energy of the ($1s^{-1}$, σ^*) state that coincides with the one of the gas phase[9]. The previous observations establish the abscence of a chemical bond between the C_2H_4 molecules and the substrate. In contrast, in all cases when a chemical bond is formed only species with π orbitals perpendicular to the surface have been observed[14]. As has been discussed in the previous Section the fine structure of the ($1s^{-1}$, π^*) state differs from the one of gas phase species. All previous findings combined to the observation of a broadening for the ($1s^{-1}$, π^*) state for the data of Fig. 1(d), indicates that for this excited state supplementary decay channels exist for the π orbitals pointing towards the surface in the abscence of an overlap between wave functions in the ground state[9].

A more precise inspection of the data of Figs. 1 and 5 indicates the existence of sharp features at the same energies as in the gas phase. For example around 287.25 eV a feature is seen assigned to a transition to a 3s Rydberg final state after gas phase measurements. The present data clearly indicate a π character as this state is only observed at grazing X–ray incidence. On the other hand the peak around 288.0 eV assigned to a transition to a 3p Rydberg state is only observed at normal X–ray incidence, an indication of σ character. A detailed quantitative analysis for two adsorbate coverages (0.2 and 0.4 ML) is given elsewhere[9] and confirms the previous observations. The previous comparison between the symmetry of these states directly determined from the experiment is different from previous assignments based only on a comparison of their energies to theoretical calculations. It is now possible to use as an input for such calculations not only the energies but the symmetry of the final state as well.

6. Conclusion

Due to improved photon energy resolution in the soft X–ray regime, small energy shifts and life–time broadenings of the order of 100 meV can be measured for the final states after core level excitation This allows for a better characterization of the chemisorption process and opens new aoportunities for the understanding for the role of low energetic excitations (inter–intra–molecular vibrations) in catalytic reactions. High energy resolution photoabsorption measurements of oriented but gas phase like molecules on a surface at low temperatures allows for a direct determination of the symmetry of Rydberg states to be used as an input to theoretical calculations.

Acknowledgement

We would like to thank M. Domke for help during the gas phase measurements. The work is supported by the BMFT, Grant Nr. 05 5KEAAB

References

*Present address: PTB am BESSY, Lentzeallee 100, D–1000 Berlin 33, Germany

1. C.T. Chen and F. Sette, Rev. Sci. Instrum. 60 (1989) 1616
2. D. Arvanitis, H. Rabus, M. Domke, A. Puschmann, G. Comelli, H. Petersen, L. Tröger, T. Lederer, G. Kaindl, K. Baberschke, App. Phys. A 49 (1989) 393
3. C.T. Chen and F. Sette, Physica Scripta T31 (1990) 119
4. P.A. Heimann, F. Senf, W. McKinney, M. Howells, R.D. van Zee, L.J. Medhurst, T. Lauritzen, J. Chin, J. Meneghetti, W. Gath, H. Hogrefe, D.A. Shirley, Physica Scripta T31 (1990) 127
5. H. Rabus, D. Arvanitis, M. Domke, A. Puschmann, L. Wenzel, G. Comelli, G. Kaindl, K. Baberschke, Physica Scripta, T31 (1990) 131
6. A. Puschmann, T. Mandel, M. Domke, C. Xue, G. Kaindl, Surf. Sci. 252/252 (1991) 353
7. H. Rabus, D. Arvanitis, M. Domke, K. Baberschke, J. Chem. Phys. 96 (1992) 1560
8. H. Rabus, D. Arvanitis, K. Baberschke, Surf. Sci. 269/270 (1992) 270
9. D. Arvanitis, J. Singh, H. Rabus, T. Lederer, K. Baberschke, Phys. Rev. B 45 (1992) 1518
10. F.X. Gaeda, H. Köppel, J. Schirmer, L.S. Cederbaum, K.J. Randall, A.M. Bradshaw, Y. Ma, F. Sette, C.T. Chen, Phys. Rev. Lett. 66 (1991) 883
11. W. Domcke, L.S. Cederbaum, J. Chem. Phys. 64 (1976) 603
12. U. Höfer, M.J. Breitschafter, E. Umbach, Phys. Rev. Lett. 64 (1990) 3050
13. M. Bertolo, W. Hansen, K. Jacobi, Phys. Rev. Lett. 67 (1991) 1898
14. D. Arvanitis, K. Baberschke, L. Wenzel, U. Döbler, Phys. Rev. Lett. 57 (1986) 3175
15. A.P. Hitchcock, A.T. Wen, E. Ruhl, J. Elec. Spec. Rel. Phen. 51 (1990) 653
16. F. Jugnet, F.J. Himpsel, Ph. Avouris, E.E. Koch, Phys. Rev. Lett. 53 (1984) 198

FINAL-STATE EFFECTS IN PHOTOEMISSION FROM METAL-SEMICONDUCTOR INTERFACES

J. KANSKI, K. KARLSSON, O. NYQVIST, P.O. NILSSON
Department of Physics, Chalmers University of Technology, S-412 96 Göteborg, Sweden,

U.O. KARLSSON
Department of Physics, Royal Institute of Technology, S-100 44 Stockholm, Sweden,

M.C. HÅKANSSON
Department of Synchrotron Radiation Research, University of Lund, S-223 62 Lund, Sweden,

S.P. SVENSSON
Martin Marietta Laboratories, Baltimore, MD 21227, USA

ABSTRACT

The effect of modified final state screening in photoemission from a semiconductor surface, which undergoes metallization, is investigated by model calculations. It is found that the screening may induce core level shifts of 0.1 - 0.2 eV, which has severe implications on photoemission studies of metal semiconductor interfaces. Results from an experiment designed to isolate the metal induced screening are in qualitative agreement with model predictions.

1. Introduction

Over the last decades photoelectron spectroscopy has become one of the most trusted techniques for detailed electron structure investigations, fundamental as well as applied ones. Depending on the nature of system under investigation, interpretation of data can be more or less straightforward. Generally, the degree of complexity increases with increased localization of the excited state. In the case of core levels, the appearance of a localized hole is accompanied by valence state excitations. These excitations may show up as distinct spectral structures, or as less pronounced modifications of the main emission. The energy of the main emission line is, of course, linked to the energy of the "final" (photohole) state, which depends on the screening properties. As long as these properties are unchanged, core level shifts can be interpreted in terms of initial state modifications, e.g. chemical shifts and band bending. It may occur, however, that the nature of the final state does change significantly as the sample conditions are changed. In such cases the measured core level shifts will include the final state modifications, and their interpretation becomes more complicated. One particular case where one has reason to suspect that such changes may be of importance is in studies of metal-semiconductor interfaces: the non-local screening (i.e. the image potential energy) of a photohole created on the semiconductor side of the interface is changed when the semiconductor surface is metallized.

Considering that much of the presently available information about metal-semiconductor interfaces (i.e. the Schottky barrier problem) is derived from core level shifts measured by photoemission, a detailed examination of these effects is well motivated. At an early stage it was found that Schottky barriers deduced from core level shifts are practically the same as those observed by electrical methods [1]. It is worthwhile to note that this conclusion was based on data obtained with submonolayer metal coverages, i.e. before it was realized that the final Fermi-level pinning is not reached until the surface layer has developed metal properties [2-5], and also before the effects of surface photovoltage [6,7] were appreciated. As a great majority of studies so far are made on nonideal (i.e. more or less disordered, intermixed and reacted) interfaces, it is justified to question the detailed equivalence between electrical and photoemission measurements. On the contrary, in the few cases where well-defined interfaces have been investigated, differences of ~ 0.2 eV have been reported between barrier heights derived from electrical and photoemission studies [8].

In this paper we present a model calculation, where the image potential energies are evaluated for the semiconductor-vacuum and semiconductor-metal interfaces. A calculation is also made for the metal-vacuum interface, in order to allow for comparison with recent experimental photoemission data from noble gas adsorbates on metal and semiconductor surfaces [9]. The parameters used in this calculation are then also used for evaluating the difference between the final state screening at the semiconductor-vacuum and semiconductor-metal interfaces. We also present results of a recent experiment aiming at isolating the effects of different image potentials.

The model calculation treats only one specific aspect of the photoemission process. A full account of the relaxation effect must, in principle, involve total energy calculations on semiconductor-vacuum and semiconductor-metal systems, with and without a core hole (ΔSCF). To our knowledge no such extensive calculations have been published so far. Model calculations of metal induced screening of impurity states [10], which pin the Fermi level, suggest that the initial state core level shifts caused by this effect are very small (< 20 meV). It should also be stressed that the subject of our interest is equally relevant whether the actual interface states responsible for the Fermi-level pinning are defect states[1] or metal-induced gap states (MIGS) [11].

2. Model calculations

The model considered here consists of two semi-infinite dielectric media, described by dielectric functions $\varepsilon_1(q)$ and $\varepsilon_2(q)$, representing the semiconductor and metal (or vacuum), respectively. The real experimental situation with the metallized semiconductor should, in principle, be modeled with two interfaces. However, as the screening length in a metal is very short (~ 0.5 Å), even a monoatomic metallic overlayer is well represented by a semi-infinite medium as viewed from the semiconductor side of the interface [12]. Approximating the core hole with a point charge Q located at a distance z_0 from the interface ($z_0 < 0$, see inset Fig. 1), we calculate the induced (image) potential in the semiconductor, ϕ_1, for the semiconductor-vacuum and semiconductor-metal systems. In order to avoid the the image potential divergence at small distances, we use wave-vector dependent dielectric functions. For a translationally invariant dielectric medium we have to solve the equation

$$-\nabla^2\phi^D = 4\pi Q\delta(\mathbf{r} + z_0\hat{\mathbf{z}}) \tag{1}$$

where
$$\phi(q) = \frac{\phi^D(q)}{\varepsilon(q)} \tag{2}$$

with the boundary conditions

$$\phi_1(q, z)\big|_{z=0} = \phi_2(q, z)\big|_{z=0} \tag{3}$$

and
$$\left.\frac{\partial \phi_1^D(q,z)}{\partial z}\right|_{z=0} = \left.\frac{\partial \phi_2^D(q,z)}{\partial z}\right|_{z=0} \tag{4}$$

The potentials ϕ^D and ϕ correspond to the **D** and **E** fields, respectively. For the semiconductor we use [13]

$$\varepsilon_1(q) = 1 + \frac{k_1^2}{c + q^2} \tag{5}$$

and for the metal
$$\varepsilon_2(q) = 1 + \frac{k_2^2}{q^2} \tag{6}$$

where k_1^{-1} and k_2^{-1} are the Thomas-Fermi screening lengths, and c is a constant determined by the static value of ε_1 (Eq. 10). In describing the dielectric properties we have neglected local-field effects, i.e. we have made spherical averaging of the screening charge density. The resulting expression for the electrostatic potential becomes (indices M and V refer to semiconductor-metal and semiconductor-vacuum interfaces):

$$\phi_1^{M,V} = B^{M,V}\left[\frac{e^{qz}}{\varepsilon_1} + \frac{k_1^2 e^{\kappa_1 z}}{2\kappa_1(\kappa_1 - q)}\right] + \frac{2\pi Q}{q}\left[\frac{k_1^2 e^{\kappa_1 z - qz_0}}{2\kappa_1(\kappa_1 + q)} + \frac{k_1^2 q e^{-\kappa_1|z + z_0|}}{\kappa_1(\kappa_1^2 - q^2)} + \frac{e^{-q|z + z_0|}}{\varepsilon_1}\right] \tag{7}$$

where
$$\kappa_1 = \sqrt{c + k_1^2 + q^2} \tag{8}$$

$$\kappa_2 = \sqrt{k_2^2 + q^2} \tag{9}$$

and
$$\varepsilon_1 = 1 + \frac{k_1^2}{c} \tag{10}$$

is the static dielectric constant. With the boundary conditions given in Eqs. (3) and (4), the coefficients $B^{M,V}$ are:

$$B^M = \frac{2\pi Q}{q}\frac{[q/\kappa_2 - 1/\varepsilon_1 + (q/\kappa_1)(1 - 1/\varepsilon_1)]e^{-qz_0} - 2(q/\kappa_1)(1 - 1/\varepsilon_1)e^{-\kappa_1 z_0}}{1/\varepsilon_1 + (q/\kappa_1)(1 - 1/\varepsilon_1) + q/\kappa_2 + 2} \tag{11}$$

$$B^V = \frac{2\pi Q}{q}\left[1 - \frac{1}{\varepsilon_1}\right]\frac{(1 + q/\kappa_1)e^{-qz_0} - 2(q/\kappa_1)e^{-\kappa_1 z_0}}{1/\varepsilon_1 + (q/\kappa_1)(1 - 1/\varepsilon_1) + 3} \tag{12}$$

The electrostatic image potential $\phi^{M,V}_{1,im}$ (ρ,z) in the semiconductor region (z < 0) is the first term in Eq. (7). To get the spatial dependence we make a Fourier transform:

$$\phi^{M,V}_{1,im}(\rho,z) = \int \frac{d^2q}{(2\pi)^2} B^{M,V} \left\{ \frac{e^{qz}}{\varepsilon_1} + \frac{1}{2}\left[1 - \frac{1}{\varepsilon_1}\right]\left[1 + \frac{q}{\kappa_1}\right] e^{\kappa_1 z} \right\} e^{iq\cdot\rho}. \tag{13}$$

where ρ = (x,y). The electrostatic image energy at the core hole (W) is obtained, finally, by building up the charge Q from infinitesimal charges dQ:

$$W_{M,V}(0,z) = \int_0^e dQ\, \phi^{M,V}_{1,im}(0,z) \tag{14}$$

To get numerical estimates of the image energy, assumptions have to be made about the screening lengths (i.e. characteristic distances beyond which the core hole is completely screened in the semiconductor and in the metal). We believe that realistic values for these quantities are ~ 2-5 Å (ionic semiconductor) and ~ 0.5 Å (metal). Using the values k_1^{-1} = 3.3 Å and k_2^{-1} = 0.5 Å, and a bulk dielectric constant ε_1 = 12 (GaAs), we obtain image energies as shown in Fig. 1.

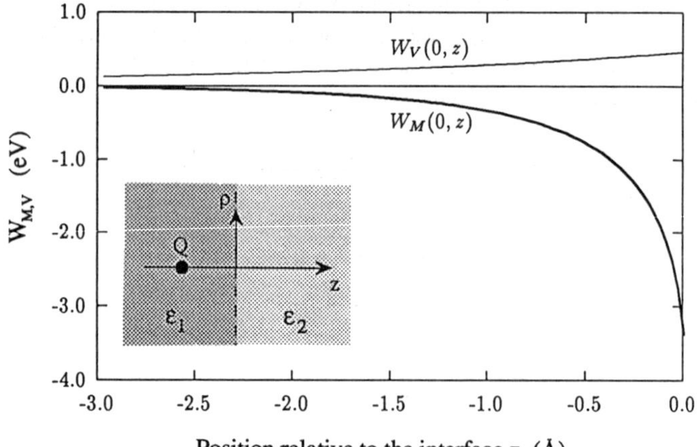

Figure 1: The electrostatic image energy for semiconductor-vacuum (W_V) and semiconductor-metal (W_M) systems. The inset describes the geometry.

Before evaluating the screening effects explicitly for photoemission from metal-semiconductor interfaces, we apply the model to a different experimental situation, namely photoemission from noble gas atoms adsorbed on semiconductor- and metal surfaces. In a recent photoemission study of Xe adsorbed on different semiconductor and metal surfaces, it was found that the 4d core level binding energy of Xe is systematically ~ 0.6 eV lower on metal surfaces, and the effect was suggested to be due to different final state screening in the two cases [9]. In the same way as above we derive an expression for the image potential, this time for the metal-vacuum (M′) and semiconductor-vacuum (S′) interfaces. (In the experiments [9] submonolayer quantities were used in order to avoid effects of Xe polarizability. This justifies the use of $\varepsilon_{Xe} = 1$.) We obtain:

$$\phi_{1,im}^{M',S'}(\rho,z) = \int \frac{d^2q}{(2\pi)^2} B^{M',S'} e^{iq\cdot\rho} \tag{15}$$

where the coefficients $B^{M'}$ and $B^{S'}$ are given by

$$B^{M'} = \frac{2\pi Q}{q} \frac{q/\kappa_2 - 1}{q/\kappa_2 + 3} e^{-2qz} \tag{16}$$

and

$$B^{S'} = \frac{2\pi Q}{q} \frac{1/\varepsilon_1 + (q/\kappa_1)(1 - 1/\varepsilon_1) - 1}{1/\varepsilon_1 + (q/\kappa_1)(1 - 1/\varepsilon_1) + 3} e^{-2qz} \tag{17}$$

The electrostatic image energy at the adatom core hole is then:

$$W_{M',S'}(0,z) = \int_0^e dQ\, \phi_{1,im}^{M',S'}(0,z) \tag{18}$$

In Fig. 2 we show the calculated core energy difference ($\Delta E = W_{S'}(0,z) - W_{M'}(0,z)$) for some different screening parameters. If we estimate the average location of the core hole by the value of the Xe atomic radius (1.2 Å), we find that the model predicts an energy difference of ~ 0.5 eV, in good agreement with experiment [9].

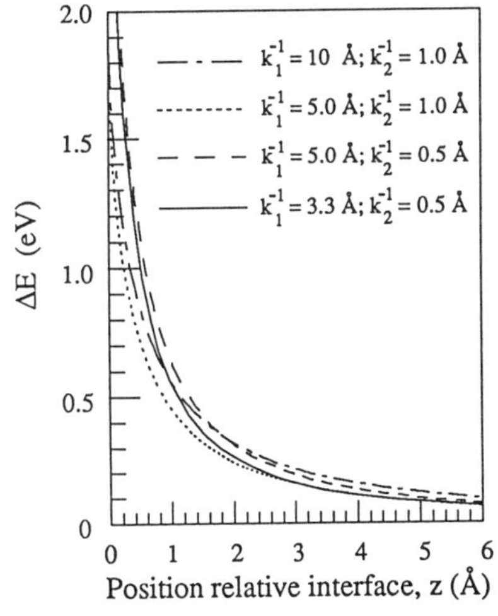

Figure 2: The core level energy difference in photoemission from atoms adsorbed on metal and semiconductor surfaces, as obtained from Eq. (18) with $\varepsilon_1 = 12$. Note the relatively low sensitivity of ΔE on the precise choice of screening parameters.

Having gained some confidence in the model, we return to the semiconductor metallization case. In a photoemission process the probability for an excited electron to escape into vacuum without any inelastic energy loss decays exponentially with increased distance to the surface. The characteristic decay length (mean free path λ) is typically 5 - 10 Å for electrons with kinetic energies ~ 10 - 100 eV. Since the screening-related shift depends on the distance between the core hole and the metal, the total effect in a photoemission spectrum has to be calculated by adding the contributions from all layers parallel with the interface:

$$I_{M,V}(E) \sim \sum_i e^{z_i/\lambda} \frac{\gamma/2\pi}{[E - E_C - W_{M,V}(0, z_i)]^2 + (\gamma/2)^2} \tag{19}$$

where E_C is the core level binding energy, z_i defines the position of the emitting core, λ is the mean free path, and γ the total broadening of the core level (for the present purpose it is not necessary to have a correct lineshape).

As a numerical example we apply Eq. (19) to Ga(3d) emission from a metallized GaAs(100) surface, in which case we assume an ideally abrupt interface with a Ga-plane facing the metal. We assume further that the first Ga layer is located halfway between the metal (defining the image plane) and the first As layer, i.e. z_1 = 1.4 Å. Using the same parameters as those in Fig. 1, we obtain photoemission spectra as shown in Fig. 3. (For simplicity we model the Ga spectrum with one component.)

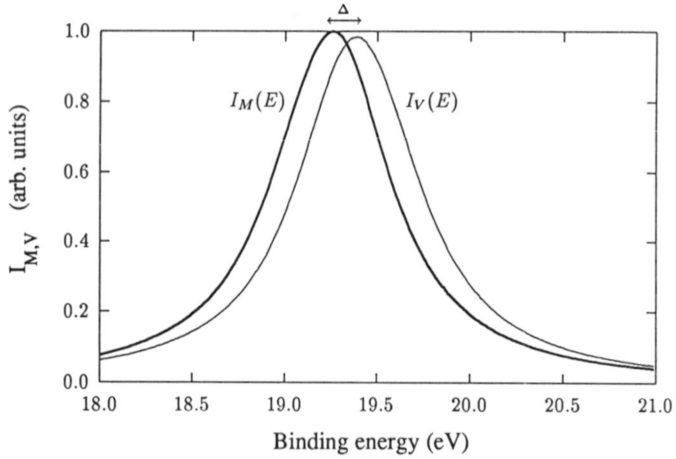

Figure 3: Simulated spectra from semiconductor-vacuum (I_V) and semiconductor-metal (I_M) systems obtained with the following parameters: k_1^{-1} = 3.3 Å; k_2^{-1} = 0.5 Å; λ = 8 Å; γ = 0.7 eV, the latter typical for XPS.

The binding energy of the core state, as observed by the peak position in photoemission, is thus reduced by an amount Δ = 0.13 eV, due to improved final state screening. To demonstrate the effect of varying surface sensitivity, we also calculate Δ as a function of λ, see Fig. 4. We notice that the shift increases with increasing surface sensitivity.

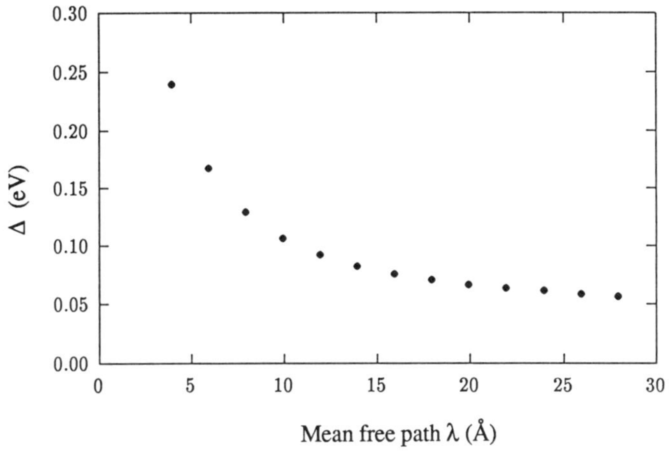

Figure 4: The core level shift as a function of the mean free path. k_1^{-1} = 3.3 Å; k_2^{-1} = 0.5 Å; γ = 0.7 eV.

3. Experimental results

The severe implications of these results to photoemission studies of the Schottky barrier phenomenon require extensive experimental tests. The first question we should ask is whether there are any signs of screening effects in previously published data. We have already mentioned the deviation between photoemission and electrical results on epitaxial Pb/Si(111) contacts, with photoemission data systematically yielding ~ 0.18 eV larger barrier heights [8]. It has been proposed that this difference could be due to structural changes at the interface when thick metal overlayers are formed [14], although there are indications that this is not the case [15]. An alternative explanation could be in terms of final state screening as discussed above: in the photoemission experiment the final state energy should be reduced with 0.1 - 0.2 eV after metallization, and this would have the same effect on the spectra as an additional band bending. Another possible sign of metal induced screening is the general observation of core level shifts of 0.1 - 0.2 eV, accompanying the appearance of metallic characteristics on semiconductor surfaces [2-5]. So for instance, in detailed studies of reacted rare earth/GaAs(110) interfaces [4] it was found that all core level energies were reduced. In this particular case an alternative interpretation was favoured (in terms of MIG-state Fermi level pinning), mainly because the Schottky barrier height deduced from the photoemission data coincided with results from electrical measurements. In view of the mentioned discrepancies for more ideal systems, this criterion can be questioned.

The metal-semiconductor interface problem is a complex one, so details like those discussed here may easily be disguised by other phenomena, like metal clustering, interdiffusion and interface chemistry. Effects of metal induced screening should be possible to identify via the predicted distance dependence (Fig. 1), in analogy with the successful studies of layer dependent screening in rare gas adsorbates [16]. Since the expected energy shifts are relatively small (~ 0.1 eV), it is essential that such investigations are performed under well controlled experimental conditions, with well defined, abrupt and nonreacted interfaces. The distance dependence can be tested in different ways. The simplest method is to compare spectra recorded at a fixed photon energy, but different emission angles. If the photon energy is chosen such that the electron mean free path is long, the relative spectral weight from the near surface region is significantly increased at large emission angles. Alternatively, by recording photoemission spectra at different photon energies, the mean free path is varied, and thus the relative spectral contribution of the outermost atomic layers can be changed. Although this latter method appears straightforward, it is connected with some practical problems. Due to the relatively slow variation of the electron mean free path with kinetic energy, quite large variation of photon energy is needed. This can be achieved using synchrotron radiation, but the required accuracy imposes high demands on a precise monochromator calibration (e.g. via a stable internal reference sample).

In the present study we have chosen an alternative method, namely to vary the semiconductor-metal distance by the use of spacer material of well controlled thickness. Such layered semiconductor structures can be prepared by means of molecular beam epitaxy (MBE). For the first studies we have chosen to use GaAs and AlAs, which are believed to form nearly atomically abrupt interfaces. In the present case our samples consisted of thick AlAs(100) grown on GaAs(100) wafers, on top of which additional GaAs layers of different thicknesses were grown to serve as spacers for the subsequent metallization. The GaAs growth rate was calibrated by intensity oscillations in reflection high energy electron diffraction (RHEED). After growth the samples were capped with Sb. (Since the MBE sample manipulator did not allow for sufficient cooling, we were unable to apply As_4 capping.) This preparation was performed at the Martin Marietta Laboratories in Baltimore, USA. The samples were then transported to the national synchrotron radiation centre MAX-lab in Lund, Sweden, where they were mounted on one sample holder in the photoelectron analyser system. Once the samples were mounted on the same holder, they were subject to exactly the same subsequent treatment, including parallel metallization. In the analysis chamber the samples were heated to desorb the Sb cap. This process was monitored via the intensity of Sb(4d) photoemission. In this way we could see that almost all Sb was removed at ~ 350 °C. The remaining Sb (similar amounts on the different samples) was quite stable, probably reflecting some As/Sb exchange in the surface layer. For metallization we used Cs, obtained from a dispenser type evaporation source. Cs was chosen for the present studies because it does not cluster on the surface and does not react chemically with GaAs. We assume that the saturated Cs overlayer on GaAs(100) formed at room temperature is metallic, just as on GaAs(110) [17]. No signs of Cs oxidation were detected.

In Fig. 5 we show Al(2p) spectra obtained before and after metallization, with two different GaAs spacer layer thicknesses, 2 and 4 monolayers (ML). To evaluate the Cs induced shifts, the spectra were analyzed using a computer aided peak fitting routine. Treating the spin-orbit components as independent structures, we obtained in each case an intensity ratio between the two peaks of 2.0, i.e. the expected branching ratio, and an energy separation of 0.41 eV. From the fitted curves it is immediately clear that the Cs induced shifts

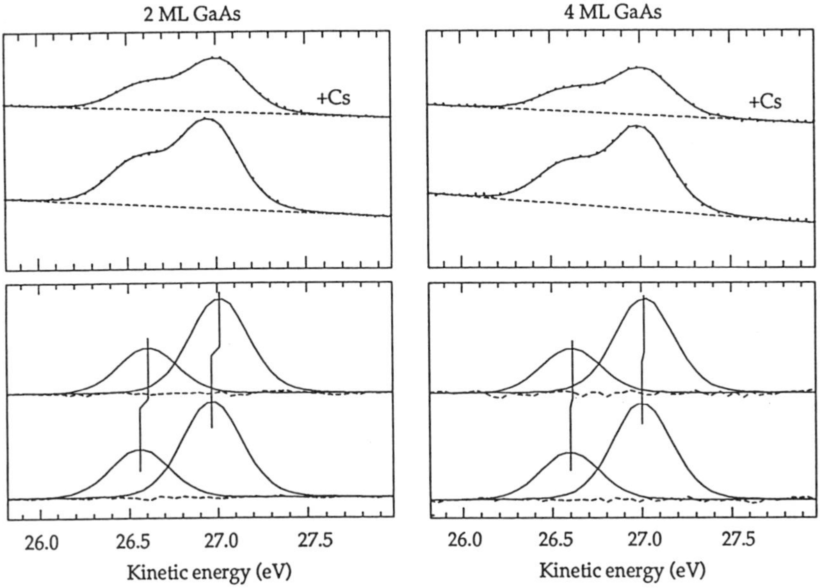

Figure 5: Al(2p) photoemission from clean and Cs covered GaAs/AlAs samples, excited with 105 eV photon energy. The data points are shown in the upper panels, together with the fitted spectra (full lines) and estimated linear backgrounds (dashed). In the lower panel we show the components in the fitted curves (full lines) and the difference between the fitted and the measured spectra (dashed lines). Fitting parameters: Γ_{Gauss} = 0.35 eV, $\Gamma_{Lor.}$ = 0.06 eV.

are larger with 2ML than with 4ML GaAs spacer, ~ 45 meV and ~ 10 meV, respectively. These values are somewhat larger than predicted by the non-local screening model (21 meV and 7 meV). The reason for the small values is, of course, the rapid decay of the image potential energy with increasing distance from the interface. According to the model, significantly larger effects would be expected for thinner spacer layers. Such investigations are in progress.

4. Conclusion

When a free semiconductor surface is metallized, a reduction of core state binding energies is expected as a result of changed final state screening. According to the model calculations presented here, the induced image energy may be as large as 0.1 - 0.2 eV. In photoemission experiments the effect should be observed via shifted and somewhat broadened core level spectra after metallization. The model is supported by a test on an experimental situation where non-local screening is more established, namely photoemission from noble gas adsorbates. The first experimental effort to single out the screening effect on a metal-semiconductor system is presented, and the results are found to be in qualitative agreement with the model predictions. This has very important implications to studies of metal-semiconductor interfaces by means of core level photoemission, as the observed shifts cannot be fully ascribed to initial state effects like metal induced band bending.

5. Acknowledgements

We wish to thank G. Le Lay for valuable comments. The generous technical assistance of the MAX-lab staff is gratefully acknowledged. This work is supported by grants from the Swedish Natural Science Research Council.

6. References

1. P.W. Chye, I. Lindau, O. Pianetta, C.M. Garner, C.Y. Su, and W. Spicer, Phys. Rev. B **18** (1978) 5545.
2. F. Schäffler and G. Abstreiter, J. Vac. Sci. Technol. B **3** (1985) 1184.
3. J. Kanski, S.P. Svensson, T.G. Andersson, and G. Le Lay, Solid State Commun. **60** (1986) 793.
4. M. Prietsch, C. Laubschat, M. Domke, and G. Kaindl, Phys. Rev. B **38** (1988) 10655.
5. C. Laubschat, M. Prietsch, M. Domke, E. Weschke, G. Remmers, T. Mandel, J.E. Ortega, and G. Kaindl, Phys. Rev. Lett. **62** (1989) 1306.
6. J.E. Demuth, W.J. Thompson, N.J. DiNardo, and R. Imbihl, Phys. Rev. Lett. **56** (1986) 1408.
7. M. Alonso, R. Cimino, and K. Horn, Phys. Rev. Lett. **64** (1990) 1947.
8. J.A. Carlisle, T. Miller, and T.-C. Chiang, Phys. Rev. B**45** (1992) 3400, and references therein.
9. G. Haugstad, A. Raisanen, L. Sorba, L. Vanzetti, X. Yu, and A. Franciosi, J. Vac. Sci. Technol. B **9** (1991) 2415, and references therein.
10. R. Ludeke, G. Jezequel, and A. Taleb-Ibrahimi, J. Vac. Sci. Technol. B **6** (1988) 1277.
11. J. Tersoff, Phys. Rev. Lett. **52** (1984) 465.
12. K. Karlsson, O. Nyqvist, and G. Wendin, Solid State Commun. **67** (1988) 339.
13. F. Bechstedt and R. Enderlein, Phys. St. Solidi (b) **94** (1979) 239; **95** (1979) 185.
14. G. Le Lay and K. Hricovini, Phys. Rev. Lett. **65** (1990) 807.
15. H.H. Weitering, D.R. Heslinga, T. Hibma, and T.M. Klapwijk, Phys. Rev. Lett. **65** (1990) 808.
16. T.-C. Chiang, G. Kaindl, and T. Mandel, Phys. Rev. B**33** (1986) 695.
17. L.J. Whitman, J.A. Stroscio, R.A. Dragoset, and R.J. Celotta, Phys. Rev. Lett. **66** (1991) 1338.

INVERSE PHOTOEMISSION OF ALKALI-METAL OVERLAYERS ON SEMICONDUCTORS

B. REIHL, R. DUDDE*, L.S.O. JOHANSSON$, and K.O. MAGNUSSON#
IBM Research Division, Zurich Research Laboratory, CH-8803 Rüschlikon, Switzerland

Alkali-metal layers on semiconductor surfaces are characterized by the strong decrease of the workfunction with coverage. Recently, the interest in these systems concentrates on questions like: Is the adsorbate system at monolayer coverage metallic or semiconducting, and does the metallization take place in the alkali overlayer or in the top layer of the semiconductor? Is the bonding ionic or covalent? What is the absolute coverage at saturation? What are the adsorption sites? Do all alkali metals behave similar on the same semiconductor surface? We try to answer some of the questions for Li, Na, K, and Cs on Si(111)2 × 1, K and Cs on Si(111)7 × 7 and on GaAs(110), and Na and K on Si(100)2 × 1 employing the techniques of direct and inverse photoemission.

1. Introduction

In contrast to transition-metal overlayers which usually intermix and form silicides,[1] alkali metals form simple overlayer systems, i.e. they do not intermix with the semiconductor at room temperature (with the exception of Li, perhaps) and can be deposited homogeneously without any clustering. In addition their monovalent character suggests relatively simple physics as compared to open-shell transition metal atoms. Employing the techniques of direct and inverse photoemission spectroscopy (UPS and IPES, respectively), we try to understand the properties of the near-monolayer covered semiconductor surfaces and to link the geometric overlayer structures with their electronic properties. IPES is the ideal tool to monitor the unoccupied conduction band states involved in charge transfer processes during metal deposition, while UPS yields the complementary information about the occupied valence and core-level states. As semiconductor substrates we have chosen the cleaved Si(111)2 × 1 and GaAs(110)1 × 1 surfaces and the annealed Si(111)7 × 7 and Si(100)2 × 1 surfaces. One drawback in choosing alkali metals is their high reactivity. They are known to be catalytic promoters for oxidation and even in a vacuum with a base pressure in the 10^{-11} mbar range, a freshly prepared alkali/semiconductor interface exhibits signs of contamination after 2 to 3 hours, this being in sharp contrast to the silicide forming metal overlayers.

The two more applied aspects of alkali/semiconductor interfaces (as photocathodes[2] and spin polarized electron sources[3]) are both connected to the strong reduction of the workfunction with coverage. As a measure for the workfunction we use the cutoff of the secondary electrons in UPS or the onset of the absorption

---

* Present address: Bundesministerium für Forschung und Technologie, Heinemannstr. 2, D-5300 Bonn 2, Germany

$ Present address: Dept. of Synchrotron Radiation Physics, University of Lund, S-22362 Lund, Sweden

Present address: Dept. of Natural Science, University of Karlstad, Box 9501, S-65009 Karlstad, Sweden

current employing the diode method. The total evaporation time from an alkali dispenser oven is used as measure for the coverage An absolute calibration of the coverage is difficult and has often created confusion leading to irreproducible results in different laboratories. We (like others) prefer to use the saturation onset of the workfunction change and name it 'saturation-monolayer coverage' and abbreviate it '1 ML'. As we shall see below this coverage is very reproducible and often goes along with a well-defined (new) LEED pattern indicating commensurate fractions of a monolayer as the absolute coverage.

2. Ionic versus covalent bonding

The nature of the bonding of the alkali-metal atoms to semiconductor surfaces, in particular to silicon surfaces, is the subject of a current controversy. It may be exemplified by describing the two extreme limits: (i) In case of a complete charge transfer of the alkali valence s electron to the semiconductor surface the bonding is *ionic*, and the metallization as a function of coverage is caused by a partial filling of the semiconductor surface-state bands. The strong dipole field created by the positive alkali ion and the negative image charge are causing the workfunction change. This picture was first forwarded for K/Si(100)2×1 by Ciraci and Batra[4] and supported for other alkali/semiconductor interfaces by several theory groups.[5-11] (ii) In the other case the charge transfer is fractional and rather small, and a weak *covalent* bonding explains the strong workfunction reduction by a polarization-dependent interaction. The metallization occurs within the alkali overlayer as soon as the valence electron orbitals can overlap as a function of coverage, hence a critical surface atom density is required in contrast to the ionic-bonding picture. This picture is favored by Ishida and Terakura[12,13] and other groups.[14,15] Some theories[16,17] also favor a mixed-type of bonding, i.e. almost ionic at low coverage, while the covalent character increases with coverage. In reality the truth can lie in between the two extreme cases (i) and (ii) outlined above, in particular when hybridization between the semiconductor surface states and alkali metal s,p states comes into play.

The amount of charge transfer as relevant parameter regarding the *ionic* versus *covalent* character of the bonding is particularly suited to be measured by the technique of inverse photoemission.[18] Empty electronic states (band states or localized levels) may get filled by the charge transfer process, and hence are no longer available as final states for IPES. This is illustrated in fig. 1. The change of the IPES signal from the clean to the metal-covered surface is a monitor of the semiconductor-metal interaction. In addition IPES provides the right surface sensitivity owing to the limited penetration depth of the incoming electrons.[18-20]

We are mainly interested in ordered surface structures, because then surface states may form bands with well-defined E(**k**) dispersion relationships in the plane of the surface. The measurement of their dispersion allows to identify the surface states, i.e. distinguish them from bulk states. In cases where they cross the Fermi energy, a necessary condition for a metallization of the interface system is fulfilled. By comparison to electronic-structure calculations we can even derive surface-atom configurations and determine the alkali adsoprtion sites as will be demonstrated below. The wavevector **k** of surface states as two-dimensional entities has only a component parallel to the surface $k_{\parallel}$ which can be related to the kinetic energy E_k and polar angle

θ of the incident electron beam by $k_{\parallel} = 0.512 \sin\theta\,(E_k)^{1/2}$, where $E_k = h\nu - \phi + E_B$ is the measurable kinetic energy of a spectral feature in IPES with binding energy E_B relative to $E_F = 0$. Energies are in electronvolts and wavevectors in reciprocal Ångströms, ϕ denotes the workfunction. Details of the UPS and IPES experiments, the sample preparation, the surface characterization, and the workfunction plots of the various alkali-metal interfaces may be found elsewhere.[18,19]

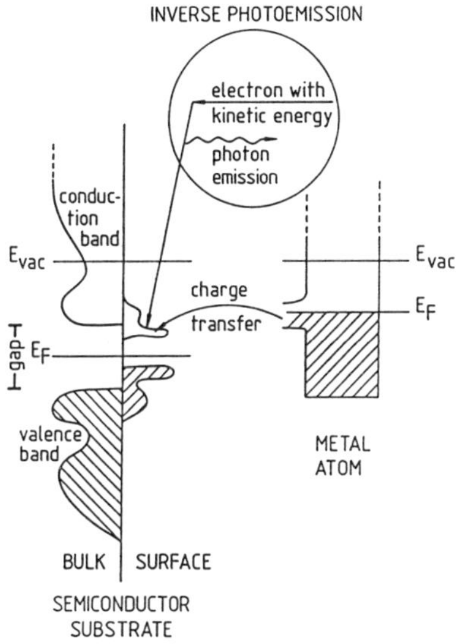

Fig. 1. Illustration of inverse photoemission as a monitor of charge-transfer processes which may occur when a metal atom interacts with a semiconductor surface with bulk and surface states.

3. Na and K on Si(100)2×1

Studies of alkali metals on the Si(100)2×1 surface started the recent interest in alkali/semiconductor interfaces. Electron-energy loss measurements[21] of the plasmon dispersion suggested[22,23] that the K chains on a Si(100)2×1 surface may form a one-dimensional metal, which is in contrast to the substrate metallization of the ionic picture.[4] A surface extended x-ray absorption fine-structure study[24] of this same surface produced a 20% larger value for the K-Si bond length than calculated. Angle-resolved photoemission measurements[25,26] of Si(100)2×1-K and -Cs showed these surfaces to be semiconducting at full coverages. From core-level photoemission Soukiassian et al.[27] derived at a covalent type of bonding with a metallization occurring within the K or Cs overlayers.

A common feature of the one-dimensional metallic K chains suggested in refs. 22,23 and the adsorption geometry of the ionic picture[4] is the absolute K coverage of 0.5, i.e. one K atom per Si dimer. In the meantime Abukawa and Kono[28] introduced the so-called double layer model with an absolute K coverage of 1 per Si surface atom, based on their x-ray photoelectron diffraction results and the angle-resolved photoemission work of Enta et al.[25,26] Ishida and Terakura[13] calculated the corresponding electronic structure and the dispersion of the surface-state bands, favoring a covalent bonding. This new surface geometry for the K atoms lacks the one-dimensionality along the dimer chains of the Si(100)2×1 surface, rather it is more isotropic with a buckling perpendicular to the dimer chains. A recent ion-scattering study[29] supports the double-layer model.

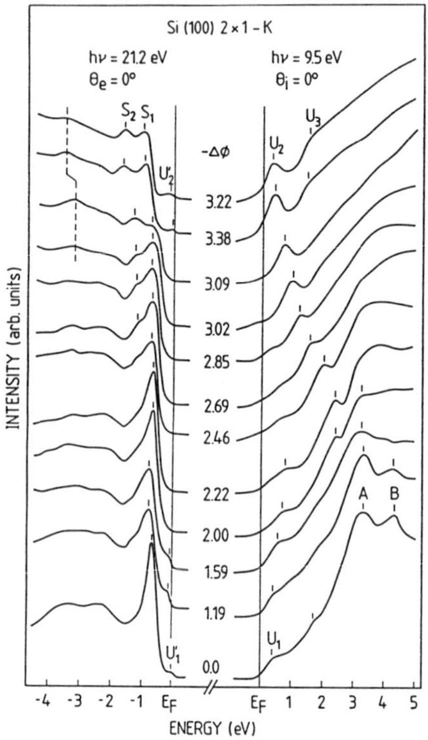

Fig. 2. Normal-emission UPS and normal-incidence IPES curves recorded on the Si(100)2×1-K surface for increasing K coverage. The coverage was indirectly controlled by the measured workfunction shift ($\Delta\phi$).

We have now studied[30] this surface and taken special effort to prepare single-domain Si(100)2×1-K surfaces by using vicinal surfaces. The evolution of the surface states at $\overline{\Gamma}$ in the surface Brillouin zone (SBZ) with increasing K coverage is shown in fig. 2. In the spectra of the clean surface ($\Delta\phi = 0.0$) contributions can be seen from the filled and empty dangling-bond surface states denoted S_1 and U_1, respectively. A small contribution from the U_1 state can also be seen in the photoemission spectrum owing to the high n-doping of the crystal. Prominent bulk structures, denoted A and B, exist above 3 eV in the IPES spectrum.[19] At a coverage corresponding to $\Delta\phi$ = -2.0 eV an empty surface state appears at 2.4 eV, denoted U_2. With increasing coverage it shifts downwards in energy until it reaches the Fermi level at the same coverage where the workfunction reaches its minimum at $\Delta\phi$ = -3.38 eV, rendering the surface metallic. Then the U_2 state becomes also visible in the UPS curves (denoted U_2').

In UPS a second surface state, denoted S_2, appears to split off from the S_1 peak for $\Delta\phi$ = -2.85 eV and move to lower energies with increasing coverage. Between $\Delta\phi$ = -3.09 eV and -3.38 eV, all UPS features shift downwards by ~ 0.3 eV, as indicated in fig. 2. This energy shift coincides with the reaching of the minimum in the $\Delta\phi$ curve and occurs quite abruptly with respect to the evaporation time (within the last 10 % of the time needed to reach the minimum). The observation of this energy shift is in good agreement with the earlier UPS valence band and K 3p core-level results of Enta et al.[25] on a two-domain Si(100)2×1-K surface. In accordance with their conclusions, we interpret it to be a band-bending shift caused by the onset of the filling of the previously empty surface state U_2.

From IPES curves (not shown) recorded for different incidence angles along the two main azimuthal directions of the rectangular SBZ we could plot the peak positions as a function of wavevector in fig. 3. The spectra[30] had been measured at a coverage corresponding to the minimum of the $\Delta\phi$ curve ($\Delta\phi_{min}$ = -3.38 eV). The U_2 state shows a strong upward dispersion away from $\bar{\Gamma}$ (cf. fig. 3). Its bandwidth is ~ 2.5 eV. A similar dispersion was also recorded in the perpendicular $\bar{\Gamma}-\bar{J}'$ direction. These dispersions were fully developed already at the coverage corresponding to $\Delta\phi$ = -3.0 eV. A surprising result is that the U_2 dispersion does not follow the periodicity of the substrate reconstruction in the $\bar{\Gamma}-\bar{J}$ direction (perpendicular to the dimer rows), as clearly seen in fig. 3. Only a faint umklapp-scattered U_2-peak is observed at $\bar{\Gamma}$ in the second SBZ.

The photon emission intensity from transitions to the U_2 state was strongly polarization dependent.[30] For s-polarized photons a strong U_2 peak was obtained, while for p-polarized photons the same peak was faint. The parabolic-like dispersion of U_2 is consistent with a K 4s and/or K $4p_z$ character, whereas the strong s/p-polarization dependence of the photon emission implies a dominant p_z character. A minor K 4s contribution can not be ruled out, but a major 4s character is very unlikely. Furthermore, meta-stable de-excitation spectroscopy by Nishigaki et al.[31] on the Si(100)2×1-K surface locates the K 4s-derived peak ~ 1 eV below the Fermi level, suggesting that the 4s level is occupied and mainly hybridized with the dangling-bond states of the Si substrate. The latter arguments are corroborated by the similar electronic structure of alkali metals on metal substrates. The measured band structure[32] of Cu(111)2×2-Na, for instance, is strikinly similar to the one of fig. 3.

A metallization of the K overlayer had been reported employing indirect experimental techniques like energy loss[21] and core-level[33] spectroscopy. The data presented above, however, is direct and conclusive evidence for overlayer metallization, since in the IPES and UPS spectra we directly observe the crossing of the Fermi level of the overlayer-derived U_2 band. This metallization occurs at the coverage corresponding to $\Delta\phi_{min}$, whereas at a slightly lower K coverage, at $\Delta\phi$ = -3.09 eV, the surface is still semiconducting (cf. fig. 2). Further evidence for a metallic overlayer is given by the parabolic-like dispersion of the U_2 band, which indicates a metallic bonding within the overlayer with strong K-K interaction both parallel and perpendicular to the dimer rows. The partial independence of the substrate periodicity for the U_2 band can be understood in the context of a weak substrate-overlayer interaction, which is consistent with a metallic overlayer. We simply observe the continuation of a mainly parabolic free-electron-like dispersion in the second SBZ, whereas the umklapp-scattered branch of the parabola is observed only as a faint structure around $\bar{\Gamma}$ in the second SBZ. This suggests that the atomic positions in the K overlayer may be close to a 1×1 periodicity.

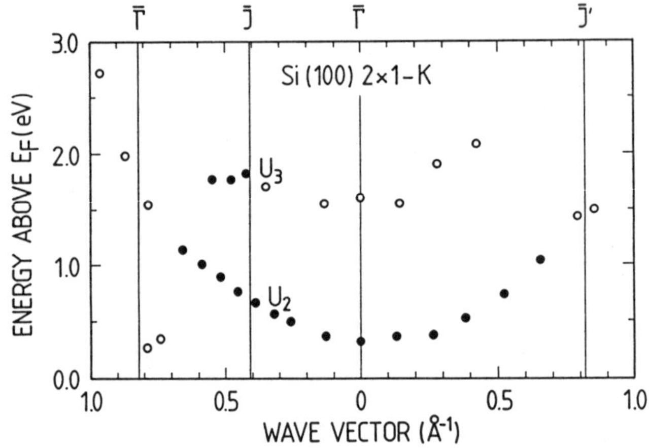

Fig. 3. The measured dispersions of the surface-state peaks in the IPES spectra of the Si(100)2×1-K surface. The K coverage corresponds to the minimum of the $\Delta\phi$ curve. While the peak positions stay above the Fermi level, the total spectral intensity of the K-induced peak crosses E_F around the $\bar{\Gamma}$ point rendering the surface metallic (see fig. 2 and ref. 30).

By comparing to calculated band structures[4,13,14] for Si(100)2×1-K, it is clear that our results are only consistent with atomic models based on 1-ML coverage, e.g. the double-layer model,[28] for which Ishida and Terakura[13] and Kobayashi et al.[14] found a semiconducting surface band structure with two filled bands derived from the Si dangling bonds and K 4s and $4p_{x,y}$ orbitals, and one empty band with mixed K $4p_z$ and 4s orbital content. The character of the empty band agrees well with the U_2 band in the present work. Our results thus give strong support to the polarized covalent Si-K bonding picture of refs. 13,14. There is perfect agreement with the calculated[14] dispersions of the empty surface states along the $\bar{\Gamma}-\bar{J}$ and $\bar{\Gamma}-\bar{J}'$ directions.

In the meantime we have also investigated sodium on single-domain Si(100)2×1 surfaces.[34] As with potassium a Na-induced empty surface-state band moves down with increasing coverage, but in contrast to K never reaches E_F. Its dispersion is weaker than the p_z band of fig. 3 for K. As a result the Si(100)2×1-Na surface stays semiconducting at saturation in agreement with Tikhov et al.,[35] but in disagreement with Soukiassian et al.[36] who observed plasmon loss peaks in photoemission. The less pronounced dispersion indicates a weaker interaction within the Na overlayer as expected from the smaller Na size as compared to K. As the dispersion is very similar for the two main symmetry directions we again favor the double-layer model for the Na adsorption as in case of K. All this is consistent with the calculations of Kobayashi et al.[14] for this overlayer system, too.

4. Li, Na, K and Cs on cleaved Si(111)

It was found from total-energy calculations[7] that K deposition on the cleaved Si(111) surface destroys the 2×1 reconstruction and forms a 1×1 surface. In contrast, we find that K deposition leaves the π-bonded chains unchanged and observe a 2×1 LEED pattern throughout the coverage regime.[37] However, cesium does destroy the 2×1 registry and forms a new $\sqrt{3} \times \sqrt{3}$ R30° surface atom arrangement at exactly the 1-ML coverage.[38] Sodium[39] and lithium[40] also destroy the 2×1 pattern and form 1×1 phases as it had been predicted[7] for K. At 1-ML coverage the Si(111)$\sqrt{3} \times \sqrt{3}$ R30° –Cs and the Si(111)1 × 1 –Na and – Li surfaces are semiconducting with a well-defined surface bandgap, whereas the Si(111)2 × 1 –K surface is metallic over a broad coverage range with a filling of the semiconductor surface bands as suggested by Ciraci and Batra.[6] In fig. 4 we show the relevant angle-resolved inverse-photoemission spectra of the latter case, i.e. for 1 ML of potassium on Si(111)2×1. As discussed in more detail in ref. 37 the K-induced surface state U'_1 in fig. 4 shifts towards the Fermi level and loses spectral intensity. Such behavior is proof for an IPES peak moving through the Fermi level as measured with a finite energy resolution,[41] hence the surface-state band crosses the Fermi level about midway along the $\bar{\Gamma} - \bar{J}$ axis, proving a metallization. Plotting the peak positions of fig. 4 (and other spectra not shown), we find that the overall shape and band width of the surface-state band is consistent with Ciraci and Batra's calculations[6] for one K atom per 2×1 unit cell, which is essentially the clean-surface band structure with the Fermi level shifted so that it cuts through the unoccupied part rendering the substrate surface metallic. Finally, we note that we also observe[37] this surface-state band to cut E_F at smaller coverages, i.e. less workfunction reductions, indicating metallicity in a broad coverage range which makes the K/Si(111) interface consistent with the ionic picture. Its ingredients were substrate metallization through filling of the surface-state bands,[6] and no critical coverage required for metallization.

The monolayer-covered Cs/Si(111) surface is quite different from the K/Si(111) case, as cesium destabilizes the π-bonded chain structure and induces a $\sqrt{3} \times \sqrt{3}$ R30° reconstruction.[38] The Cs-induced surface-state features exhibit only little dispersion as function of θ and never approach the Fermi level for all SBZ symmetry directions measured (data not shown). This holds true for the unoccupied surface band dispersion as obtained by IPES as well as the occupied surface band as derived from UPS. Their combination yields an indirect surface bandgap of 1.5 eV at the $\bar{M}'$ point. All these observations are at variance with previous predictions and experiments of the Cs/Si(111)2×1 interface system.

Recently we have also investigated lithium and sodium on the cleaved Si(111) surface.[39,40] The workfunction minima are less pronounced (-2.40 eV for Li and -2.7 eV for Na) as compared to K and Cs on the same surface. At about half of the saturated-monolayer coverage the LEED pattern change from 2×1 to 1×1 which implies the existence of yet another alkali phase on the same silicon surface. Angle-resolved UPS and IPES measurements (not shown) reveal that at 1-ML coverage both interfaces are semiconducting. In fig. 5 we compare our measured

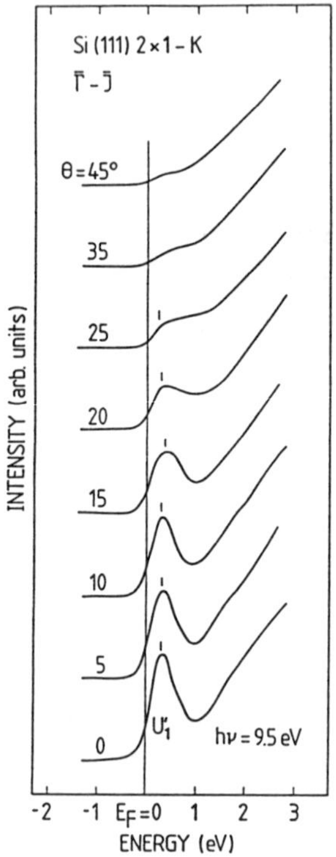

Fig. 4. Angle-resolved inverse photoemission spectra at $h\nu = 9.5$ eV for different incidence angles θ, probing states along the $\overline{\Gamma} - \overline{J}$ line of the Si(111)2×1-K surface Brillouin zone (from ref. 37).

energy dispersion of the Na-induced unoccupied (U'_1) and occupied (S) surface state features to existing band-structure calculations which we have split for the two different azimuths. Along $\overline{\Gamma} - \overline{K}$ we have reproduced the bands from a first-principles pseudopotential calculation[5] which finds $\Delta\phi = -2.7$ eV in perfect agreement with our value and an ionic bond of the Na 3s with the 3p dangling bonds of the Si surface. Amongst the various atom configurations this calculation favors the threefold hollow site. The latter is in contrast to the calculations by Ossicini et al.,[10] which favor the threefold filled site with a charge transfer of 0.14 electrons per atom indicating a covalent bonding. Their electronic structure results are based on the linear muffin-tin orbital method in the atomic-sphere approximation and are shown in fig. 5 along the $\overline{\Gamma} - \overline{M}$ azimuth. Clearly their empty surface state band does not follow our measured dispersion. Recently, ab-initio molecular dynamics simulations using a plane-wave expansion have been performed for both Na[15] and Li[40] covered Si(111). These calculations simulate temperature annealing with a random displacement of atoms in order to find the equilibrium structure. The authors also find the 1×1 surface structure with the threefold filled site for Li and the threefold hollow site for Na as the most stable configurations with the lowest total energy. Their surface state dispersions calculated for the two different adsorption sites agree almost perfectly with our measured dispersions for Li and Na, respectively.

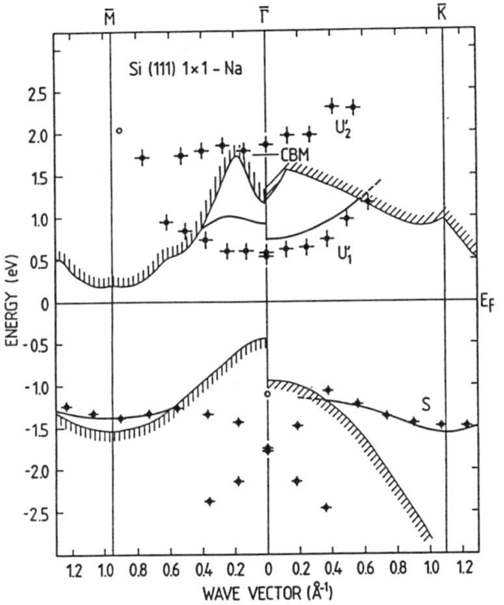

Fig. 5. The energy dispersion $E(k_{\parallel})$ of the occupied and empty surface-state features of 1 ML Na on Si(111) as measured by angle-resolved UPS and IPES, respectively. The crosses indicate the estimated uncertainty. Along $\overline{\Gamma}-\overline{K}$ (right side) we compare to the calculated bands of Northrup (ref. 5), along $\overline{\Gamma}-\overline{M}$ (left side) to those of Ossicini et al. (ref. 10).

5. K and Cs on annealed Si(111)

The alkali/semiconductor systems presented so far could inherently undergo only a semiconductor/metal transition as a function of alkali coverage. We recall that Cs/GaAs(110) (ref. 42) and K/Si(111) (chap. 4) indeed showed such transitions, while all other systems stayed semiconducting partly accompanied by strong surface structural changes. Now the annealed Si(111)7×7 surface offers the possibility to study the reverse transition from a metallic to a semiconducting surface, as it exhibits a surface state which renders this surface metallic.[43] Using STM Hamers et al.[44] identified the surface state S_1 at the Fermi level and the lowest empty surface state U_1 as originating from the adatom dangling bonds in the dimer-adatom-stacking fault model,[45] while the occupied S_2 state belongs to the restatom dangling bonds.

Changes observed[46,47] by UPS and IPES after successive evaporations of K or Cs are initially very similar. The observed downward shift of S_1 away from E_F occurs already after small amounts of K or Cs deposition, and after 1 min of evaporation the surfaces appear to be semiconducting. Simultaneously we observe the reduction in intensity of the unoccupied U_1 state, indicating in both cases a filling of the empty adatom dangling-bond states. At the minima of the workfunction (1 ML) the surfaces are clearly non-metallic. Upon further evaporation of Cs beyond the minimum of the workfunction the electronic structure changes. The Cs-induced empty surface states U_1'' and U_2'' shift down in energy, U_1'' even enough to yield considerable intensity at the Fermi level. At the same time UPS also shows enhanced intensity at E_F. We conclude that over-cesiation, in contrast to over-exposure of potassium, drives the semiconducting surface towards metallicity again. We suggest the driving force to be the Cs-Cs interaction which at a continued flow of Cs atoms onto the already saturated surface can change the surface ordering and thereby the direction of charge transfer between Cs and Si. An important parameter is the size of the Cs atom in relation to the distance between Cs adsorption sites.

All in all, Cs on Si(111)7×7 undergoes a metal-to-semiconductor and then a semiconductor-to-metal transition with increasing coverage. From the electronic-structure changes the latter seems to be similar to Cs on GaAs(110) where the metallization occurs within the Cs overlayer and not in the semiconductor surface.[42] Photoemission of the Cs 4d core levels employing synchrotron radiation at MAX-Lab, Lund support this conclusion, as the lineshape of the 4d emission becomes significantly asymmetric above 1-ML coverage. This is well known[48] for photoemission from metals and reflects final-state effects with multielectron excitations involving band states above and below E_F. Hence the Cs atoms start to feel a metallic local environment when the coverage is going beyond 1 ML. All this points to a metallization of the Cs overlayer and not of the silicon substrate.

Acknowledgment: We thank M. Tschudy for his competent assistance.

References

1. *Formation of Semiconductor Interfaces*, ed. by P. Perfetti S.U. Campisano, G. Margaritondo, and O. Bisi, *Appl. Surf. Sci.* **56-58** (1992).
2. W.E. Spicer, *Appl. Phys.* **12** (1977) 115.
3. B. Reihl, M. Erbudak, and D.M. Campbell, *Phys. Rev.* **B19** (1979) 6358; D.T. Pierce, R.J. Celotta, G.C. Wang, W.N. Unertl, A. Galeijs, C.E. Kuyatt, and S.R. Mielczarek, *Rev. Sci. Instrum.* **51** (1980) 478.
4. S. Ciraci and I.P. Batra, *Phys. Rev. Lett.* **56** (1986) 877.
5. J.E. Northrup, *J. Vac. Sci. Technol.* **A4** (1986) 1404.
6. S. Ciraci and I.P. Batra, *Phys. Rev. Lett.* **58** (1987) 1982.
7. I.P. Batra and S. Ciraci, *Phys. Rev.* **B37** (1988) 8432.
8. P.S. Bagus and I.P. Batra, *Surf. Sci.* **206** (1988) L895.
9. R. Ramirez, *Phys. Rev.* **B40** (1989) 3962.
10. S. Ossicini, C. Arcangelia, and O. Bisi, *Phys. Rev.* **B42** (1990) 7671.
11. J. Hebenstreit, M. Heinemann, and M. Scheffler, *Phys. Rev. Lett.* **67** (1991) 1031.
12. H. Ishida, K. Terakura, and M. Tsukada, *Solid State Commun.* **59** (1986) 356.
13. H. Ishida and K. Terakura, *Phys. Rev.* **B40** (1989) 11519.

14. K. Kobayashi, Y. Morikawa, K. Terakura, and S. Bügel, *Phys. Rev.* **B45** (1992) 3469.
15. I. Moullet, W. Andreoni, and M. Parrinello, *Phys. Rev.* **B46** (1992) 1842.
16. Y. Ling, A.J. Freeman, and B. Delley, *Phys. Rev.* **B39** (1989) 10144.
17. T. Kato, K. Ohtomi, and M. Nakayama, *Surf. Sci.* **209** (1989) 131.
18. B. Reihl et al., *Appl. Surf. Sci.* **56-58** (1992) 123.
19. B. Reihl, K. O. Magnusson, J. M. Nicholls, P. Perfetti, and F. Salvan, *NATO Advanced Studies Institute Series* **B195** (1989) 397.
20. F.J. Himpsel, *Surf. Sci. Rep.* **12** (1990) 1.
21. T. Aruga, H. Tochihara, and Y. Murata, *Phys. Rev. Lett.* **53** (1984) 372.
22. M. Tsukada, H. Ishida, and N. Shima, *Phys. Rev. Lett.* **53** (1984) 376.
23. H. Ishida, N. Shima, and M. Tsukada, *Phys. Rev.* **B32** (1985) 6246.
24. T. Kendelewicz, P. Soukiassian, R.S. List, J.C. Woicik, P. Pianetta, I. Lindau, and W.E. Spicer, *Phys. Rev.* **B37** (1988) 7115.
25. Y. Enta, T. Kinoshita, S. Suzuki, and S. Kono, *Phys. Rev.* **B39** (1989) 1125.
26. Y. Enta, S. Suzuki, S. Kono, and T. Sakamoto, *Phys. Rev.* **B39** (1989) 5524.
27. P. Soukiassian, M.H. Bakshi, Z. Hurych, and T.M. Gentle, *Surf. Sci.* **221** (1989) L759.
28. T. Abukawa and S. Kono, *Phys. Rev.* **B37** (1988) 9097.
29. A.J. Smith, W.R. Graham and E.W. Plummer, *Surf. Sci.* **243** (1991) L37.
30. L.S.O. Johansson and B. Reihl, *Phys. Rev. Lett.* **67** (1991) 2191.
31. S. Nishigaki et al., *Surf. Sci.* **231** (1990) 271.
32. R. Dudde, L.S.O. Johansson, and B. Reihl, *Phys. Rev.* **B44** (1991) 1198.
33. E.M. Oellig and R. Miranda, *Surf. Sci.* **177** (1986) L947; E.M. Oellig, E.G. Michel, M.C. Asensio, R. Miranda, J.C. Duran, A. Munoz, and F. Flores, *Europhys. Lett.* **5** (1988) 727.
34. L.S.O. Johansson and B. Reihl, to be published.
35. M. Tikhov, G. Boishin, and L. Surnev, *Surf. Sci.* **241** (1991) 103.
36. P. Soukiassian, M.H. Bakshi, Z. Hurych, and T.M. Gentle, *Surf. Sci.* **221** (1989) L759.
37. B. Reihl and K.O. Magnusson, *Phys. Rev.* **B42** (1990) 11839.
38. K.O. Magnusson and B. Reihl, *Phys. Rev.* **B39** (1989) 10456.
39. B. Reihl, S.L. Sorensen, R. Dudde, and K.O. Magnusson, *Phys. Rev.* **B46** (1992) 1838.
40. I. Moullet, W. Andreoni, M. Parrinello, H. Bernhoff, and B. Reihl, to be published.
41. R. Clauberg, K.H. Frank, J.M. Nicholls, and B. Reihl, *Surf. Sci.* **189** (1987) 44.
42. K.O. Magnusson and B. Reihl, *Phys. Rev.* **B40** (1989) 7814.
43. F.J. Himpsel, D.E. Eastman, P. Heimann, B. Reihl, C.W. White, and D.M. Zehner, *Phys. Rev.* **B24** (1981) 1120.
44. R.J. Hamers, R.M. Tromp, and J.E. Demuth, *Phys. Rev. Lett.* **56** (1986) 1972.
45. K. Takayanagi et al., *J. Vac. Sci. Technol.* **A3** (1985) 1502.
46. K.O. Magnusson and B. Reihl, *Phys. Rev.* **B41** (1990) 12071.
47. K.O. Magnusson, S. Wiklund, R. Dudde, and B. Reihl, *Phys. Rev.* **B44** (1991) 5657.
48. S. Hüfner and G.K. Wertheim, *Phys. Rev. Lett.* **35** (1975) 53.

PHOTON STIMULATED REACTIONS IN ADSORBED AND CONDENSED ATOMS AND MOLECULES STUDIED BY ION DESORPTION

G. DUJARDIN, L. HELLNER, L. PHILIPPE AND M.J. BESNARD-RAMAGE.
Laboratoire pour l'Utilisation du Rayonnement Eletromagnétique (LURE)
Bâtiment 209d, Université Paris-Sud, 91405 Orsay, France

and

Laboratoire de Photophysique Moléculaire
Bâtiment 213, Université Paris-Sud, 91405 Orsay, France

ABSTRACT

Ion desorption from adsorbed and condensed atoms and molecules is shown to be a powerful probe of new ionization processes involving the simultaneous excitation of several electrons. The basic issues of ion desorption are discussed in hydrogenated Si (111) surfaces and multilayers of condensed argon on platinum.

1. INTRODUCTION :

For photon energies higher than 10-15 eV, the interaction of photons with atoms and molecules on surfaces is dominated by ionization processes, i.e. the formation of singly or multiply charged ions. Photoionization processes are usually studied by photoemission techniques based on the detection and the kinetic, angle and spin analysis of ejected electrons. We will show in this paper that the study of ion desorption may lead to the discovery of new ionization processes which cannot or could hardly be studied by photoemission of electrons. Furthermore, the detailed analysis of desorbed ions may also provide informations on the dynamics of produced ions which cannot be derived from photoemission studies.

Examples from two particular systems will be considered ; adsorbed hydrogen on silicon and multilayers of condensed argon on platinum. These systems illustrate two extreme situations. On the hydrogenated Si (111) surface, an hydrogen atom is strongly chemisorbed (dissociation energy ≃ 3.5 eV) to each surface silicon atom whereas for multilayers of condensed argon, surface argon atoms are only weakly bound (physisorbed ; dissociation energy ≃ 0.012 eV) to underlying argon atoms.

According to the work by Jennison and Emin [1], it is generally believed that single ionization in a crystal or at the crystal surface cannot produce any ion desorption because of a rapid delocalization of the excitation. For example, the ejection of an electron from an electronic band having a width of 0.5 eV would give rise to an ionization with a localization time of the order of 10^{-15} s which is much too short to allow any desorption or self trapping of the excitation [1] to occur. Single ionization will be shown indeed to induce no ion desorption from hydrogenated silicon and from condensed argon whereas multiple excitation processes do produce ion desorption. However we will argue that there is no need to consider any delocalization and that energetic reasons are sufficient to explain this lack of ion desorption by single ionization.

2. ION DESORPTION FROM HYDROGENATED Si (111) SURFACES :

The hydrogenated Si (111) surface is prepared by wet chemical etching of a clean oxidized silicon surface by a dilute HF solution [2]. This provides a non reconstructed surface, well characterized by infrared spectroscopy [2], where an hydrogen atom is bound to each surface silicon atom.

The experimental set-up used for recording the intensity of mass selected ions as a function of the excitation photon energy has been previously described [3]. We show in figure 1 the ion yield for H^+ desorption from the Si (111)-H surface by photon excitation in the 15-45 eV valence energy range. The threshold energy for formation of H^+ is found at ≃ 20 eV. The low intensity signal below 20 eV is considered to be due to higher orders of the light transmitted by the monochromator. The results in figure 1 are in very good

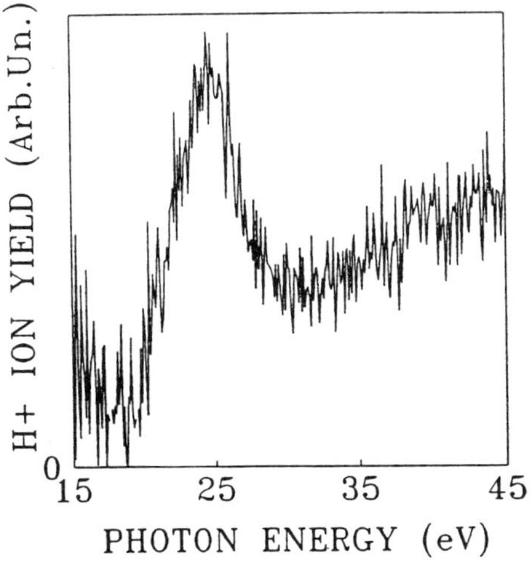

Figure 1 :
H^+ ion Yield from Si (111)-H

agreement with those obtained with electron impact by Madden et al. [4] and with photon impact by Takakuwa et al [5]. The assignment of the electronic processes giving rise to the observed feature in figure 1 is difficult since no electronic process is known to occur in this 15-45 eV energy range. Previous photoemission experiments on the Si (111)-H surface [6, 7] have shown that all the observed valence bands including the two surface states have binding energies relative to the botton of the conduction band smaller than 13 eV. It was suggested by Madden et al. [4] that the threshold energy at 20 eV would be related to a two valence electron ejection. In such a case, one would expect the ion desorption signal to behave, as a function of the photon energy, like the double photoionization cross section [3, 8], i.e. to increase over more then 20 eV above the threshold energy. The resonant behavior of the H^+ ion yield curve in figure 1, which could not be seen by electron impact [4], is then very unlikely due to double ionization. Takakuwa et al. [5] argued that the 3s ionization threshold energy of gas phase SiH_4 [9] is very close to the observed threshold energy of H^+ desorbed from Si (111)-H. We emphasize however that such a 3s ionization does not exist in Si (111)-H and in clean Si (111) because of the hybridization of the 3s and 3p valence orbitals. These considerations lead us to conclude that the electronic process giving rise to the H^+ ion desorption cannot be ascribed to a single valence excitation and most probably corresponds to an ionic satellite state excitation in which two valence holes are created, one electron being ejected from the surface and the second one being excited to an unoccupied valence band. This is very similar to what is observed in argon layers (see section III) in which the broad resonant shape of the ion yield is believed to be related to a Cooper minimum in the cross section for formation of ionic satellite states. Wether the same explanation is valid for Si (111)-H has to be confirmed by further calculations.

Jennison and Emin [1] discussed the conditions for observing H^+ ion desorption from Si (111)-H. The lack of ion desorption by single valence hole ionization was explained by a localization time of the excitation much too short to allow any desorption or self trapping of the excitation. By contrast, the two valence hole excitation having a longer localization time was expected to produce ion desorption. We want to emphasize however that alternative reasons may also explain the H^+ ion desorption results. The energy needed to desorb an H^+ ion is the sum of the Si-H binding energy at $\simeq$ 3.5 eV [10], the ionization energy of the hydrogen atom at 13.6 eV plus the electron affinity of the surface. For an electron being captured in the conduction band of the surface, the electron affinity of the surface is expected to lie between 0 and $\simeq$ -3.5 eV [11]. It follows that the minimum energy to desorb an H^+ ion is at 13.6 eV. This explains that single valence ionization which is known to occur at much lower excitation energies cannot produce ion desorption.

3. ION DESORPTION FROM ARGON LAYERS ON PLATINUM :

Rare gas layers on a metal are especially interesting systems for studying ion desorption mechanisms. They have indeed a reduced number of well known electronic excited and ionized states [12]. It was shown only very recently that various electronic excitations, including the formation of ionic satellite states [3], doubly ionized states [3], core excited states [13, 14] and exciton pair states [15, 16], may induce ion desorption from multilayers of condensed argon. We note that rare gas (sub) monolayers on a metal show no ion desorption [14] because of a rapid neutralization by the metal, but rare gas multilayers, in which the top monolayer has a much reduced interaction with the metal, do show ion desorption.

Figure 2 shows the ion yield for desorption of Ar^+, Ar_2^+ and Ar_3^+ from 4 monolayers of condensed argon on polycrystalline platinum over the 20-100 eV photon energy range. These curves are recorded using the experimental set-up described in ref. 3. The non zero background of Ar^+ signal around 20 eV is due to the 4th order of the light transmitted by the monochromator.

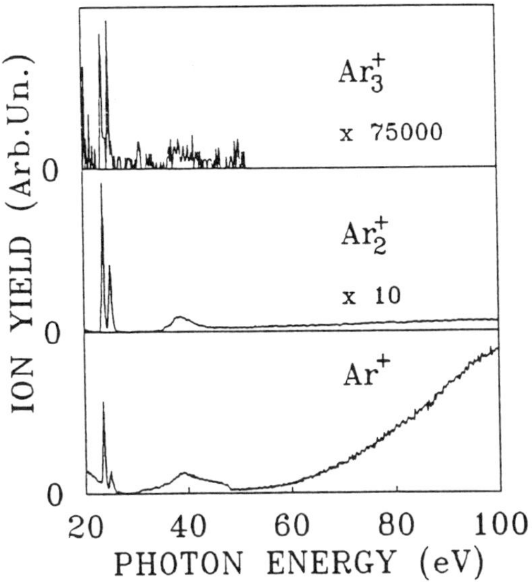

Figure 2 :
Ar^+, Ar_2^+ and Ar_3^+ ion yields from 4 monolayers of condensed argon on platinum.

Three markedly different features are seen in ion yield curves of figure 2. The two resonances at about 23.5 eV and 25.5 eV have been previously ascribed to surface and bulk exciton pairs respectively [15, 16]. We note that the desorption of Ar_3^+ trimer ions, which has never been reported before, is observed here with a very weak intensity relative to Ar^+ ions. The second feature observed for photon energies between 35 eV and 50 eV has been assigned to the formation of ionic satellite states [3]. This will be discussed into more details in the following. The last feature which gives rise to the large increasing of the Ar^+ ion yield beyond 50 eV is due to the primary formation of doubly charged Ar^{++} ions [3]. We emphasize that the observed ion desorption occurs in an energy range 20-100 eV where no electronic process have been detected by photoabsorption [17] or by photoemission [12], except the 3s ionization of condensed argon which is known to have a threshold at 28.4 eV [18]. However, as seen in figure 2, this latter process seems to produce no ion desorption. We note in particular that single 3p ionization occuring around 13.8 eV [12] produces no detectable ion desorption [3, 15].

Let us now examine into more details the ion desorption following the formation of satellite states of Ar^+ ions. We see in fig. 3a and 3b the ion yield curves for desorption of Ar^+ and Ar_2^+ ions from a 50 monolayer sample. In order to identify the processes giving rise to the Ar^+ and Ar_2^+ ion yield curves in figure 3, we show, on the same scale, the photoemission spectrum recorded for the same sample (figure 3c) as well as low resolution (figure 3d) and high resolution (figure 3e) photoelectron spectra recorded in gas phase argon [19]. The gas phase photoelectron spectra have been shifted to lower binding energies in order to get the same binding energy for the $3s^{-1}$ band in gas phase and condensed argon. We note that spectra 3c and 3d are quite similar although the binding energies of bands 1,3,4, and 5 relative to the binding energy of the $3s^{-1}$ band are slightly higher in condensed argon than in gas phase argon.

Looking carefully at the ion yield curves of figure 3, one can distinguish in the Ar^+ curve three features having threshold energies associated with the formation of bands 1,2 and 3 in the photoelectron spectra whereas the Ar_2^+ curve exhibits only two features correlated to bands 2 and 3. According to the detailed assignment of the gas phase satellite states given by Kossman et al. [19], one can then also assign the sattelite states which produce the desorption of Ar^+ and Ar_2^+ ions from condensed argon.

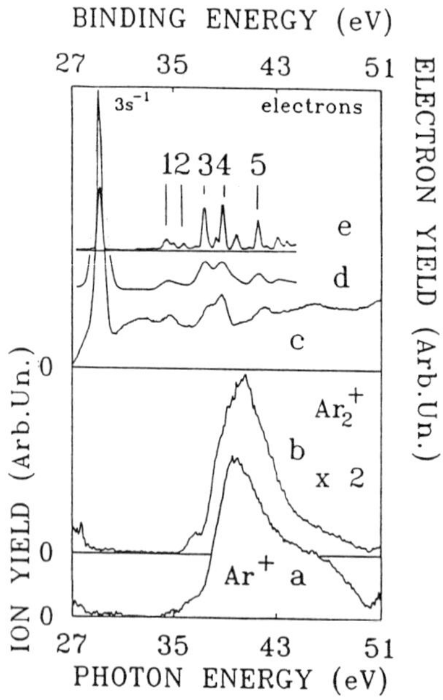

Figure 3 :
Ar^+ (curve a) and Ar_2^+ (curve b) ion yields from 50 monolayers of condensed argon on platinum. The photoemission spectrum (curve c) from the same sample is recorded at a photon energy of 80 eV. The gas phase photoelectron spectra (curves d and e) are taken from ref. 19.

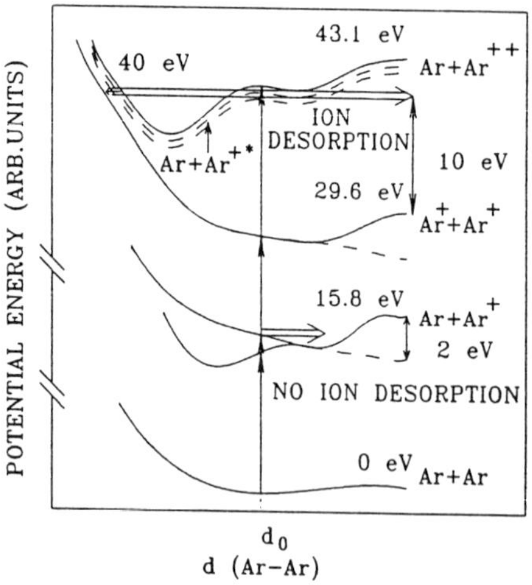

Figure 4 :
Schematic potential energy curves accounting for the dynamics of doubly charged Ar^{++}, satellite state $(Ar^+)^*$ and singly charged Ar^+ ions interacting with a neighbour Ar argon atom.

The mechanism for ion desorption from multilayers of argon produced by double photoionization and by ionic satellite state formation is schematically shown in figure 4. The basic idea is that a doubly charged Ar^{++} ion or a singly charged $(Ar^+)^*$ ion in a satellite state have enough internal energy to ionize a neighbour argon atom leading to the formation of a pair of ions. Due to the strong Coulomb repulsion between the two positive Ar^+ charges, the ejected Ar^+ ions can get enough kinetic energies to overcome the energy barrier which prevent the positive ions to be desorbed. This energy barrier of about 2 eV originates from the lower ionization energy of condensed argon (≃ 13.8 eV) as compared to that of argon in the gas phase (≃ 15.8 eV). By contrast, the interaction of a singly charged Ar^+ ion with a neutral neighbour Ar atom is either attractive or weakly repulsive [20] so that is this case the Ar^+ ions would not have enough kinetic enery to overcome the 2 eV energy barrier. This simple desorption model explains why double photoionization and ionic satellite state formation produce ion desorption whereas single ionization do not produce any ion desorption. A more quantitative test of this model, based on the measurement of kinetic energies of desorbed ions, is in progress in Orsay.

4. CONCLUSION

We have seen that single ionization do not produce any ion desorption from Si (111)-H or from argon layers on platinum. It was previously believed that a rapid delocalization of the electron hole was necessary to explain this result. In fact, the electron hole may be localized at various defect sites on the surface so that the delocalization of the excitation may be not the correct explanation. Moreover it turns out that simple energetic considerations clearly prevent single ionization to produce ion desorption from Si (111)-H and argon layers on platinum.

This lack of ion desorption by single ionization has important consequences. Other minority ionization processes, involving the simultaneous excitation of several electrons, are not seen in usual experiments such as photoabsorption or photoemission experiments where signals originating from single ionization are by for the dominant one. On the contrary, in ion desorption studies, such minority ionization processes can be clearly observed since they are the only processes giving rise to ion desorption. For this reason, new ionization processes such as formation of ionic satellite states, double photoionization, formation of exciton pairs could be observed in Si (111)-H and argon layers on platinum. Much efforts are now necessary in oder to study in details the dynamics of the ion desorption following these excitation processes.

5. ACKNOWLEDGEMENTS

We thank P. Dumas for his participation to the work on Si (111)-H and for fruitfull discussions. We also acknowledge support from the Science program of the E.E.C.

6. REFERENCES :

1. D.R. Jennison and D. Emin, Phys. Rev. Lett. **51**, (1983) 1390.
2. P. Dumas, Y. Chabal and G.S. Higashi, Phys. Rev. Lett. **65**, (1990) 1124.
3. G. Dujardin, L. Hellner, M.J. Besnard-Ramage and R. Azria, Phys. Rev. Lett. **64**, (1990) 1289.
4. H.H. Madden, D.R. Jennison, M.M. Traum, G. Margaritondo and N.G. Stoffel, Phys. Rev. B **26**, (1982) 896.
5. Y. Takakuwa, M. Nogawa, M. Niwano, H. Katakura, S. Matsuyoshi, H. Ishida, H. Kato and N. Miyamoto, Jap. J. Appl. Phys. **28**, (1989) L.1274.
6. T. Sakurai and H.D. Hagstrum, Phys. Rev. B **12**, (1975) 5349.
7. P. Dumas, presented at VUV 10, Paris (1992).

8. G. Dujardin, M.J. Besnard, L. Hellner and Y. Malinovitch, Phys. Rev. A **35**, (1987) 5012.
9. R.A. Rosenberg, J. Vac. Sci. Technol. A3, (1986) 1463.
10. G.W. Trucks, K. Raghavachari, G.S. Higashi and Y.J. Chabal, Phys. Rev. Lett. **65**, (1990) 504.
11. R.S. Becker, G.S. Higashi, Y.J. Chabal and A.J. Becker, Phys. Rev. Lett. **65**, (1990) 1917.
12. N. Schwentner, E.E. Koch and J. Jortner, in Electronic Excitation in Condensed rare gases (Springer, Berlin, Heidelberg, 1985).
13. G. Rocker, P. Feulner, R. Scheuerer, L. Zhu and D. Menzel, Phys. Sci. **41**, (1990) 1014.
14. D. Menzel, Appl. Phys. **A51**, (1990) 163.
15. Y. Baba, G. Dujardin, P. Feulner and D. Menzel, Phys. Rev. Lett. **66**, (1991) 3269.
16. T. Schwabenthan, R. Scheuerer, E. Hudel and P. Feulner, Solid State Commun. **80**, (1991) 773.
17. B. Sonntag, in Rare Gas Solids, Vol. II, edited by M.L. Klein and J.A.Venables (Academic, New York, 1977) p. 1083.
18. H.N. Biester, M.J. Besnard, G. Dujardin, L. Hellner and E.E. Koch, Phys. Rev. Lett. **59**, (1987) 1277.
19. H. Kossman, B. Krassig, V. Schmidt and J.E. Hansen, Phys. Rev. Lett. **58**, (1987) 1620.
20. H.U. Böhmer aand S.D. Peyerimhoff, Z. Phys. **D3**, (1986) 195.

PHOTOEMISSION AND NEXAFS STUDIES OF ORGANIC MOLECULAR AND POLYMERIC MATERIALS

Kazuhiko SEKI

Department of Chemistry, Faculty of Science, Nagoya University, Furocho, Chikusa-ku, Nagoya 464-01, JAPAN

Recent progress in the elucidation of electronic and geometrical structures of electronically functional organic materials UPS and NEXAFS is reviewed, taking examples mainly from author's group.

1. Introduction

Recently many electronically functional organic materials attract attention for their potential applications in various fields such as molecular devices, color display, solar cell, organic ferromagnets, and photography. Elucidation of their electronic and geometrical structures form basis for understanding various organic systems and also for improving their functions. UV photoelectron spectroscopy (UPS) is a powerful tool of studying the electronic structure of matter, in particular the occupied electronic states, and has been also applied to organic materials[1-4]. On the other hand, the study of unoccupied states has become possible by the development of novel techniques such as near edge X-ray absorption spectroscopy (NEXAFS)[5,6], inverse photoemission spectroscopy (IPES)[7,8], and electron transmission spectroscopy (ETS)[9]. NEXAFS has advantage of the possibility of also studying the molecular orientation in the specimen (e.g. ultrathin films), which is important when we consider the construction of molecular systems, by using polarization dependence of absorption.

In this talk, we review recent progress in this field, taking example mainly from the work of our group by using UPS and NEXAFS, also mentioning related works by other workers.

2. Organic Semiconductors

The search for organics with good electrical conduction has been performed in two ways. One was the combination of two components such as charge-transfer (CT) complexes and conjugated polymers doped with electron acceptors or donors. This approach was successful and has led to the discovery of organic conductors and even organic superconductors. Our recent studies in this field will be described in the next section. The other attractive approach is single component system. Although this has not reached conductors or superconductors, single component system has an advantage of being generally more stable and easier to characterize than binary systems.

In single-component systems, the number of intrinsic charge carriers are limited by energy gap much larger than the thermal energy. One approach to overcome this difficulty is to increase intermolecular Ointeractions in order to reduce the energy gap and increase the band width. Introduction of calcogen atoms is effective to increase intermolecular interaction. We already reported[10-12] that tetrakis(alkylthia)tetrathiafulvalenes (TTC_n-TTFs) (**Fig. 1a**) show variation in ionization threshold of solid I_s from 4.6 to 5.1_5 eV for the carbon number n in each alkyl chain n = 1-18, while gas phase ionization energy is constant within 0.1 eV. The variatin of I_s follow that of conductivity from 10^{-10} to 10^{-5} S/cm, and was ascribed to strong intermolecular interaction through short S..S contacts, the shortest being 3.57 A for n = 9, which is smaller than twice of van der Waals radius (3.70 A).

Recently Yamashita et al. synthesized new organic semiconductor bis[1,2,5]thiazolo-*p*-quinobis(1,3-dithiole) (BTQBT) (**Fig. 1b**)with conductivity of 10^{-3} S/cm^{-1} and activation energy of 0.21-0.24 eV at room temperature.[13,14] Its good conductivity is ascirbed to the strong intermolecular interaction caused by short S..S contacts (3.26 A), which is even smaller than that in TTC_9-TTF. The UPS data of BTQBT[15], shown in **Fig. 2**, revealed again a small gap of 0.2_6 eV between the valcence band maximum (VBM) and Fermi level E_F, which agrees well with the observed activation energy. The simulated UPS spectrum by MNDO calculation shown in Fig. 2 is in fair agreement with the observed spectra, and the calculated HOMO has large contribution from the S atoms in short contact. Band calculation using the crystal structure indicates a fairly large width of 0.7 eV for HOMO-derived band, supporting possibility of band conduction proposed on the Hall mobility measurements. This

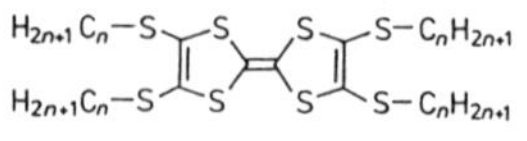

TTC_n-TTF

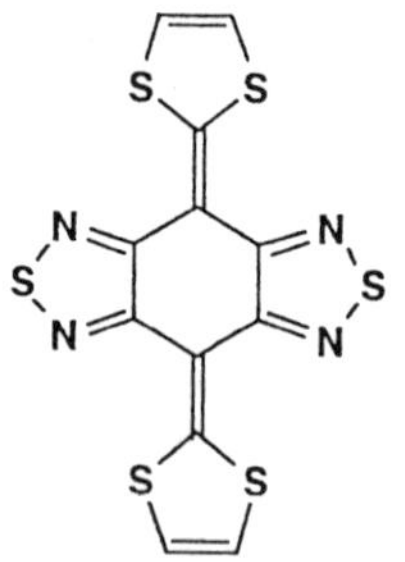

BTQBT

Fig. 1. Molecular structures of TTC_n-TTF (a) and BTQBT (b).

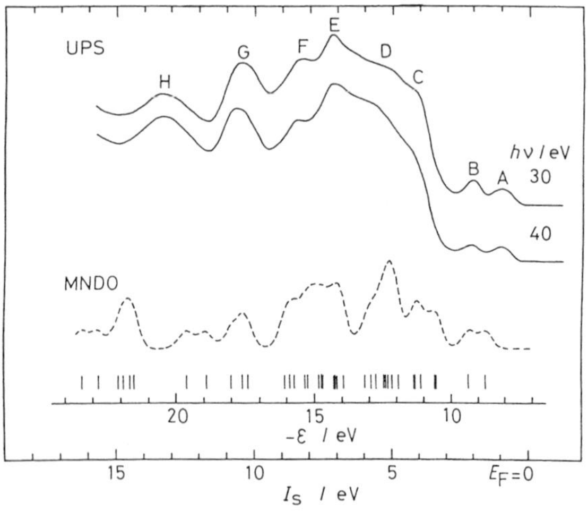

Fig. 2 UPS spectra of BTQBT and simulated spectrum by MNDO calculation. Note the small gap (0.2 eV) between Fermi level (E_F) and valence band maximum (right hand onset of peak A).

case demonstrates usefulness of UPS for studying organic semiconductors.

Recently Hasegawa et al.[16] studied angular dependence of UPS spectra of BTQBT carefully evaporated on graphite and MoS_2. The observed dependence was properly explained by assuming almost flat molecular orientation on the surface. This is also in agreement with the preliminary results of polarization dependence in NEXAFS.

Another possible way of reducing band gap is the use of radical systems including unpaired electron. Typical example is Lu phthalocyanine,[17-19] for which UPS spectra[20,21] show a small gap between VBM and E_F, and its interpretation is a matter of recent controversy.

3. Conducting Polymers and Molecular Complexes

As mentioned above, conducting polymers has been one of the major subjects of UPS studies of organics. Many polymers, most of which have extended π-conjugated system, are doped with acceptors (e.g. iodine, ClO_4^-, and AsF_5) or donors (alkali metals) and become conducting.

Among them, polyacetylene (PA) is the simplest prototype polymer. It is prepared by polymerizing acetylene on Ziegler-Natta type catalyst at low temperature. The polymer at this stage is in the *cis*-form, and converted to *trans*-form by raising temperature or by doping. The conductivity of ca. 10^3 S/cm[22] by the original method by Shirakawa et al.[23] (S-PA) was improved to ca. 10^5 S/cm for so-called new PA (N-PA) by modifying preparative details.[24]

The UPS measurements of PA was difficult due to its extreme sensitivity to air. For example, expected peak at VBM was often smeared out. Recently, spectra of good quality were obtained by Kamiya et al.[25] and Rasmussen et al.[26] In **Fig. 3**, the spectra of both *cis*- and *trans*-formes of N-PA are shown. They were measured by using glove boxes for sample preparation and incorporation into UPS spectrometer. The broken lines are simulated spectra by ab initio MO calculations of long polyenes. The spectra correspond well with the calculation, and the feature at VBM is now clearly seen. The spectrum of *trans*-form agrees fairly well with those of *trans*-S-PA by Swedish group.[26]

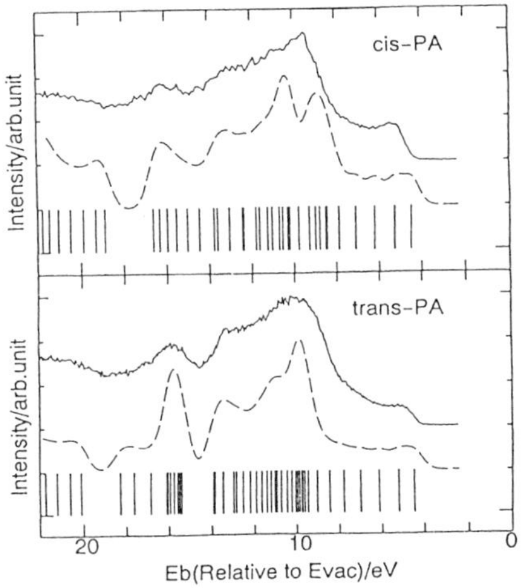

Fig. 3 UPS spectra of undoped cis- and *trans*-polyacetylenes. The broke lines are simulated spectra by using vertical bars from MO calculations.

The change of electronic structures of such polymers on doping is another interesting subject. In **Fig. 4** we show the change of the UPS spectrum by both p-type (ClO_4^- by immersion in CH_3CN solution of $Cu(ClO_4)_2$) and n-type (potassium evaporation) doping.[27,28] Before doping, the gap between VBM and E_F (determined by the Fermi edge of gold) is 0.7 eV, in good agreement with optical bandgap of 1.4 eV. By light doping with ClO_4^-, E_F shifts to larger binding energy by 0.7 eV. By further doping to 8 %, E_F shifts further (0.9 eV vs. undoped state), and crosses VBM, indicating a metallic state. This transition to metallic statec at heavily doped stage corresponds to the change of other properties such as magnetic susceptibility at 5-6 %. On n-type doping, similar change of spectra, but in this case the shift of E_F to smaller binding energy (0.7 and 0.9 eV for 6.5 and 10 %, respectively), was observed. Such a shift of E_F relative to VBM was explained by considering the formation of various defects such as soliton (neutral radical) or polson (combined states of soliton and polaron - charged radical), as shown by calculated density of states for charged polyene chain with counter ions.[29] Similar change to metallic state by p-type doping, with shift of E_F, was also reported for polythiophene[30] and polypyrrole,[31] studies of n-type doping were performed for small model systems of poly(*p*-phenylene) and polythiophene.[32] Charge transfer complexes of molecular solids are also important systems, in particular for realizing superconductors. Thypical example is alkali-doped states of fullerenes. **Figure 5** depicts spectra of C_{60} and higher fullerenes (C_{70}, and C_{84}) and their alkali metal complexes at their states of maximum conducivity by doping, determined with *in situ* monitoring of conductivity change by van der Pauw method.[33-36] The results, in combination with inverse photoemission studies of C_{60} with K doping,[35] indicates rather complex change of electronic structure. The rigid band model of LUMO filling with excess electron from alkali metal completely fails, and VBM of higher fullerenes do not cross E_F even at the stage of maximum conductivity. We, however, do not go into the details of this subject, since several talks in this conference deals with it.

Studies of other complexes have also been reported. In particular, detailed studies of the metal com plexes of DCNQI (dicyano-*p*-quinodi-

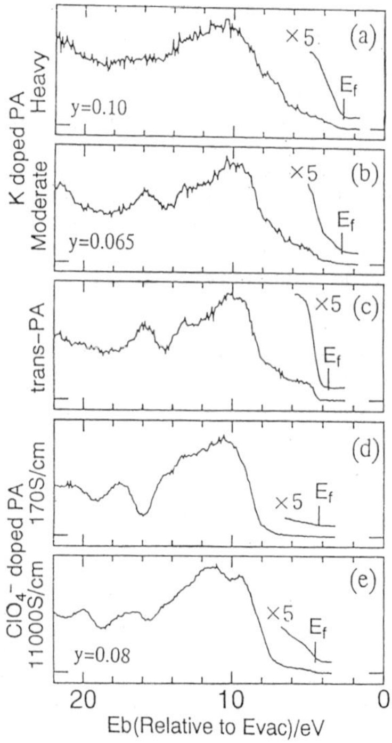

Fig. 4. Change of UPS spectra of trans-polyacetylene by n-type (K) and p-type (ClO_4) doping. Note the shift of Fermi level (E_F) relative to VBM.

imine) derivatives, including observation of band dispersion relation,[37] is notable.

4. Si- and Ge-Based Polymers

Besides carbon-based π-conjugated polymers, there is growing interest in polysilanes and polygermanes, which are con-ducting polymers without unsaturated bonds. They consists of main chain formed by Si or Ge sequence and C-derived side groups. In **Fig. 6** the whole valence spectra of poly(methylpropylsilane) $(SiCH_3C_3H_7)_n$ (PMePrSi)[38] is shown in comparison with spectra of building blocks. These spectra, of much better quality than those reported previously,[39] show many fine structures, and comparison with related compounds demonstrates that the spectra can be in the first approximation well understood as an overlap of the spectra of main chain and side groups, with the topmost part of the valence band being formed by the polysilane back bone. The ionization threshold I_s of polysilane (5.7 eV for PMePrSi) is much smaller than that of carbon analogue (alkyl chain), as seen by comparison with the onset of propane in Fig. 6. This difference, due to smaller electronegativity of Si than that of C, makes p-type doping possible even without unsaturated bonds. In polymer with aromatic pendants (e.g. poly(methyl-phenylsilane)), the σ-π interaction between the main chain and pendant reduces I_s further to 5.3 eV.

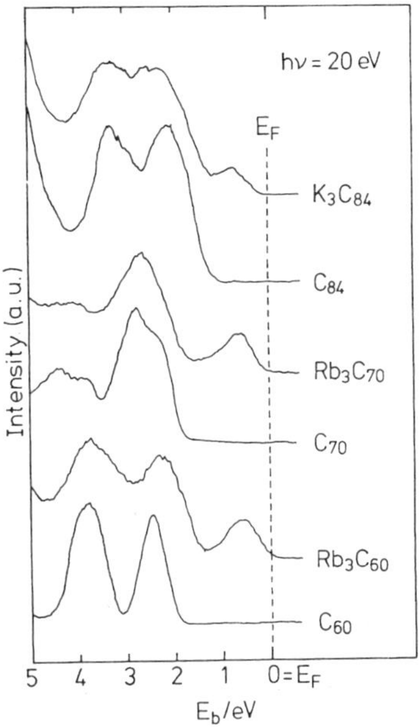

Fig. 5 UPS spectra of fullerenes and their K-doped states at maiximum electric conductivity. (a) C_{60}, (b) C_{70}, and (d) C_{84}.

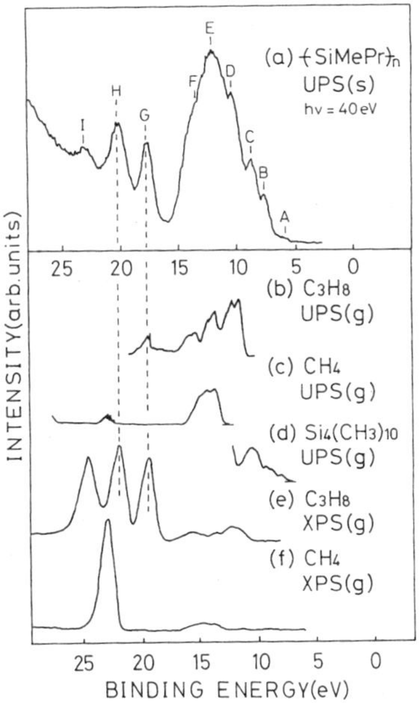

Fig. 6. UPS spectra of poly-(methylpropylsilane) at hν = 40 eV compared with spectra of related compounds.

5.3 eV. Polygermanes give fairly similar spectra with polysilanes.

Polysiloxane is another major class of Si-based polymers, with Si-O-Si sequence in the main chain. The UPS spectrum of poly(dimethylsiloxane) $(Si(CH_3)_2O)_n$ is shown in **Fig. 7**.[41] The value of I_s is 8.3 eV, which is much higher than that of poly(dimethylsilane) (5.9 eV), reflecting the effect of electron-withdrawing oxygen. This corresponds to the difference in three dimensional silicon (5.35 eV) and SiO_2 (10.6 eV).

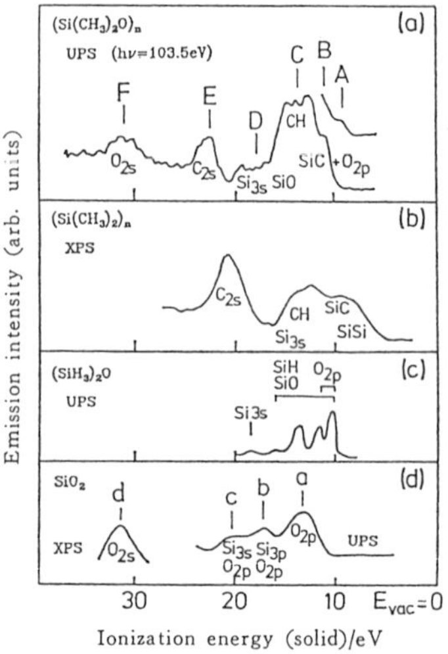

Fig. 7 UPS spectra of poly(dimethylsiloxane) at hν = 103.5 eV compared with spectra of related compounds.

5. Dye/AgBr and other Interfaces

The organic/organic and organic /inorganic interfaces are important for various phenomena such as electron transfer. In **Fig. 8** the change of UPS spectra of AgBr evaporated film by depositing one of the merocyanine dyes shown in **Fig. 9**, little by little, until no further change become observed.[42] The abscissa is retarding voltage between the sample and electron energy analyzer. The location of E_F, determined by the Fermi edge of silver substrate, is also indicated. The right hand onset corresponds to VBM, and the left hand cutoff corresponds to the energy of the vacuum level. By dye deposition, we see the growth of a peak at ca. 6 V by HOMO of the dye, and also a large lowering of the vacuum level of ca. 1 eV. From such data and the measurements of optical absorption edges, interfacial energy diagram between AgBr and 5 dyes shown in **Fig. 10** was derived. The relative locations of the bottom of the conduction band of AgBr and LUMO of dyes were in excellent correspondence with the observed ability of dyes for spectral sensitization, which is a phenomena that sensitive region of silver halide in photography to the absorption region of the adsorbed dye on silver halide grains. When electron is excited from HOMO to LUMO of the dye and LUMO is above the bottom of the conduction band of AgBr, electron can be injected into the conduction band, causing additional sensitivity of AgBr. Previous trials[43,44] for constructing interfacial diagrams, from independent measurements of AgBr and dye, failed in showing correspondence with the sensitizing ability of dyes due to the neglect of the shift of vacuum level. Thus the present study removed one of the difficulties of the electron transfer mechanism of spectral sensitization.

NEXAFS spectroscoy is also effective in studying the orientation of adsorbed dye on silver halide. In **Fig. 11** the S K-edge NEXAFS spectra of two merocyanine dyes are shown for dye powder and adsorbed state on

AgCl.[45] For dye A, where the sulfur atom is in the aromatic ring, S1s -> π^* transition is not well separated from transitions to σ^* states, unlike in the case of dye B with thioketone (=S) group, where the former is clearly separated from the latter. When adsorbed on AgCl, the spectrum of dye A is not much affected, while the spectrum of dye B is significantly changed, with the intensity of S1s -> π^* transition being reduced. Thus dye B should be adsorbed on AgCl at the site of =S group, possibly changing from the neutral form to zwitterionic form with O^-

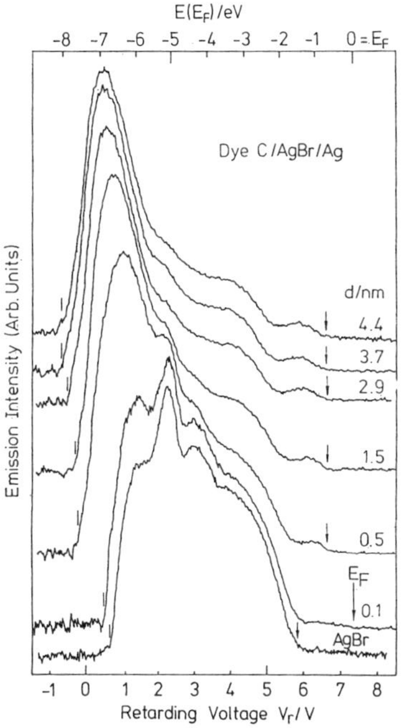

Fig. 8 Change of UPS spectrum of AgBr by stepwise deposition of dye C in Fig. 9.

Fig. 9. Structures of merocyanine dyes.

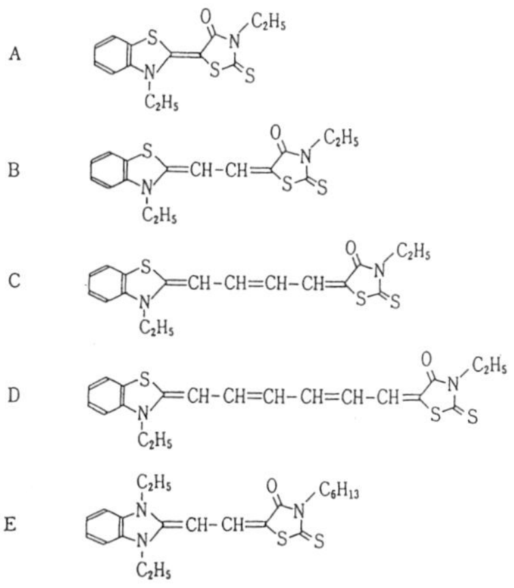

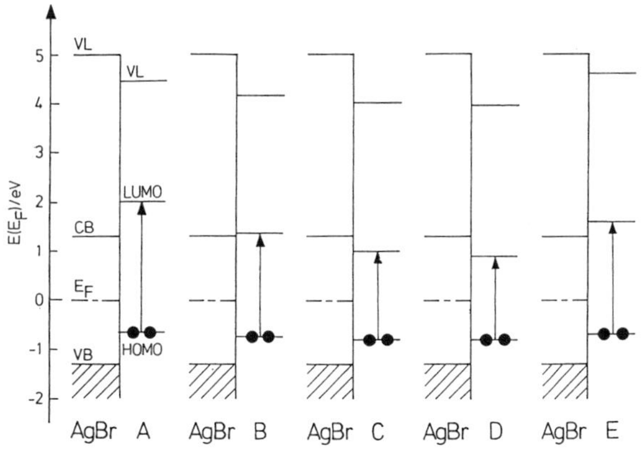

Fig. 10 Energy level diagram of AgBr/dye interfaces for dyes in Fig. 9. VL: Vacuum level, CB and VB: bottom of conduction band and top of valence band of AgBr, LUMO and HOMO: lowest unoccupied and highest occupied molecular orbitals of dye, E_F: Fermi level.

part adsorbed on Ag^+ ion.

The alignment of the dye molecule with the negative site on Ag halide may partly contribute to the formation of electric double layer detected as the lowering of vacuum level in UPS study of dye/AgBr system. Electron transfer between the dye and AgBr to reach Fermi level alignment is another possibility. In this relation, we mention detailed studies of Fermi level alignment between C_{60} and various metals and semiconductors.[46]

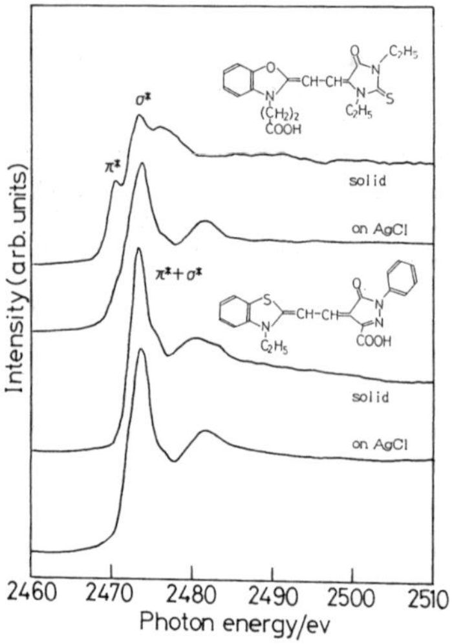

Fig. 11 S K-edge NEXAFS spectra of merocyanine dye solid and in adsorbed state on AgCl. Dye with thioketone (=S) group is perturbed by adsorption, while dye with thia (-S-) group does not.

5. Concluding Remarks

Here we could not menetion interesting subjects such as chemical reaction in thin films, where systems such as polyimide formation by thermal treatment of Langmuir-Blodgett films,[47] photopolymerization of diacetylene[48] and distyrilpyrazine,[49] and covalent bonding of phenylenediamine on Si(111) surface[50] have been studied.

Including these studies, recent studies of organic materials has become increasingly sophisticated in both sample manipulation and measurement techniques. This should be due to the effort for attacking even highly intractable materials (e.g. polyacetylene) and also to the real fusion of the fields of organic materials science and surface science. This trend, with the efforts of organic chemists to create atractive new compounds, will make the future of this field to be even more fruitful.

6. Acknowledgements

The author thanks H.Fujimoto, K. Kamiya, S. Hasegawa, T. Miyamae, J. Tanaka, C. Tanaka, H. Inokuchi, Y. Yamashita, A. Yuyama, S. Narioka, H. Ishii, M. Fujino, H. Isaka, M. Fujiki, K. Takeda, N. Matsumoto, T. Sugano, M. Kobayashi, H. Yanagi, T. Araki, T. Ohta, T. Takahashi, T. Morikawa, S. Hino, S. Suzuki, K. Kikuchi, I. Ikemoto, S. Suzuki, Y. Achiba, S. Watanabe, T. Tani, and N. Ueno for their kind collaboration.

7. References

1. W.D. Grobman and E.E. Koch, in *Photoemission in Solids*, ed. L. Ley and M. Cardona (Springer, Berlin, 1979), vol. 2, p. 261.

2. H. Inokuchi, K. Seki, and N.Sato, *Phys. Scripta* T17 (1987) 120.
3. K. Seki, in *Optical Techniques to Characterize Polymer Systems*, ed. H. Baessler (Elsevier, Amsterdam, 1989).
4. W. Salaneck, *Rep. Prog. Phys.* **54** (1991) 1215.
5. D.A. Outka and J. Stoehr, *Springer Ser. Surf. Sci.* **7**, ed. R. Vanse low and R. Howe (Springer, Berlin, 1988).
6. G. Tourillon, D. Guay, A. Fontaine, R. Garrett, and G.P. Williams, *Faraday Disc. Chem. Soc.* **89** (1990) 275.
7. M.L.M. Rocco, K.H. Frank, P. Yannoulis, and E.E. Koch, *J. Chem. Phys.* **93** (1990) 6859, and references therein.
8. H.M. Meyer III, T.J. Wagener, J.H. Weaver, M.W. Feyereisen, and J. Almloef, *Chem. Phys. Lett.* **164** (1989) 527.
9. N. Ueno, K. Seki, K. Sugita, and H. Inokuchi, *Phys. Rev.* **B43** (1991) 2384, and references therein.
10. H. Inokuchi, G. Saito, P. Wu, K. Seki, T.B. Tag, T. Mori, K. Imaeda, T. Enoki, K. Inaka, and N. Yasuoka, *Chem. Lett.* (1986) 1263.
11. K. Seki, T.B. Tang, T. Mori, Wu P.J., G. Saito, and H. Inokuchi, *J. Chem. Soc. Faraday Trans. 2* **82** (1986) 1067.
12. H. Yamamoto, K. Seki, H. Inokuchi, and G. Saito, *J. Chem. Soc. Fara day Trans. 2* **83** (1987) 2151.
13. Y. Yamashita, S. Tanaka, K. Imaeda, and H. Inokuchi, *Chem. Lett.* (1991) 1213.
14. K. Imaeda, Y. Yamashita, Y. Li, T. Mori, H. Inokuchi, and M. Sano, *J. Mater. Chem.* 2 (1992) 115.
15. H. Fujimoto, K. Kamiya, S. Tanaka, T. Mori, Y. Yamashita, H. Inokuchi, and K. Seki, *Chem. Phys.* in press, and Th53 of this conference.
16. S. Hasegawa, S. Tanaka, Y. Yamashita, H. Inokkuchi, H. Fujimoto, K. Kamiya, K. Seki, and N. Ueno, to be publlished.
17. J.J. Andre, K. Holczer, P.Petit, M.-T. Rioux, C. Clarisse, R. Even, J. Simon, M.Fourmigue, *Chem. Phys. Lett.* **115** 1985 463.
18. M. Maitrot, G. Gouillaud, B. Boudjema, J.J. Andre, H. Strzelecka, J. Simon, and R. Even, *Chem. Phys. Lett.* **133** (1987) 59.
19. M. Bouvet and J. Simon, *Chem. Phys. Lett.* **172** (1990) 299.
20. M. R. Fahy, H. Fujimoto, A.J. Dann, H. Hoshi, H. Inokuchi, Y. Ma ruyama, and M.R. Willis, *Phys. Scripta*, **41** (1990) 550.
21. J. Buffler, M. Abraham, M. Bouvet, J. Simon, and W. Goepel, *J. Chem. Phys.* **95** (1991) 8459.
22. C.K. Chiang, C.R. Fincher, Jr., Y.W. Park, A.J. Heeger, H. Shiraka wa, E.J. Louis, S.C. Gan, A.G. MacDiarmid, *Phys. Rev. Lett.* **39** (1977) 1098.
23. T. Ito, H. Shirakawa, and S. Ikeda, *J. Polym. Sci. Chem. Ed.* 12 (1974) 11.
24. H. Naarman and N. Theophelou, Syn. Metals, 22 (1987) 1; J.Tsukamoto, A. Takahashi, and K. Kawasaki, *Jpn. J. Appl. Phys.* 29 (1990) 251.
25. K. Kamiya, H. Inokuchi, M. Oku, S. Hasegawa, C. Tanaka, J. Tanaka, and K. Seki, *Syn. Metals.* **41-43** (1991) 155.
26. J. Rasmussen, S. Stafstroem, M. Loeglund, W.R. Salaneck, U.O. Karls son, D. B. Swansson, A.G. MacDiarmid, and G.A. Arbuckle, *Syn. Metals* **41-3** (1991) 1365.
27. T. Kamiya, M. Oku, J. Tanaka, K. Seki, and H. Inokuchi, to be pub lished.

28. T. Miyamae, K. Kamiya, S. Hasegawa, J. Tanaka, K. Seki, and H. Inokuchi, to be published.
29. C. Tanaka and J. Tanaka, *Syn. Metals* **41-43** (1991) 3709; *Mat. Res. Symp. Proc.* **247** (1992) 577.
30. R. Lazzaroni, M. Loeglund, S. Stafstroem, W.R. Salaneck, and J.L. Bredas, *J. Chem. Phys.* **93** (1990) 4433.
31. P. Baetz, D. Schmeisser, and W. Goepel, *Phys. Rev.* **B43** (1991) 9178.
32. F.P. Netzer, M.G. Ramsey, and D. Steinmueller, *Syn. Metals* **41-43** (1991) 1343; M.G. Ramsey, D. Steinmueller, and F.P. Netzer, *Phys. Rev.* B42 (1990) 5902; M.G. Ramsey, F.P. Netzer, D. Steinmueller, D. Steinmueller-Nethl, and D.R. Lloyd, *J. Chem. Phys.* in press.
33. T. Takahashi, T. Morikawa, S. Sato, H. Katayama-Yoshida, A. Yuyama, K. Seki, H. Fujimoto, S. Hino, S. Hasegawa, K. Kamiya, H. Inokuchi, K. Kikuchi, S. Suzuki, I. Ikemoto, and Y. Achiba, *Physica* **C185-189** (1991) 417.
34. T. Takahashi, T. Morikawa, S. Hasegawa, K. Kamiya, H. Fujimoto, S. Hino, K. Seki, H. Katayama-Yoshida, H. Inokuchi, K. Kikuchi, S. Suzuki, I. Ikemoto, and Y. Achiba, *Physica* **C190** (1992) 205.
35. T. Takahashi, S. Suzuki, T. Morikawa, H. Katayama-Yoshida, S. Haseg awa, H. Inokuchi, K. Seki, K. Kikuchi, S. Suzuki, I. Ikemoto, and Y. Achiba, *Phys. Rev. Lett.* **68** (1992) 1232.
36. S. Hino, K. Matsumoto, S. Hasegawa, K. Kamiya, H. Inokuchi, T. Mori kawa, T. Takahashi, K. Seki, K. Kikuchi, S. Suzuki, I. Ikemoto, and Y. Achiba, *Chem. Phys. Lett.* **190** (1992) 169.
37. D. Schmeisser, P. Baetz, W. Goepel, W. Jaegermann, Ch. Pettenkofer, H. Wachtel, A.Jimenez-Gonzales, J.U. von Schutz, H.C. Wolf, J. Tabor ski, V. Wustenhagen, E. Umbach, P. Erk, H. Meixner, and S. Huenig, *Ber. Bunsenges. Phys. Chem.* **95** (1991) 1441 and references therein.
38. A. Yuyama, S. Narioka, H. Ishii, S. Hasegawa, K. Seki, M. Fujino, H. Isaka, M. Fujiki, and N. Matsumoto, to be published.
39. G. Loubriel and J. Zeigler, *Phys. Rev.* **B33** (1986) 4203.
40. K. Takeda, M. Fujino, K. Seki, and H. Inokuchi, Phys. Rev. B36 (1987) 8129; K. Seki, T. Mori, H. Inokuchi, and K. Murano, *Bull. Chem. Soc. Jpn.* **61** (1988) 351.
41. T. Sugano, K. Seki, T. Ohta, H. Fujimoto, and H. Inokuchi, *J. Chem. Soc. Jpn.* (1990) 594 (in Japanese).
42. K. Seki and T. Tani, *Proc. Int. Conf. Polymer/Solid Interf.* (Adam Hilger, London, 1992), in press; and Tu16 of this conferernce.
43. I.A. Akimov, V.M. Bentsa, F.I. Vilesov, and A.N. Terenin, *Phys. Stat. Solidi* **20** (1967) 771.
44. R.C. Nelson, *J. Opt. Soc. Am.* **51** (1961) 1186.
45. T. Araki, K. Seki, S. Narioka, K. Horiuchi, Y. Takata, T. Yokoyama, T. Ohta, S. Watanabe, and T. Tani, Th23 of this conference.
46. T.R. Ohno, Y. Chen, S.E. Harvey, G.H. Kroll, and J.H. Weaver, *Phys. Rev.* **B44** (1991) 13747.
47. Th.Schedel-Niedrig, M.Keil, H.Sotobayashi, T. Schilling, B.Tesche, and A.M. Bradshaw, *Ber. Bunsengers. Phys. Chem.* **95** (1991) 1385.
48. K. Seki, I. Morisada, K. Edamatsu, H. Tanaka, H. Yanagi, T. Yokoya ma, and T. Ohta, *Phys. Scripta* **41** (1990) 172.
49. M. Abraham, J. Duettig, M. Schreck, R. Lege, R. Reich, D. Oelkrug, and W. Goepel, *J. Chem. Phys.* **94** (1991) 3235.
50. Th. Kugler, M. Abraham, and W. Goepel, *Surf. Sci.* **260** (1992) 64.

HIGH ENERGY ELECTRON LOSS AND UV PHOTOEMISSION SPECTROSCOPIES OF H/SI(111)-(1x1) SURFACES

P.DUMAS

Laboratoire Utilisation Rayonnement Electromagnetique
Centre Universitaire Paris-Sud F-91405 Osay Cedex
and LASIR-CNRS F-94320-Thiais

ABSTRACT

Unreconstructed, ideally hydrogen terminated silicon (111) surfaces are investigated by vibrational and photoemission spectroscopies.Very good qualitative and quantitative agreement is obtained between the experimental vibrational and valence band dispersion curves, and the theoretical calculations.The mechanism responsible for the energy loss spectrum in the electron scattering process at the H/Si surfaces, is dominated by impact scattering.High resolution core-level spectroscopy reveal that the lifetime broadening (< 40 meV)is very close to the value obtained for SiH_4 gas phase.Surface Si 2p core levels are clearly observed when varying the escape depth of the photoelectrons.Vibrational fine structure including both bending and stretching Si-H modes are also resolved.

1.Introduction:

Hydrogen termination of clean silicon surfaces is an interesting chemisorption system, because hydrogen is the simplest atom that can saturate dangling bonds at the surface of a semiconductor.Hydrogen also plays a crucial role in the formation of polycrystalline and amorphous silicon and its presence has a profond influence on the electronic properties. It is therefore important to investigate thechemisorption of H on silicon surfaces to help understand the properties of hydrogen in the bulk of crystalline, microcrystallites or amorphous silicon.

Traditionnally, hydrogen is adsorbed on a silicon single crystal surface by exposing atomic H to the clean surface, prepared in ultra high vacuum environment [1].H atoms saturate the dangling bonds, quench the surface states and induce well-defined bonding states[2] .However, the main problem in obtaining useful data is the atomic imperfection of the H-covered surfaces. The Si(111) 7x7 surface has a strong reconstruction which is not eliminated upon atomic H exposure.In case of cleaved Si(111)-(2x1), H lifts up the reconstruction, but the interface is only locally ordered ((1x1) arrangement), and the long range order is disrupted by steps and defects [3].

Consequently, all the previous vibrational and electronic measurements of hydrogen terminated silicon single crystal surfaces are characterized by spectra that are featureless and difficult to interpret.

Hydrogen termination of silicon surfaces can also be obtained by wet chemical etching [4], which in fact leads to remarkable interfacial properties[5].Early, observations of (1x1) low-energy electron diffraction patterns of H-passivated surfaces[6a], and of the various vibrational losses in electron energy loss spectroscopy[6b], have suggested that HF-prepared silicon surfaces may be unreconstructed with ideal H termination.Although such surfaces were shown to be atomically rough with various hydride termination [7], recent use of buffered HF have shown that flat, unreconstructed Si(111) surfaces can indeed be prepared [8]. .Among their remarkable properties, these surfaces are stable in air , and can be introduced into an ultra high vacuum chamber, without any appreciable loss of hydrogen or contamination[9].As a result, a number of studies have been performed with conventional surface science tools : Infrared absorption [9] and Sum Frequency generation [10] to study vibrational dynamics, Scanning Tunneling Microscopy [11],Energy Electron Loss spectroscopy [12], Inelastic He atom Scattering[13] and Inverse photoemission[14] .

In this paper,vibrational and electronic properties of H/Si(111)-(1x1) surfaces are presented.The high perfection of the surfaces leads to unusually narrow lines of both the vibrational and the valence band features ,making it possible to gain new insights into the physical properties of H/Si interface.The paper is organized as follows : in section 2, the preparation procedure and the analysis of the structural and chemical quality of the sample details are reviewed ; experimental details are briefly presented in section 3; the vibrational properties of the H/Si(111)-(1x1) surface are described in section 4, with an emphasis on the nature of the different surface phonon modes, and on the electron scattering mechanism at the H/Si interface . The valence band and core level spectra obtained with a high resolution photoemission spectrometer, using a synchrotron radiation as the photon source, are shown and analyzed in section 5.

2.H/Si(111)-(1x1) surfaces: extent of the structural and chemical perfection:

2.1.Preparation and handling:

The general preparation of the samples, involving chemical oxidation and HF etching, has been previously described in the literature[8,9,15].However , since minor variations in the details of the procedure can make substantial differences in the resulting morphology, the chemical composition and the vibrational and electronic band lineshape , the procedure which produces the best samples[15] is summarized here.The samples (n-doped, $\rho \approx 150$ W.cm) are first thermally oxidized at 1050°C, and then post annealed during 15 minutes in an inert gas , at the same temperature.Post annealing is important to make nearly atomically smooth Si/SiO_2 interfaces[16].Samples are then degreased first in hot trichloethylene, room temperature acetone and methanol sequentially and second in hot (80°C) solution of NH_4OH:H_2O_2:H_2O (1:1:4). After a thorough rinsing in deionized water(D.I.) (18.2 Ω.cm), the thick oxide is removed in a buffered HF solution (pH=5; time=2 minutes). The samples are rinsed again in deionized water before being etched during 4 minutes in a 40% NH_4F solution (pH≈8). After this last etching step,the samples are briefly rinsed in D.I.(≈10 seconds), to remove most of the unwanted chemicals physisorbed on the surface (SiF_4,NH_4OH,HF,and others).Good quality deionized water is required to avoid any contamination , such as hydrocarbons or oxide, during the last rinsing step.

Introduction into a UHV chamber requires a careful evacuation procedure (slow evacuation ≈ 10 Torr/mn and clean pumping) [9]. No further treatment of the sample (cleaning, baking ...) is needed after introduction into UHV.

2.2 .Extent of the perfect H terminated domains:

Infrared absorption spectroscopy (using a multiple internal reflexion technique[17]) is a powerful tool to analyze the quality of the samples right after their preparation and prior to introduction in the UHV chamber. The best samples exhibit a single band, polarized strickly perpendicular to the surface, with a frequency of 2083.9 cm-1 at room temperature, and a half width at full maximum of 0.71 cm-1[15] .A detailed analysis of the lineshape, involving the dynamical dipole coupling within finite size domains[18] , can be used to deduce the average domain size (2.10^4 Si-H units or 600 Angstroms domains in one dimension for the best samples). Furthermore, a slight asymetry towards higher frequency of the Si-H stretch mode , is resolved at low temperature [19], and assigned to Si-H stretch next to point defects .Depending on the nature of the defects, their concentration is estimated to be between 0.01% and 1% [19] .

2.3.I mpurity content:

Figure 1 displays the electron loss spectrum of H/Si(111)-(1x1) surfaces, taken with a primary energy of 5eV, at the center of the Surface Brillouin Zone (SBZ).No hydrocarbon contamination(CHx stretch modes in the 2850-2960 cm-1 frequency range), no oxide (SiOx stretch:1000-1100 cm-1), no NH2 or NH4F or NH4OH (1500,2200 and 3200 cm-1 regions)can be detected , at least up to the detection limit of the HREELS technique (<1%). This is in total departure from the usual contamination observed in previous HREELS, for chemically etched silicon samples, introduced in UHV [20] , and still substantially cleaner than the samples used in more recent EELS studies [21].

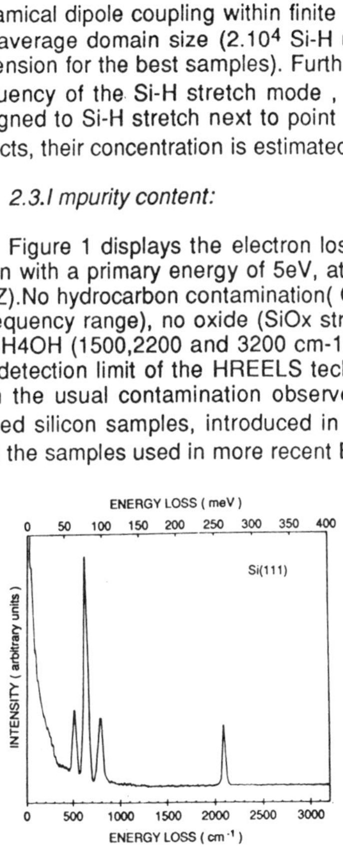

Fig.1.Specular electron-energy-loss spectrum of the H/Si(111)-(1x1) surfaces.The primary electron energy is 5 eV, the incidence and reflection angles are both 60°, and the electron trajectory, projected onto the surface, is along the <110> direction

The four losses, observed on Fig.1 , are assigned to the Lucas mode (520 cm-1 or 64.5 meV), the monohydride bending (or wagging) mode (636 cm-1 or 79 meV), a Fermi resonance-induced loss (795cm-1 or 98 meV) and the monohydride stretching mode (2084cm-1 or 258 meV)[23] , respectively.The only contreversial assignment is that of the 795 cm-1 (or 98 meV) mode which has been assigned to an impurity [21] (Si-OH mode). This origin is not supported by the observation that: (1) it is resolution limited and maintains a constant intensity ratio with the bending mode;(2) its intensity, compared to that of the elastic peak would require with this assignment a concentration of hydroxyl groups at the surface over 5%, which is not consistent with the defect concentration determined by Infrared absorption spectroscopy[22] ;(3)it appears intrinsic to perfectly hydrogen terminated Si(111) surfaces [12,23], i.e. it is not observed on the atomically rough H/Si(111) and on the H/Si(100) .The Fermi resonance origin will be discussed in the next section.

3.Experimental:

The apparatus and experimental procedure for the HREELS study have been described in details elsewhere[12,23].We simply recall here that the spectrometer is a Leybold-Hereaus (ELS 22) with a rotating analyser and the UHV chamber is equipped with a load-lock chamber. The count rate in the specular beam is close to 5 10^5 Hz , with a 4 meV resolution.

The photoemission spectra have been obtained with the French-Swiss beam line SU3 on Superaco, at LURE. The undulator beam line is equipped with a P.G.M. monochromator. The combined resolution of the PGM and the VSW analyser is 45 meV at 110 eV photon energy.Samples are in the horizontal plane, and the incident photons impinge on the surface at grazing incidence (80°) with the electric field in the surface plane.The analyser is set at 45° from the horizontal plane.The UHV chamber is equipped with a load lock chamber.

4.Surface phonons :

Previous electron loss spectroscopy on H -terminated Si(111) and Si(100) surfaces [24] have identified surfaces losses , but the complexity of the surface reconstruction has prevented theorical calculations of the surface phonon dispersion and thus comparison with experimental studies .H/Si(111)-(1x1) surfaces, with their long range order, constitute an unique system to study experimental dispersion curves and to compare with existing theoretical approaches.In addition, the electron scattering mechanism can be carefully investigated thanks to the well defined geometry of the Si-H bond (oriented strickly perpendicular to the surface).

4.1 .Phonons at the H/Si(111)-(1x1) surfaces:

Electron loss spectra have been recorded, as a function of primary energy and off-specular angle.Fig. 2 and Fig.3 display the spectra, recorded at different off-specular angles, for primary energy of 5 eV and 9 eV respectively.All spectra are taken along the <110> azimuth.

Spectra taken at 9 eV primary energy exhibit an additional loss, located at 390 cm-1, at the Γ point of the SBZ.This mode has never been observed before on the H/Si surfaces. Its detection depends both on the primary energy (Ep> 9eV for θ_{inc}=60°) and on the angle of incidence (θ_{inc}>60° for Ep=5eV [25]).This mode is mainly impact in character. It is also detected on rough Si(100) surfaces, but with less intensity[25] .All these observations indicate that it is a bulk phonon transition[21,25] .

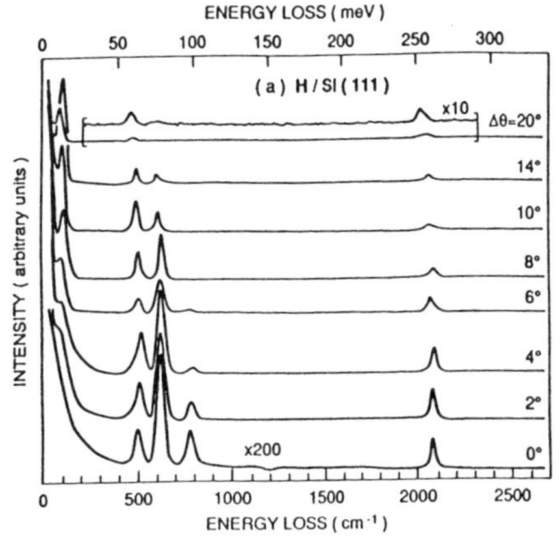

Fig.2.Off specular energy loss spectra of H/Si(111)-(1x1) surfaces, for a primary electrons of 5 eV. The angle of incidence is kept constant (60°), and the reflection angle departs of ΔΘ from the specular reflection angle.The electron trajectory, projected onto the surface plane, is oriented along the <110> direction.

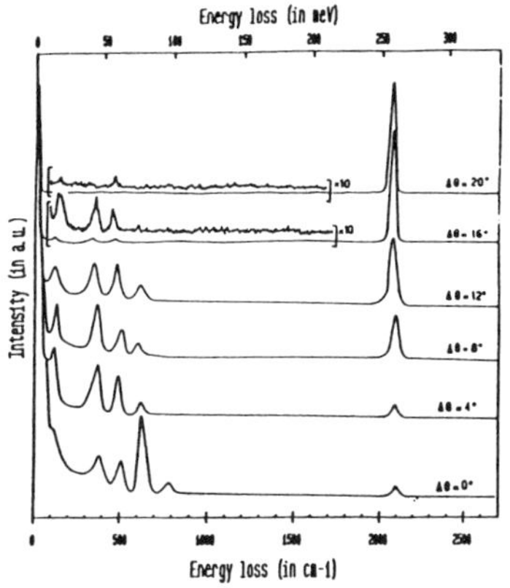

Fig.3. Same as Fig.2, but in this case, the primary electron energy is 9 eV.

In spectra taken both at 5 and 9 eV, a mode appears at low frequency (≈110 cm-1), and shifts upward away from Γ.It has also a strong impact character, but is only observed along the Γ-K direction of the SBZ.It has to be associated with the S8 surface band predicted by the bond charge calculations 26 (Fig.4).

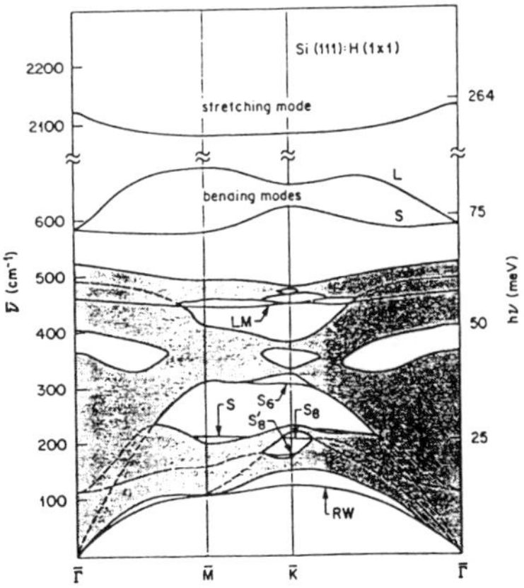

Fig.4.Dispersion relations of surface-projected bulk modes (shaded area) and surface modes (solid lines) along the borders of the surface Brillouin zone (from ref.27)

The dispersion of the various losses has been recorded only around the Γ point,(Ep<10eV - 9% of the SBZ unit vector , due to the poor transmission of our spectrometer for higher energy).Within this range, a measurable dispersion is observed for most features .The dispersion curve has now been extended by C.Stuhlmann et al.[21] to the zone boundary. Of particular interest is a mode at 330 +/- 20 cm-1 at the K point of the SBZ (S6 phonon)which has been invoked as possibly responsible for the 795 cm-1 loss.This interpretation requires a strong interaction(mixing) between the S6 two-phonon band and the Si-H bending mode, and is supported by the following observations:(1) at K, the S6 phonon involves the motion of the top silicon atoms completely in the surface plane, with a corresponding motion of the hydrogen atoms in the surface plane, but with over twice the amplitude of the silicon (i.e. this mode involves a large bending of the Si-H bond[26]); (2) the two phonon band has an energy (650 cm-1) , which is essentially identical to that of the band associated with the unperturbated Si-H bending vibration.Therefore, a strong anharmonic coupling is expected between the S6 two-phonon mode and the bending mode, similar to a Fermi resonance in molecular spectroscopy. This explanation is still tentative[12] , which makes the assignment of the 795 cm-1 mode contreversial.

4.2 .Electron energy loss mechanism:

The observation of an intense bending mode in the HREEL spectrum (dipole forbidden because its dipole moment is strickly parallel to the surface) and the intensity behavior of the loss peaks at 9 eV primary energy (the intensity of the stretch loss is higher than that of the elastic peak at $\Delta\Theta=20°$), clearly indicate that impact scattering is the dominant mechanism at H/Si surfaces.The apparent dipole behaviour at 5 eV primary energy, as suggested by the proportional decrease of the various loss intensities compared to that of the elastic peak, may be due to diffraction effects, previously identified in the case of hydrogen on W(100)[27].

5. Valence and core-level surface states:

Ultraviolet photoelectron spectroscopy has been widely used to investigate the electronic properties of the H-exposed silicon surfaces. Si(111)7x7 surfaces have been investigated the most.They suffer a substantial rearrangement of the atomic structure when exposed to hydrogen , exhibiting similarities with the Si(111)-(1x1)-H surface[28] .However, accurate measurements of the surface states dispersion and accurate determination of the core level surface components are not easy due to the short range order of such surfaces.

Due to their simplicity, ideally H-terminated Si(111) surfaces are the more appropriate system to study electronic properties at H/Si interfaces.The unprecedented narrow linewidths recorded in both the valence band and the core level spectra illustrate the structural homogeneity of the surface[35].

5.1.Valence band region:

On fig.5 are displays the valence band spectra, recorded with 17.5 eV photon energy for emission angles ranging from 35° to 55° along the Γ-K-M direction of the (1x1) SBZ .

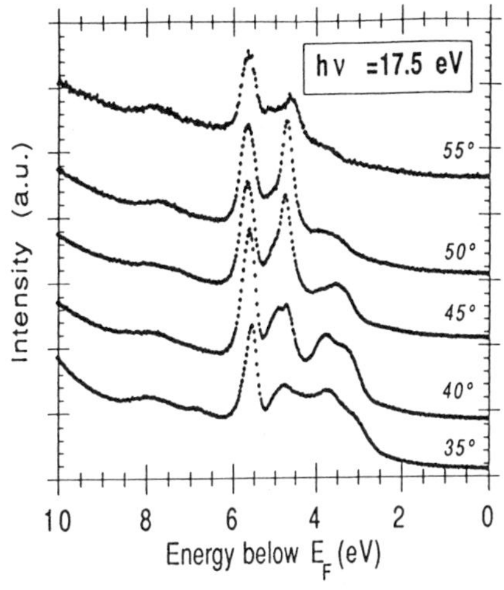

Fig.5.Photoemission spectra recorded with 17.5 eV photon energy for various angles of emission along the Γ-K-M direction of the 1x1 SBZ.

Among the observed features, two narrow bands at 4.5 and 5.8 eV below the fermi level, with linewith less than 350 meV, are clearly evidenced. They have both a flat dispersion.They correspond to surfaces states.In order to make a comparison with theory, an approach involving first a determination within the local density approximation (LDA) of the surface geometrical structure via total energy minimization was performed [34] .The computation of the surface states energies was then achieved using a quasiparticle self-energy approach [34].Figure 6 shows the quasiparticle surface states energies as compared to the angle-resolved photoemission data.The agreement between theory and experiment is excellent.The width of the surface states indicates that the lineshape of the valence spectra of hydrogen adsorbed on reconstructed silicon surfaces[28] , or that of ideally hydrogen terminated Si(111) obtained by removal of In adlayer by atomic hydrogen[29] (typically 500 meV), has a strong inhomogeneous contribution .

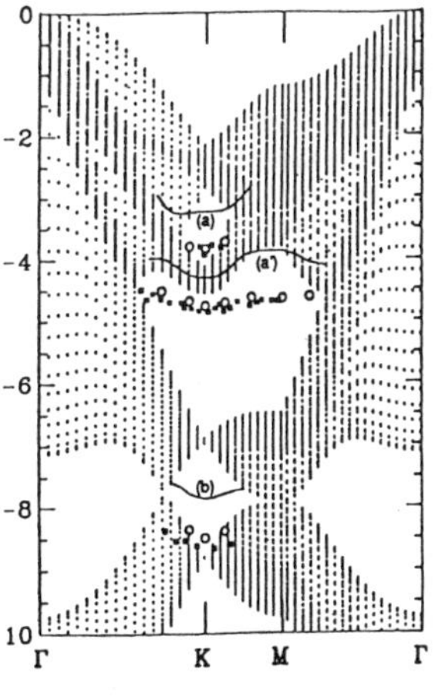

Fig.6.GW self-energy calculations (open circles) and experimental (filled squares)surface band structure for the H/Si(111)-(1x1) surfaces[34].The photons are shined onto the surface at 80° incidence angle, with polarisation along the surface plane.

5.2 .Core level region:

High quality core level spectra are obtained on the H/Si(111)-(1x1) surfaces and give precise information on the surface core level shifts .This is clearly demonstrated on Fig.7 showing Si2p core level spectra, recorded with photon energy ranging from 108 eV to 135 eV . A surface state appears around 0.12 eV higher binding energy when decreasing the escape depth of the photoelectrons. It is best seen at low temperatures in a spectrum taken at 130eV photon energy (Fig.8).In this spectrum, bulk and surface component are well separated. The surface state (hereafter labelled S2) is assigned to the outmost top silicon atom directly bonded to the hydrogen (top layer silicon).

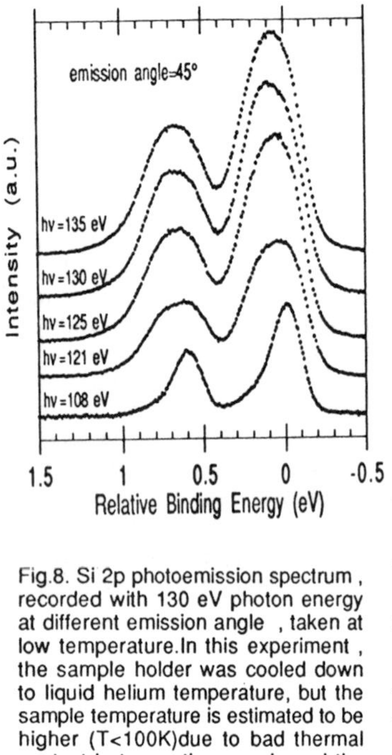

Fig.7.Si2p photoemission spectra from a H/Si(111)-(1x1) surface, at room temperature, with photon energies from 108 to 135 eV.The emission angle is 45°.

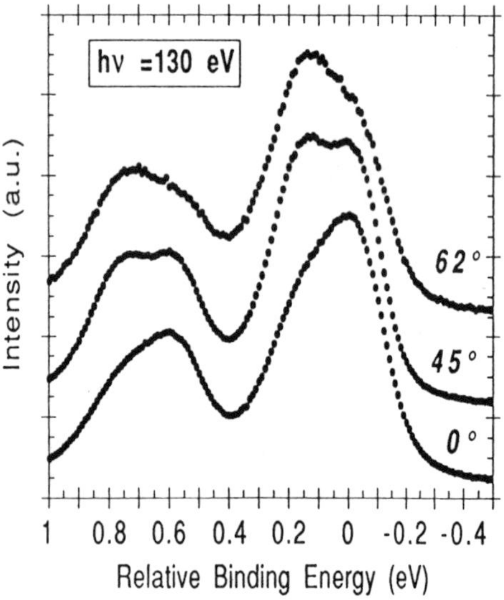

Fig.8. Si 2p photoemission spectrum , recorded with 130 eV photon energy at different emission angle , taken at low temperature.In this experiment , the sample holder was cooled down to liquid helium temperature, but the sample temperature is estimated to be higher (T<100K)due to bad thermal contact between the sample and the cryostat.

Substraction of spectra, taken at two different photon energy, after normalizing to the bulk component , reveals the presence of an additional surface state, located around 100 meV lower binding energy (Fig 9.).Such a state (labelled S1 in the following) is tentatively assigned to the second layer

Substraction of spectra, taken at two different photon energy, after normalizing to the bulk component , reveals the presence of an additional surface state, located around 100 meV lower binding energy (Fig 9.).Such a state (labelled S1 in the following) is tentatively assigned to the second layer silicon atoms.Considering an adsorbate -induced shift of 1.8 eV per electron[30] , an effective charge difference between the outmost top and the backbonded silicon atoms of 0.12eV is calculated.

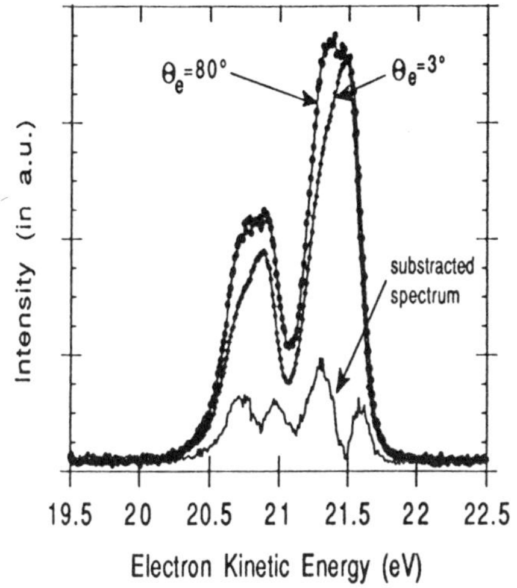

Fig.9.Si2p photoemission spectra, recorded with 125eV photon energy, at two different emission angles (normal and glazing angle).The dashed curve is the result of substraction of the 2 spectra, after normalizing to the bulk component.

The high electron kinetic energy part of the bulk sensitive $Si2p_{3/2}$ half band, taken at 108eV photon energy,is a direct measure of the total experimental half width of the bulk component . In the present case, this total width is 160 +/-10 meV . The contributions to the total width come from a Lorentzian lifetime broadening, a Gaussian phonon broadening and a Gaussian instrumental resolution(45 meV in the present case). Thus, it appears , without need of deconvolution, that both the photon and lifetime broadening are below the recent published value (160meV and 95+/- 25 cm-1 respectively[31]). Fitting the core level spectra taken at 108 eV photon energy with Voigt function lines gives values of 35+/-10 meV and 130+/-10 meV for the lifetime and phonon broadening of the bulk component respectively.

This lifetime broadening is the lowest value obtained on H chemisorbed on silicon surfaces, and it is close to that obtained for SiH_4 molecules (45 meV)[32] . It is lower than that extracted from the previous measurements of H-exposed silicon surfaces, and indicates that reconstruction in the near-surface region contributes significantly to the broadening of Si2p peaks, and consequently in the determination of the core hole lifetime by curve fitting .

UV-photoemission spectra of molecules in the gas phase usually exhibit a well-resolved multiple vibrational fine structure[33] .However, in many cases, the vibrational splitting remains unresolved when molecules or atoms are adsorbed on a solid surface, due to the general broadening of the lines. Vibrational components are present in case of core level spectroscopy of H/Si(111)-(1X1) surfaces. According to the high resolution electron energy loss spectra presented in the previous section,two Si-H motions are considered: a bending and a stretching modes, which frequency of 79.5 meV and 260 mev respectively in the electronic ground state.

Fig 10 shows a deconvoluted surface sensitive core level spectrum, , for 130eV photon energy.The curve fitting parameters are reported on Table 1.

Component	SCS (eV)	Gaussian width (eV)
Bulk	0	0.13 +/-0.01
S1	-0.1 +/-0.01	0.14 +/-0.01
S2	0.13 +/- 0.01	0.18 +/- 0.01
Bend	0.25 +/- 0.01	0.18 +/- 0.02
Stretch	0.42 +/- 0.01	0.18 +/- 0.01

TABLE 1: Curve fitting parameters for the components used to fit the Si 2p core-level spectra.The spin-orbit split, branching ratio, and Lorentzian width are 0.606, 0.51 and 0.035 eV respectively.
S1= backbonded atoms
S2= outmost top silicon atoms directly bonded to hydrogen.
Instrumental resolution at 130 eV photon energy=0.065 eV

The vibrational splittings of the bending (120 +/-20 meV)and stretching (300 +/- 20 meV) modes are increased by 50% and 10% respectively compared to their value in the electronic ground state. The increase of the Si-H stretching frequency mode is in good agreement with the result obtained for SiH4 gas [34], and corresponds to a shortening of the Si-H bond. The much larger increase of the bending mode is not understood at present.

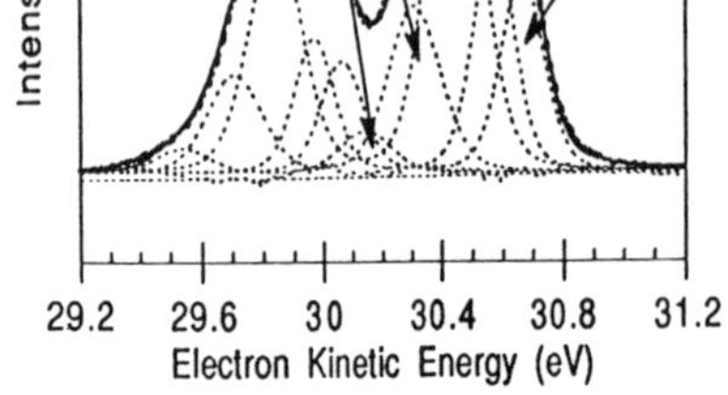

Fig.10.Curve fitting of a core-level spectra, with 130eV photon energy and emission angle of 45°.Circles are data points; the solid line is the fit to the data. The dashed lines are the individual peaks determined by the fit, using parameters reported on Table 1.The instrumental resolution is 65 meV.

6. Conclusions:

Despite their preparation procedure,ideally hydrogen terminated Si(111) surfaces, chemically elaborated in basic HF solutions, are exempt of impurity, and have long range perfect domains termination. After introduction in an UHV environment , such systems provide new physical understanding of vibrational and electronic properties at the H/Si surfaces.

High energy electron loss spectroscopy reveals surface and bulk phonons, and dispersion of all the observed modes are in very good agreement with previous theoretical calculation, using bong charge model. Due to the well defined structure of the adsorbate-substrate bond , impact scattering mechanism is clearly evidenced, and confirmed by the energy and off specular dependence of the loss spectra.

High resolution photoemission spectroscopy has allowed a complete description of the experimental valence band dispersion curve.The linewidth of the surface states , especially observed in the vicinity of the K point of the surface Brillouin zone ,is less than 350 meV. The agreement with theoretical calculations involving GW self-energy formalism is excellent.

Core-level peaks have a lifetime broadening (<45 meV), very close to the value of Si-H in the gas phase.Surface core-level peaks are easily observed when varying both the photon energy and the photoelectron emission angle.One is assigned to backbonds (100 +/- 10 mev towards lower binding energy) and the other to outmost top silicon atoms bonded to the hydrogen (150+/- 20 meV towards higher binding energy).Vibrational replica of the bending and the stretching Si-H modes are also resolved.

7. Acknowledgements:

It is a pleasure to acknowledge all my collaborators in this work, Y.J.Chabal,Y.Petroff,K.Hracovini,J.E.Bonnet,P.A.Thiry,P.Thiry, A.Talehb-Ibrahimi,G.Indelkhofer , R.Gunther, X.Blase,Xuejun Zhu and S.G.Louie.

Y.J.Chabal, my long time collaborator, and Y.Petroff deserves special thanks for the critical reading of this manuscript and for the helfull discussions.

8. References:

1.Y.J.Chabal,G.S.Higashi and S.B. Christman, *Phys.Rev.* B28(1983)4472; Y.J. Chabal, *Surf. Sci.* 168(1986)594.
2.T.Sakurai and H.D.Hagstrum , *Phys. Rev.* B12 (1975)6160.
H.Ibach and J.E.Rowe , *Surf. Sci.* 43 (1974) 481.
K.Fujiwara, *Phys.Rev.*B24 (1981) 2240.
3.U.Harten, J.P.Toennies, Ch.Woll,L.Miglio,P.Ruggerone,L.Colombo and G.Benedek, *Phys. Rev.*B38(1988)3305.
4. W.Kern , *Semicond.Int.*94(April 1984).
5.E.Yablonovitch, D.L.Allara, C.C.Chang, T.Gmitter and T.B.Bright, *Phys.Rev.Lett.* 57(1986)249.

6a.P.O.Hahn, *Mater.Res.Soc.Symp.Proc.*54(1986)645.
b.M.Grunder and R.Schultz *in "Deposition and growth limits for Microelectronics"*, edited by G.W.Rubloff and G.Lucovsky, AIP Conf.Proc.N°167 (1988) pp.329-337.
7.Y.J.Chabal, G.S.Higashi, S.B.Christman and V.A.Burrows *J.Vac.Sci.and Tech.A* 7 (1989) 2104.
8. G.S.Higashi, Y.J. Chabal, G.W.Trucks and K.Raghavachari, *Appl.Phys.Lett.* 56(1990)656.
9.P.Dumas, Y.J.Chabal and G.S.Higashi, *Phys.Rev.Lett* 65(1990)1124.
10. P.Guyot Sionnest, P.Dumas , Y.J.Chabal and G.S.Higashi , *Phys.Rev.Lett.* 64 (1990)2156 ; P.Guyot Sionnest, P.Dumas and Y.J.Chabal , *J.Elec.Spectr. Related Phenomena* 54/55 (1990)27.
11.R.S.Becker,G.S.Higashi,Y.J.Chabal and A.J.Becker, *Phys.Rev. Lett.* 65 (1990)1917.
12.P.Dumas and Y.J.Chabal , *Chem Phys Lett* 181 (1991)537.
13.R.B.Doak,Y.J.Chabal,G.S.Higashi and P.Dumas *J.Electron Spectr.Related Phenomena* 54/55 (1990)291.
14. S.Bouzidi,F.Coletti,J.M.Debever,P.A.Thiry,P.Dumas and Y.J. Chabal *Phys.Rev.B* 45(1992)1187.
15.P.Jakob,P.Dumas and Y.J.Chabal Appl.Phys.Lett.59(1991)2968.
16. P.O.Hahn and M.Henzler , *J.Appl.Phys.* 52 (1981)4122;*J.Vac.Sci.Technol.* A2(1984)574;J.M.Gibson,M.Y.Lanzerotti and V.Elser , *Appl.Phys.Lett.*55(1989)1394;A.Ogura, *J.Electrochem.Soc.*138(1991)807.
17. Y.J.Chabal *Surf. Sci. Reports* 8 (1988)211.
18. P.Jakob, Y.J.Chabal and K.Raghavachari *Chem Phys.Lett.* 187(1992)325.
19. P.Jakob,Y.J.Chabal,K.Raghavachari, P.Dumas and S.B.Christman to be published.
20. D.Graf,M.Grunder,R.Schulz and L.Muhlhoff *J.Appl.Phys.* 68 (1990)515 ; D.Graf,M.Grunder,R.Schulz *J.Vacuum Sci.Technol.* A7 (1989) 808 ; M.Grunder *J.Vacuum Sci .Technol.* A5 (1987) 2011.
21.C.Stuhlmann, G.Bogdanyi and H.Ibach *Phys.Rev B* 45 (1992)6786.
22.P.Dumas, Y.J. Chabal and P.Jakob *Appl.Surf.Sci.* (to be published)
23.P.Dumas , Y.J.Chabal and P.Jakob ,*Surf. Sci.*269/270(1992)867.
24.H.Froitzheim, H.Lammering and H.L.Gunter *Phys.Rev.B* 27(1983) 7429 ; H.Froitzheim, U.Kohler and H.Lammering *Surf.Sci.*149 (1985)537.
25. P.Dumas and Y.J.Chabal *J. Vac.Sci.Tech.* B (july 1992)
26. L.Miglio, P.Ruggerone, G.Benedek and L.Colombo, *Physica Scripta* 37(1988)768.
27.W.Ho, R.F.Willis and E.W.Plummer, *Phys.Rev.*B21(1980)4202.
J.P.Woods and J.L.Erskine, *Phys.Rev.Lett.* 55(1985)2595.
J.P.Woods,A.D.Kulkarni,J.L.erskine and F.W.de Wette , *Phys.Rev.*B36(1987)5848.
28.C.J.Karlsson,E.landemark,L.S.O.Johansson,U.O.Karlsson and R.I.G.Uhrberg *Phys.Rev.*B41(1990)1521.
K.Mortensen,D.M.Chen,P.J.Bedrossian,J.A.Golovchenko and F.Besenbacher, *Phys.Rev.*B43(1991)1816.
29.E.Landemark,C.J.Karlsson and R.I.G. Uhrberg, *Phys.Rev, Rapid Comm.*44(1991)1950.

30.F.J Himpsel, B.S.Meyerson, F.R.Mc Freely, J.F.Morar, A.Taleb-Ibrahimi and J.A.Yarmoff in .*"Photoemission and Absorption Spectroscopy of Solids and Interfaces with Synchrotron Radiation"*,"Enrico Fermi"Course CVIII, edited by M.Campagna and L.Rosei (North Holland,Amsterdam, to be published).
31.G.K.Wertheim,D.M.Riffe,J.E.Rowe and P.H.Citrin, *Phys.Rev.Lett.*67(1991)120.
32.J.D.Bozek,G.M.Bancroft,J.N.Cutler and K.H.Tan ,*Phys.Rev.Lett.*65(1990)2757.
33. D.W.Turner, C.Baker, A.D.Baker and C.R.Brundle, *Molecular Photoelectron Spectroscopy* (Wiley, London,1970).
34.K. Hracovini, R.Gunther, A.Talheb-Ibrahimi, G.Indlekhofer,P.Thiry, P.A.Thiry,Y.J.Chabal,P.Dumas,Y.Petroff,S.G.Louie ... submitted

EXCITED STATES IN ELECTRONIC STRUCTURE CALCULATIONS

STEVEN G. LOUIE

Department of Physics, University of California at Berkeley, and Materials Sciences Division, Lawrence Berkeley Laboratory, Berkeley, CA 94720 USA

ABSTRACT

A first-principles quasiparticle approach to the electronic excitation energies in crystals and at surfaces is described. The quasiparticle energies are calculated within the GW approximation for comparison with photoemission and other spectroscopic experiments. Applications of the method to bulk semiconductors and the Si(111)2×1, Ge(111)2×1, and H/Si(111) surfaces are presented. In both cases, significant self-energy corrections arising from many-electron effects to the excitation energies are found. Using atomic positions from total energy minimization, the calculated excitation energies explain quantitatively the experimental spectra. This approach thus provides an *ab initio* means for analyzing and predicting results from spectroscopic probes.

1. Introduction

Understanding the electron excitation spectra of condensed matter such as those measured in photoemission experiments requires detailed knowledge of the electronic structure. Electron-electron interaction effects, moreover, often lead to a significant renormalization of the properties of the electrons from those of the independent-particle picture. Nevertheless, experimental spectra from spectroscopic measurements may often be understood in terms of excitation between quasiparticle states of the interacting electron system. In this paper, we give a brief review of a method[1] for calculating the quasiparticle energies from first principles and discuss several selected applications.

The method is based on an expansion of the electron self-energy operator to first order in the dressed electron Green's function and the dynamically screened Coulomb interaction with local field effects included. This is the so-called GW approximation.[2] *Ab initio* calculations using this approach have successfully explained results from optical, photoemission, scanning tunneling, and other spectroscopic measurements for a variety of systems including bulk crystals, surfaces, interfaces, and small clusters.[3] Here we focus the discussion on the application to semiconductors and their surfaces. The calculated quasiparticle energies are in general accurate to within ~ 0.1 eV of experimental values. The predictive capability of the approach is of particular value in many studies. For example, it is found that self-energy corrections to surface-state energies and band dispersions are substantial and that electron-hole (excitonic) interactions may play a significant role in the optical response of some surfaces such as the Si(111)2×1 surface.

2. Self-energy Green's Function Approach to Quasiparticle Energies

It is the transitions between the quasiparticle states which are measured in most spectroscopic experiments.[2] Exchange and correlation effects can significantly modify the quasiparticle properties from those calculated using one-electron theories. The neglect of these many-electron effects in standard band structure calculations in the past has often given rise to rather severe discrepancies between theory and experiment. For example, although state-of-the-art local density functional (LDA) calculations yield very accurate structural properties, the LDA eigenvalues fail to give accurate band gaps and related excited-state properties.[1] Typically, the minimum gap in bulk semiconductors is underestimated by 30 - 100% in comparison to experiment leading to the so-called "band gap problem" in semiconductors.

The quasiparticle properties of an interacting electron system may be obtained using a Green's function approach. The electron or hole energies and wavefunctions are given by[2]

$$[T + V_{ext}(\mathbf{r}) + V_H(\mathbf{r})]\psi(\mathbf{r}) + \int dr'\, \Sigma(\mathbf{r},\mathbf{r}';E^{qp})\psi(\mathbf{r}') = E^{qp}\psi(\mathbf{r}) \ , \tag{1}$$

where T is the kinetic energy operator, V_{ext} the external potential due to the ions, V_H the average electrostatic Hartree potential, and Σ the electron self-energy operator respectively. The self-energy operator contains the effects of exchange and dynamical correlations. In general, it is nonlocal, energy-dependent, and nonHermitian with the imaginary part giving the lifetime of the quasiparticles.

In the GW approximation,[2] the self-energy operator is expanded to first order in the screened Coulomb interaction:

$$\Sigma(\mathbf{r},\mathbf{r}';E) = \frac{i}{2\pi}\int d\omega\, e^{-i\delta\omega}\, G(\mathbf{r},\mathbf{r}';E-\omega)W(\mathbf{r},\mathbf{r}';\omega) \ . \tag{2}$$

Here, G is the crystalline Green's function and W the dynamically screened Coulomb interaction. Our approach[1] is to make the best possible approximations for G and W, calculate Σ, and then obtain the quasiparticle energies. Inclusion of local fields in the screened Coulomb interaction is found to be crucial for obtaining accurate quasiparticle energies because the off-diagonal elements of the dielectric matrix in Fourier space distinguish the variations in screening properties at the different regions in the crystal.

Since the quasiparticle energies E_{qp} and wavefunctions ψ enter into G, the quasiparticle energies together with Σ and G must be obtained self-consistently. In practice, the electron Green's function is constructed initially using the LDA Kohn-Sham eigenfunctions and eigenvalues, and subsequently updated with the quasiparticle spectrum from Eq. (1). The dynamical response matrix used to construct the screened Coulomb interaction W is typically obtained in two steps. First, the static dielectric matrix is calculated as a ground-state property from the LDA. Second, the dielectric matrix is extended to finite frequencies using a generalized plasmon pole model employing exact dispersion and sum rule relations. To simplify further the calculation, the self energy is calculated only for the valence electrons. The exchange and correlation interactions between the valence and core electrons are treated via a pseudopotential formulation with these effects included at the LDA level. This last approximation has been shown to be very accurate in general. The only exceptions are some selected gaps in a few semiconductors with unusually shallow and polarizable core states. For these cases, realistic core-valence interactions may be successfully included using a core-polarization potential.[4]

The present quasiparticle approach has been applied to a range of crystals including semiconductors, ionic insulators, and the simple metals. Accurate band gaps, optical transition energies, and photoemission spectra have been obtained. These calculations have also been extended to various surfaces, interfaces, superlattices, and clusters[3] and to the study of pressure induced insulator-metal transitions.[5] A first-principles approach is of particular importance to the latter group of applications since these systems are often experimentally less well-characterized. The validity of the GW approximation to the d-band metals and other highly correlated electron systems however remains an open subject at this time.

3. Band Gaps and Photoemission Spectra of Bulk Semiconductors

Table 1. Comparison of calculated band gaps (in eV) with experiment. Core polarization and relaxation effects are included for Ge, AlAs, and GaAs.

	LDA	Present Theory	Expt.[a]
diamond	3.9	5.6	5.48
Si	0.5	1.16	1.17
Ge	-0.26	0.73	0.74
GaAs	0.12	1.42	1.52
AlAs	1.28	2.01	2.24
LiCl	6.0	9.1	9.4

[a]See Refs. 1 and 4.

A major success of the quasiparticle method described in Sec. 2 has been its application to the *ab initio* calculation of the band gaps and excitation energies in semiconductors and insulators.[1] The calculated band gaps of some selected crystals are presented in Table 1 together with the experimental values and the LDA Kohn-Sham gaps. In general, the LDA gaps significantly underestimate the experimental values. Hartree-Fock calculations, on the other hand, usually overestimate the gaps by several folds because of neglect of electron correlations. As seen in Table 1, with the excitation energies properly interpreted as transition between quasiparticle states, the calculated quasiparticle gaps are in excellent agreement with experiment. The present results were obtained with only input being the atomic number of the constituent elements and the crystal lattice parameters. The use of a self-consistent crystal Green function and the inclusion of local fields (the full dielectric matrix) and dynamical screening effects are all important factors in describing accurately the self-energy effects.

As mentioned above, core-valence interactions can sometimes affect the value of certain gaps in materials with very shallow and highly polarizable core states. In Table 1, the values for the minimum band gap of Ge, GaAs, and AlAs were calculated with these effects included through a core-polarization potential.[4] The largest change is in the direct gap of GaAs which amounts to ~ 0.4 eV. The effects of core polarization on the gaps of the other materials listed in Table 1 are however typically about 0.1 eV or less.

In addition to the band gaps, the quasiparticle approach yields excellent optical transition energies and band dispersions. In general, the calculated transition energies (neglecting excitonic effects) are within about 0.1 - 0.2 eV of the observed spectral features in optical measurements. This agreement is comparable to empirical methods in which several parameters are used to fit to the experimental spectra. A comparison of the calculated quasiparticle band structure of Ge with results from angle-resolved direct[6] and inverse[7] photoemission measurements is depicted in Fig. 1. In general, the agreement between theory and experiment is well within the experimental and theoretical error bars. The calculated unoccupied conduction band states were in fact predictions of the theory. Recently, Ortega and Himpsel[7] have performed detailed direct and inverse photoemission studies and obtained similarly good agreement with our predicted quasiparticle band structures for a number of other materials.

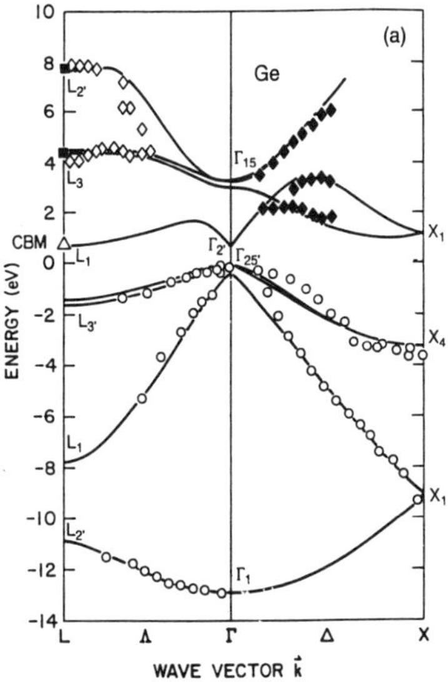

Fig. 1. Calculated quasiparticle energy bands of Ge compared to data from direct (Ref. 6) and inverse (Ref. 7) photoemission experiments.

Si (111) 2 x 1 Surface

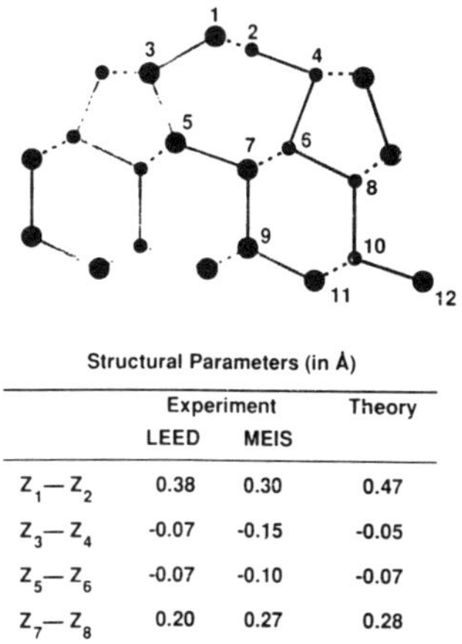

Structural Parameters (in Å)

	Experiment LEED	Experiment MEIS	Theory
$Z_1 - Z_2$	0.38	0.30	0.47
$Z_3 - Z_4$	-0.07	-0.15	-0.05
$Z_5 - Z_6$	-0.07	-0.10	-0.07
$Z_7 - Z_8$	0.20	0.27	0.28
$Z_9 - Z_{10}$	0.13	0.14	0.15

Fig. 2. Structure of the Si(111) 2×1 surface. Experimental data are from Ref. 18 (LEED) and Ref. 19 (MEIS).

4. Semiconductor Surfaces

In this section we describe the extension of the first-principles quasiparticle approach to semiconductor surfaces. We use, as examples, the (111)2×1 surface of Si and Ge and the H/Si(111) surface to illustrate some recent efforts[8-10] in this area. To compare with experimental excitation spectra, we calculate also the quasiparticle energies associated with the surface states. The quasiparticle studies complement structural studies: Atomic coordinates obtained in total energy calculations are used as input for the calculation of the excitation energies.

The calculations are carried out using a repeated slab geometry. Typically, about a dozen layers of the crystal are used with a vacuum region equivalent to several layers separating the slabs. The surface structure is determined by minimizing the total energy in a self-consistent LDA *ab initio* pseudopotential calculation.

4.1. The Si(111) and Ge(111)2×1 Surfaces

The Si(111)2×1 surface is one of the most studied semiconductor surfaces both theoretically and experimentally. The reconstructed structure has an interesting π-bonded chain of surface atoms.[11] Its electronic structure is also quite intriguing. The surface-state band gap measured from optical processes[12-14] appears to be significantly different from that measured in photoemission[15,16] and tunneling experiments[17] indicating that electron-hole (excitonic) interactions may be considerably larger at this surface than in the bulk.

Figure 2 shows the calculated minimal energy structure[8] of Si(111)2×1 surface as compared with structural parameters determined from LEED[18] and medium energy inelastic scattering (MEIS)[19] measurements. z_i is the coordinate of the ith atom along the surface normal. The positions of the surface atoms agree very well with experiments[18-20] and with previous theoretical calculations.[21] Another interesting feature of this surface is that there is a rather large buckling of the surface chain atoms.

The surface electronic structure of Si(111)2×1 is dominated by a pair of π and π* bands of surface states.[8] These surface bands arise from the dangling bonds on the 3-fold coordinated surface atoms and are dispersive only along the chain direction resulting in one-dimensional-like bands. Figure 3 compares the calculated quasiparticle surface-state energies with angle-resolved direct[15] and inverse[16] photoemission data. Both the position and the dispersion of the surface states are given accurately by the quasiparticle theory. The overall agreement between theory and photoemission experiment is better for the occupied surface states than for the empty states which have larger experimental error bars. A subsequent analysis of the effects of the limited resolution in the inverse photoemission experiment[16] indicates that the true energy of the π* state at $\bar{J}$ could be lower than that indicated in Fig. 3 by as much as 0.15 eV yielding a surface band gap in good agreement with the theoretical value of 0.62 eV. (The LDA surface-state gap is only 0.27 eV for this surface.) Scanning tunneling spectroscopy[17] also gives an direct surface energy gap of 0.6 eV in agreement with our theoretical prediction. However, the measured onset energy[12-14] for electron-hole creation is significant smaller than the calculated quasiparticle surface-state gap. (See Fig. 4.) At low temperature, the optical gap is measured[13] to be only 0.47 eV in significant disagreement with theory and with photoemission and tunneling results. Also observed is that there is only one asymmetric peak in the experimental optical absorption spectrum which is quite different from a two-peak structure predicted by the joint quasiparticle density of states of the π and π^* bands.

The situation with the Ge(111)2×1 surface is very similar to that of the Si(111)2×1 surface. The calculated quasiparticle surface-state energies[9] show good agreement with results from direct and inverse angle-resolved photoemission experiments. For the surface-state band gap, a value of 0.65 ± 0.2 eV is inferred from combined direct[22] and inverse[23] photoemission results, and a value of 0.65 eV is obtained from scanning tunneling spectroscopy.[17] These values are in accord with our theoretical value of 0.67 eV. On the other hand, optical measurements from undoped samples[24,25] and direct photoemission from heavily n-doped (to occupy the surface conduction states) samples[26] appear to give a band gap around 0.5 - 0.55 eV. This difference between the surface band gap measured by optical processes and that by single-particle processes is very similar to the situation in Si(111)2×1. There is an added complexity in the case of the Ge(111)2×1 surface. The energy minimum corresponding to the buckled chain geometry is much more shallow than that of the Si case. Since the surface-state band gap is quite sensitive to the amount of buckling,[9,27] any theoretical uncertainty in this structural parameter would lead to a change in the gap value.

The observed discrepancies in the quasiparticle and the optical gap indicate that there may be enhanced excitonic effects[8,28] of order of 0.1 eV in the surface optical data. This possibility has been suggested because of the quasi-1D dispersion

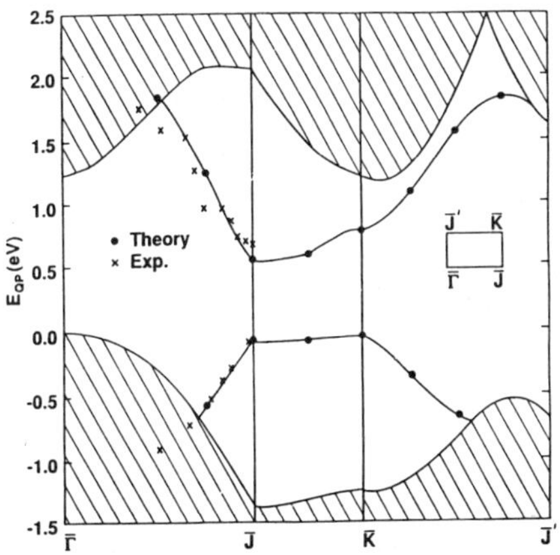

Fig. 3. Quasiparticle surface-state bands of Si(111)2×1 compared to photoemission (Ref. 15) and inverse photoemission (Ref. 16) data.

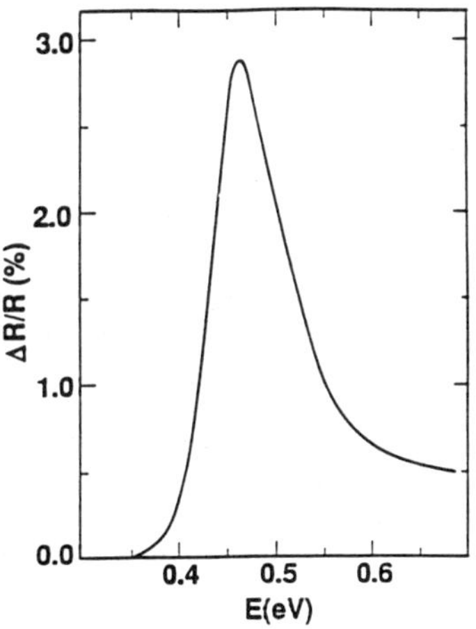

Fig. 4. Experimental differential reflectivity spectrum for the Si(111)2×1 surface (after Ref. 13).

and reduced screening of the surface states of this particular surface. We performed a model calculation to estimate the effects of the electron-hole interaction. The exciton energy spectrum and wavefunctions were obtained within a model which employs the calculated surface-state quasiparticle bands and a model statically screened electron-hole interaction.[8,29]

Figure 5 depicts the calculated $\alpha(E)$ for the Si(111)2×1 surface for light polarized along the chains. Curve (b) corresponds to a complete neglect of the electron-hole interaction. In this case, $\alpha(E)$ exhibits two peaks corresponding to the critical points in the joint density of states. Curve (a) is that from an excitonic spectrum for which the lowest energy exciton had a binding energy of 0.13 eV obtained with a electron-hole Coulomb interaction screened by a dielectric constant taken to be 6.5. The absorption spectrum is dominated by the ground-state exciton, and the spectrum exhibits a single asymmetric peak. The large shift in oscillator strength as the electron-hole interaction is included is a result of both the reduced screening and the one-dimensional nature of the electronic structure near the edge ($\bar{J}\,\bar{K}$) of the surface Brillouin zone. The experimental differential reflectivity, which is closely related to $\alpha(E)$, has been measured by Ciccacci *et al.*[13] and is shown in Fig. 4. It is clear that the absorption spectrum obtained by neglecting the electron-hole interaction cannot explain the data. Spectrum (a) in Fig. 5 which corresponds to an exciton binding energy of 0.13 eV, on the other hand, reproduces quite well the experimental value.

It should be emphasized that the calculated quasiparticle energies correspond to the energy needed to create an isolated electron or hole in the solid. Thus, it is consistent that these energies describe well the photoemission and tunneling data but not necessarily the optical data unless electron-hole interaction effects are small.

4.2. The H/Si(111)1x1 Surface

The recent development of a chemical method[30] for preparing ideally hydrogen terminated Si(111) surfaces which are highly stable and easily transportable has renewed great interest in this system. Because of their structural perfection over large areas, remarkably sharp features in the valence and core level photoemission spectra have been obtained.[10] The surface states on this surface have been examined in several calculations.[31,32] However, discrepancies in their energies and dispersions as large as 1 eV were found between previous theories and experiment. We find that a quantitative understanding of the electronic structure requires a full quasiparticle calculation even for this simple chemisorption system.

From total energy minimization, the Si-H bond length is found to be 2.87 a.u. with the first silicon layer slightly relaxed inward while the relaxations of the deeper layers are negligible. Figure 6 depicts the calculated quasiparticle surface-state bands as compared to those from LDA calculation as well as the measured surface-state bands from photoemission.[10] Well-defined surface states are found in each local gap of the projected band structure near K and M. Comparison of the LDA results to the quasiparticle energies reveal two important features. First, the self-energy corrections to the LDA surface-state energies are exceptionally large: the 0.76 eV correction for the level (a') at M is larger by a factor 2 to 3 as compared to typical previous self-energy corrections on occupied surface states.[8,9] This is related to the

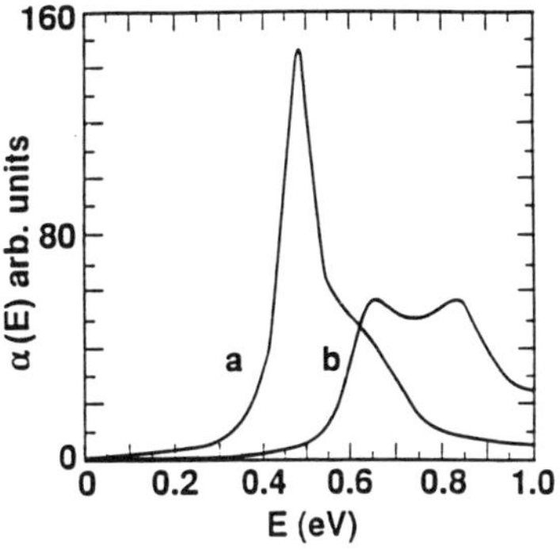

Fig. 5. Calculated surface-state optical absorption spectra obtained (a) with and (b) without excitonic correlations.

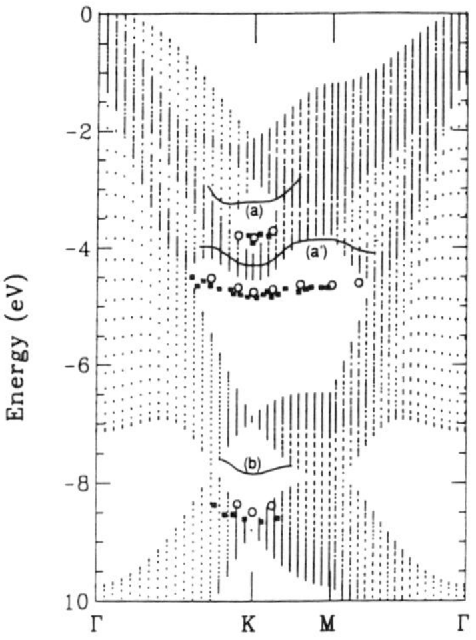

Fig. 6. Calculated surface-state bands of H/Si(111): LDA (full lines) and GW (open circles). The photoemission data (black dots) are from Ref. 10.

very localized hydrogen 1s orbital forming the surface states in the present system. Second, the self-energy effects give rise to a sizable correction to the band dispersion of the surface states. In particular, the LDA surface state band (a′) shows a 0.42 eV dispersion going from M to K which does not appear in the experimental data. This discrepancy is completely removed in the quasiparticle self-energy formalism. This again may be understood from the sensitivity of the nonlocal self-energy operator to the localization of the electron wavefunction. The surface state (a′) is much more localized at M than at K.

Figure 6 shows that the agreement between the present theory and experiment is excellent, including the location of the pockets in the projected bulk density of states. The self-energy approach yields an impressive improvement in the energy location and dispersion of all the surface states. Even the state (a′) at K, which in the LDA treatment is a surface resonance, is successfully extracted by the self-energy operator from the continuum of bulk states.

5. Summary

We have described in this brief review a first-principles method for calculating the quasiparticle energies in solids. Several recent studies are presented to illustrate the applicability of this approach to the excited-state properties of semiconductors and their surfaces. Specific examples discussed include the bulk crystals and the Si(111)2×1, Ge(111)2×1, and the H/Si(111) surfaces. In general, agreement to within

~ 0.1 eV is achieved between the calculated results and data from optical, photoemission, scanning tunneling, and other spectroscopic measurements. Self-energy corrections to standard self-consistent field theories such as the LDA or Hartree-Fock methods are shown to be quite substantial for both bulk and surface systems. It is also found that electron-hole (excitonic) interactions play an important role in the optical response of some surfaces such as the Si(111)2×1 surface.

Acknowledgements

This work was supported by National Science Foundation Grant No. DMR91-20269 and by the Director, Office of Energy Research, Office of Basic Energy Sciences, Materials Sciences Division of the U.S. Department of Energy under Contract No. DE-AC03-76SF00098.

References

1. M. S. Hybertsen and S. G. Louie, Phys. Rev. Lett. **55** (1985) 1418; Phys. Rev. B **34** (1986) 5390.
2. L. Hedin, Phys. Rev. **139** (1965) A796; L. Hedin and S. Lundqvist, Solid State Phys. **23** (1969) 1.
3. S. G. Louie, in *Electron Materials: A New Era of Materials Sciences,* eds. J. R. Chelikowsky and A. Franciosi (Springer-Verlag, Berlin, 1991), p. 23.
4. E. L. Shirley, X.-J. Zhu, and S. G. Louie, Phys. Rev. Lett. **69** (1992) 2955 .
5. H. Chacham, X.-J. Zhu, and S. G. Louie, Europhys. Lett. **14** (1991) 65; Phys. Rev. B **46** (1992) 6688.
6. A. L. Wachs *et al.,* Phys. Rev. B **32** (1985) 2326.
7. D. Straub, L. Ley, and F. J. Himpsel, Phys. Rev. B **33** (1986) 2607; J. E. Ortega and F. J. Himpsel, to be published.
8. J. E. Northrup, M. S. Hybertsen, and S. G. Louie, Phys. Rev. Lett. **66** (1991) 500.
9. X.-J. Zhu and S. G. Louie, Phys. Rev. B **43** (1991) 12146.
10. K. Hricovini, R. Gunther, P. Thiry, A. Taleb-Ibrahimi, G. Indlekofer, J. E. Bonnet, P. Dumas, Y. Petroff, X. Blase, X.-J. Zhu, S. G. Louie, Y. J. Chabal, and P. A. Thiry, to be published.
11. K. C. Pandey, Phys. Rev. Lett. **47** (1981) 1913; **49** (1982) 223.
12. M. A. Olmstead and N. M. Amer, Phys. Rev. Lett. **52** (1984) 1148.
13. F. Ciccacci, S. Selci, G. Chiarotti, and P. Chiaradia, Phys. Rev. Lett. **56** (1986) 2411.
14. J. Bokor, R. Storz, R. R. Freeman, and P. H. Bucksbaum, Phys. Rev. Lett. **57** (1986) 881.
15. R. I. G. Uhrberg, G. V. Hansson, J. M. Nicholls, and S. A. Flodstrom, Phys. Rev. Lett. **48** (1982) 1031; F. J. Himpsel, P. Heimann, and D. Eastman, Phys. Rev. B **24** (1981) 2003; F. Houzay, G. M. Guichar, R. Pinchaux, and Y. Petroff, J. Vac. Sci. Technol. **18** (1981) 860.

16. P. Perfetti, J. M. Nicholls, and B. Reihl, Phys. Rev. B **36** (1987) 6160; A. Cricenti, S. Selci, K. O. Magnusson, and B. Reihl, Phys. Rev. B **41** (1990) 12908.
17. R. M. Feenstra, Surf. Science **251**, (1992) 401.
18. F. J. Himpsel, P. M. Marcus, R. M. Tromp, I. P. Batra, M. Cook, F. Jona, and H. Liu, Phys. Rev. B **30** (1984) 2257.
19. R. M. Tromp, L. Smit, and J. F. van der Veen, Phys. Rev. Lett. **51** (1983) 1672.
20. H. Sakama, A. Kawazu, and K. Ueda, Phys. Rev. B **34** (1986) 1367.
21. J. E. Northrup and M. L. Cohen, Phys. Rev. Lett. **49** (1982) 1349; J. Vac. Sci. Technol. **21** (1982) 333; Phys. Rev. B **27** (1983) 6553.
22. J. M. Nicholls, G. V. Hansson, U. O. Karlsson, R. I. G. Uhrberg, R. Engelhardt, K. Seki, S. A. Flodström, and E. E. Koch, Phys. Rev. Lett. **52** (1984) 1555; J. M. Nicholls, G. V. Hansson, R. I. G. Uhrberg, and S. A. Flodström, Phys. Rev. B **27** (1983) 2594; F. Solal, G. Jezaquel, A. Barski, P. Steiner, R. Pinchaux, and Y. Petroff, Phys. Rev. Lett. **52** (1984) 360.
23. J. M. Nicholls and B. Reihl, Surf. Sci. **218** (1989) 237.
24. M. A. Olmstead and N. M. Amer, Phys. Rev. B **29** (1984) 7048; M. A. Olmstead, Surf. Sci. Rep. **6** (1986) 159.
25. S. Selci, A. Cricenti, F. Ciccacci, A. C. Felici, C. Goletti, Y. Zhu, and G. Chiarotti, Surf. Sci. **189/190** (1987) 1023; S. Nannarone, P. Chiaradia, F. Ciccacci, R. Memeo, P. Sassaroli, S. Selci, and G. Chairarotti, Solid State Comm. **33** (1980) 593.
26. J. M. Nicholls, P. Martensson, and G. V. Hansson, Phys. Rev. Lett. **54** (1985) 2363.
27. N. Takeuchi, A. Selloni, A. Shkrebtii, and E. Tosatti, Phys. Rev. B **44** (1991) 13611; E. Artacho, X.-J. Zhu, and S. G. Louie, to be published.
28. R. Del Sole and A. Selloni, Phys. Rev. B **30** (1984) 883.
29. F. Bassani and G. Pastori Parravicini in: *Electronic States and Optical Properties in Solids* (Pergamon, New York, 1975).
30. G. S. Higashi, Y. J. Chabal, G. W. Trucks, and K. Raghavachari, Appl. Phys. Lett. **56** (1990) 656 ; G. S. Higashi, R. S. Becker, Y. J. Chabal, and A. J. Becker, Appl. Phys. Lett. **58** (1991) 1656.
31. J. A. Appelbaum and D. R. Hamann, Phys. Rev. Lett. **34** (1975) 806; K. C. Pandey, T. Sakurai, and H. D. Hagstrum, Phys. Rev. Lett. **35** (1975) 1728; K. M. Ho, M. L. Cohen, and M. Schlüter, Phys. Rev. B **15** (1977) 3888.
32. M. B. Nardelli, F. Finocchi, M. Palummo, R. Di Felice, C. M. Bertone, F. Bernardini, and S. Ossicini, Surf. Sci. **269/270** (1992) 879.

VII. MAGNETIC DICHROISM

MAGNETIC CIRCULAR DICHROISM IN ANGLE-RESOLVED PHOTOEMISSION FROM FERROMAGNETIC SURFACES IN THE X-RAY AND VUV REGIME

C.M. Schneider and J. Kirschner
Max-Planck-Institut für Mikrostrukturphysik, Weinberg 2, 0-4050 Halle/Saale, Germany

D. Venus
Physics Department and Institute for Materials Research, McMaster University, Hamilton, Canada, L8S 4M1

ABSTRACT

Angle-resolved photoemission experiments from ferromagnetic surfaces using circularly polarized x-ray and VUV radiation show pronounced magneto-dichroic effects. This "magnetic circular dichroism" (MCD) appears in photoemission from valence bands and core levels. It is attributed to the interplay of spin-orbit coupling and exchange interaction in the occupied electronic states. The effect of MCD in photoemission has potential applications as a new analytical tool in investigations of surface magnetism.

1. Introduction

The interaction of polarized light with magnetic matter has long been of considerable interest in materials science. Classical examples are the magneto-optical Faraday and Kerr effects in the visible range of light, which are widely used nowadays to study the magnetic properties of solids and thin films. These effects are specific forms of the more general phenomenon of "magnetic dichroism", which refers to a situation where the absorption of light in a magnetic sample depends on both the state of light polarization and the magnetization. Magnetic circular dichroism (MCD), in particular, corresponds to a different absorption of left and right hand circularly polarized light in a magnetic material.

Studies of magnetic dichroism and particularly MCD have long been restricted to the range of visible light. It was only through the development of dedicated synchrotron radiation facilities that highly intense and almost continuously tunable light sources, reaching far into the UV and X-ray regime became available. Elliptically polarized light with a high degree of circularity may be obtained in the simplest case by using the off-plane radiation[1] of a synchrotron source. The access to circularly polarized VUV and X-ray radiation recently stimulated novel experiments on magnetic materials, which led to the discovery of magneto-dichroic effects in the VUV and X-ray regime.

Magnetic circular dichroism may be observed in a number of different experimental configurations From the underlying physical mechanisms these fall into two main classes. On the one hand, there are various types of photoabsorption experiments, in which the transmitted light or a secondary signal (e.g. total yield) are measured[2,3]. MCD in this case means that the photoabsorption cross section depends on the polarization of the incident light and the sample magnetization. These techniques probe the spin character of the unoccupied electronic states close to the Fermi level and represent the high energy counterpart of classical magneto-optics. On the other hand, the emitted photoelectrons can be analyzed with energy and angular resolution. These photoemission experiments probe the spin character of the occupied electronic states and yield information complementary to the photoabsorption method. It turns out that the intensity spectra in these photoemission experiments depend on the relative orientation of the sample magnetization, photon spin and the crystalline axes[4,5]. By analogy to the optical phenomenon this effect is termed as magnetic circular dichroism in photoelectron spectroscopy (MCDUPS, MCDXPS). It should, however, not be confused with the so-called circular dichroism in the angular distribution of photoelectrons (CDAD) which is similar in appearance, but has a completely different origin[6]. In the following, we will concentrate on magnetic circular dichroism in angle-resolved photoemission from solid surfaces.

2. Spinpolarization of Photoelectrons and the Physical Origin of MCD

Magnetic circular dichroism in photoemission appears in a conventional intensity measurement as the difference of two intensity spectra, recorded at opposite light helicities or sample magnetizations. The quantitative measure of this difference is the intensity asymmetry distribution A(E)

$$A(E) = \frac{I_+(E) - I_-(E)}{I_+(E) + I_-(E)} \tag{1}$$

where $I_+(E)$ and $I_-(E)$ denote the intensity spectra taken at positive and negative helicity, respectively. Although MCD is observed in intensity measurements, it occurs due to the electron spin and spin-dependent mechanisms in the photoemission process. MCDUPS and MCDXPS can be interpreted basically as the excitation of photoelectrons from a spin-polarized electron system in the presence of spin-orbit coupling. This situation may be rigorously treated only within a fully relativistic one-step photoemission theory. A qualitative understanding, however, can be developed in the framework of the three-step model of photoemission by combining aspects of spin-polarized photoemission from ferromagnets and nonmagnetic materials. Extensive work in both fields has been carried out in the past, and the relevant key aspects may be summarized in the following for the example of band states.

2.1 Spinpolarized Photoelectrons from Ferromagnets

The initial state involved in the photoemission process in ferromagnets is spin-polarized. Other than in non-magnetic materials all electronic states are split due to the influence of exchange interaction, forming energy bands of opposite spin character. The spin quantization axis is defined by the remanent sample magnetization $\underline{M}$ within a ferromagnetic domain. The spin of the electrons in a majority (minority) band is oriented antiparallel (parallel) to the direction of M. The energetic separation between the spin-up and spin-down bands, the so-called exchange-splitting, generally depends on E and k.

When spin-orbit coupling is neglected, spatial and spin parts of the electronic wave functions are decoupled, and each set of bands can be treated independently. The bands are characterized by their spatial (single group) symmetry. The electronic excitations due to the incident light may be described by non-relativistic dipole selection rules[7] within the spin-up or spin-down set of bands. These considerations have been successfully applied to the interpretation of photoemisson data from ferromagnets. Majority and minority electrons excited from bands of a given spatial symmetry carry the information of the spin character of the initial state. Provided a sufficient energy resolution is available, these transitions give rise to individual spectral features in spin-integrated photoemission. A subsequent spin-analysis will reveal their opposite spin character. These principles form the basis of a large number of valence band structure analyses of 3-d and 4-f ferromagnetic surfaces and thin films[8]. Similar formalisms have been successfully employed in spin-resolved core level photoemission from ferromagnets[9].

It is important to note that for a given geometry, excitation with unpolarized, linearly or circularly polarized light yields essentially the same intensity and spin polarization spectra. A magnetization reversal leaves the energy distribution curves unaffected, but reverses the spin direction in the spectral features and thus the spin polarization spectrum. As a consequence, the idealized case of a ferromagnet without spin-orbit coupling cannot give rise to magneto-dichroic effects.

2.2 Spinpolarized Photoelectrons from Nonmagnetic Materials

In contrast to ferromagnets each energy band in a non-magnetic solid contains electrons of both spin states. The emission of spin-polarized photoelectrons from such materials requires a non-zero spin-orbit interaction in the initial or final state, or in both. Compared to the non-relativistic case spin-orbit coupling changes the electronic states in several respects. First, bands of different non-relativistic symmetry may belong to the same relativistic (double group) symmetry character, which gives rise to additional hybridizations. Second, certain degeneracies in the band structure are lifted, resulting in a splitting of the corresponding electronic states. The wavefunctions of a given band state have different spatial parts for the spin-up and spin-down electrons. If we limit the discussion to bulk electronic states, only excitation with circularly polarized light will yield different transition matrix elements for the two spatial wave functions and thus a preferential population of one spin character in the final state. In the most symmetric experimental set-up (normally incident light and emitted photoelectrons), the spin quantization axis is defined by the photon spin (or equivalently the direction of incidence), which explains why this process is called "optical orientation". The spin-polarization in the final state may be qualitatively described by relativistic selection rules[10].

A reversal of the helicity of the incident light reverses the sign of the spin polarization, but leaves the intensity spectrum essentially unchanged. This also explains, why the spin-polarization disappears with unpolarized or linearly polarized light, which may be regarded as a superposition of right and left circularly polarized components. In particular situations, however, where the breaking of space inversion symmetry by the surface becomes important, excitation even with unpolarized light will produce spin-polarized photoelectrons[11]. These cases, which require a treatment within a one-step photoemission theory will be omitted in the following. Finally, the phenomenon of optical orientation is not limited to band states, but has its analogue in photoemission from localized atomic states and core levels where it is well-known as Fano effect[12].

Experimental results have been reported for a variety of metals, ranging from high-Z materials such as Pt down to Cu. Surprisingly, large photoelectron spin-polarization due to optical orientation has been found even in Cu, where relativistic effects are usually believed to be negligible[13]. These data lead to the important conclusion that spin-orbit coupling must have a significant influence on the valence electronic states of 3d ferromagnets. The picture developed in section 2.1 is thus oversimplified and applies only under certain conditions.

2.3 Magnetic Circular Dichroism in Photoemission from Ferromagnets

Due to spin-orbit coupling electronic states of a certain single-group symmetry split, others may hybridize. This process may even mix bands of opposite spin character in a ferromagnet[14]. Furthermore, the coupling of the spin to the spatial part of the wavefunction leads to spin dependent transition matrix elements. By analogy to optical orientation in non-magnetic materials, a similar effect should occur in ferromagnets by excitation with circularly polarized light.

The situation is more complex, however, as the electronic states are already spin-polarized due to the magnetic exchange field. This results in two competing spin quantization axes: the sample magnetization $\underline{M}$ aligns the spins along a certain spatial direction, and the photon spin (or equivalently, helicity of the circularly polarized light) defines a preferred orbital orientation through the helicity dependence of the transition matrix elements. Spin-orbit coupling links these two independent orientations. MCD arises from the difference in the photoexcitation probabilities when these two mechanisms prefer the same or different spin orientations along a common quantization axis. As states of opposite spin character are split in energy (in contrast to relativistic paramagnets), not only the spin polarization spectrum (as in the case of relativistic paramagnets) but also the shape and structure of the photoemission intensity spectrum depends on the relative orientation of these competing quantization axes, i.e., the relative spatial orientation of the sample magnetization and the light helicity. These differences in the energy distribution curves give rise to a non-zero intensity asymmetry A, when comparing spectra recorded at opposite helicities but otherwise identical experimental parameters. Since in a ferromagnet the interplay of spin-orbit coupling and exchange interaction is present in both valence bands and core levels, magneto-dichroic effects must be expected in photoemission from both types of electronic states.

3. Experimental Results

3.1 MCD in Core Level Photoemission - Localized electronic levels

Early investigations of the 2p and 3p core levels in 3d ferromagnets revealed unusually large line widths[15]. As one of the possible origins an experimentally unresolved exchange splitting of the core lines was discussed. In order to check this hypothesis, the first experiments on MCD in x-ray photoemission have been performed on the 2p levels of iron[4]. The results obtained from a Fe(110) surface are reproduced in fig. 1a. This experiment was carried out at a grazing incidence geometry, in order to have a large component of the photon spin along the sample magnetization M. This geometry has been found favorable in photoabsorption experiments. The upper panel shows the total intensity curve I in the region of the spin-orbit split $2p_{1/2}$ and $2p_{3/2}$ lines of iron, averaged over the two magnetization states of the sample. The inset displays, on enlarged scales, the peaks of the partial intensities I_+ and I_- for the two magnetization states separately. Obviously the peaks in I_+ and I_- appear at different

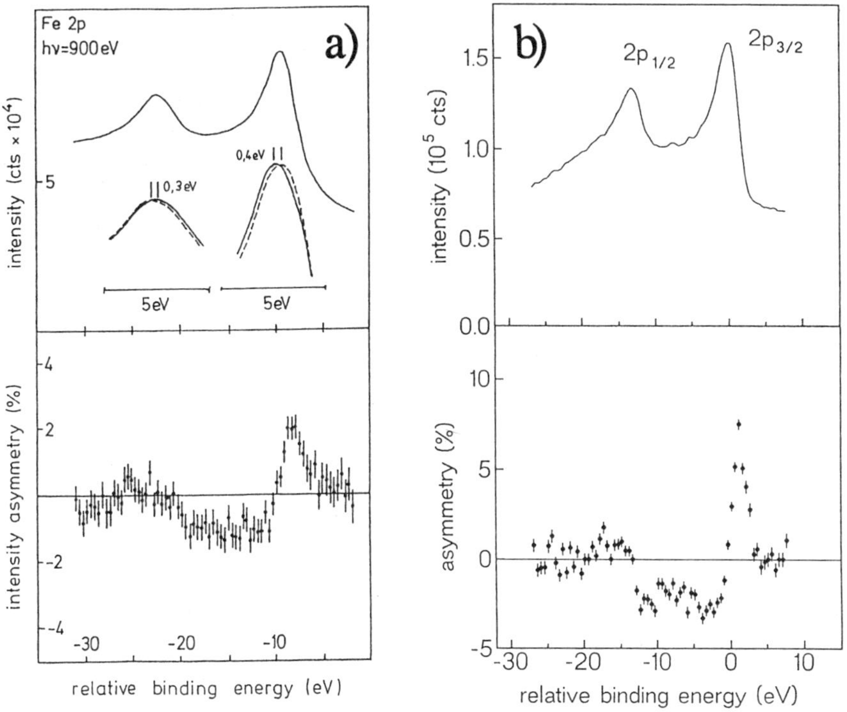

Fig. 1: MCD signal from the 2p levels of iron obtained at grazing incidence (a) and normal incidence (b). The sample was Fe(110) with the emission direction being close to [111].

binding energies, giving rise to a non-zero magneto-dichroic signal A(E) (lower panel), which is calculated according to eq. 1. A reversal of the photon helicity reverses the asymmetry curve. This finding can be interpreted by assuming both the $2p_{1/2}$ and $2p_{3/2}$ states to split into spin-polarized sublevels due to exchange interaction. This sublevels are characterized by definite azimuthal quantum numbers m_J and energy eigenvalues $E(m_J)$, and have different excitation probabilities for left- and right-hand circularly polarized light. The energy splittings are 0.3 ± 0.2 and 0.4 ± 0.2 eV for the $2p_{1/2}$ and $2p_{3/2}$ lines, respectively, and thus much smaller than the spin-orbit splitting (~10 eV). In the case of the 2p levels the exchange interaction can be treated as perturbation, which allows a qualitative interpretation of the MCD signal[4]. The qualitative interpretation has been substantiated by fully relativistic photocurrent calculations based on absorption models[16]. These nicely reproduced the binding energy dependence of the experimental asymmetry distribution, but predicted absolute values of A(E) a factor of 2-3 larger than found experimentally. The above calculation treated the core exchange splitting as a ground state property, whereas other approaches attribute it to the coupling of the core hole to the polarized valence electrons[17].

At present the calculations on MCDXPS are still based on absorption models. This is a serious drawback, since the angular resolution in a photoemission experiment, and thus a more or less defined electron emission direction, is not taken into account. In particular in photoemission from crystalline materials, the emission direction may play an important role in the occurrence and the size of the magneto-dichroic asymmetry A(E). This is demonstrated by the results displayed in fig. 1b, again showing the intensity spectrum (upper panel) and the magneto-dichroic asymmetry (lower panel) of the Fe 2p levels.

The data have been recorded with normally incident light of 886eV photon energy and a photoelectron take-off angle of 45° with respect to the surface normal. Photon spin $\underline{q}$ and sample magnetization $\underline{M}$ are orthogonal in this particular geometry. According to absorption models the MCD signal should disappear under these conditions. The measured peak-to-peak intensity asymmetry A, however, is about a factor of 2 larger than the value at grazing incidence (fig. 1). This result reflects the influence of final state selection effects, which can only be understood by taking into account the emission direction of the electrons with respect to the crystalline axes of the sample. Further experiments show an angular dependence of the MCD signal in angle-resolved photoemission, which is more complicated than expected from photoabsorption theories[17]. So far, the existing theories may explain the basic mechanism of MCDXPS, but they are still inadequate to model the details.

3.2 MCD in Valence Band Photoemission - Itinerant electronic states

In the case of the 2p levels described above the energetic level splittings due to exchange interaction are an order of magnitude smaller than those introduced by spin-orbit coupling. Considering the spin-resolved photoemission results from Cu[13], the opposite situation is found in the valence bands of the 3d ferromagnets. This raised the question, whether or not an analogue of MCDXPS

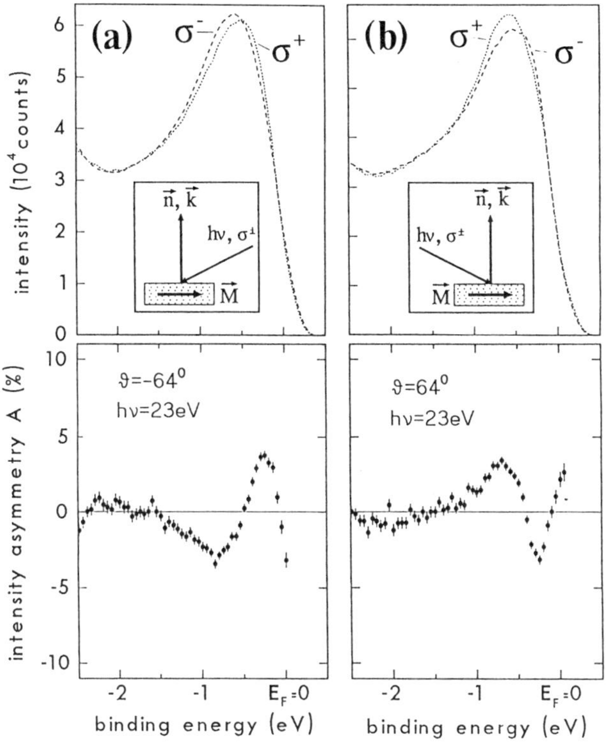

Fig. 2: MCD signal from the 3d valence bands of cobalt. The sample consisted of 5 monolayers of fcc-Co epitaxially grown on a Cu(100) substrate. In (b) the direction of incidence has been reversed compared to (a).

in the valence band range might be observable with circularly polarized VUV light. Such experiments have been performed at the 6.5m Normal Incidence Monochromator at BESSY. The samples consisted of Co ultrathin films of typically 5 monolayers (ML) thickness epitaxially grown on a Cu(001) substrate. Preliminary spin-resolved photoemission measurements indicated a bulk like dispersion of the electronic states in these films, in good agreement with fully relativistic band structure calculations[19]. Typical results obtained in normal emission at a photon energy of 23 eV are displayed in fig. 2. A geometry with normal emission has been chosen to minimize spin effects due to transmission through the surface[20]. The peak below the Fermi energy (upper panels) corresponds to emission from the Co d-bands of predominantly Δ_5 symmetry. For a given geometry and direction of magnetization the intensity spectrum clearly depends on the helicity of the incident light. This gives rise to the dichroic asymmetry A (lower panels) which exhibits a pronounced plus/minus feature in

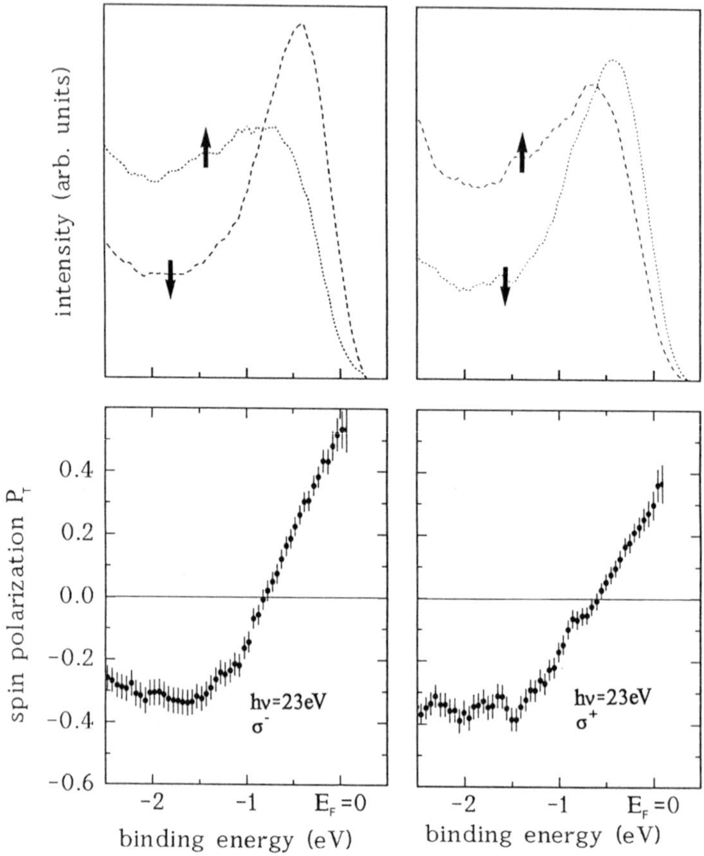

Fig. 3: Spin-resolved photoemission from fcc-Co films. Partial intensities (upper panels) of majority and minority character are marked by (↑) and (↓), respectively. Lower panels: corresponding spectra of the spin polarization component parallel to the surface along the remanent sample magnetization.

the region of the dominant peak. A change of the direction of incidence from ϑ to $-\vartheta$ (corresponding to a reversal of the component of the magnetization along the photon spin) leads to a reversal of the asymmetry distribution A(E) (fig. 2b). Further investigations showed that the operations $\underline{M} \rightarrow -\underline{M}$ and $\vartheta \rightarrow -\vartheta$ produce the same result within the experimental uncertainty, and the asymmetry A disappears at normal incidence ($\vartheta = 0°$), where both the directions of light incidence ($\underline{q}$) and electron emission ($\underline{k}$) are perpendicular to $\underline{M}$. These results prove the existence of magnetic circular dichroism in valence band photoemission (MCDUPS). Similar experiments on MCD from Fe have been reported very recently[21].

It has been stated that MCD in photoemission has its origin in spin-dependent mechanisms. This can be shown directly only in a spin-resolved experiment. An example is given in fig. 3 for the system fcc-Co/Cu(100). Fig. 3a (3b) corresponds to the excitation with left (right) hand circularly polarized light of 23 eV photon energy. The experimental geometry and parameters are the same as in fig. 2b. The lower panels display the spin polarization spectrum of the emitted photoelectrons. From the sign of the spin polarization one deduces the peak below the Fermi energy to be of predominantly minority spin character, followed by a broader distribution of majority electrons at higher binding energies. This can be seen more clearly in the partial intensities (upper panels) which represent the contributions with majority (↑) and minority spin (↓) character to the spin-integrated intensity. In going from fig. 3a to fig. 3b the change of the light helicity introduces subtle changes in the spin polarization spectrum due to spin-dependent processes in the photoexcitation. Again these changes show up more clearly in the partial intensities where the minority spin peak at the Fermi level has increased, but the majority peak has decreased. Since the two partial intensities sum up to the spin-integrated intensity distribution I(E), the height and energetic position of the dominant peak in I(E) is different for left (σ^-) and right hand (σ^+) circularly polarized light, causing a magneto-dichroic asymmetry essentially like the one shown in fig. 2b. This result proves magnetic circular dichroism in photoemission to be indeed due to spin-dependent excitation mechanisms which arise from the interplay of spin-orbit coupling and exchange interaction in the electronic structure. A qualitative interpretation of the above finding requires spin-orbit coupling to be regarded as a perturbation on the exchange split Co d-bands. Spin-orbit coupling splits the Δ_5 bands of each spin character by a small amount. The important point is that these split states differ slightly in the spatial part of the electronic wave functions, leading to a dependence of the corresponding photoexcitation probabilities on the light helicity. As a consequence, the partial intensity of a certain spin character changes upon excitation with σ^-- or σ^+-light, finally giving rise to the apparent energy shifts observed in the spin-integrated intensities. In contrast to the situation described in sect. 3.1, the main effect comes from the difference in the spatial part of the wave functions rather than the energetic splitting of levels.

4. Applications of MCD in Angle-Resolved Photoemission

The observation that magnetic circular dichroism appears in both UV and X-ray photoemission suggests MCD to be used as a new and versatile tool in surface and thin film magnetism. Since the information about spin of the excited photoelectron and thus the magnetism of a sample (e.g. exchange splitting, direction of $\underline{M}$) can be extracted from rather simple intensity measurements, MCD may in many cases replace the explicit and time-consuming spin polarization analysis. Detailed investigations of magnetic circular dichroism can give a deeper insight into the interrelation between spin-orbit and exchange interaction in ferromagnets. The elemental specificity inherent to XPS experiments can be used to link chemical and magnetic information on a local scale. This opens a way to study the magnetism of the individual constituents in alloys, interfaces in layered systems or sublattices in antiferromagnetic compounds.By exploiting the angular dependence of MCDXPS, one might even extract the spatial orientation of the magnetic moments. On the basis of appropriate theoretical approaches, MCD could be developed into a method to determine absolute values of the magnetic moments.

Magnetic circular dichroism in UV and x-ray photoemission has a potential application as a new contrast mechanism in photoelectron microscopy to image magnetic domains. Since the MCD signal (intensity asymmetry) depends on the relative orientation of the magnetization and the photon spin, photoemission from domains which are magnetized parallel and antiparallel to the photon spin will give rise to a photoemission yield which varies spatially across the sample surface. This spatial intensity distribution can be translated into a black and white contrast to display an image of the domain pattern on a television screen. As an advantage over scanning methods, this technique will permit the investigation of magnetization dynamics at ferromagnetic surfaces.

5. Acknowledgements

This work was supported by the Bundesminister f. Forschung u. Technologie, grant No. 05 413 AXI 7.

6. References

1. *Application of Circularly Polarized Light using Synchrotron and Ordinary Sources,* ed. F. Allen and C. Bustamente (Plenum, New York, 1985)
2. G. Schütz, W. Wagner, W. Wilhelm, P. Kienle, R. Zeller, R. Frahm, G. Materlik, *Phys. Rev. Lett.* **58**, 737 (1987)
3. C.T. Chen, F. Sette, Y. Ma and S. Modesti, *Phys. Rev.* B 43, 7262 (1990)
4. L. Baumgarten, C.M. Schneider, H. Petersen, F. Schäfers, J. Kirschner, *Phys. Rev. Lett.* **65**, 492 (1990)
5. C.M. Schneider, M.S. Hammond, P. Schuster, A. Cebollada, R. Miranda, J. Kirschner, *Phys. Rev.* B **44**, 12066 (1991)
6. G. Schönhense, *Physica Scripta* T**31**, 255 (1990)
7. J. Hermanson, Sol. State Commun. **22**, 9 (1977)
8. E. Kisker, in *Polarized Electrons in Surface Physics,* ed. R. Feder (World Scientific, Singapore, 1985)
9. C. Carbone and E. Kisker, Sol. State Commun. **65**, 1107 (1988)
10. M. Wöhlecke and G. Borstel, in *Optical orientation,* eds. F. Meier and B.P. Zakharchenya (North Holland, Amsterdam, 1984)
11. J. Kirschner, R. Feder and J. Wendelken, *Phys. Rev. Lett.* **47**, 614 (1981)
12. J. Kessler, *Polarized Electrons* (Springer, Berlin, 1985), p. 144
13. C.M. Schneider, J.J. de Miguel, P. Bressler, P. Schuster, R. Miranda and J. Kirschner, *J. Electron Spec. Rel. Phen.* **51**, 263 (1990)
14. L. Fritsche, J. Noffke and H. Eckhardt, *J. Phys.* F **17**, 943 (1987)
15. J.C. Fuggle and S.F. Alvarado, *Phys. Rev.* A **22**, 1615 (1980)
16. H. Ebert, L. Baumgarten, C.M. Schneider and J. Kirschner, *Phys. Rev.* B **44**, 4406 (1991)
17. G. v.d. Laan, J. Phys.: *Cond. Matt.* 3, 1015 (1991)
18. C.M. Schneider, D. Venus and J. Kirschner, *Phys. Rev.* B **45**, 5041 (1992)
19. C.M. Schneider, P. Schuster, M.S. Hammond, H. Ebert, J. Noffke and J. Kirschner, *J. Phys.: Cond. Matt.* **3**, 4349 (1991)
20. H.P. Oepen, K. Hünlich and J. Kirschner, *Phys. Rev. Lett.* **56**, 496 (1986)
21. J. Bansmann, M. Getzlaff, C. Westphal, F. Fegel and G. Schönhense, *Surf. Science* (1992), in press

SOFT X-RAY LINEAR DICHROISM

Maurizio Sacchi
Laboratoire pour l'Utilization du Rayonnement Electromagnetique
Université Paris Sud, F-91405 Orsay

ABSTRACT

Polarization dependent X-ray absorption spectroscopy in the soft X-ray energy range can be used to investigate anisotropic properties in an element selective mode. We report examples of recent applications of linear dichroism to the study of magnetic (ferro- and anti-ferro) ordering and of crystal field effects.

1. Introduction

X-ray absorption spectroscopy (XAS) is a widespread technique in physics and chemistry that allows to study electronic and structural properties with some peculiar advantages, notably the element selectivity. In the dipole approximation, the selection rule $\Delta l = \pm 1$ determines the symmetry of the final electronic state, given the initial one: only the transitions $(s \rightarrow p), (p \rightarrow s, d), (d \rightarrow p, f)$, etc. will be observed. In the soft X-ray region, the absorption process usually presents very high cross-sections (up to about $1Å^2$) which allow to perform experiments on very small amounts of material in relatively short times. Finally, XAS can be made surface sensitive using appropriate detection schemes, e.g. total electron yield, partial electron yield or total reflection.

Performing XAS with polarized X-rays extends the range of information available to this technique. The use of polarized light introduces a specific symmetry in the probe that can be tuned to that possibly existing in the sample in order to make it observable. Circularly polarized light can then be used to show the existence of unidirectional asymmetries along one axis, as in ferromagnetic materials or chiral molecules, while the presence of an axial symmetry is better investigated by using linearly polarized light.

Both linearly and circularly polarized light can be obtained from a storage ring. The simplest configuration involves the use of a bending magnet, emitting radiation which is linearly polarized in the orbit plane and elliptically polarized out of plane, with circular components of opposite helicity above and below. New insertion devices, like asymmetric wigglers and elliptical and crossed undulators, are now tested in several storage rings and will probably represent the new standard in the third generation synchrotron radiation sources. The high flux (10^9-10^{13} ph./sec.) over an extended energy range (1-10^4 eV) with tunable polarization makes the synchrotron radiation from storage rings the ideal source for X-ray dichroism studies [1].

2. The beamline SA-22 at SuperACO

All the experiments that will be presented and discussed in this paper as examples of linear dichroism have been performed on the beamline SA-22 of the 800 MeV positron storage ring SuperACO at LURE (Orsay). The unfocussed radiation from a bending magnet goes through a system of vertical and horizontal slits (typical divergence 25x50 μrad.) and through a 1 μm thick carbon filter that attenuates the low energy radiation. The monochromator consists of two Beryl crystals and gives a resolving power of about 3000 over the energy range 800$\rightarrow$1550 eV. The upper energy limit is set by the presence of Al and Si in Beryl, which introduce absorption edges at about 1560 and 1840 eV. The entire beamline is in ultra high vacuum (UHV), the pressure ranging from 2×10^{-10} to 2×10^{-9} in the different stages. The measuring chamber is a standard UHV system, equipped with Auger and LEED analyzers. Evaporation and sputtering facilities are available, the sample temperature can be varied between 10 K and 1200 K, and magnetic fields can be applied in situ. XAS measurements are performed by collecting the total electron yield from the sample.

3. Classical applications of X-ray linear dichroism

The exploitation of the linearly polarized in-plane synchrotron radiation dates back to more than 10 years ago [2], and has found two main applications that are related to the so called ***search light effect***: the study of the orientation of adsorbed molecules and of the distribution of empty states in low dimensionality systems. In the dipole approximation and one electron picture, the absorption cross section is proportional to $| < \Psi_i|\epsilon \cdot \mathbf{r}|\Psi_f > |^2$ where ϵ is the polarization vector of the incident light and $|\Psi_{i,f} >$ are the initial and final states for the transition. This determines a dominant sensitivity of XAS to those states that project along the polarization direction, with a specific angular dependence that involves the symmetry of the Ψ wavefunctions.

Linear dichroism has been largely exploited to determine the spatial distribution of the empty states in metal/semiconductor and metal/metal interfaces [3]. An example is shown in Fig.1 of the incommensurate quasi-5x5 reconstruction of the Cu/Si(111)7x7 interface [4]. The Cu 2p edges ($L_{2,3}$) are excited with light linearly polarized parallel and perpendicular to the interface plane. For coverages below 1.5 ML (where the q-5x5 reconstruction is observed) there is a clear and strong polarization dependence: spectra with ϵ in the interface plane (normal incidence) show a higher integrated intensity, with a sharp edge followed by three well resolved structures. When polarization is oriented along the [111] direction

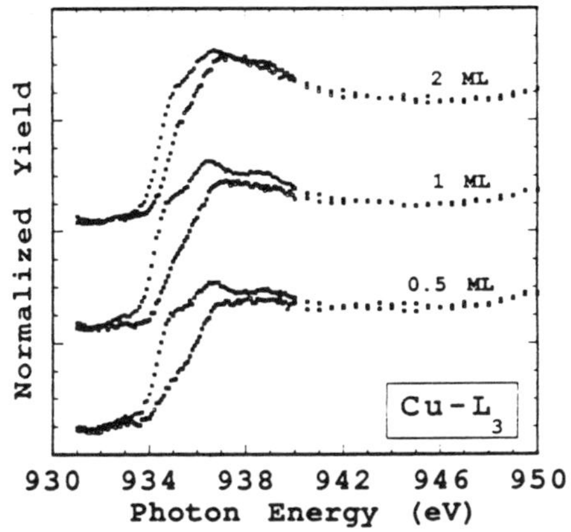

Fig.1 Cu L_3 absorption edge for the quasi-5x5 Cu/Si(111) interfaces at different Cu coverages. Filled squares : polarization vector in the surface plane (normal incidence of the photons). Empty squares : polarization vector 10° off the surface normal (grazing incidence). All the spectra are as collected.

of the substrate (grazing incidence), the edge rise is smoother and featureless, suggesting a lower density of empty states out of the interface plane. The $L_{2,3}$ edges correspond to transitions from a $2p$ core level towards states of s and d symmetry. Empty states close to the Fermi level in metallic copper are expected to be mainly of s character, but the matrix elements for $p \rightarrow d$ transitions are about 50 times larger than for $p \rightarrow s$, allowing to observe very small contributions of d character. The data in Fig.1 have been tentatively interpreted as a sign of a selective strong hybridization of the Cu d states in the interface plane [5]. A weak hybridization along the [111] direction also agrees with the model assuming a *frustrated interface* derived on the basis of STM observations [6]. One remark is worth making: the lowest investigated coverage (0.25 ML) corresponds to about 1×10^{14} atoms·cm^{-2}, and the data have been taken in less than one hour. This gives a more quantitative basis to the initial statement of ***high sensitivity*** for soft X-ray linear dichroism. Considering that metallic copper does not show the highest cross section in this energy range and that the unfocussed radiation from a bending magnet is not the most intense source of photons, a large margin for improvement is expected.

4. Perturbed level scheme and X-ray dichroism

In the last few years there has been a steady increase of interest in X-ray dichroism [7-10]. Attention has been focussed on the use of circularly polarized light for studying ferromagnetic ordering [10]. One shouldn't overlook the fact that linear dichroism can have, with a slightly different approach, a much more general use than the *classical applications* we have previously referred to. It is worth mentioning again that, for example, the first experiments on X-ray dichroism in rare earths used linearly polarized light to investigate magnetic properties [9]. In the following, a simple description of this different approach is presented.

A free ion is in spherical symmetry and cannot give rise to polarization dependent XAS. This is an obvious statement that can be partly reversed to give a more interesting one: if an ion shows polarization dependent absorption it is not in spherical symmetry, hence no longer free. In other words, X-ray dichroism is the manifestation of the presence of a perturbing field acting on the absorbing ion.

Let us express this in a more formal way for the case of a weak perturbation (first order perturbation theory). The absorption cross-section $\sigma_J(\omega)$ for a free ion of total momentum J can be separated in three partial terms corresponding to fixed ΔJ values in the transition [7,11]

$$\sigma_J(\omega) = \Sigma_K \sigma_{JK}(\omega) = \Sigma_K \frac{1}{2J+1} | < J||O_1||J+K > |^2 \; \delta(\hbar\omega + E_\circ - E') \qquad (1)$$

where $K = \Delta J = J' - J$ can take the values 0,±1. The σ_{JK} functions have been calculated in the atomic approach for all the RE ions [9]. We can also write the total cross section as: $\sigma_J = \Sigma_{pi} \sigma_J^{pi}$ where

$$\sigma_J^{pi}(\omega) = \Sigma_K \begin{pmatrix} J & 1 & J+K \\ M_J^i & p & M_J' \end{pmatrix}^2 | < J||O_1||J+K > |^2 \; \delta(\hbar\omega + E_\circ - E') \qquad (2)$$

$\begin{pmatrix} j_1 & j_2 & j_3 \\ m_1 & m_2 & m_3 \end{pmatrix}$ being the 3j-symbols for the composition of angular momenta [12]. $\sigma_J^{pi}(\omega)$ is the contribution from a specific state $|JM_J^i>$, summed over the possible ΔJ values and referring to a specific polarization of the light: right circular (p=+1), left circular (p=-1) or linear parallel (p=0) with respect to the quantization axis **q**. Introducing a weak perturbation V determines a new level scheme defined by $(H_\circ + V)| \; \Phi^i_J > = \mathrm{E}_i \; | \; \Phi^i_J >$ with $|\Phi^i_J> = \Sigma_j a_{ij} |JM_J^j>$. Indicating with s the partial cross-sections for the perturbed ion, we obtain [11]

$$s_J^{ip} = \Sigma_K s_{JK}^{ip} = \Sigma_j |a_{ij}|^2 \Sigma_K \begin{pmatrix} J & 1 & J' \\ M^j & p & -(M^j+p) \end{pmatrix}^2 | < J||O_1||J' > |^2 = \Sigma_j |a_{ij}|^2 \sigma_J^{ip} \quad (3)$$

where s_J^{ip} is the absorption cross section for the ion in its perturbed level $|\Phi^i_J>$ induced by a photon of polarization p. The total cross-section for a given value of

p is now a function of temperature, according to the distribution $n_i(T)$ of the ions over the different $|\Phi^i_J>$ eigenstates:

$$n_i(T) = \frac{e^{\frac{-E_i}{kT}}}{\Sigma_m e^{\frac{-E_m}{kT}}}$$

(the occupation of states higher in energy than the $\{|\Phi^i_J>\}_i$ set is not considered). Hence

$$s^p_J(T,\omega) = \Sigma_i n_i(T) s^{ip}_J(\omega) = \Sigma_j \ [\Sigma_i \frac{e^{\frac{-E_i}{kT}}}{\Sigma_m e^{\frac{-E_m}{kT}}} |a_{ij}|^2] \ \sigma^{jp}_J(\omega) \qquad (4)$$

The dichroism and its temperature dependence are entirely determined by the set of parameters $\{a_{i,j}, E_i\}$ which defines the perturbed level scheme.

5. Examples of soft X-ray linear dichroism applied to the investigation of perturbing fields

5.1. Ferromagnetic ordering.

Ferromagnetism is, in XAS studies, the domain of circular dichroism, due to the defined handedness of the field. But ferromagnetic ordering also defines an axis of symmetry and can be studied by using linearly polarized light [7,9]. Here we report part of the results obtained on $Fe_{1-x}Tb_x$ amorphous films, showing net magnetization normal to the film plane according to Kerr rotation measurements [13]. The spectra in Fig.2 refer to a 1000 Å film with $x = 0.2$, sandwiched between two silicon nitride layers 125 Å thick and supported by a silicon substrate. The spectra were taken on the as-prepared samples, with no external magnetic field applied during the measurements. The slight difference observed at room temperature (RT) between the $\epsilon \parallel H_{tot}$ and $\epsilon \perp H_{tot}$ spectra become more relevant as the temperature is lowered. The M_5 absorption edge of Tb is well understood from atomic calculations [14], where the ensemble of the transitions from the $^7F_6$ ground state of the trivalent ion into all the dipole allowed final states can be separated in subsets according to the variation of the total momentum J ($\Delta J = 0, \pm 1$). For example, the transitions characterized by $\Delta J = -1$ are grouped together to form the right-hand peak (at about 1242 eV) of the principal double structure, while the peak on the left (1240.5 eV) is mainly formed by $\Delta J = 0$ transitions. The presence of a magnetic field H breaking the spherical symmetry around the ion introduces a quantization axis and splits the degenerate ground state of the free ion in $2J + 1 = 13$ Zeeman levels, ordered in energy for increasing M_J values from $-J$ to J. The separation in energy $\mu g H$(μ is the Bohr magneton and g the Landé factor) among different Zeeman levels is of the order of meV, and is not resolvable in the XAS spectrum. On the other hand, from the simple model discussed above one finds that for $M_J = -6$ about 90% of the intensity of the $\Delta J = 0$ set of

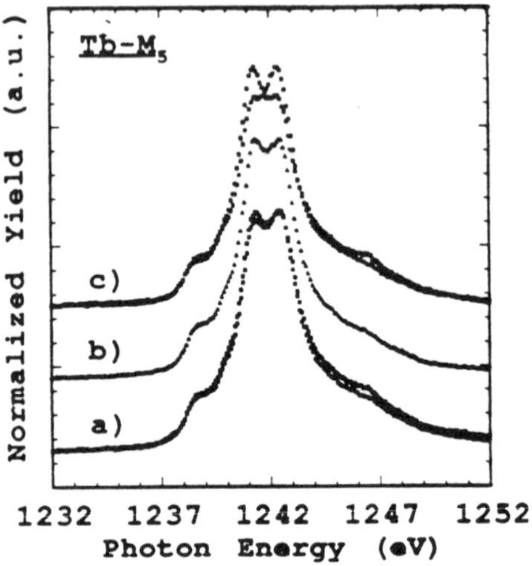

Fig.2 Tb M_5 edge in $TbFe_4$: (a) at RT for $\alpha = 12^\circ$ (open squares) and $\alpha = 90^\circ$ (filled squares), (b) unpolarized spectrum for a polycrystalline reference sample, (c) as (a) at T = 35 K.

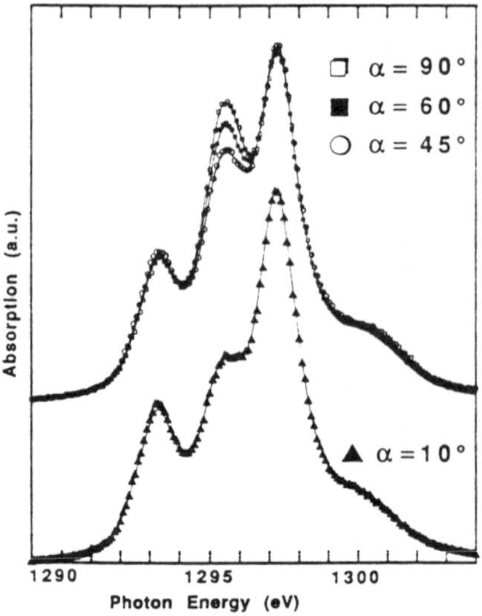

Fig.3 The dichroism at the Dy M_5 edge in the Dy/Si(111) interface (symbols) is compared to the model calculation for the Dy^{3+} ion in $C_{\infty v}$ symmetry (lines). The free parameter γ is adjusted on the $\alpha = 10^\circ$ spectrum.

transitions is associated with $\Delta M_J = 0$, while all the $\Delta J = -1$ set is characterized by $\Delta M_J = +1$. With $\epsilon \parallel H$ we excite transitions with $\Delta M_J = 0$ but not with $\Delta M_J = \pm 1$, hence we expect the low energy peak to be more intense. The opposite is true for $\epsilon \perp H$. Assuming, as shown by Kerr rotation measurements, that the axis of net magnetization is perpendicular to the film surface, one can readily understand the low temperature data of Fig.2. From the detailed angular and T dependence, the orientation of the magnetic moments of Tb is determined to follow a canted structure with an aperture angle of about 30 deg. and the axis perpendicular to the surface. The total energy spread of the perturbed levels is about 13 meV. If we assume that the perturbation is only represented by a magnetic field acting on the Tb site and determining a Zeeman splitting, its intensity is determined to be 12±2 T.

5.2. Ordering at a surface.

Eq.(4) makes no specific reference to the type of perturbation that originates the dichroism. An ion in the solid state already experiences a symmetry lower than spherical, due to the crystal field. An extreme situation is reached at a surface, where a description in terms of purely axial symmetry ($C_{\infty v}$) is a fairly good approximation [15]. As an example, let us consider the Dy/Si(111) interface [16]. This system has been extensively studied also in the epitaxial regime, showing that the D_{6h} symmetry of the Dy ion in $DySi_{2-x}$ gives rise to strong linear dichroism at low temperature (~ 100 K). Here we will only consider the case of a thin deposit (0.4 ML) annealed at 600 C for one minute. In the surface layer of the silicide structure, Dy has C_{6v} symmetry (main axis along the [111] direction of the substrate), that we approximate with $C_{\infty v}$. The perturbed level scheme of Dy ($J=\frac{15}{2}$) in $C_{\infty v}$ is defined by the parameters $a_{i,j}$ and E_i of eq.(4) given by

$$a_{ij} = 2^{-\frac{1}{2}}(\delta_{ij} + \delta_{j,2(J+1)-i})^{\frac{1}{2}} \quad ; \quad \Delta E_{ij} = \gamma(M_i^2 - M_j^2)$$

γ is the only free parameter in this model: its sign determines the ground state (either $|\pm\frac{1}{2}>$ or $|\pm\frac{15}{2}>$), its value the total energy spread ($\Delta E_{tot} = 56\gamma$). The Dy M_5 edge is reported in Fig.3 as a function of the angle α between ϵ and the [111] direction (symbols). By fitting the calculations on the $\alpha = 10^\circ$ spectrum, the sign and value of γ can be determined: $|\pm\frac{1}{2}>$ is the lowest lying level and $\Delta E_{tot} = 24$meV ($\gamma = 430\mu eV$). Without introducing any further parameters, the entire angular and T-dependence of the absorption can now be calculated and compared to the experiment. An example for the angular dependence at RT is shown in Fig.3 where the data for $\alpha = 45^\circ, 60^\circ$ and 90° are compared with the calculations for $\gamma = 430\mu$eV, as determined by the fitting procedure at $\alpha = 10^\circ$. The excellent agreement confirms in this case the validity of the $C_{\infty v}$ approximation for the symmetry at the Dy site.

5.3. Double RE layers.

As an example of a problem that stresses the peculiar characteristics of X-ray dichroism, let us consider the effect of the presence of a second RE layer on the magnetic ordering of a Dy submonolayer [17]. Fig.4_a shows the linear dichroism observed in 0.4 ML of Dy on Ni(110) deposited at 50 K, annealed at 500 K and measured again at 50 K. The detailed analysis of the spectra gives a picture similar to that obtained for Dy/Si: the Dy ions have their magnetic moment randomly oriented in the interface plane (i.e. the ground state is $|\pm\frac{1}{2}>$). Holding the substrate at 50 K, an additional layer of Ho was deposited on top, and the Dy M_5 edge measured again as a function of the polarization to surface geometry (Fig.4_b). The sample still shows dichroism, with reduced intensity. The most interesting point, however, is the reversal of the polarization dependence. The magnetic moment of Dy is now preferentially oriented out of the interface plane, with an average M_J of $\frac{11}{2}$. In this experiment both the high sensitivity (submonolayer, total measuring time less than one hour) and element selectivity (multicomponent system) of X-ray dichroism are indispensable and fully exploited.

5.4. Antiferromagnetic ordering in NiO.

Extensive calculations have been recently performed to include transition metal ions in the description of magnetic and crystal field dichroism [18-20]. For TM,

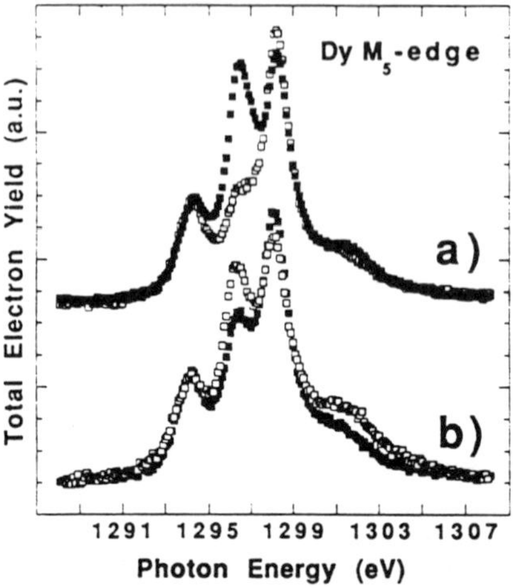

Fig.4 Dy M_5 spectrum for 0.4 ML Dy/Ni(110) measured at 50 K before (a) and after (b) the deposition of an extra Ho layer. Filled and open squares correspond to the polarization vector oriented in the surface plane and perpendicular to it, respectively.

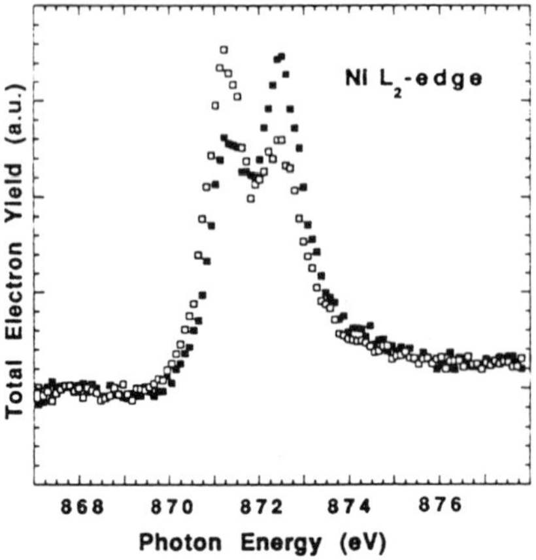

Fig.5 Ni L_2 absorption edge in a 17 ML NiO/MgO sample, taken with the polarization vector in plane (open squares) and 15° off the surface normal (filled squares).

the edges of interest are mainly the $L_{2,3}$, dominated by the $2p \rightarrow 3d$ transitions, hence extremely sensitive to magnetic and crystal fields. Examples of linear dichroism induced by lowering the symmetry at the surface in CaF_2 [19] and by the antiferromagnetic ordering in Fe_2O_3 [20] have been reported.

An experiment on NiO thin layers epitaxially grown on MgO(100) has been recently performed in LURE, in collaboration with the Materials Science Center in Groningen [21]. NiO has a cubic structure: the crystal field at the Ni site influences to a large extent the absorption spectrum [19,22], but has too high symmetry to induce linear dichroism. The surface layer in a truncated NiO(100) crystal has $C4v$ symmetry, and in this configuration the Ni^{2+} ion is expected to show dichroism. Moreover, NiO orders antiferromagnetically below 525 K, with antiparallel orientation of the Ni magnetic moments in adiacent (111) planes.

XAS spectra have been taken as a function of the angle α between the polarization vector and the [100] direction normal to the film plane, of the temperature in the range 100→600 K, and of the layer thickness (from 3 to 50 monolayers). A typical result is shown in Fig.5 for 17 ML at RT with $\alpha = 15°$ and $\alpha = 90°$. The ensemble of the results from this study [21] shows that:

i) at a fixed temperature, linear dichroism in NiO layers depends on the thickness (e.g. below 6ML there is no dichroism at RT).

ii) for a given sample, the dichroism increases by lowering the temperature. The T-dependence is always reversible and reproducible.

iii) for the thickest sample (50 ML) the dichroism vanishes exactly at the Néel temperature for the antiferromagnetic ordering in bulk NiO.

By comparing the angular and temperature dependent data to calculations, the origin of the linear dichroism has been identified as the occurence of antiferromagnetic order, the surface effects being comparatively small and negligible above 100 K. As a consequence, it has been possible, starting from XAS measurements, to trace the thickness dependence of the Nèel temperature in NiO(100) epitaxial layers.

6. Conclusions

Soft X-ray dichroism offers new access to the investigation of local anisotropies and perturbing fields, and its characteristics of element selectivity and high sensitivity make it a unique tool. The use of polarized light avoids the limitations of energy resolution which are implicit in an XAS experiment (core hole lifetime and instrumental broadening), so that ground state properties of the ion on the *meV* energy scale can be investigated.

Acknowledgements The X-ray dichroism project developed at LURE is part of a long term collaboration with the Universities of Nijmegen and Groningen, and is partly supported by the EC contract n. SC1-CT91-0630.

References

[1]. *Rayonnement Synchrotron Polarisé, Electrons Polarisés et Magnétisme* eds. E.Beaurepaire, B.Carrière and J.P.Kappler (I.P.C.M.S. Strasbourg, 1989)

[2]. see for example D.Norman, *J.Phys.* **C19** (1986) 3273 and references therein

[3]. J.Stöhr, in *X-ray Absorption: principles, Applications, Techniques of EXAFS, SEXAFS and XANES*, eds. R.Prins and D.C.Koningsberger (Wiley, New York, 1986), and references therein

[4]. M.Sancrotti, M.Sacchi, O.Sakho, G.Rossi, *Phys.Rev.* **B44** (1991) 1958

[5]. S.Iacobucci, M.Sancrotti, M.Montagnoli, M.Sacchi, G.Rossi *(unpublished)*

[6]. K.Mortensen, *Phys.Rev.Lett.* **66** (1991) 461

[7]. B.T.Thole, G. van der Laan, G.A.Sawatzky, *Phys.Rev.Lett.* **55** (1985) 2086

[8]. J.B.Goedkoop, B.T.Thole, G.van der Laan, G.A.Sawatzky, F.M.F.de Groot, J.C.Fuggle, *Phys.Rev.* **B37** (1988) 2086 ;
H.Ebert, G.Schütz, W.M.Temmermann, *Solid State Commun.* **4** (1990) 475 ;
B.T.Thole, P.Carra, F.Sette, G.van der Laan, *Phys.Rev.Lett.* **68** (1992) 1943

[9]. G.van der Laan, B.T.Thole, G.A.Sawatzky, J.B.Goedkoop, J.C.Fuggle, J-M.Esteva, R.C.Karnatak, J.P.Remeika, H.A.Dabkowska, *Phys.Rev.* **B34** (1986) 6529 ;
J.B.Goedkoop, M.Grioni, J.C.Fuggle, *Phys.Rev.* **B43** (1991) 1179

[10]. G.Schütz, W.Wagner, W.Wilhelm, P.Kienle, R.Zeller, R.Frahm, G.Materlik, *Phys.Rev.Lett.* **58** (1987) 737 ;
C.T.Chen, F.Sette, Y.Ma, S.Modesti, *Phys.Rev.* **B42** (1990) 7262

[11]. M.Sacchi and L.Reining, *unpublished*

[12]. L.D.Landau and E.M.Lifsits, *Course of Theoretical Physics* Vol.III. (Pergamon, Oxford, 1984)

[13]. M.Sacchi, R.J.H.Kappert, J.C.Fuggle and E.E.Marinero, *Appl.Phys.Lett.* **59**(1991) 872

[14]. B.T.Thole, G.van der Laan, J.C.Fuggle, G.A.Sawatzky, R.C.Karnatak, J-M.Esteva, *Phys.Rev.* **B32** (1985) 5107

[15]. e.g. I.Peschel and P.Fulde, *Z.Phys.* **259** (1973) 145

[16]. M.Sacchi, O.Sakho, G.Rossi, *Phys.Rev.* **B43** (1991) 1276 ;
M.Sacchi, O.Sakho, F.Sirotti, G.Rossi, *Appl.Surf.Sci.* **56-58** (1992) 1

[17]. R.J.H.Kappert, J.Vogel, M.Sacchi, J.C.Fuggle (*unpublished*)

[18]. G.van der Laan and B.T.Thole, *Phys.Rev.* **B43** (1991) 13401

[19]. F.M.F.de Groot, J.C.Fuggle, B.T.Thole, G.A.Sawatzky, *Phys.Rev.* **B41** (1990) 928 and *ibid.* **B42** (1990) 5459 ;
F.M.F.de Groot, *Ph.D. Thesis, University of Nijmegen (unpublished)*

[20]. P.Kuiper, B.G.Searle, P.Rudolf, L.H.Tjeng, C.T.Chen, *(unpublished)*

[21]. C.Levelut, J.Vogel, M.Sacchi, D.Alders, S.Peacor, M.Takano, G.A.Sawatzky, (*unpublished*)

[22]. J.van Elp, B.G.Searle, G.A.Sawatzky, M.Sacchi, *Solid State Commun.* **80** (1991) 67
J.van Elp, *Ph.D. Thesis, University of Groningen (unpublished)*

MAGNETIC DICHROISM THEORY OF CORE LEVEL SPECTROSCOPY

GERRIT VAN DER LAAN
Daresbury Laboratory, Warrington WA4 4AD, United Kingdom

ABSTRACT

Magnetic dichroism in x-ray absorption and photoemission are interpreted in a many-electron approach. In x-ray absorption the integrated circular dichroism over a given edge allows one to determine the ground state expectation value of the orbital angular momentum. In core level photoemission the spin polarization and magnetic dichroism give information about the correlation between the valence spin, core hole orbital momentum and core hole spin. These measurements are important for the understanding of the anisotropic properties of transition metal and rare earth materials.

1. Introduction

Knowledge of the magnetic anisotropy in magnetic bulk materials, multilayers, thin films and surfaces is of large technological importance. Consequently, the separation of the orbital and spin contributions to the magnetic moment has become a major challenge in neutron and x-ray scattering research. In neutron scattering the two contributions can be separated by fitting the measured form factors to a suitable model. In the x-ray scattering from antiferromagnets the magnetic superlattice reflections can be decoupled from the structural Bragg peaks. [1,2] In ferro- and ferri-magnets the charge and magnetic Bragg peaks coincide and the latter are swamped, but a first order interference term exists if the structure factor is complex, or if the radiation source is circularly polarized. The key to the separation of the spin and orbital magnetization lies in the different dependence on the Bragg angle, which can be tuned by the photon energy. [3] However, attempts to implement a quantitative separation experimentally have so far been inconclusive.

In this paper a general model for magnetic dichroism in x-ray absorption and photoemission in the presence of atomic interactions between the hole created and the valence holes is discussed. [4-14] We can distinguish three fundamental spectra in x-ray absorption: the isotropic, linear and circular dichroism spectrum; and eight fundamental spectra in photoemission: the isotropic spectrum, spin spectrum, orbit spectrum (magnetic circular dichroism), spin-orbit spectrum, spin-orbit magnetic-quadrupole spectrum, anisotropic spectrum (magnetic linear dichroism), anisotropic spin magnetic dipole spectrum and anisotropic spin magnetic octupole spectrum. We will show that these spectra provide important information about the orbital and spin part of the magnetic moment, which cannot be obtained by other methods.

2. X-ray absorption

Recently, it was shown that the spin and orbital momentum can be separated using circular magnetic x-ray dichroism (CMXD). Thole *et al.* [6] have derived a powerful sum rule which states that the integrated absorption signal for right- minus left circularly polarization is proportional to the orbital magnetic moment per hole. This can be illustrated for e.g. the Ni $2p$ CMXD spectrum of ferromagnetic nickel. This spectrum consists of two peaks corresponding to excitations from the $2p_{3/2}$ and $2p_{1/2}$ level, respectively, into the unoccupied $3d$ states above the Fermi level which are exclusively of minority spin character. Without $3d$ spin-orbit interaction the $2p_{3/2}$ and $2p_{1/2}$ peaks have negative and positive intensities of equal magnitude, thus the integrated signal is zero. However, in the presence of $3d$ spin-orbit interaction the $2p_{3/2}$ peak increases and the $2p_{1/2}$ peak decreases in absolute intensity. The integrated signal predicts an orbital moment of 0.053 μ_B, which is in good agreement with neutron diffraction and local spin density calculations. [12]

Similarly, the integrated linear dichroism measures the quadrupole moment $Q_{zz} = \langle L_z^2 \rangle - \frac{1}{3} l\ (l+1)$ of the ion. A positive value of Q_{zz} means high values of $|m|$ are occupied which makes the ion flat in the x,y plane. For a negative value it is elongated along the z-axis. This quantity is non-zero not only in ferro- and antiferromagnets, but also in spatially anisotropic media even if non-magnetic (crystal fields).

The core excitation makes CDMXD element and site specific. The dipole selection rule allows us to resolve energy separations in the ground state in the order of a few meV, despite the fact that the excitation energy is about 1 keV and the energy resolution 100 meV. [13] The technique is very sensitive and can be used on a sub-monolayer level. Interesting results have been obtained for dipole transitions to localized empty states such as the $2p \rightarrow 3d$ and $3p \rightarrow 3d$ in $3d$ transition metal compounds and the $3d \rightarrow 4f$ in rare earth compounds, which agree well with theoretical predictions. [4,5,14]

Additional information can be obtained from the spin-orbit split branching ratio. The isotropic branching ratio is proportional to the angular dependent part of the ground state spin-orbit interaction. It has been shown that this can be used to assess the local magnetic moments of magnetically unordered materials. [15] An increase in the magnetic moment is accompanied by an increase in the branching ratio. In CDMXD the branching ratio is a sum over various terms such as S_z and tensor products of $l.s$. [16] In the case that $\langle L \rangle = 0$, such as in Gd f^7, the branching ratio of the MXD is proportional to $\langle S_z \rangle$.

3. X-ray photoemission

The experimental proof of magnetic x-ray dichroism in 1986 [13] was followed by research in related phenomena in the x-ray region, such as resonant magnetic x-ray scattering, [17] magnetic Kerr effect [18] and Faraday rotation. [19] In 1990 Baumgarten *et al.* [20] observed circular dichroism in x-ray photoemission (CDXPS) in the Fe $2p$ level of ferromagnetic iron. In core hole photoemission the excited electron has no interaction with the remaining system and the observation of dichroism requires, apart from spin-orbit coupling, an interaction between the core level and the polarized valence levels. This is a fundamental difference compared to CDMXD where the electrostatic interaction is not essential when the excited electron is subjected to the Pauli exclusion principle.

In a general setup for polarized XPS circularly or linearly polarized light is shone on a magnetically ordered sample and the intensity of the emitted electrons is measured for both up- and down spin as a function of energy. We do not consider here the angular distribution or the final state interactions of the photoelectrons. Thus, for some fixed magnetization M of the sample we obtain six spectra $I_{q\sigma}$, where q = -1, 0 and 1 for left, Z- and right circularly polarized light and $\sigma = \uparrow$ and $\downarrow$ for spin up and down, respectively. Reversal of the magnetization produces no additional spectra because, in an obvious notation $I_{-q\text{-}\sigma\text{-}M} = I_{q\sigma M}$.

These six spectra are all more or less different, and for each material we might be satisfied by a theory which reproduces these spectra. However, analysis shows that we can assign specific differences between the spectra to specific physical effects, thus allowing to draw definite conclusions. This is possible when we take the linear combinations I^{xy} (x = 0, 1, 2 and y = 0, 1) defined in Table I. It may be checked that a spectrum with x = 0 can be measured using isotropic light, and x = 1 measures circular and x = 2 linear dichroism. When y = 0, no spin polarization needs to be measured while y = 1 denotes the difference spectrum for spin up and down.

Table I. The six fundamental photoemission spectra I^{xy}, which are linear combinations of the primitive spin polarized spectra. x = 0, 1, 2 denotes isotropically, circularly and linearly polarized radiation, respectively. y = 0, 1 denotes without and with spin polarization measurement, respectively. z = 0, 1, 2, 3 denotes the value $\langle M^z \rangle$ of the magnetic moment M in the ground state (minus the isotropic value).

I^{xy}	combination of primitive spectra	z	significance
I^{00}	$I_{1\uparrow} + I_{0\uparrow} + I_{-1\uparrow} + I_{1\downarrow} + I_{0\downarrow} + I_{-1\downarrow}$	0	isotropic spectrum
I^{01}	$I_{1\uparrow} + I_{0\uparrow} + I_{-1\uparrow} - I_{1\downarrow} - I_{0\downarrow} - I_{-1\downarrow}$	1	spin spectrum
I^{10}	$I_{1\uparrow} - I_{-1\uparrow} + I_{1\downarrow} - I_{-1\downarrow}$	1	orbit spectrum (MCD)
I^{11}	$I_{1\uparrow} - I_{-1\uparrow} - I_{1\downarrow} + I_{-1\downarrow}$	0,2	spin-orbit spectrum
I^{20}	$I_{1\uparrow} - 2I_{0\uparrow} + I_{-1\uparrow} + I_{1\downarrow} - 2I_{0\downarrow} + I_{-1\downarrow}$	2	anisotropic spectrum (MLD)
I^{21}	$I_{1\uparrow} - 2I_{0\uparrow} + I_{-1\uparrow} - I_{1\downarrow} + 2I_{0\downarrow} - I_{-1\downarrow}$	1,3	anisotropic spin magnetic spectrum

The atomic photoemission process is depicted in Fig. 1. It shows a system consisting of a valence shell and a core shell with momenta S_v, L_v and L_c, S_c respectively. In a localized model there are exchange interactions between S_v and S_c; Coulomb interactions between L_v and L_c and spin-orbit interactions between L and S of the same shell. The exchange field polarizes the valence spin S_v of the ground state which means that only some of the magnetic sublevels m are populated giving non-isotropic values for $\langle M \rangle$, $\langle M^2 \rangle$, $\langle M^3 \rangle$, etc. The electric polarization vector q of the light acts on the core orbital momentum L_c, producing polarized core hole (and photoelectron) orbitals, and the photoelectron spin polarization σ measured is the same as that of the core hole spin S_c. We indicate the multipole moments x, y and z of the polarizations by 0, 1, 2, 3 for isotropic, dipole, quadrupole and octupole moments, respectively. For a spectrum to be observable interactions connecting the non-zero values of x, y and z in Fig. 1 are required.

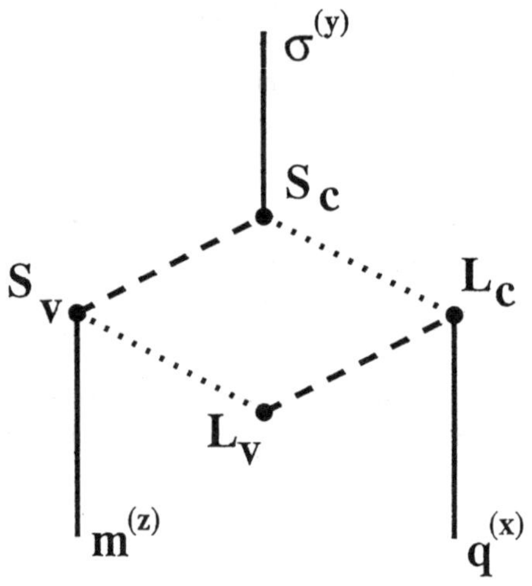

Fig. 1. Schematic picture for spin polarized photoemission with spin-orbit interactions (dotted lines) and electrostatic interactions (dashed lines). $L_{v(c)}$ and $S_{v(c)}$ denote the orbital and spin moments of the valence (core) state, respectively; x, y and z are the polarizations of the electric vector q, the spin σ and the magnetic moment m. For a polarization spectrum to be observable we need interactions connecting the non-zero values of x, y and z.

The physical picture in Fig.1 can be formulated in an exact way. It can be shown that a spectrum I^{xy} is non-zero only when a z-th moment has been induced in the atoms of the sample by the exchange field, where z must be in the range $|x - y| \ldots (x + y)$ and $x + y + z$ is even. When this is the case the spectrum measures the amount of correlation (alignment) of the x-th moment of the orbital momentum of the core hole with the y-th moment of the core hole spin and the z-th moment induced in the ground state.

This correlation is different for each final state, thus we obtain information on the character of each final state. Moreover, when such a correlation between moments is observed this indicates the presence and effects of the specific interactions which cause this correlation.

The first fundamental spectrum, I^{00}, is the familiar isotropic spectrum. It measures the parentage of the ground state to the final states. This paper mainly analyses those polarized spectra where one of the x, y, z is zero and the other two are 1. For qualitative purposes we can say that the spectrum I^{01} (z = 1) measures $S_c.S_v$, i.e. the alignment of the core hole spin with the valence spin. Likewise, I^{10} (z = 1) measures $-L_c.S_v$ and I^{11} (z = 0) measures $-L_c.S_c$, the core hole spin orbit coupling.

In I^{01}, the electric vector q is isotropic (x = 0) and cannot produce polarized core hole orbitals, but the sample has a magnetic moment (z = 1) and we measure the spin polarization (y = 1). To couple the m and σ polarization requires at least exchange interaction between S_v and S_c. The exchange interaction directly couples the spins of the core hole and valence holes; neither spin-orbit coupling nor circular polarization is needed to polarize the core hole spin and thus the photoelectron spin.

In I^{10} (magnetic circular dichroism) no spin polarization is measured (y = 0), but m is polarized (z = 1) and the difference in total emission for right and left circular polarization is measured (x = 1). The required coupling between L_c and S_v is always indirect: the photon spin creates a core hole with polarized orbital momentum, which is coupled by electrostatic and spin-orbit interaction to the spin of the valence electrons. [8-9] Fig.1 shows that there are two possible ways to do this.

The I^{11} spectrum has two z contributions: 0 and 2. It is clear from Fig. 1 that in the z = 0 contribution core hole spin-orbit interaction is required, but no interaction with the valence electrons is needed. The photon spin acts on (polarizes) the electron spin via the spin-orbit coupling. Since no interaction is needed between the core and valence levels this spectrum can also be obtained in non-magnetic atoms or in the one-electron model.

The cases in which x, y or z are larger than 1 are too complicated to allow a simple explanation of the behaviour of each peak in the spectrum. We will only discuss the conditions under which these spectra are non-zero.

The I^{20} spectrum measures the difference in emission using perpendicular and parallel linearly polarized light (x = 2), hence magnetic linear dichroism. The spin is not detected. Since z = 2 (quadrupole polarization), this spectrum is present both in ferro- and antiferromagnets, where $\langle M \rangle = 0$, but $\langle M^2 \rangle - J(J+1)/3 \neq 0$ when J or $S > 1/2$. In the latter case it measures a many-electron effect. A quadrupole can also be induced by a low-symmetry crystal field (keeps M and $-M$ degenerate). The crystal field polarizes L_v and of course this moment couples in a specific way to the other moments. The importance of this spectrum is that it is relatively simple to measure.

Finally, in the I^{21} spectrum we need a magnetic dipole or octupole moment to measure spin polarization using linearly polarized light. Spin-orbit as well as Coulomb interactions are needed. In theory the z = 1 and 3 contributions can be determined separately when we vary the magnetic polarization, e.g. by temperature dependence, giving information on the effects of both $\langle M \rangle$ and $\langle M^3 \rangle$.

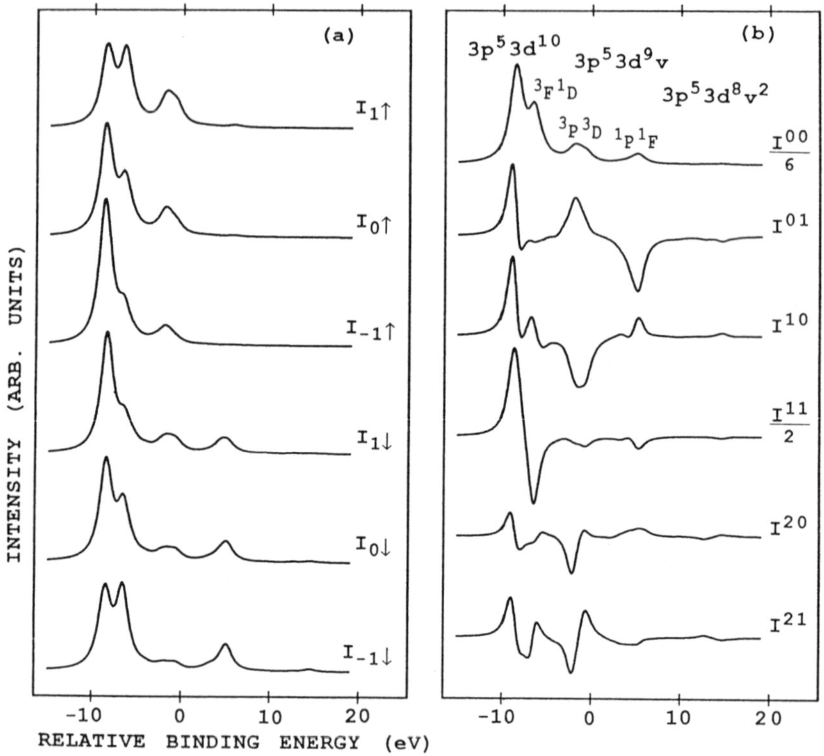

Fig. 2. Ni 3*p* spin polarized photoemission of Ni metal, (a) primitive and (b) fundamental spectra.

4. Ni 3*p* CDXPS

We will now demonstrate the theory on the 3*p* XPS spectra of ferromagnetic Ni metal. Fig. 2 shows the primitive spectra for *d* emission. The fundamental spectra shown are those for *s* emission, which are stronger by a factor 1, -2 and 10 than those for *d* emission for $x = 0$, 1 and 2, respectively. The spectra were calculated using an Anderson impurity model. [11] A ground state consisting of a mixture of d^8, d^9 and d^{10} with weights of 8, 44 and 48 %, respectively [9] gives a good agreement with the experimental 2*p* and 3*p* isotropic XPS [12] and magnetic x-ray dichroism (MXD) absorption spectra.[21,22]

In the 3*p* XPS the core spin-orbit coupling is small and the core valence Coulomb interactions dominate. The spin spectrum I^{01} clearly shows a negative low-spin $^1P^1F$ and positive high-spin $^3P^3D$ peak. In the I^{10} spectrum the alignment of L_c with S_v is produced indirectly. By valence spin-orbit coupling L_v is anti-parallel to S_v. Next, core-valence Coulomb interactions couple L_c to L_v; in the F, D and P states they are parallel, weakly- and strongly antiparallel respectively. In the $^1P^1F$ peak the effects of P and F almost cancel, in the $^3P^3D$ peak the ^3P gives the net parallel alignment of L_c and S_v and in the low energy structure ^3F gives the net antiparallel peak. This indirect coupling gives a net alignment of S_c and L_c even in the absence of core spin-orbit interaction, giving the effect that I^{11} is to some extend a product of I^{01} and I^{10}. But the *real* spin-orbit interaction in p^5d^{10}, which has no Coulomb interactions, is more effective, giving the dispersive curve at the low energy side.

5. Conclusion

Polarized x-ray absorption and photoemission spectroscopy of transition metal-, rare earth- and actinide magnetic systems greatly increases the amount of information on the magnetic moment of the ground state and on the interaction of this moment with the spin and orbit of the hole created.

References

1. P.M. Platzman and N. Tzoar, *Phys. Rev.* **B 2** (1970) 3356.
2. F. deBergevin and M. Brunel, *Phys. Lett.* **A 39** (1972) 141.
3. M. Blume, *J. Appl. Phys.* **57** (1985) 3614; M. Blume and D. Gibbs, *Phys. Rev.* **B 37** (1988) 1779.
4. B.T. Thole, G. van der Laan, and G.A. Sawatzky, *Phys. Rev Lett.* 55 (1985) 2086.
5. G. van der Laan and B.T. Thole, *Phys. Rev.* **B 43** (1991) 13401.
6. B.T. Thole, P. Carra, F. Sette and G. van der Laan, *Phys. Rev. Lett.* **68** (1992) 1943.
7. T. Jo, *Synchrotron Radiation News* **5** (1992) 21.
8. G. van der Laan, *J. Phys. Condens. Matter* **3** (1991) 1051.
9. G. van der Laan, *Phys. Rev. Lett.* **66** (1991) 2527.
10. B.T. Thole and G. van der Laan, *Phys. Rev.* **B 44** (1991) 12424.
11. B.T. Thole and G. van der Laan, *Phys. Rev. Lett.* **67** (1991) 3306.
12. G. van der Laan and B.T. Thole, *J. Phys. Condens. Matter* **4** (1992) 4181.
13. G. van der Laan, B.T. Thole, G.A. Sawatzky, J.B. Goedkoop, J.C. Fuggle, J.M. Esteva, R. Karnatak, J.P. Remeika, and H.A. Dabkowska, *Phys. Rev.* **B 34** (1986) 6529.
14. J.B. Goedkoop, B.T. Thole, G. van der Laan, G.A. Sawatzky, F.M.F. de Groot and J.C. Fuggle, *Phys. Rev.* **B 37** (1988) 2086.
15. B.T. Thole and G. van der Laan, Phys. Rev. **B 38** (1988) 3158.
16. P. Carra, B.T. Thole, M. Altarelli, and X. Wang, submitted to *Phys. Rev. Lett.*
17. D. Gibbs, D.R. Harshman, E.D. Isaacs, D.B. McWhan, D. Mills and C. Vettier, *Phys. Rev. Lett.* **61** (1988) 1241.
18. J. Bonarski and J. Karp, *J. Phys. Condens. Matter* **1** (1989) 9261.

19. D.P. Siddons, M. Hart, Y. Amemiya and J.B. Hastings, *Phys. Rev. Lett.* **64** (1990) 1967.
20. L. Baumgarten, C.M. Schneider, H. Petersen, F Schäfers, and J. Kirschner, *Phys. Rev. Lett.* **65** (1990) 492.
21. C.T. Chen, F. Sette, Y. Ma, and S. Modesti, *Phys. Rev.* **B** 42 (1990) 7262.
22. T. Koide, T. Shidara, H. Fukutani, K. Yamaguchi, A. Fujimori, and S. Kimura, *Phys. Rev.* **B 44** (1991) 4697.

MAGNETIC CIRCULAR DICHROISM IN X-RAY ABSORPTION AND CORE-LEVEL PHOTOEMISSION OF Fe/Cu(001)

J.G. TOBIN and G.D. WADDILL

Chemistry and Materials Science Department, University of California, Lawrence Livermore National Laboratory, P.O. Box 808, Livermore, CA 94550 USA

D.P. PAPPAS

Naval Research Laboratory, Washington, DC 20375 USA

ABSTRACT

We report the first observation of magnetic circular dichroism in both x-ray absorption and core-level photoemission of ultra thin magnetic films using circularly polarized x-rays. Iron films (1–4 ML) grown on a Cu(001) substrate at 150 K and magnetized perpendicular to the surface show dramatic changes in the $L_{2,3}$ branching ratio for different x-ray polarizations. For linearly-polarized x-rays perpendicular to the magnetic axis of the sample the branching ratio was 0.75. For films $\geq$ 2 ML, this ratio varied from 0.64 to 0.85 for photon spin parallel and anti-parallel, respectively, to the magnetic axis. This effect was observed either by changing the x-ray helicity for a fixed magnetic axis, or by reversing the magnetic axis for a fixed x-ray helicity. Our observation can be analyzed within a simple one-electron picture, if the raw branching ratios are normalized so that the linear value becomes statistical. Furthermore, warming the films to ~300 K eliminated this effect, indicating a loss of magnetization in the film over a temperature range of ~30 K. Finally, reversing the relative orientation of the photon spin and the magnetic axis from parallel to anti-parallel allowed measurement of the exchange splitting of the Fe 2p core levels which was found to be ~0.2 eV. These results are consistent with earlier studies, but the use of off-plane circularly-polarized x-rays from a bending magnet monochromator offers ~2 orders of magnitude greater intensity than typical spin-polarization measurements.

One of the basic thrusts of the investigation of nanoscale magnetic structures, whether it be ultrathin monolayer films, multilayers, or clusters, is the establishment of structure-property relationships. To be more specific, the determination of the interconnections between geometric, electronic, and magnetic structure is the fundamentally important by intermediate goal, with an eye towards the ultimate manipulation and application of this understanding. Ordinarily, efforts to measure nanoscale magnetic properties in conjunction with atomistic geometric and electronic structures runs headlong into the same problem: The magnetic perturbation tends to be a small component of the overall effect. In contrast to this, here we report giant circular dichroism in the near-edge core-level x-ray absorption of a near-monolayer metal film.[1] An example of our data is shown in Fig. 1. The essential effect is the relative amplification of the $2p^{1/2}$ peak where the magnetization and helicity are parallel, regardless of whether the magnetization is into or out of the surface plane. (Because the magnetization and x-ray incidence are normal to the surface, we call it perpendicular dichroism.) This is a direct measurement of the spin polarization and the density of the unoccupied states near E_F in a ferromagnetic system.

The temperature dependence of the perpendicular magnetization can also be followed with this technique. Figure 2 illustrates the behavior for 4 ML of Fe/Cu(001). Here, we are using branching ratio (BR) as a measure of the dichroism. BR is defined in Eq. (1). I is the integrated intensity of the white line peak at each edge jump.

$$BR = I(2p_{3/2}) / \left[I(2p_{3/2}) + I(2p_{1/2}) \right] \tag{1}$$

We have developed a simple, one-electron picture to analyze our results, which can be summarized into the closed form analytic expression shown in Eq. (2).

$$BR = \frac{3 + 2(1-\alpha)\beta + 2(1-\beta)\alpha}{6} \tag{2}$$

α is the spin down alignment of the unoccupied 3d states and β is the degree of right circular polarization. ($\beta = 100\%$ for right $\beta = 0\%$ for left, and $\beta = 50\%$ for linear.)

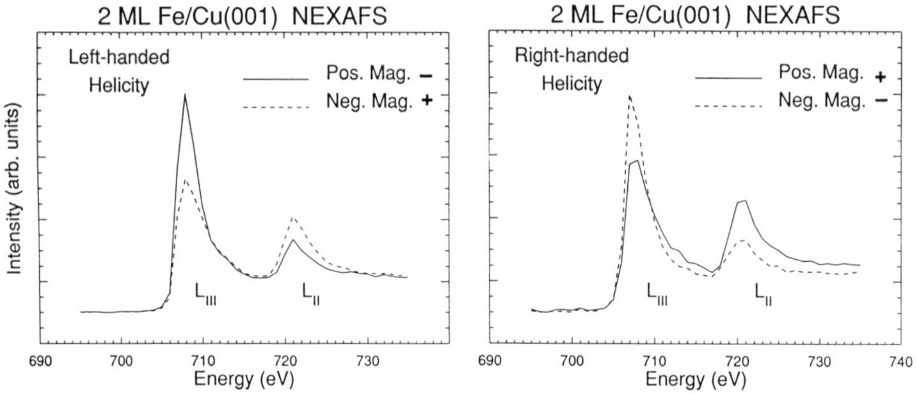

Figure 1. The near-edge x-ray absorption fine structure (NEXAFS) dichroism of 2ML of Fe/Cu(001). These are plots of absorption versus photon energy. The upper panel shows the effect of reversing the magnetization while maintaining the left-handed helicity of x-rays. Similarly, for the lower panel and right-handed helicity x-rays. Samples are perpendicularly magnetized either into (pos. mag.) or out of (neg. mag.) the surface. The symbol +(-) means that helicity and magnetization are parallel (anti-parallel). The −2p3/2 peak is at the LIII edge and the 2p1/2 peak is at the LII edge. The spectra were normalized to each other by equating the pre-edge intensity, at energies below approximately 700 eV.

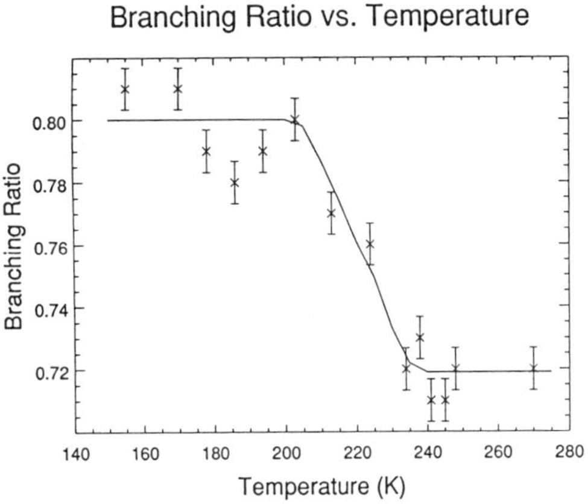

Figure 2. The temperature dependence of the magnetization of 4 ML Fe/Cu(001) using NEXAFS magnetic circular dichroism as the probe. Because of the imperfect placement of the thermocouple on the sample plate, the sample temperature may be slightly higher than that shown. Error bars of approximately ± 1% are included for comparison but the true time error estimate is probably higher (about 4%). A branching ratio near 0.72 corresponds to an unmagnetized sample.

To use a one-electron picture, it is necessary to "normalize-out" the non-statistical many-body effects. To do this we multiply all BR values by $^2/_3/BR_{EXP}(lin)$. This allows us to concentrate upon the effects of helicity variation. The results of our analysis, assuming $2\mu_B$/Fe atom ($\alpha = 25\%$ and 75%) is shown in Table I.

Table 1.

P vs M	BR_{EXP}	BR'_{EXP}	BR_{TH}
+	0.645	0.585	0.60
–	0.83	0.75	0.73
lin	0.74	0.67	0.67

P = polarization
M = magnetization direction
P vs M: relative directions of helicity and magnetization
+ = parallel
– = anti-parallel
BR = branching ratio
BR_{EXP} = raw experimental results
BR'_{EXP} = experimental results, normalized to the linear statistical prediction
BR_{TH} = atomic theory prediction, 2mB/Fe atom, $\alpha = 25\%$, 75% and b = 10%, 90% pol.

Finally, we have also used MCD with core-level photoemission to measure the exchange splitting of the 2p3/2 and 2p1/2 peaks of 4ML of Fe/Cu(001)[2]. The spectra are shown in Fig. 3. The exchange splitting of 2p3/2 peak is 0.22 ± 0.10 eV. The apparent change in the spin-orbit splitting between the two peaks is 0.33 ± 0.14 eV. There are less than those observed in the bulk (see Ref. 2 and References therein).

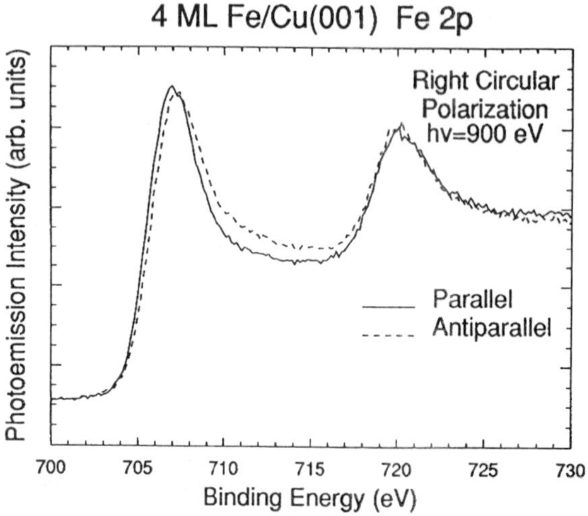

Figure 3 Fe 2p spectra taken with hv = 900 eV and positive photon helicity. The solid lines are for parallel orientation of photon spin and sample magnetization and the broken lines for an anti-parallel orientation. The two orientations were achieved by fixing the photon helicity and reversing the magnetization of the sample. The spectra are as collected and show the differences in binding energy for the 2p3/2 and 2p1/2 levels for the two helicity and magnetization orientations.

We have performed magnetic circular dichroism (MCD) experiments upon Fe/Cu(001), using both core-level photoemission and near-edge absorption fine structure. A simple, one electron picture worked well to explain normalized branching ratio measurements. We plan to continue and expand this work.

Acknowledgments

Work performed under the auspices of the U.S. Department of Energy by the Lawrence Livermore National Laboratory under contract No. W-7405-ENG-48. D.P. Pappas was supported by IBM Almaden when this experimental work was done. These measurements were made on Beamline 8–2 at the Stanford Synchrotron Radiation Laboratory. Beamline 8–2 is part of the UC/National Laboratories Participating Research Team (PRT)3. The authors wish to thank Ms. Karen Clark for supporting this work.

References

1. J.G. Tobin, G.D. Waddill, and D.P. Pappas, Phys. Rev. Lett., 68, 3642 (1992) and references therein.

2. G.D. Waddill, J.G. Tobin, and D.P. Pappas, Phys. Rev. B, 46, No. 1 (1992).

3. L.J. Terminello, G.D. Waddill, and J.G. Tobin, Nucl. Instrum. Meth. (1992), in press.

SPIN POLARIZED PHOTOEMISSION STUDIES OF INTERFACIAL AND THIN FILM MAGNETISM

P D. JOHNSON, Y. CHANG and N.B. BROOKES †
Physics Dept., Brookhaven National Lab., Upton, NY, 11973, USA.

ABSTRACT

Spin polarized photoemission has been used to study the magnetic interfaces formed when Cr and Ag are deposited on a Fe(001) substrate. In the case of Ag/Fe(001) a minority spin interface state is observed to step towards the Fermi level in well defined discrete steps dependent on the thickness of the overlayer. For the Cr overlayers with the d-bands lying at similar binding energy to that of the substrate d-bands, it becomes more difficult to unravel the origin of the different interface states. Using a Stoner type model within the tight-binding framework it is possible to calculate the spin resolved interface states and further to examine their behavior as the overlayer films are extended from the monolayer regime to thicker coverages. For Cr overlayers on Fe the film is shown to slowly develop the antiferromagnetic structure characteristic of the bulk material. In the case of Ag/Fe(001) we show how the discrete steps in the binding energy of the interface state reflect the coupling to the continuum states within the overlayer film.

1. Introduction

Magnetic thin films and multilayers are currently the subject of considerable research activity.[1] Of particular interest are the transition metal multilayers which may show ferromagnetic or antiferromagnetic coupling between adjacent ferromagnetic layers dependent on the thickness of the intervening spacer layer.[2] The related giant magnetoresistance properties[3] and the large enhancement of the Kerr rotation make these systems particularly interesting technologically. The exact mechanism by which the adjacent magnetic layers couple is still not understood.

Straightforward application of RKKY theory with a period length λ given by π/k_F will not apply. Here the free electron Fermi wavevector k_F is defined by $k_F=(3\pi^2N)^{1/3}$ where N is the number of electrons per unit volume. As an example, if we consider a multilayer grown in the (001) direction with monovalent copper as the intervening layer RKKY theory would predict a period length λ = 2.31Å, only slightly larger than the interlayer spacing of 1.8Å. In many multilayers on the other hand two periods of oscillation in the coupling have been observed, a long period corresponding to an intervening layer with thickness of the order of 5-6 layers and a short period of the order of 2-3 layers.

Theories have therefore considered the effects of the finite thickness of the intervening layer and also of the non-spherical Fermi surfaces associated with this layer.[4-8] The introduction of periodicity into the intervening layer with associated reciprocal lattice vector has been shown to result in a longer period length for the coupling. Several of these theories also suggest that the coupling will reflect the presence of states localized in the interfaces between the layers.

In this paper we examine localized interface states and further examine how such states couple to states within the adjacent layers. In particular we examine the interfaces formed when Cr and Ag are deposited on a Fe(001) surface, antiferromagnetic coupling having previously been observed in both Cr/Fe[9-11] and Ag/Fe multilayers.[12] For both these systems we are able to identify interface states with a well defined spin. Further, the binding energies of these states are found to reflect the thickness of the overlayer film. In the case of the silver film we are able to show how these magnetic or spin polarized states evolve into the continuum or "quantum well" states of the overlayer and thereby potentially allow, as has been suggested elsewhere,[13] direct coupling of the two interfaces either side of the intervening layer in a multilayer.

2. Experimental Details

The primary experimental technique used in this study to examine the interfacial electronic states is spin polarized photoemission. In all of the experiments detailed in this paper the incident photon energy is 52 eV and the light is p-polarized. Described in more detail elsewhere,[14] the photoemitted electrons are first energy and momentum analyzed with a conventional hemispherical analyzer and then spin analyzed with a low energy scattering detector of the type described by Unguris et al.[15] Photons are provided by the U5 undulator located on the UV ring at the National Synchrotron Light Source.

In both experiments the Fe(001) substrate, cut in the form of a picture frame crystal, was cleaned using standard cycles of argon bombardment followed by annealing to temperatures of the order of 600°C. The silver and chromium films were evaporated from a tungsten basket at a rate of approximately 0.25 ML per minute. In both cases the lattice matches are excellent, being 2.86 and 2.87Å for bcc Cr and bcc Fe and 2.88Å the appropriate spacing for fcc Ag. The substrate was held at room temperature in the case of silver deposition and 70°C for the chromium deposition. Auger Electron Spectroscopy (AES) ratios were measured as an estimate of the coverage and the evaporations were also monitored using a quadrupole mass spectrometer. Low Energy Electron Diffraction (LEED) measurements showed a good sharp p(1x1) pattern at all coverages up to approximately 3 ML for the Ag and up to thicknesses of the order of 10 ML for the Cr. In neither overlayer case was any evidence found for intermixing.

3. Results

Figure 1 shows a series of spin integrated spectra recorded from an Fe(001) surface with Cr deposited to different thicknesses. The clean surface spectrum shows a peak at a binding energy of 2.4 eV which has previously[16] been shown to be a minority spin surface resonance. The presence of a Cr monolayer, 1.6Å, causes this state to move

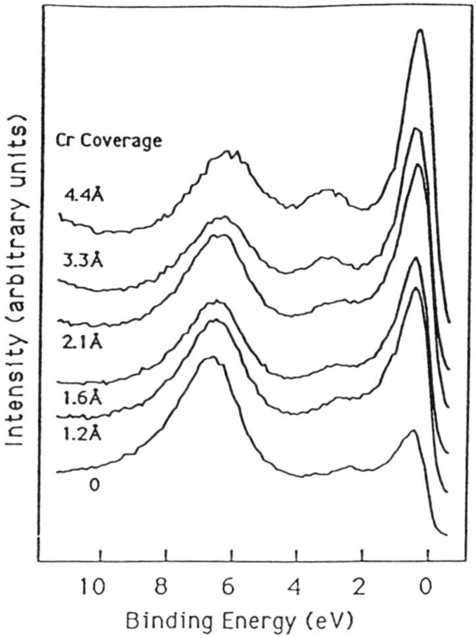

Fig. 1. Spin integrated photoemission spectra recorded from Cr deposited on Fe(001) as a function of Cr coverage. The incident photon energy is 52 eV and the angle of incidence corresponds to p-polarized light.

to a slightly higher binding energy and with increasing thickness the state evolves into a surface state associated with the Cr surface. The spectra in fig. 1 also show an increase in the emission intensity of the peak in the vicinity of the Fermi level.

Figure 2 shows a spin resolved spectrum obtained from the Fe(001) surface with approximately one monolayer of Cr deposited. The spectrum clearly shows that the peak derived from the surface resonance still displays minority spin character. The increase in the intensity at the Fermi level also reflects new states of minority spin character. A spin polarized photoemission spectrum at this photon energy from the clean Fe(001) surface shows the majority peak to be the much stronger component. Both the minority spin state near E_F and the peak at a binding energy of 2.7 eV characterize the interface between the Fe substrate and the Cr overlayer. Spin polarized photoemission spectra recorded from the same surface with approximately two monolayers of Cr show a reduction of the intensity of the minority spin peak near the Fermi level such that the majority peak at higher binding energy again becomes the more intense. The minority spin peak near the Fermi level would therefore appear to characterize the monolayer coverage.

Figure 3 shows spin integrated spectra recorded from the the Fe(001) surface but now with silver deposited to different thicknesses. Described in greater detail

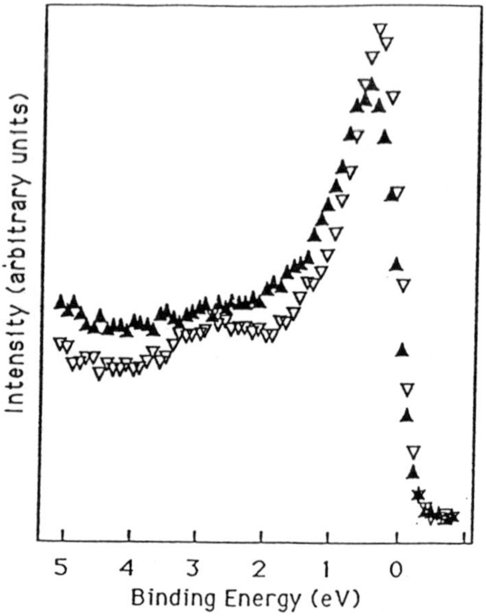

Fig. 2. Spin polarized photoemission spectra recorded from approximately one monolayer (1.6Å) of Cr deposited on Fe(001). The incident photon energy is 52 eV and the angle of incidence corresponds to p-polarized light. The minority spin states are indicated by open triangles, the majority spin states by closed triangles.

elsewhere,[17] the spectra show a series of peaks in the region from 0 to 2 eV below the Fermi level with well defined binding energies dependent on the thickness of silver deposited. Spin resolved studies of these states, which characterize the interface between the iron and silver, show them to be entirely of minority spin character.

4. Calculations and Discussion

In order to better understand the photoemission spectra presented in the previous section, we have developed a spin dependent tightbinding scheme in a slab formulation that allows us to calculate the electronic structure of these systems. Such a calculation is particularly useful in interpreting spectra from systems where the substrate and overlayer have overlapping d bands as in the case of Cr overlayers on Fe.

These tight-binding calculations were carried out using an approach based on the use of a Hubbard Hamiltonian. Thus

$$H = \sum_{\mathbf{k}} E(\mathbf{k}) n_{\mathbf{k}} + \frac{U}{N} \sum_{\mathbf{k},\mathbf{k}'} n_{\mathbf{k}\uparrow} n_{\mathbf{k}'\downarrow} \qquad (1)$$

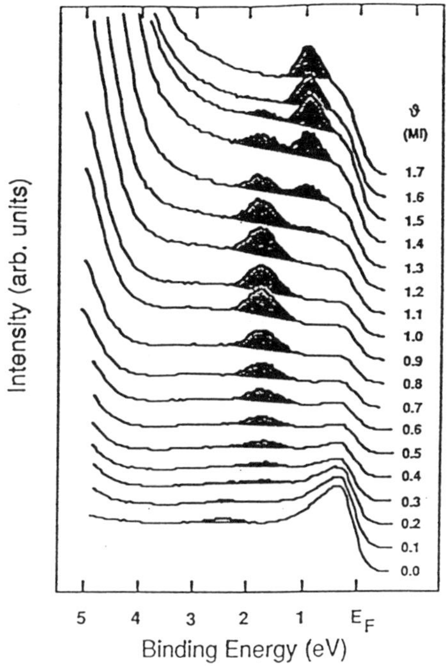

Fig. 3. Photoemission spectra recorded along the surface normal from a series of Ag coverages deposited on Fe(001). The different coverages are indicated. The incident photon energy is 52 eV and the angle of incidence corresponds to p-polarized light.

where the first term reflects the paramagnetic bandstructure and the second term represents the modification due to an on-site spin dependent potential, U. This "effective exchange" potential leads to a spin dependent splitting of the electronic states with the resulting formation of local moments. With the assumption that the electronic states are described by Bloch states a calculation of the susceptibility shows that U may be associated with the Stoner parameter.[18] It has been demonstrated elsewhere that the the latter parameter is essentially an atomic property showing little variation from one environment to another. Indeed in a first principles FLAPW calculation for a five layer slab of vanadium it was found that the effective Stoner parameter was identical for the surface and central layers.[19]

Our approach is to assume in the case of the transition metals, that the moments are carried almost entirely by the more localized d electrons. Thus we take the parameters associated with a tight-binding fit to the appropriate paramagnetic band structure, split the on-site spin dependent energies for the d-blocks by an amount Δ and, in the case of a non-orthogonal fit, make appropriate adjustments to the associated off-diagonal elements. We then integrate the spin-dependent densities of states up to the

Fermi-level to obtain the resulting layer dependent moments. A self-consistent solution is sought such that for each layer

$$\Delta_l = U_l m_l \qquad (2)$$

where Δ_l is the layer dependent splitting introduced into the d-block, U_l the layer dependent Stoner parameter and m_l the calculated moment for each layer. Seeking the solution to equation (2) is equivalent to find a local minimum in the total energy of the system.

Initially a small ferromagnetic moment is applied to every site to move away from the trivial solution m_l=0. Charge neutrality for the system as a whole is maintained throughout the calculation. Two center tight binding parameters are taken from the compilation of Papaconstantopoulos[20] and values for U were taken from Local Spin Density calculations of the susceptibility and related Stoner parameters.[21,22] The total density of states was calculated for each overlayer system by summing over 28 k-points evenly distributed throughout the irreducible surface Brillouin zone.

Shown in figure 4, a calculation for an 11 layer Fe(001) slab using this method yields an average internal magnetic moment of 2.2 ± 0.05 μB. The surface layer in the (001) plane has an enhanced moment of 2.94 μB. These results are in excellent agreement with the results of a full-potential linearized-augmented-plane-wave (FLAPW) calculation.[23] Also shown in fig. 4, a 13 layer Cr(001) slab moved from the

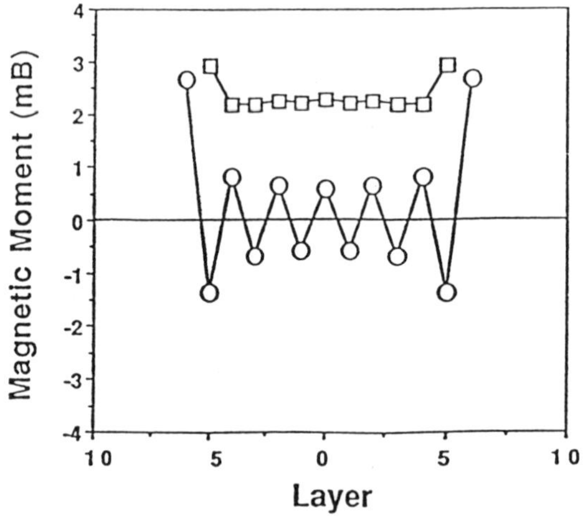

Fig. 4. Tight binding calculations of the magnetic structure of an 11 layer Fe (001) slab (squares) and a 13 layer Cr (001) slab (circles). The sources for the different parameters are given in the text. The different layers are labelled out from the center of the slabs.

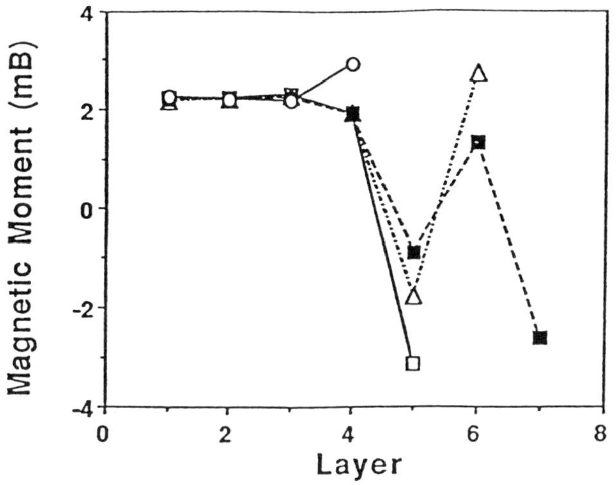

Fig. 5. Tight binding calculations of the magnetic structure of 1 (□), 2 (Δ), and 3 (■) layers of Cr deposited on a Fe (001) substrate. The clean surface moments are indicated by the open circles. The different layers are labelled out from the center of the slab with layer 4 corresponding to the iron surface.

initial ferromagnetic configuration through to an antiferromagnetic alignment of adjacent (001) planes. In order to achieve a bulk moment of 0.6 μB in agreement with experiment it was found to be necessary to set U = 53 mryd, the mean of two first principles calculations of the Stoner parameter for Cr.[21,22] As in previous calculations the surface showed an enhanced moment of 2.3 μB.[24,25] Interestingly the decay of the absolute value of the moments away from the surface is more gradual in Cr than in Fe where the surface enhancement is apparently screened within one layer.

Turning to the growth of thin films we consider first a Cr film grown on the Fe(001) substrate. The chromium-iron interaction parameters were taken as the mean of the chromium and iron parameters. Where required the scaling scheme of Anderson et al. was used.[26] The on site energies were adjusted to align the Fermi levels of the two metals. The lattice constants were taken as the bulk values and for the interfacial separation the mean of the two elemental interlayer spacings was used. However, as noted above the lattice match for Fe and Cr is excellent.

In figure 5 we show the calculated magnetic moments for the Fe/Cr system. We start out with the clean iron (001) surface and introduce, layer by layer, three monolayers of Cr. In agreement with earlier tight binding calculations[24] the initial Cr layer reduces the surface iron moments by 40% from 2.9 μB to 1.9 μB. The initial Cr layer is antiferromagnetically aligned to the substrate with a large moment of 3.1 μB. Each new layer of Cr aligns itself antiferromagnetically with the previous layer. The "buried" Cr layers slowly reduce their moments towards the value associated with bulk chromium. However the newly deposited surface layers always show an enhanced moment reflecting the presence of the solid-vacuum interface or more importantly the reduced number of nearest neighbors. Discussed elsewhere[27] the deposition of-only a single layer of Fe on these three Cr layers immediately results in the magnetization profile for the Cr that has previously been found in several earlier multilayer calculations for this system.[28]

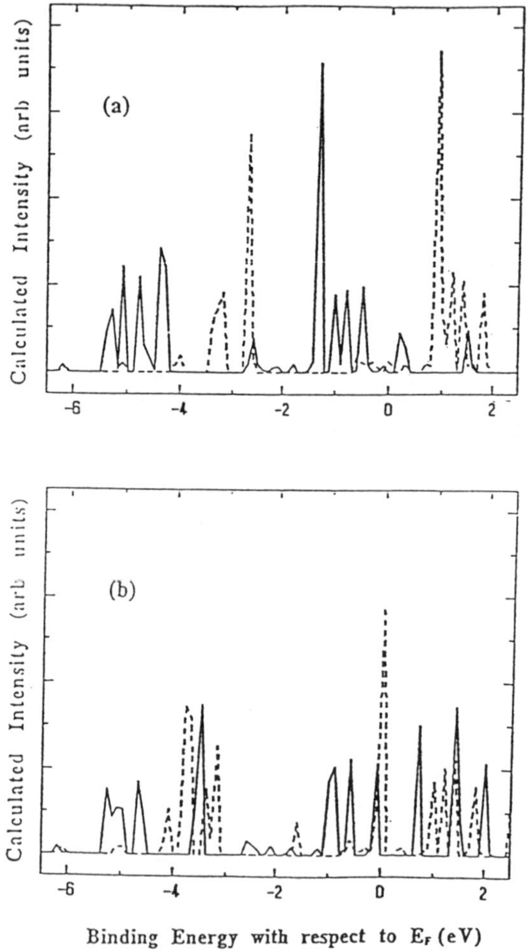

Fig. 6. Spin-dependent densities of states in a region around Γ, the center of the Brillouin zone, to a radius $k_{||}$=0.1Å for (a) the clean Fe(001) surface and (b) an Fe(001) surface with one monolayer of Cr deposited. Minority spin states are indicated by the dashed lines and majority spin states by the solid lines.

In order to make a more direct comparison with the photoemission spectra recorded along the surface normal, we show in figure 6, the calculated density of states in a narrow region around Γ, the Brillouin Zone center. The layer dependent eigenvectors, corresponding to the different eigenvalues are weighted by factors reflecting the relative photoionization cross-sections for iron and chromium and for the mean free path of the photoemitted electrons. For the present calculations a representative mean free path of 4.5 Å was used.

The upper panel in fig. 6 shows the calculated spectra from a clean Fe(001) surface and the lower panel the same but now from a Fe(001) surface with a deposited monolayer of Cr. In both cases the incident light is modelled as p-polarized. By examining the eigenvectors producing the different peaks in these "spectra" we can identify the origin of the structure within the photoemission spectra of fig. 2. It should be noted however that the model does not include any final state scattering, which may modify the relative peak heights in the figure.

The calculated "spectrum" in fig. 6 shows a strong minority surface feature at 2.7 eV binding energy for the clean surface. For the Cr covered surface this peak is shifted to slightly higher binding energy and dispersed over an energy range from 3.0 to 4.0 eV binding energy. This shift to higher binding energy is in agreement with the experimental observations shown earlier in fig. 2. The calculated spectra also show that the presence of the Cr leads to a strong minority feature being induced right at the Fermi level again in agreement with the experimental observations discussed earlier. Both the peak at the Fermi level and the peak at higher binding energy are of d_{z^2} symmetry, the former having slightly more Fe character and the latter slightly more Cr character.

We now use the same procedure to analyze the Ag/Fe(001) system. Elsewhere we have discussed the layer by layer development of the interface states and shown how these states coupled to the sp states in the overlayer.[17] Indeed it was the finite size effects associated with the latter overlayer that lead to discrete stepping in the binding energy of the interface state. Here we demonstrate this idea more clearly by comparing the experimental observations with tight binding calculations of silver slabs of different thicknesses and for thinner films the same silver slabs deposited on an Fe(001) substrate.

In any calculation of silver of finite thickness we will sample each of the bands in the bulk band structure n times where n is the number of layers in the slab. In momentum space therefore each band will be sampled at intervals given by $k_{BZ}/(n-1)$ where k_{BZ} is the zone boundary momentum. In fig. 7 we illustrate this by examining the calculated eigenvalues with eigenvectors of predominantly sp_z symmetry in the region of the Fermi level for silver slabs of different thicknesses. As the thickness of the slab increases the eigenvalues, indicated by the dashed lines, will converge because we sample the sp band with increasing frequency.

We also show in the figure our spin dependent calculations of these same states for 1,2 and 3 monolayers of silver placed on a Fe(001) substrate. The states now display magnetic character with both minority and majority spin states appearing in the calculation. A first principles FLAPW calculation of the monolayer Ag/Fe(001) system also finds both majority and minority spin interface states.[29] In agreement with the results of the same calculation our model shows that the deposition of the initial

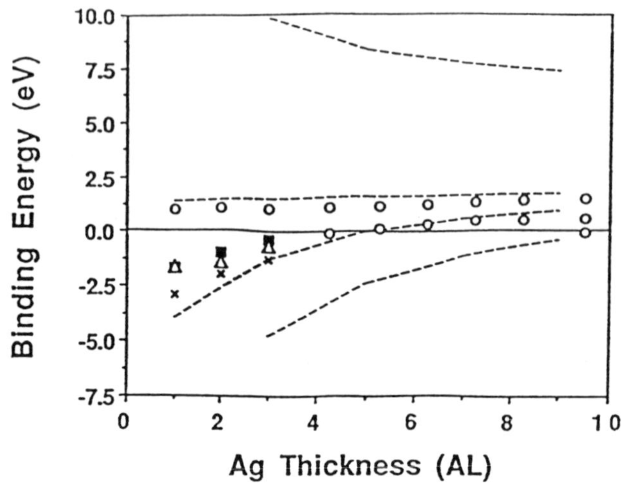

Fig. 7. Calculated states of sp symmetry for silver slabs of different thicknesses (dashed lines). Calculated minority spin states (Δ) and majority spin states (x) for silver layers deposited on an Fe(001) substrate. Experimentally observed minority spin states observed in photoemission studies (■) (ref. 17) and the same states (o) observed above the Fermi level in an Inverse Photoemission study (ref. 13)

monolayer reduces the surface moments of the iron from 2.94 μB to 2.5 μB. With subsequent deposition of silver layers the Fe surface moments show little change in our calculation.

The experimental observations from our spin polarized photoemission study of the interface states and the results of an Inverse Photoemission study[13] of the same states observed above the Fermi level as they evolve into quantum well states of the silver are also shown in fig. 7. Examination of the figure shows that good agreement is obtained between both sets of experimental observations and the results of the calculation. Both the experiment and the model suggest that the minority spin state crosses the Fermi level at a coverage of between three and four layers of silver.

The minority spin states dominate the experimental spectrum with only the smallest indication of a possible majority state for the silver monolayer coverage. An examination of the eigenvectors in our tightbinding model suggests that the minority spin states are more localized in the interface region than the majority spin states which display weight back into the bulk of the Fe. This result is not inconsistent with the idea discussed elsewhere that the presence of a band gap at E_F in the even symmetry states of minority spin character for the Fe substrate should produce a dominance of minority spin polarization for the states in the silver.[13]However we note that a study of the same states in copper films grown on a fcc Co substrate shows minority spin polarization even though there is no appropriate band gap spanning the Fermi level for the substrate.

5. Conclusions.

Our studies have shown that when one of the two materials is ferromagnetic, well defined magnetic states may exist in the interface between two metals. Further these states may couple to states in the adjacent layers effectively polarizing the latter. If such polarized states are responsible for the magnetic coupling in transition metal multilayers it will be necessary, in systems showing more than one periodicity, to examine in greater detail the polarization and dispersion of the states at points other than the center of the Brillouin Zone.

6. Acknowledgements.

The author would like to acknowledge useful discussions with M. Weinert. This work has been supported by the Department of energy under contract DE-AC02-76CH00016 and by the NSF under contract No. DMR-86-03304.

7. References.

† Present Address: ESRF, BP 220, F38043 Grenoble Cedex, France.

1. *Magnetic Surfaces, Thin Films, and Multilayers*, ed. S.S.P. Parkin, H. Hopster, J-P Renard, T. Shinjo and W. Zinn, MRS Proceedings **231**, (Pittsburg,1992).
2. S.S.S. Parkin, p. 145 in ref. 1.
3. M.N. Baibich, J.M. Broto, A. Fert, F. Nguyen Van Dau, F. Petroff, P. Etienne, G. Creuzet, A. Friederich, and J. Chazelas, *Phys. Rev. Lett.* **61** (1986) p. 2472.
4. Y. Wang, P.M. Levy and J.L. Fry, *Phys. Rev. Lett.* **65** (1990) p. 2732.
5. D.M. Deaven, D.S. Rokhsar and M. Johnson, *Phys. Rev* **B44** (1991) p. 5977.
6. P. Bruno and C. Chappert, *Phys. Rev. Lett.* **67** (1991) p. 1602.
7. R. Coehoorn, *Phys. Rev.* **B44** (1991) p. 9331.
8. F. Herman, J. Sticht, and M. Van Schilfgaarde, p. 195 in ref. 1.
9. P. Grunberg, R. Schreiber, Y. Pang, M.B. Brodsky, and H. Sowers, *Phys. Rev. Lett.* **57** (1986) p. 2442.
10. S.S.P. Parkin, N. More and K.P. Roche, *Phys. Rev. Lett.* **64** (1990) p. 2304.
11. J. Unguris, R.J. Celotta and D.T. Pierce, *Phys. Rev. Lett.* **67** (1991) p. 140.
12. J. Unguris, D.T. Pierce and R.J. Celotta, to be submitted.
13. J.E. Ortega and F.J. Himpsel, *Phys. Rev. Lett.* **69** (1992) p. 844.
14. P. D. Johnson, S. L. Hulbert, R. Klaffky, N. B. Brookes, A. Clarke, B. Sinkovic, N.V. Smith, R. Celotta, M. H. Kelly, D.T. Pierce, M. R. Scheinfein, B. J. Waclawski and M.R. Howells, *Rev. Sci. Inst.* **63** (1992) p. 1902.
15. J. Unguris, D.T. Pierce, R.J. Celotta, *Rev. Sci. Instrum.*, **57** (1986) p. 1314 ; M.R. Scheinfein, D.T. Pierce, J. Unguris, J.J. McClelland, R.J. Celotta, and M.H. Kelley, *Rev. Sci. Instrum.* **60(1)** (1989) p. 1.
16. N. B. Brookes, A. Clarke, P. D. Johnson, and M. Weinert, *Phys. Rev.* **B41** (1990) p. 2643.
17. N. B. Brookes, Y. Chang, and P. D. Johnson, *Phys. Rev. Lett.* **67** (1991) p. 354; N.B. Brookes, Y. Chang, P.D. Johnson, A. Clarke, and N.V. Smith, to be submitted.
18. J.M. Ziman, *Principles of the Theory of Solids* (Cambridge University Press, Cambridge, 1972), p. 339.

19. S. Ohnishi, C.L. Fu and A.J. Freeman, *J. Magn. Mater.* **50** (1985) p. 161.
20. D.A. Papaconstantopoulos, *Handbook of the Band Structure of Elemental Solids* (Plenum Press 1986).
21. J.F. Janak, *Phys. Rev.* **B 16** (1977) p. 255.
22. O.K. Andersen, O. Jepsen and D. Glotzel, in *Highlights of Condensed Matter Theory*, edited by F. Bassani, F. Fumi, and M.P. Tosi (North-Holland, Amsterdam, 1985), p. 154.
23. S. Ohnishi, A.J. Freeman and M. Weinert, *Phys. Rev.* **B 28** (1983) p. 6741.
24. R.H. Victora and L.M. Falicov, *Phys. Rev.* **B 31** (1985) p. 7335.
25. C.L. Fu and A.J. Freeman, *Phys. Rev.* **B 33** (1986) p. 1755.
26. O.K. Andersen and O. Jepsen, *Physica* **B 91** (1977) p. 317.
27. P.D. Johnson, to be published.
28. F. Herman, J. Sticht, and M. Van Schilfgaarde, *J. Appl. Phys.* 69 (1991) p.4783.
29. S. Ohnishi, M. Weinert and A.J. Freeman, *Phys. Rev* **B 30** (1984) p. 36
30. Y. Chang, K. Garrison and P.D. Johnson, to be published.

VIII. ELECTRON DIFFRACTION AND HOLOGRAPHY

Surface and Near-surface Atom Imaging with Photoelectron Holography

L. J. Terminello
Lawrence Livermore National Laboratory
Livermore, CA 94550
and
J. J. Barton and D. A. Lapiano-Smith
IBM Research
Yorktown Heights, NY 10598

Abstract

Near-surface atoms from a Cu(001) surface were imaged using 3p photoelectron holograms. The angle-dependent electron diffraction patterns that constitute an electron hologram were Fourier transformed to create a three dimensional, real-space image of the scattering atoms. We have made use of a multiple-wavenumber, phased summing method to improve the atom imaging capabilities of exerimental photoelectron holography using the Cu(001) prototype system.

1 Introduction

The recently developed atom imaging technique of photoelectron holography has generated much interest as a probe of single crystal materials and, eventually, more challenging abrupt interfaces. Suggested by Szöke,[1] and subsequently formulated by Barton,[2] the electron holography technique has undergone a series of refinements to improve the quality and reliability of the atom images. Most of these improvements — such as scattering factor compensation, reduced angular windowing, and Gaussian elimination methods[3-5] — have been demonstrated mainly on simulated data, but some of the experimental demonstrations to date of the holography technique have also benefited.[6-10] However, most of these methods require *a priori* knowledge of the system being studied to achieve image improvement, particularly the suppression of conjugate atom images.

Another method for improving the quality of holographically derived atom images was developed by Barton.[11, 12] He showed with simulated holograms that a third-axis transform-approximation — a phased-sum of multiple-wavenumber holograms

— can suppress twin images and multiple scattering artifacts that are a natural consequence of hologram inversion. We show in this work the application of this image improvement method on experimental holograms and have chosen the Cu(001) bulk system using the photoejected Cu 3p core-state as the coherent wave source for atom illumination. This method is well served by vacuum ultra-violet and soft x-ray synchrotron radiation because of the tunability of the monochromatic light. One attractive feature of the multiple-wavenumber, phase-summing formalism is that it requires only experimentally observable information (the hologram and its energy) to create a higher fidelity atom image.

2 Background

The electron scattering physics that produces a photoelectron hologram is experimentally the same as photoelectron diffraction.[13] However, when measured as an electron angular distribution, the diffraction (interference) between the core-level electron measured directly from the emitter (i.e., the reference beam), and the same photoelectron wave that has scattered off neighboring atoms (i.e., the object wave), form a true hologram. The electron scattering physics that forms a hologram is shown schematically in Fig. 1. A hologram's chemical specificity is a consequence of using element-specific core-level photoelectrons, and oxidation-state and interfacial sensitivity can be achieved when high-resolution, synchrotron-radiation photoelectron spectroscopy is used to select the oxidation-state or interfacially shifted core-level electron.

The elegance of the holography technique permits the multiple-angle photoelectron interference pattern to be two-dimensional Fourier transformed to produce a real-space, three-dimensional intensity map of the electron-scattering atoms neighboring the photoemitter.[5–8] Even without transforming the holograms, the electron forward scattering, which is implicit in patterns measured from buried systems, yields important structural information. The powerful combination of synchrotron radiation and photoelectron holography also enables higher resolution images of atoms to be created through multiple-energy filtering of the resultant images.

3 Experimental

We measured copper 3p photoelectron holograms from a clean, Cu(001) crystal face using well established cleaning and annealing procedures.[14] The surface was checked with core-level photoelectron spectroscopy for contaminants, and the Cu valence-band photoemission features were used to orient the sample with respect to the analyzer. All photoemission spectra were recorded using an Eastman ellipsoidal mirror analyzer that has been described previously.[15] This angle-resolving, energy-band-pass electron analyzer characterized the Cu sample as an angle- integrating detector and also measured the 3p photoelectron holograms in an angle-resolved mode.

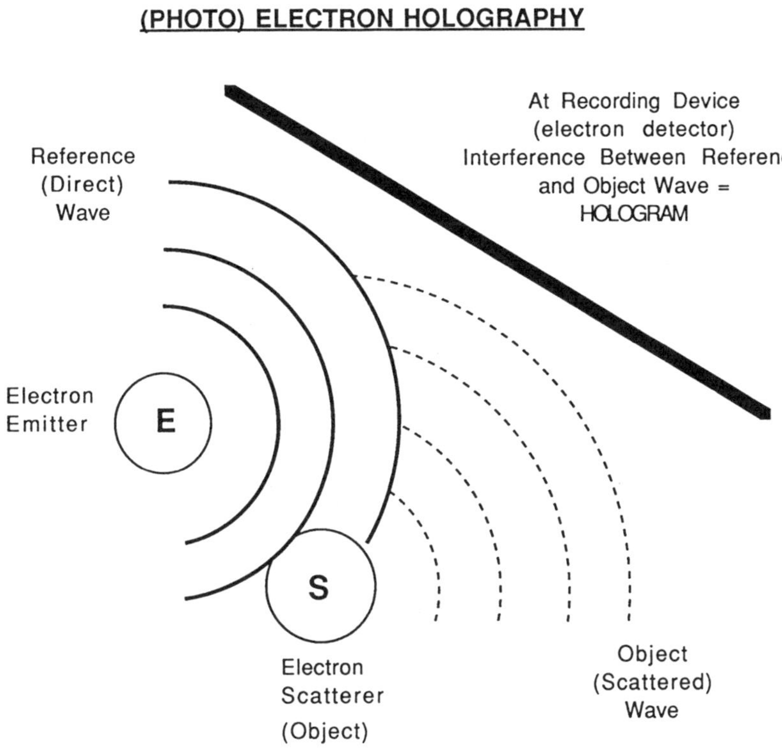

Figure 1: A schematic that illustrates the diffraction process that forms an electron holography is shown. A localized electron is ejected from an atom (electron emitter) at some fixed kinetic energy and propagates towards an angle resolved electron detector. Part of that same electron wave scatters off of neighboring atoms *en route* to the electron detector. The interference between these two waves contains the structural information in the form of the path length difference incurred during the scattering process.

Measuring the Cu 3p multiple-energy electron angular distribution patterns at several energies mandated the use of monochromatic synchrotron radiation as the excitation source so that photoelectron holograms over a wide kinetic energy range could be obtained. We conducted these measurements at the National Synchrotron Light Source on the IBM/U8 beamline[16] This facility provided the 316 – 560 eV photons needed for ejecting the Cu 3p core electrons. Electron angular distributions were measured at nine kinetic energies ranging from 244 to 477 eV. This range had several experimental advantages over a higher energy which could produce better resolution:[6] the cross section for Cu 3p photoemission is well matched with the flux throughput of the monochromator, and non-forward scattering intensities are large in this region of k-space.[17] This implies that electron back scattering atoms as well as forward scatterers may be observed through this choice of energy range.

4 Results

From the raw electron angular distribution patterns, holograms were extracted using methods that have been described elsewhere.[18] Three of the holograms measured in this work are pictured in Fig. 2 where χ_1 was measured at 244 eV, χ_2 at 322 eV, and χ_3 at 477 eV. Fourier transforms were used to invert all nine holograms into three-dimensional, real-space, atom intensity. The numerical inversions of the three holograms pictured in Fig. 2 are shown schematically below their respective source data. The visible slice in each case is a cut through the Fourier volume taken through the first layer of electron forward scatterers. Note the four high intensity spots around the center (the surface normal) that can be assigned to the four scatterers. Also note that going from lower to higher kinetic energy, the lateral resolution of each atom image improves.

Phase-summing of the nine Cu 3p hologram reconstructions was achieved using the simple, multiple-energy transform approximation presented by Barton.[12] Briefly, this approximation can be described as:

$$A(\mathbf{r}) = \sum_i F_i(k_i, \chi_i, \mathbf{r}) e^{-ik_i\mathbf{r}} \tag{1}$$

where the resultant real-space ($\mathbf{r}$) scattering atom intensity, A($\mathbf{r}$), is given by a sum over the individual atom reconstructions $F_i(k_i, \chi_i, \mathbf{r})$ — which were computed from the holograms $\chi_i(k_i)$ measured at each wave vector value k_i, times exp(-$ik\mathbf{r}$) — the phase term that isolates the real, single-scattering contribution to the reconstruction. The phase-dependence as a function of wavenumber is different for the conjugate image than the true image thereby leading to suppression of the twin image. The result of the nine-energy phased sum is shown as a volume contour at 70 % of maximum in Panel A of Fig. 2. The five atoms directly above the emitter are visible in this image and are the highest intensity features. When compared to the ideal atom image of the Cu(001) lattice in panel B, and to the 477 eV result in Panel C, it is apparent

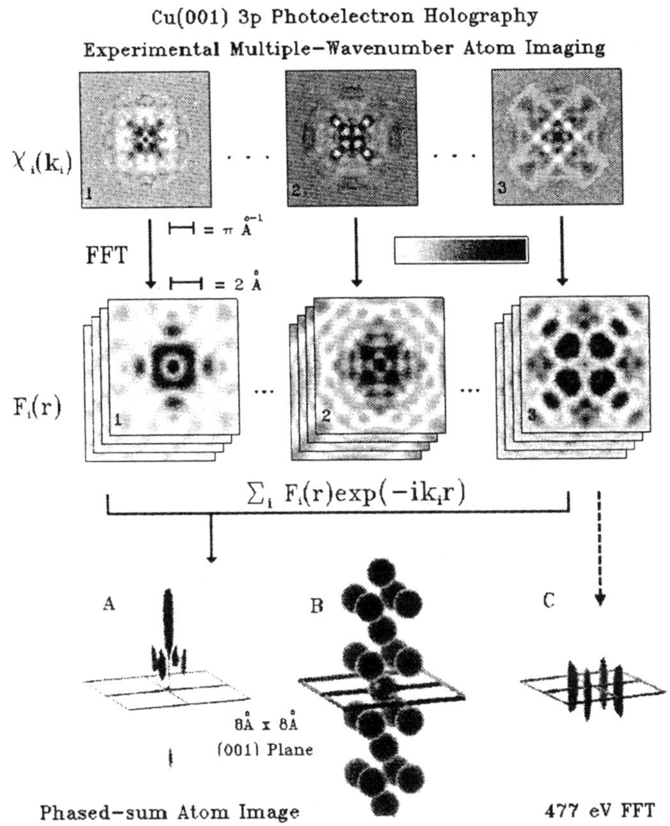

Figure 2: Several steps in the experimental multiple-wavenumber atom imaging method are shown. A series of holograms (Cu 3p $\chi_i(k_i)$) measured at several wavenumbers (k_i, 1 = 244 eV, 2 = 322 eV, 3 = 477 eV) are numerically inverted to form a series of real space atom images $F_i(r)$. The amplitude of a nine wavenumber phase-summed reconstruction for Cu(001) is shown in panel A. This iso-contour volume rendering has an 8 × 8 Å reference plane parallel to the (001) face passing through the emitter and is shown next to an ideal atom rendering of the Cu(001) lattice (Panel B) with an identical reference plane. The volume contour depiction of the atom image intensity for 3 is shown in Panel C.

that the quality of the atom imaging has been improved mostly through suppression of the conjugate image. In the phase-summed image we also note that the conjugate image is approximately 0.1 the intensity of the real image.

Scattering atoms appear shifted from their nominal lattice location in the multiple-energy (A) and all single-energy images[6, 8, 19] because of the additional phase introduced by the scattering physics. This has been observed in earlier experimental electron holography studies[6] and Fourier-transformed, energy-dependent, photoelectron diffraction.[13, 20] Forward scattering amplitudes are greater than back scattering intensities throughout the structurally-useful range of wavenumbers and determines why atoms that lie between the emitter and the detector dominate the Fourier reconstruction. In our multiple-energy, phase-summed image, we obtain a radial shift of 0.4 Å for the first layer atoms, and a 0.9 Å shift for the second layer atom located directly above the emitter.

5 Conclusions

It is not clear that the intensity below the emitter in Panel A is entirely attributable to the twin image of the second layer, (forward scattering) atom above the emitter, or to the second layer (back scattering) atom below the emitter; the expected intensities for both should be the same. A more conclusive demonstration of back scattering atom imaging would be afforded by an adsorbate overlayer system. We have performed such an experiment on the c(2x2)Cl/Cu(001) overlayer system where we measured the Cl 2p photoelectron hologram at 220 eV and numerically inverted the result to image the five atom Cu cluster that constitutes the four-fold hollow Cl adsorption site. This result is currently being published. Coupling of the multiple-wavenumber, phase-summing method with this adsorbate system imaging demonstration, we are now prepared to use photoelectron holography on challenging abrupt interfaces, such as buried heterojunctions.

6 Acknowledgements

We would like to thank C. Costas and J. Yurkas for technical assistance with the experiment. This work was conducted under the auspices of the US Department of Energy by the Lawrence Livermore National Laboratory under contract No. W-7405-ENG-48, and was conducted at the National Synchrotron Light Source, Brookhaven National Laboratory, which is supported by the Department of Energy (Division of Materials Sciences and Division of Chemical Sciences of Basic Energy Sciences) under Contract No. DE-AC02-76CH0016.

References

[1] A. Szöke. *Short Wavelength Coherent Radiation: Generation and Applications.* AIP Conference Proceedings No. 147, American Institute of Physics, New York, 1986.

[2] J. J. Barton. *Phys. Rev. Lett.*, 61:1356, 1988.

[3] B. P. Tonner, Zhi-Lan Han, G. R. Harp, and D. K. Saldin. *Phys. Rev. B*, 43:14423, 1991.

[4] S. Y. Tong, C. M Wei, T. C. Zhao, H. Huang, and Hua Li. *Phys. Rev. Lett.*, 66:60, 1991.

[5] G. S. Herman, S. Thevuthasan, Y. J. Kim, T. T. Tran, and C. S. Fadley. *to be published.*

[6] G. R. Harp, D. K. Saldin, and B. P. Tonner. *Phys. Rev. Lett.*, 65:1012, 1990.

[7] Z. L. Han, S. Hardcastle, G. R. Harp, H. Li, X. D. Wang, J. Zhang, and B. P. Tonner. *Surf. Sci.*, 258:313, 1991.

[8] S. Thevuthasen, G. S. Herman, A. P. Kaduwela, R. S. Saiki, Y. J. Kim, and C. S. Fadley. *Phys. Rev. Lett.*, 67:469, 1991.

[9] H. Huang, Hua Li, and S. Y. Tong. *Phys. Rev. B*, 44:3240, 1991.

[10] C. M. Wei, T. C. Zhao, and S. Y. Tong. *Phys. Rev. Lett.*, 65:2278, 1990.

[11] J. J. Barton and L. J. Terminello. In S. Y. Tong, M. A. Van Hove, X. Xide, and K. Takayanagi, editors, *Structure of Surfaces III, Milwaukee, WI, USA*, page in press. Springer-Verlag, 1991.

[12] J. J. Barton. *Phys. Rev. Lett.*, 67:3106, 1991.

[13] L. J. Terminello, X. S. Zhang, Z. Q. Huang, S. H. Kim, A. E. Schach von Wittenau, K. T. Leung, and D. A. Shirley. *Phys. Rev. B*, 38:3879, 1988.

[14] C. C. Bahr, J. J. Barton, Z. Hussain, S. W. Robey, J. G. Tobin, and D. A. Shirley. *Phys. Rev. B*, 35:3773, 1987.

[15] D. E. Eastman, J. J. Donelon, N. C. Hien, and F. J. Himpsel. *Nucl. Instrum. and Meth.*, 172:327, 1980.

[16] F. J. Himpsel, Y. Jugnet, D. E. Eastman, J. J. Donelon, D. Grimm, G. Landgren, A. Marx, J. F. Morar, C. Oden, R. A. Pollack, J. Schneir, and C. Crider. *Nucl. Instrum. Methods Phys. Res.*, 222:107, 1984.

[17] J. J. Barton and D. A. Shirley. *Phys. Rev. A*, 32:1019, 1985.

[18] L. J. Terminello and J. J. Barton. *to be published.*

[19] D. Hardcastle, Z.-L. Han, G. R. Harp, J. Zhang, B. L. Chen, D. K. Saldin, and B. P. Tonner. *Surf. Sci.*, 245:L190, 1991.

[20] J. J. Barton, C. C. Bahr, S. W. Robey, Z. Hussain, E. Umbach, and D. A. Shirley. *Phys. Rev. B*, 34:3807, 1986.

VALENCE PHOTOELECTRON DIFFRACTION: FROM PARTIAL DENSITIES OF STATES TO FERMI SURFACES

J. OSTERWALDER, A. STUCK, T. GREBER*, P. AEBI, L. SCHLAPBACH
Institut de Physique, Université de Fribourg,
CH-1700 Fribourg, Switzerland

and

S. HÜFNER
Fachbereich Physik, Universität des Saarlandes,
D-6600 Saarbrücken, Germany

ABSTRACT

Angular distributions of *energy-integrated* valence photoelectron spectra are measured and compared to those of localized core emission. A striking similarity is found at high energies (1200eV) *and* at very low energies (12eV), where direct transition effects dominate the spectra. This suggests a *sum rule* for angle-resolved ultraviolet photoemission which dictates, for a single energy spectrum, how much total intensity can be distributed over a set of direct transition peaks.

1. Introduction

Photoemission from core levels and from valence levels is usually treated very differently. In core emission, the well-localized initial state forms a prerequisite for chemical sensitivity, multiplets, strong relaxation effects and also for the common photoelectron diffraction (PD) in the final state.[1-3] The latter phenomenon has been very successfully described by theories considering coherent superposition of a photoelectron source wave emitted from a single atom within the surface, with all the waves that have been singly or multiply scattered by nearby atoms.[3-6] The measurement of core-level anisotropies from single-crystal surfaces, particularly at x-ray energies, has since developed into a widely used technique for structure determination in surface science.[3]

On the other hand, valence photoemission usually involves delocalized initial states, and should in principle be considered as a coherent emission from many lattice sites. Here, the theoretical approach has been to treat both initial and final states as Bloch waves with wave numbers $\mathbf{k}_i$ and $\mathbf{k}_f$, and to use energy and momentum conservation in order to locate direct transitions (DT) between such states within the band structure. By tracing the measured energy positions of such transitions as a function of photon energy and/or emission angle the band structure can in principle be mapped out.[7,8] Photoelectron diffraction effects are considered implicitly, if the final-state band structure is known exactly.

It is noteworthy that in angle-resolved core emission one usually measures *intensities* while in valence emission one measures energy *positions* as a function of angle. Relative intensities of DT peaks within the same spectrum have only been evaluated qualitatively in order to draw some conclusions about the symmetries of the initial and final electronic states (matrix element effects).[7,8] Only very recently, some work has been published presenting measured angular distributions of absolute intensities within narrow energy windows close to the Fermi energy[9] or at higher binding energies.[10]

In this paper we try to combine these rather different viewpoints to provide a description of the angular intensity distributions for both core and valence photoelectrons. We will begin our discussion with high-energy photoemission of highly delocalized valence electrons. The measuring conditions here are such as to effectively average initial states over the entire Brillouin zone (BZ). The energy spectra are therefore essentially representative of the density of states, disregarding cross section effects, and the spectral shape varies little or not at all with emission angle. We find that the angular distribution of the energy-integrated valence band intensity is essentially identical to that of a near-lying core level, and is thus determined by final-state scattering only.

We then present similar experiments at rather low energies of the order of 10eV. Quite surprizingly, the energy-integrated 3d band intensity of copper shows again a very similar angular distribution as that calculated for emission from a localized d orbital, although the individual energy-resolved spectra vary quite dramatically with angle as direct transitions disperse up and down. Apparently, the energy-integrated emission reflects the atomic orbitals of which the initial band states are composed. The so-far missing connection between core and valence emission is thus established, at least for relatively localized states as found in the copper 3d band.

2. Experimental

For both x-ray and UV excited photoemission experiments we follow a brute force approach to measure angular scans or the full hemispherical (2π) intensity distributions, which has proven to be very accurate, stable and quite successful. While the highly angle-resolving hemispherical analyzer is at fixed position relative to the excitation source, the sample is mounted on a two axis goniometer, permitting computer-controlled rotation about both a polar and an azimuthal axis. In the case of 2π scans, data acquisition proceeds sequentially in an azimuthal way, starting near grazing emission, and decreasing the polar angle by 2° after each completed 360° scan, thus covering a total of about 3600 angular settings.[11] At each setting, a spectrum over an arbitrary energy range can be measured and integrated. Depending on the signal strength acquisition times are between 2 and 30 hours.

3. Valence Emission at X-Ray Energies

Aluminium is a nearly-free electron metal with highly delocalized valence states. In Fig. 1 we give the angle-resolved

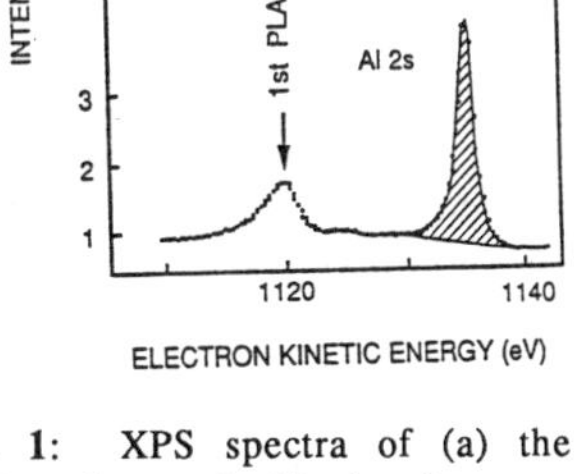

Fig. 1: XPS spectra of (a) the valence band region and (b) the 2s core level of Al, recorded along [101] direction of a clean Al(001) single crystal. The shaded areas represent the intensities that were calculated at each angular setting to obtain the curves in Fig. 2.

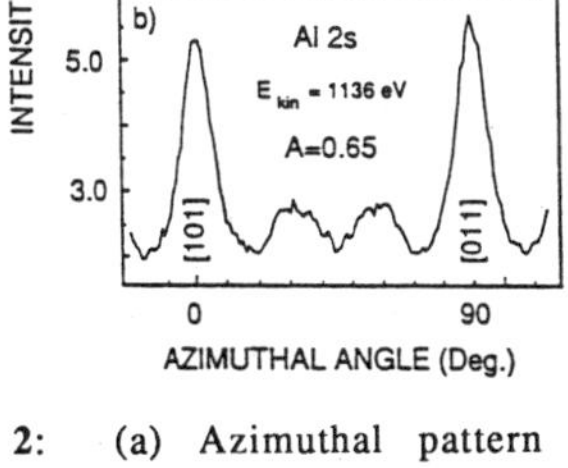

Fig. 2: (a) Azimuthal pattern of the integrated valence band measured at a polar angle of 45° from Al(001). The angular resolution ΔΩ was better than 3° full acceptance cone. (b) Same as (a) except that here 2s electrons were measured.

valence band and 2s core spectra, excited by Mg Kα radiation of 1253.6eV. Fig. 2 shows the angular variation of the integrated signals along an azimuthal scan passing through <110> nearest-neighbor directions of this (001) surface. The significant result of this experiment is the striking similarity of the two curves in shape and also in absolute anisotropies $A=I_{max}-I_{min}/I_{max}$, which are 0.65 in both cases.[12] The two emission processes thus produce very similar final states as far as the photoelectron diffraction is concerned, which necessarily must have the character of the emission from a localized orbital. Hole-state localization [12] and localization by BZ averaging [13] have been put forward to explain this phenomenon.

Independently of the mechanism behind, this finding is rather useful if applied to ordered alloys and compounds. As an illustrative example let us consider valence emission from $AuCu_3(001)$ (Fig. 3). Here, the contributions of Au and Cu to the valence band are rather

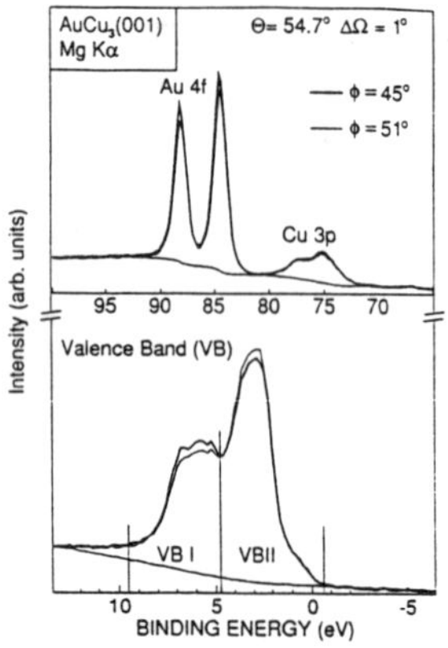

Fig. 3: Au 4f and Cu 3p core-level and valence-band spectra of $AuCu_3(001)$ at two azimuthal angles, recorded at a fixed polar angle of 54.7° off normal. The two angles chosen show strong diffraction differences between Au and Cu.

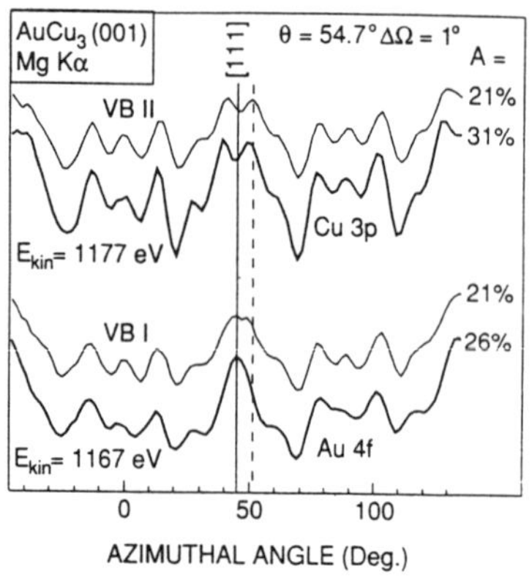

Fig. 4: Comparison of the Cu 3p and Au 4f core levels and associated valence band diffraction patterns, 54.7° off normal. Intensities have been integrated as indicated in Fig. 3.

well separated in energy,[14] and the angular variation of both types of emission, marked as VB I (Au) and VB II (Cu) in Fig. 3, respectively, can be measured independently. Fig. 4 shows azimuthal intensity curves passing through the [111] direction, which exhibit small but significant differences between VB I and VB II emission. These same differences are also found in the azimuthal curves for the nearby Cu 3p and Au 4f signals, permitting a straightforward assignment of elemental emission in the valence band. Evidently, the valence emission patterns again correspond to those expected from localized sources, either of the Au or Cu type. The x-ray excited valence spectrum can thus be viewed as simply an incoherent superposition of a Au spectrum and a Cu spectrum. On this basis, Stuck et al.[15] have demonstrated that a channel-by-channel evaluation of diffraction differences, which are measured with the core levels, can be used to deconvolute the measured spectrum into the partial densities of states.

Aluminium and $AuCu_3$ represent cases in which the BZ averaging is essentially complete at x-ray energies, mainly due to the limited angular resolution and a high level of phonon-assisted indirect transitions, where momentum conservation involves additional phonon momenta.[16] Recently, Herman et al.[17] carried out a similar experiment with tungsten, where the lattice vibrates much less at room temperature and valence spectra still show remnants of direct transitions. They find small differences in azimuthal angle scans of W 4f and integrated valence emission that do, however, not correlate with the appearance of a distinct DT feature in the spectra. Still, the authors attribute these differences to "remnant direct-transition selection rules" *and* to the different angular momentum character of the initial states, i.e. 4f and mainly 5d, respectively. Friedman and Fadley[18] had previously shown by means of single-scattering-cluster (SSC) calculations using proper spherical-wave final states that such angular momentum effects exist even in high-energy photoelectron diffraction.

4. Valence Emission at UV Energies

In order to further assess the effects of direct transitions on the angular distribution of the energy-integrated valence spectra, we have chosen to perform experiments at much lower energies where direct transitions dominate the spectra. Copper represents the one element that has been the most extensively studied by means of angle-resolved photoemission, and there exists a wealth of experimental and theoretical data on its band structure.[8] In particular, the upper and lower bounds of the fully occupied 3d band are well known to be at 2.0eV and 5.6eV binding energy, respectively. In Fig. 5 we give a series of angle-resolved ultraviolet photoelectron spectra (He I, hν=21.2eV) from Cu(111), taken with an angular resolution of the order of 3° full acceptance cone. The spectra have been measured at 5° intervals over one symmetry-equivalent stretch of 120° in azimuthal angles ϕ, at an arbitrary polar angle of 66° off normal. The origin of the ϕ-axis is defined by the ΓLUX plane, and the parallel component $\mathbf{k}_{\|}$ of $\mathbf{k}$ is close to the zone boundary at the X point for ϕ=0°, which is, however, of no importance for our

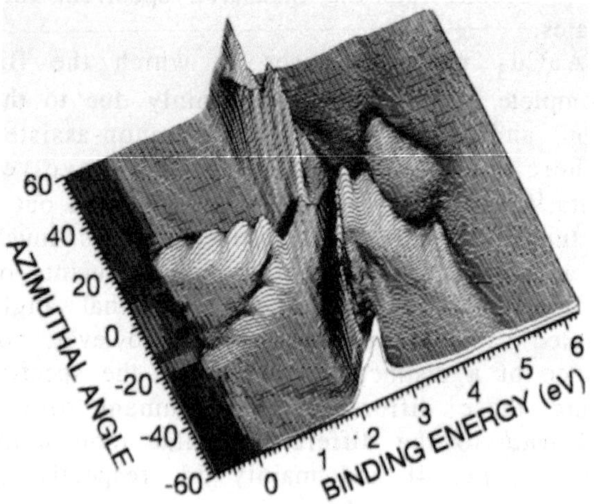

Fig. 5: Series of electron spectra from Cu(111), excited by He I radiation (hν=21.2eV). Spectra were taken every 5° of azimuthal emission angle, at a polar angle of 66° off normal. The origin of the azimuthal angle scale lies in the ΓL U X plane. Sinusoidal oscillations on the parabolic sp band (0-2eV) are an artifact due to the two-dimensional interpolation of the data to produce this graph.

present reasoning. The purpose of Fig. 5 is simply to demonstrate that the spectra are essentially composed of direct transition peaks that show rather pronounced dispersion effects. The free-electron-like parabola of the sp band can be clearly seen between the Fermi energy and 2eV binding energy. In the d band, direct transitions are found moving over the entire bandwidth.

To demonstrate the effect of direct transitions on energy integrated angular distributions, we have divided the d band binding energy region into three arbitrary windows of equal widths: from 2-3.2eV (window 1), from 3.2-4.4eV (window 2) and from 4.4-5.6eV (window 3). In the upper part of Fig. 6 we give the full angular distributions as measured with these three energy windows. Clearly they are all rather different, indicating that the anisotropies are

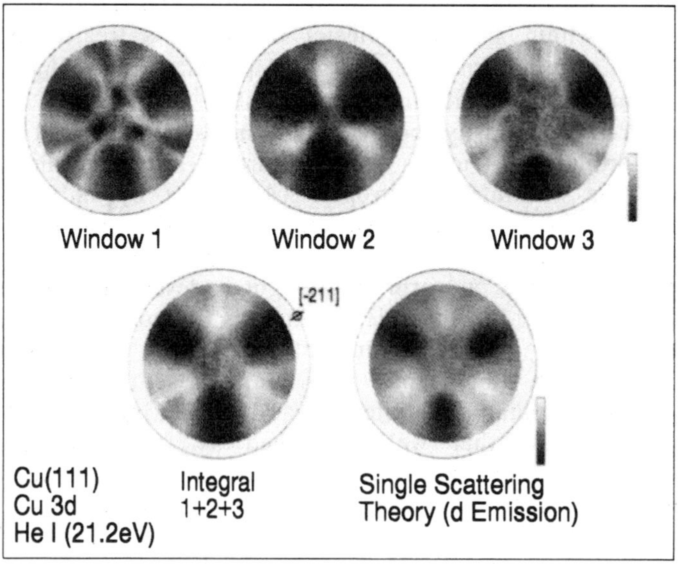

Fig. 6: Complete angular distributions of Cu 3d electrons from Cu(111), shown in the stereographic projection, with intensities given in a linear grey scale. To obtain these patterns, energy spectra have been integrated over different binding energy ranges: window 1 (2.0-3.2eV), window 2 (3.2-4.4eV), window 3 (4.4-5.6eV) and integral (2.0-5.6eV). A calculated pattern for emission from a localized d orbital (single scattering) is also shown.

largely determined by direct transition peaks moving from one window to the next. However, there are also some similarities, and in particular the wide and deep intensity depressions around the [-211]-type azimuths are common to all windows.

Now that we have established that DT effects indeed do influence angular distributions, we proceed to consider the angular distribution of the complete d band, which is given simply by the sum of the three separate energy windows. Contributions due to the spectral background are contained in this sum and are neglected in the further discussion. We compare this d band distribution to the one obtained by means of an SSC calculation, using a localized d orbital as initial state. The spherical-wave scattering code of Friedman and Fadley[18] has been used, which considers emission into proper p and f final state waves that are coupled by atomic photoemission radial matrix elements and phase shifts,[19] and with quantization axes corresponding to the photon polarization vectors. The scattering within a limited cluster of atoms is treated using the Rehr-Albers approximation,[20] with partial wave phase shifts obtained from standard muffin-tin potentials.[21]

The strong similarity between the measured d band and the simulated d orbital distributions (bottom part of Fig. 6), which is also reflected in the absolute anisotropies (A differs by only 15%), is remarkable for two reasons. First, single scattering calculated at a mean band kinetic energy of only 12.5eV with respect to the vacuum level appears to give a surprizingly good description of angular distributions at these low energies, which is in contrast to common belief. We do not want to comment further on this point here. Second, and more importantly for our present purposes, this comparison shows that the integrated d band intensity shows a distribution that is rather well described by final-state scattering from a localized source, and which reflects the orbital symmetry of the atomic states that enter the tight-binding description of the d band. This latter point is corroborated by SSC calculations carried out for the same energy but using s or p orbitals as initial states (not shown here). The angular distributions differ very strongly for each case, i.e. at these low energies the solid, through final-state scattering, acts as a very efficient angular momentum filter.

We have repeated the same experiment for Cu(111) using HeII radiation at a photon energy of 40.8eV (not shown). The mean band kinetic energy is here 32.5eV, and we find again the very striking similarity between the fully integrated d band anisotropy and a SSC calculation using d emission. Also here, the solid discriminates very strongly between different angular momentum initial states.

5. Fermi Surfaces

It is well known that angle-resolved photoemission permits in principle to measure Fermi surfaces. One has to determine over a sufficiently large area in angle and energy space the Fermi energy crossings of the bands. With conventional experimental setup this is a very tedious procedure. The situation is more favourable for two-dimensional electronic states such as e.g. surface states.[22] Here, the momentum component $\mathbf{k}_{||}$ parallel to the surface is the only relevant quantum number, and it is conserved modulo a vector of the surface reciprocal lattice. It is easily determined by the electron energy and emission angle. The usual procedure is then to follow the dispersion of a surface state as a function of $\mathbf{k}_{||}$, and the surface Fermi surface is then obtained by connecting all $\mathbf{k}_{||}$ points where the surface state cuts the Fermi energy. Recently, Santoni et al.[9] have demonstrated an alternative way for obtaining the same information: Using a two-dimensional display-type analyzer they mapped the Fermi surface of layered graphite directly by measuring the absolute intensity within a narrow energy window at the Fermi energy. Naturally, bright structures (high intensities) will appear on the display at $\mathbf{k}_{||}$ locations where a direct transition moves through the energy window.

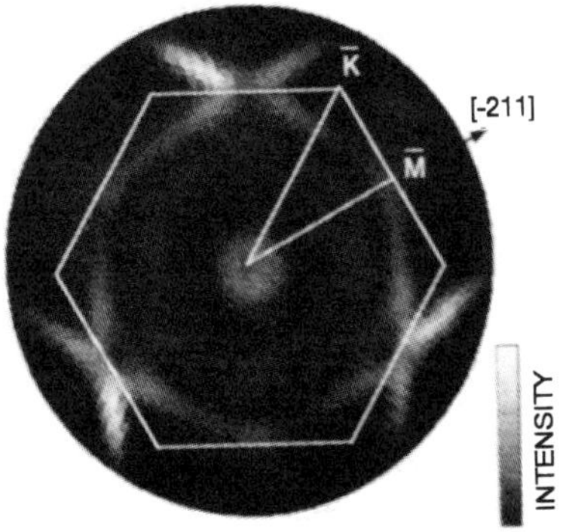

Fig 7: Fermi surface scan from Cu(111), excited by He I radiation (21.2 eV). Absolute intensities within a narrow energy window ($\pm$0.2 eV) around the Fermi energy have been mapped over all emission angles. The data is presented in the parallel projection. The surface BZ is also indicated.

We have adopted this procedure, using our sequential angle-scanning acquisition mode for obtaining intensity displays. In Fig. 7 we give such a Fermi surface scan for Cu(111). Note that the angular projection is here different from that of Fig. 6 in order to permit a linear measurement of $\mathbf{k}_{||}$. Well defined sets of lines are seen in this Figure. The interpretation of these lines is not as straightforward as in the two-dimensional cases, because we try to map a three-dimensional Fermi surface, which is well known for copper[23] onto a two-dimensional display. Still, the lines must be identified with direct transitions, and energy conservation demands that the initial states lie on the Fermi surface. Furthermore, the lines appear to show the proper translational symmetry in reciprocal space, as indicated by the surface BZ in Fig. 7. We thus conclude that these lines represent some reasonably well defined cut through the Fermi surface of copper. We have further confirmation for this interpretation from a similar measurement on Cu(001). It thus appears that such angular patterns should be useful for obtaining direct information on Fermi surface topology also in three-dimensional systems. The small, ringlike structure in the center of Fig. 7, by the way, represents the surface Fermi surface of the well-known surface state of Cu(111).[7]

6. Discussion and Conclusions

We have no reason to expect that the correspondence of energy-integrated valence emission to emission from a localized source is limited to the case of copper. It should be a rather general effect, at least for fully occupied bands. Let us therefore further discuss the consequences of this rather surprizing finding.

We consider a single, angle-resolved energy spectrum, such as one of those depicted in Fig. 5. Direct transitions dominate this spectrum. If we now want to learn something about the intensities of these DT peaks, we realize that for each emission angle we know that the energy integral over the band is equal to a number that can be calculated with relative ease. A multiple scattering calculation for the same localized orbital emission should give these numbers even more quantitatively. Of course this gives only the integral intensity added over the whole d band in the case under consideration. In fact, our experiment suggests a *sum rule*, which connects the absolute intensities of DT peaks in an angle-resolved photoemission spectrum, independently for all emission angles. If we are dealing with only one band (see below) the local final-state scattering calculation should give us directly the intensity of the single direct transition.

In the case of the Fermi surface scans (Fig. 7) the absolute intensities along the observed lines are modulated by what is usually attributed quite generally to matrix element effects. In view of our new results on energy-integrated band intensities, we may, however, consider these intensity modulations from a different view. In copper, the Fermi surface is determined solely by the simple sp band, which disperses rather strongly as shown e.g. in Fig. 5. Since there is only one band, only one direct transition can be detected at each angle, which thus takes the full intensity as dictated by the above mentioned sum rule. The intensity modulations should thus follow those obtained by a local scattering calculation using the appropriate initial state symmetry. This point is currently under investigation.

7. Acknowledgements

We greatfully acknowledge experimental help by D. Naumovic and R. Fasel, as well as fruitful discussions with H. Castella and D. Baeriswyl. Special thanks go to D. Friedman and C. S. Fadley for sharing with us their spherical-wave scattering code. This work was supported by the Fonds National Suisse and the NFP 30.

8. References

* Current address: Fritz-Haber Institut der Max-Planck Gesellschaft, Faradayweg 4-6, D-1000 Berlin 33, Germany.

1. C. S. Fadley, in *Electron Spectroscopy, Theory, Techniques, and Applications*, eds. C. R. Brundle and A. D. Baker (Pergamon Press, 1978), Vol. II, Chap. 1.
2. A. Kotani, in *Core-Level Spectroscopy in Condensed Systems*, eds. J. Kanamori and A.Kotani (Springer, Berlin 1988) p. 3.
3. C. S. Fadley, in *Synchrotron Radiation Research: Advances in Surface and Interface Science*, ed. R. Z. Bachrach (Plenum, New York 1992) Chap. 11.
4. A. Liebsch, Phys. Rev. Lett. **32** (1974) 1203; and Phys. Rev. B **13** (1976) 544.
5. C. H. Li, A. R. Lubinsky, and S. Y. Tong, Phys. Rev. B**17** (1978) 3128.
6. J. J. Barton and D. A. Shirley, Phys. Rev. B **32** (1985) 1906.
7. E. W. Plummer and W. Eberhardt, Adv. Chem. Phys. **49** (1982) 533.
8. R. Courths and S. Hüfner, Phys. Rep. **112** (1984) 53.
9. A. Santoni, L. J. Terminello, F. J. Himpsel, and T. Takahashi, Appl. Phys. A **42** (1991) 299.
10. F. J. Himpsel, L. J. Terminello, D. A. Lapiano-Smith, E. A. Eklund, and J. J. Barton, Phys. Rev. Lett. **68** (1992) 3611.
11. J. Osterwalder, T. Greber, A. Stuck, and L. Schlapbach, Phys. Rev. B **44** (1991) 13764.
12 J. Osterwalder, T. Greber, S. Hüfner, and L. Schlapbach, Phys. Rev. Lett. **64** (1990) 2683.
13 D. D. Sarma, W. Speier, and J. F. van Acker, Phys. Rev. Lett. **66**

(1991) 2834 (Comment); and J. Osterwalder, T. Greber, L. Schlapbach, and S. Hüfner, Phys. Rev. Lett. **66** (1991) 2835 (Reply).

14. G. S. Sohal, C. Carbone, E. Kisker, S. Krummacher, A. Fattah, W. Uelhoff, R. C. Albers, and P. Weinberger, Z. Phys. B **78** (1990) 295.
15. A. Stuck, J. Osterwalder, T. Greber, S. Hüfner, and L. Schlapbach Phys. Rev. Lett. **65** (1990) 3029.
16. R. C. White, C. S. Fadley, M. Sagurton, and Z. Hussain, Phys. Rev. B **34** (1986) 5226.
17. G. S. Herman, T. T. Tran, K. Higashiyama, and C. S. Fadley, Phys. Rev. Lett. **68** (1992) 1204.
18. D. J. Friedman and C. S. Fadley, J. Electron Spectrosc. Relat. Phenom. **51** (1990) 689.
19. S. M. Goldberg, C. S. Fadley, and S. Kono, J. Electron Spectrosc. Relat. Phenom. **21** (1980) 285.
20. J. J. Rehr, R. C. Albers, Phys. Rev. B **41** (1990) 8139.
21. J. B. Pendry, *Low Energy Electron Diffraction* (Academic Press, London 1974).
22. S. D. Kevan, Phys. Scr. **T31** (1990) 32.
23. D. Shoenberg, D. J. Roaf, Phil. Trans. Roy. Soc. **255** (1962) 85.

IX. FLUORESCENCE

Soft X-ray Fluorescence Spectroscopy Excited by Synchrotron Radiation: Threshold Effects

D.L. Ederer and K.E. Miyano
Tulane University
New Orleans, LA. 70118, USA

W.L. O'Brien,* T.A. Callcott, Q.-Y. Dong, and J.J. Jia
University of Tennessee
Knoxville TN, 37996 USA

D.R. Mueller
National Institute of Standards and Technology
Gaithersburg, Md, 20899, USA

E.T. Arakawa
Oak Ridge National Laboratory
Oak Ridge, TN, 37930, USA

ABSTRACT

X-rays have been used as a scientific tool for almost 100 years. The presence of x-ray fluorescence has been detected almost since the day x-rays were discovered. This paper will be a report on the valence band emission intensity as a function of the energy of the photon excitation in the 100-200 eV energy band. Emission measurements from silicon and several wide band gap insulators will be presented.

1. Introduction

X-rays have been used as a scientific tool for almost 100 years[1]. Fluorescence x-rays have been detected in the hard x-ray region almost since their discovery. The use of the soft x-ray region of the spectrum for condensed matter research has lay dormant almost until the middle of the twentieth century, when Skinner used them to obtain information about the band structure of solids[2] Skinner showed how the soft x-ray portion of the spectrum can be utilized for experiments that map out the localized density of states, through transitions between the valence and the core level. This map is modified by the dipole selection rule and one obtains a selected localized density of states of one of the elements in the compound under investigation. Thus, one obtains complimentary information to that obtained by photoelectron spectroscopy.

Soft x-ray fluorescence emission is especially powerful because the shallow core levels have a natural width that is about a factor of ten less than deeply bound levels. Thus the valence band spectrum is not unduly broadened by the width of the core level. Furthermore the fluorescence photons are not affected by electric and magnetic fields, thus insulating samples may be examined without experimental complications. They produce less damage than other probes, and photons scatter less than electrons and ions, and are more bulk sensitive than photoemission. Fluorescence is sensitive to the electronic structure around selected elemental species facilitating the analysis of multicompound samples. X-rays obey the dipole selection rule, and the emission from a specimen also conveys information about the angular momentum symmetry of the electronic states.

The yield hasn't changed in the intervening years since Skinner's seminal work, but technology has advanced. High sensitivity multi-wavelength detectors and better diffraction gratings are available.[3] These in combination with high brightness synchrotron sources make photon-in photon-out a viable and exciting scientific technique. Spectra that took Skinner hours to record on photographic plates by electron beam sources are now obtained in a few minutes using photons for excitation. The advent of bright synchrotron sources permitted their use for excitation. All these qualities more than offset the low florescent yields that dominate x-ray fluorescence in the 50-1000 eV photon energy range.

2. Instrumentation

2.1 Soft x-ray excitation scheme

At present we employ a spectrometer mounted on a beam line at the National Synchrotron Light Source (NSLS)[4]. The spectrometer incorporates toroidal gratings which have a gain of 10 in efficiency over spherical gratings, and a position sensitive detector further enhances the sensitivity of the instrument by another factor of 100.

At NSLS, an electron beam, white light, or monochromatic radiation from the storage ring can be used to excite the sample as shown in Fig. 1.

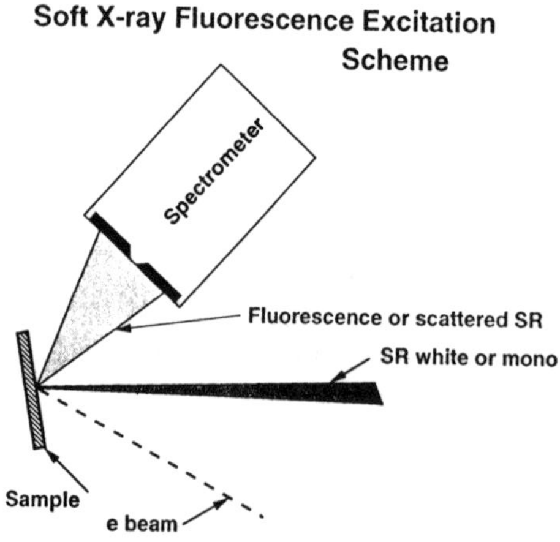

Figure 1

Soft x-ray fluorescence can be excited by an electron beam or by a broad band of radiation from a synchrotron source. A monochromator can also be used to provide a narrow band of excitation.

When white light is used for excitation, the reflected beam from the sample can be detected by the fluorescence spectrometer, and provide a measure of the absorption coefficient of the sample. At near normal incidence, the reflectivity is given by:

$$R=((n-1)^2+k^2)/((n+1)^2+k^2)$$

For values of the refractive index, n, near 1, the reflectivity is proportional to k^2, the square of the extinction coefficient. Thus, one can measure a quantity proportional to the unoccupied density of states by a measurement of the reflectivity, and the occupied density of states can be determined from the intensity of the valence band emission.

The monochromatized beam from the recently installed variable line spaced grating[5] provides a flux to the sample of about 10^{13} photons per second at 100Å in a 0.4 eV bandwidth with an electron current of 500 mA. in the ring.

2.2 Variable line spaced monochromator

This monochromator has several unusual features that are especially important for fluorescence spectroscopy. The layout of the monochromator is shown in Fig. 2. It has fixed input and output axes, good throughput, simple optics, and scans wavelength by the translation and rotation of a plane mirror. The average density of the rulings on the plane

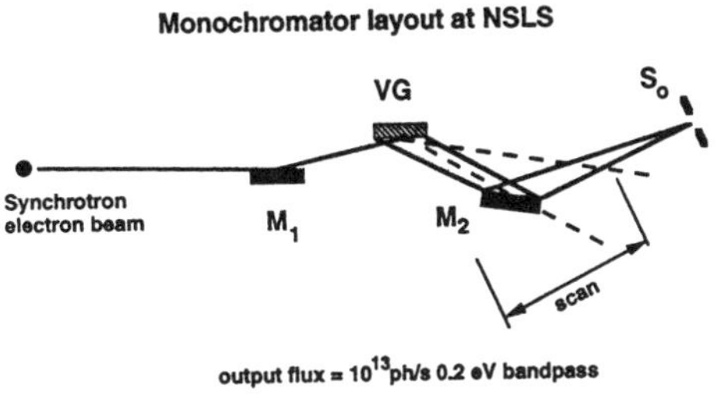

Figure 2 (above)

At the NSLS there is no entrance slit. The synchrotron beam defines the band width of the monochromator. A bent cylinder, M_1, focuses the radiation on the sample. The plane grating, VG, with a grove spacing that increases linearly, disperses the light and the scanning mirror, M_2, illuminates the sample, S_o, with monochromatic light.

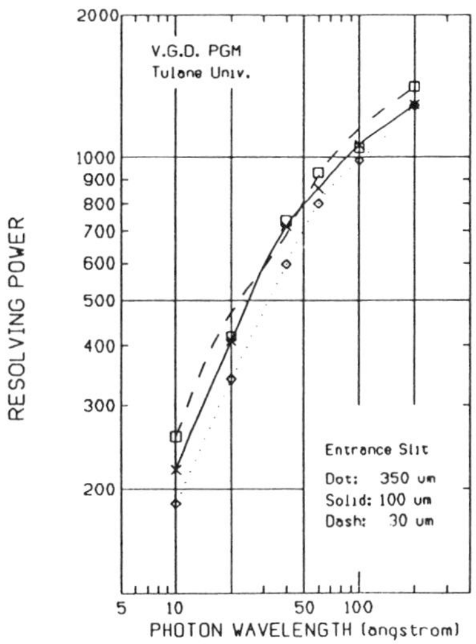

Figure 3 (to the left)

The resolution of the monochromator with a grating having variable spaced grooves. The slit width changes a factor of ten and the resolution changes about 20%.

grating is 1000 lines per mm. The monochromator achieves a resolving power of about 500 at 100Å with a 250 mm exit slit. Finally by ray tracing the optical system we were able to show that the resolving power is not a strong function of the entrance slit width as is shown in Fig. 3. Here a change in the entrance slit width by a factor of four results in a 20% change in the resolving power. This translates into high throughput at bandwidths that are especially suitable for solid state experiments. A complete report on this instrument can be found in Ref. 5

3. Results

3.1 Emission from barium fluoride

Fig 4 is the spectrum recorded during white light excitation of barium fluoride. In this spectrum the features at photon energies greater than 90 eV are due to change in the reflectivity of the sample, caused by the rapidly changing absorption coefficient that occurs when the threshold energy for the removal of a 4d electron is exceeded. The structure observed is in good agreement with measurements of the absorption coefficient of barium fluoride[6], and is very similar to the absorption of doubly ionized barium.[7] The double peaked structure at a photon energy of 102 eV corresponds to transitions between the 4d initial state and a 5f final state in the continuum that is highly localized. The structure in the fluorescence emission at photon energies near 75 eV is due to the filling of the 4d hole produced by the excitation photons whose energy is greater than the energy to produce a 4d hole. The emission from states of p and f symmetry in the valence band appear between 82 and 85 eV photon energy. Two bands appear because of the 2.4 eV spin orbit split 4d core level.

Figure 5a illustrates emission spectra excited by monochromatic phonons of the energies indicated. The spectra are presented following correction for spectrometer response and background; and following division by a factor of photon energy cubed to correct for the energy dependence of the photon density of states within the dipole approximation. The curves have been offset among the vertical axis for clarity. the incident photon beam was characterized by an energy spread corresponding to a full width at half maximum of ~0.8 eV. Several features of the spectra should be noted. The spectra show contributions from the decay of the $4d_{3/2}$ and $4d_{5/2}$ holes that differ in relative intensity as function of excitation energy. Since the excitation of the initial core vacancy occurs on a time scale much faster than its subsequent decay, the excitation and emission process are in-dependent, and changes in the relative intensity of the $4d_{3/2}$ and $4d_{5/2}$terms reflect the energy dependence of the relative photoionization cross sections of the $4d_{3/2}$ and $4d_{5/2}$ levels.

Figure 5b. is a comparison of spectra excited with incident energies of 108 and 134 eV. The relative intensity of the $4d_{3/2}$ and $4d_{5/2}$ contribution changes with photon energy, however, the spectra do not exhibit evidence of multiplet structure due to f-occupancy, which would split the 4d-5p transitions into a series of lines over an energy range of about 5 eV. For excitations greater than about 115 eV, however, the emission spectra exhibit greater intensity in the region near 77 eV than those spectra excited at lower photon energies. The added intensity likely reflects satellite emission arising from the simultaneous excitation of 4p and 5p electrons.

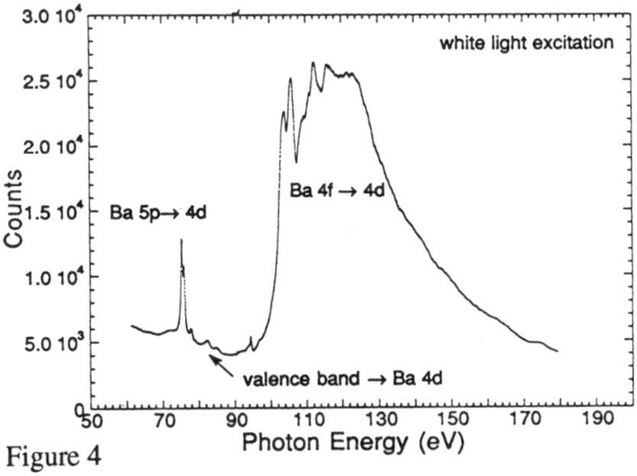

Figure 4

Soft x-ray emission from barium floride excited by white light from the NSLS storage ring. The structure at photon energies greater than 95 eV is due to the increased reflection from the barium fluoride.

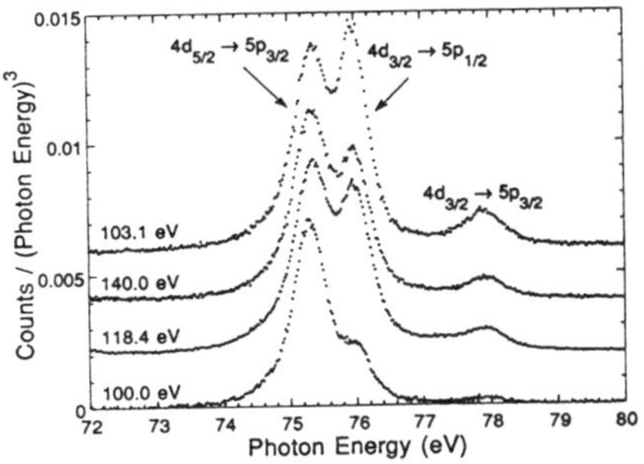

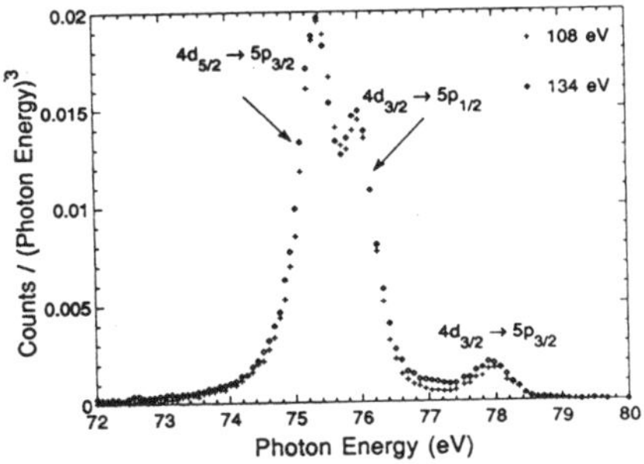

Figure 5a

Intensity of core-core transitions between the 5p and 4d shell in barium fluoride at the photon energies indicated on the left hand side of the graph.

Figure 5b

Intensity of transitions between 5p and 4d shells at 108 and 134 eV photon energy illustrating the possibility of multielectron satellite emission.

The difference in the intensity for the core-core transitions for the two different excitation energies is a measure of the branching ratio for the different cross sections to produce a $4d_{5/2}$ or a $4d_{3/2}$ hole for the two energies. If correlations are small, this branching ratio is a constant as a function of the energy, equal to the statistical weight of the states. In the case of barium 4d photoionization, the value of the ratio is 1.5. An inspection of Fig. 5a clearly reveals that the ratio is not constant, indicating that electron correlations are a factor in the photoionization of barium fluoride. Additionally if the holes are produced by circularly polarized light, it would be possible to observe the dichroic behavior of the unoccupied states through a recording of the emission intensity from the spin-orbit split core-core transitions. Furthermore, it would also be possible to observe dichroic x-ray fluorescence from the occupied states through transitions of the valence electrons to the polarized core hole . [8,9]

3.2 The excitation of silicon $L_{2,3}$ emission vs. photon energy

When a photon excites an electron from a core state to a state in the conduction band, the resulting emission is expected to be independent of the excitation energy. The phenomena that can disturb this behavior are the screening produced by the excitation of an exciton, Raman processes, many electron interactions and phonon relaxation. The soft x-ray spectral wavelength band is especially suited for studies of these processes, for which we will show examples in this and following sections. Silicon is a good candidate for such an investigation because the band structure is well known,[10] and it is easy to obtain high quality samples. The first investigation,[11] made several years ago using a monochromator with a band width of 4 eV, yielded the spectra shown in Fig. 6a. The clearly visible amplitude variation of the emission peaks labeled a,b,and c is shown in Fig.6b. The largest changes take place near the threshold and slowly the spectrum takes on the same form as one excited by electron beam bombardment or by high energy photons. In the paper[11] the variation was attributed to a many body scattering effect that takes place near threshold excitation energy.

More recently K emission spectra from, a single crystal diamond, were obtained[12] using undulator produced narrow band photon excitation near threshold. The resulting variation of the x-ray emission as a function of excitation energy was interpreted as a resonant scattering phenomena, by which the photon momentum and the crystal momentum were conserved. The authors proposed that the electrons were excited by the photons to unoccupied states of specific values of the electron wave vector, **k,** in the Brillouin zone having a high density according to the band structure. If the valence band also had a high density of states, There would be an enhanced emission at a band energy corresponding to the high density of states. If this model is one of the factors that produce a variation of the intensity of emission as a function of photon energy, the emission from an amorphous sample would show little or no change as the photon excitation is changed, because of the lack of crystalline order. We conducted measurements again on amorphous and crystalline silicon[13] using the new monochromator with a narrow band pass to produce excitation photons in a 0.2 eV band pass. The results are shown in Fig. 7. The amorphous sample shows no emission dependence as a function of the excitation energy. On the other hand emission from the crystalline sample shows a pronounced variation as a function of the excitation energy. Excitation energy at 99.7, 100.2 and 101.2 eV occur near high density points X and L in the Brillouin zone. Emission from the valence band should be enhanced at these symmetry points, which is indicated by the arrows labeled 1, for the point of X symmetry and by 2 for the point of L symmetry. Enhancement in the emission is clearly visible for these symmetry directions. The emission spectrum from the valence band gradually resembles the electron beam. spectrum as the excitation energy increases.

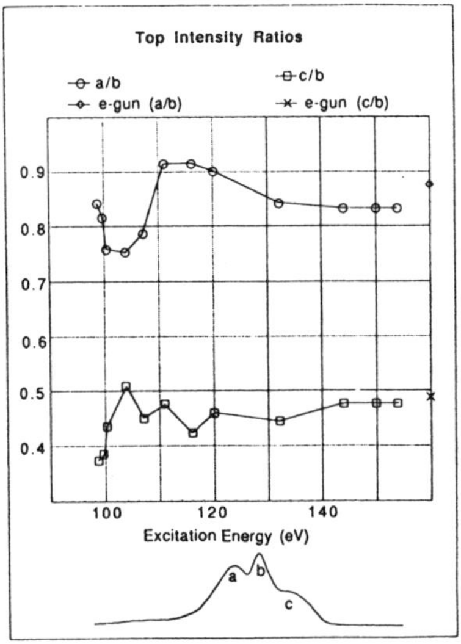

Figure 6

Peak heights as a function of excitation energy. The peaks are denoted according to the scheme below the graph. The cross and the diamond symbols placed at 160-eV excitation refer to the spectrum excited by 2-keV electrons.

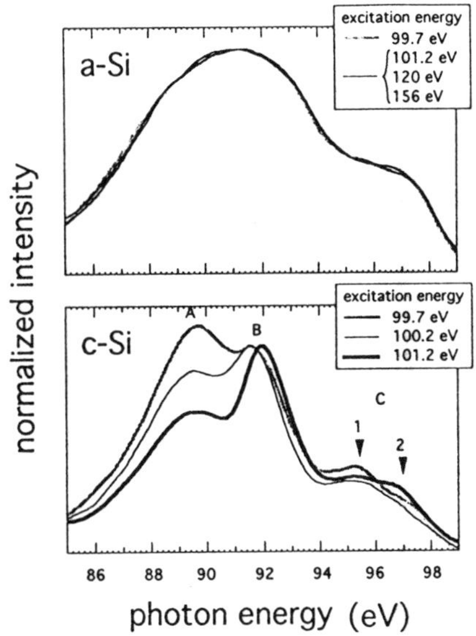

Figure. 7

Si $L_{2,3}$ soft X-ray emission spectra are plotted for amorphous and crystalline silicon samples with $h\nu_e$ as a parameter. The a-Si data are normalized to equal area, while the c-Si data are normalized to the intensity of peak B. The dotted line near peak A of the crystalline data marks the energy position of this peak for higher $h\nu_e$.

It has been suggested[12] that these enhancements be used to map out the density of states in certain crystalline substances. If this is a viable suggestion there will be yet another reason to use soft x-ray emission as a research tool.

3.3 Electron screening and spectator decay: Emission from B_2O_3

Our group has observed soft x-ray valence band emission from a number of insulating compounds, such as MgO and SiO_2. We have studied intermediate coupling,[14] temperature effects,[15] and phonon relaxation[16] in these compounds. Insulators such as these also have a core exciton as a bound state. In B_2O_3 we have studied[17] the change in the valence emission band as affected by the screening produced by the exciton. Fig. 8 is the result of these measurements. Emission spectra from B_2O_3 are shown as a function of excitation energy, indicated by the numbers ranging from 193.7 to 217 eV. The lowest energy of excitation, 193.7 eV, corresponds to that required for the excitation of a core exciton. At this photon excitation energy, one observes resonant fluorescence from the exciton (omitted for clarity) and emission from the valence band. At resonance with the exciton, emission from the valence band occurs while the exciton acts as a "spectator". At a slightly higher photon energy, off resonance with the exciton, one observes emission from the valence band. The band shifts significantly in energy (almost 2 eV) and changes shape. The shift is due partially to the change in initial and final state screening produced by the absence of the electron excited to the bound excitonic state. The change in shape is possibly due to the resonant photon scattering phenomena described in the preceding paragraph, and the large difference in the phonon coupling is also partially responsible for the difference in shape. Off resonance the valence emission spectrum remains relatively unchanged as a function of the photon energy, until hv=202 eV when an inelastic collision of the outgoing electron can excite one of the valence electrons into the excitonic state. Then one observes emission from both the exciton and from the valence band.

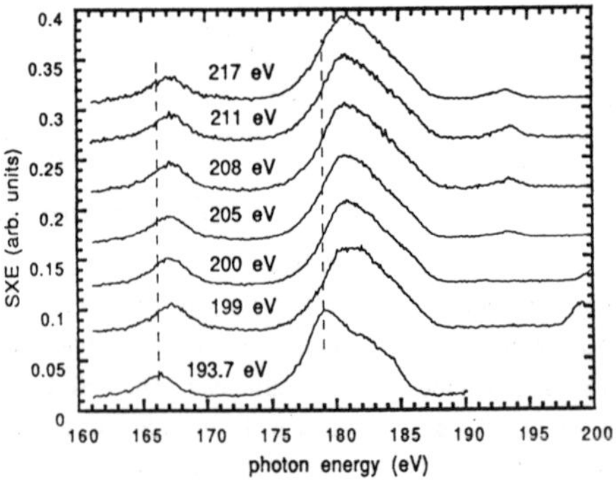

Figure 8.

Boron K soft x-ray emission spectra from B_2O_3 for different photon excitation energies.

The branching ratio for radiative decay of the boron K exciton can be determined. Linear combinations of the normal and spectator SXE spectra were compared to the SXE spectra obtained at energies greater than 210 eV. From these comparisons it was determined that 12% of the valence SXE features were due to spectator emission. The relative intensity of the exciton peak to the intensity of the two main emission peaks is 0.033. Thus the branching ratio for radiative decay of the core exciton to that of the valence band is 0.033/0.12=0.28. The large value of the branching ration suggests that the oscillator strength of the exciton is comparable to the integrated number of electrons in the valence band.

3.4 Phonon Relaxation and Broadening: Emission from MgO

In the last section we showed how photon-in photo-out could be used to study electronic relaxation in insulating crystals. In this section we want to show how SXE can be used to study phonon relaxation, or Stokes shift, in insulating compounds. The main ideas of phonon relaxation were put forth in two papers.[18,19] Measurements of this type were carried out on metals[20,21] using data from SXE and SXA from the same spectrometer. In the case of metals the shift is less than 0.1 eV, where as for MgO the Stokes shift is of the order of 0.5 eV. The large value of the shift enabled us to avoid significant errors due to calibration and use two spectrometers. One was used to obtain a measurement of the valence band spectrum by SXE and a second one was used to obtain a measurement of the same spectrum by photoelectron spectroscopy (PES). No knowledge of the band gap is required for this analysis

In SXE the valence band maximum of a semiconductor can be obtained relative to a core electron binding energy. In PES the binding energy of a core electron can be obtained relative to the valence band maximum. This is illustrated in the cartoon shown as Fig. 9a. On the left hand side of the figure a core hole is filled by the emission of a soft x-ray photon hν, leaving a hole in the valence band. On the right had side of the figure the same information can be obtained by the a measurement of the kinetic energy

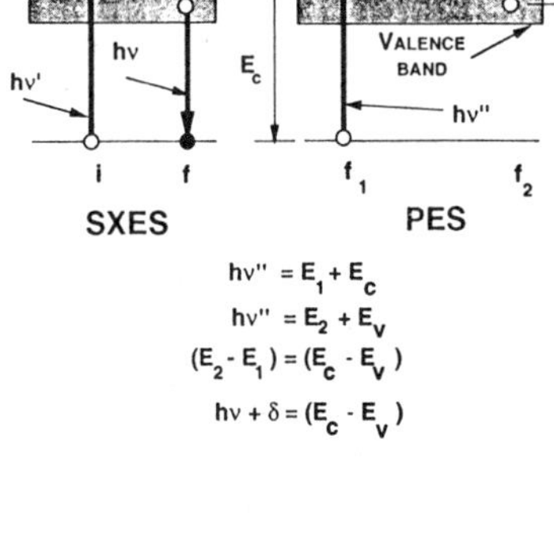

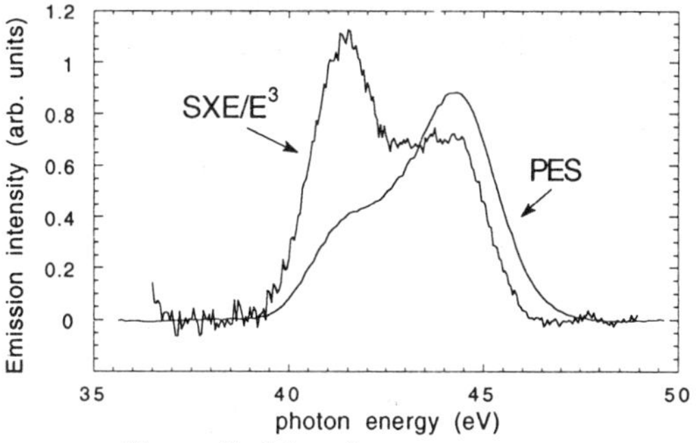

Figure 9b (above)

Comparison of MgO $L_{2,3}$ soft x-ray emission spectrum (SXE) to photoemission spectrum (PES). The SXE has been divided by E^3 to approximate the density of states.

Figure 9a (to the left)

Cartoon illustrating the way PES and SXE can be used to obtain the phonon relaxation δ.

difference between electrons originating from a core photoionization process and electrons originating from a valence band photoionization process. Now if final state electronic screening is complete and if phonon relaxation effects are unimportant, the valence band maximum obtained by either measurement should yield the same result. This is because the final state of the core PES event is the initial state of SXE, and the final state of the conduction band PES event is equivalent to the final state of the SXE transition.

A shift will be observed if phonon relaxation is important relative to the lifetime of the excited atom. In the case of these insulating compounds we assume electronic relaxation is complete. The results of the measurements are shown in Fig 9b. In this figure the photoelectron spectra were determined by measuring the valence band and the $L_{2,3}$ core electrons simultaneously. The valence spectra was shifter by the binding energy of the L_2 and the L_3 core electrons, and convoluted using the statistical weights of the core levels. In this manner the two spectra should align at the valence band maximum, if phonon relaxation is unimportant. There are major differences between the SXE and the PES spectra. The overall line shape is different and the valence band obtained by PES is shifted 0.5 eV higher in energy relative to the valence band spectrum obtained by SXE. The difference in line shape is due to the dipole selection rule, which projects out states of s and d symmetry in the valence band, while PES samples the entire symmetry range. The energy shift between the two spectra is due to a Stokes shift.

Mahan[22] has shown that core electron excitations couple most strongly with longitudinal optical phonons at X. For MgO these phonons have an energy of 0.06 eV. In an independent measurement[15] we have determined the phonon relaxation energy for MgO to be 1.5 eV. This large shift in MgO is due to the longer core hole lifetime, faster phonon relaxation and larger phonon relaxation energy.

4. Discussion and the Future: Circular Dichroic X-ray Fluorescence

In the series of examples put forth in the preceding paragraphs we have illustrated a number of new applications of photon excited soft x-ray spectroscopy, namely resonant photon scattering, absorption branching ratios, electron and phonon relaxation, and we have suggested that soft x-ray dichroic fluorescence will play an important role in the development of new materials. Strange *et al* [8] have shown recently that the density of majority and minority valence band spin states can be measured by the x-ray emission anisotropy between the valence and core holes prepared by the absorption of a circularly polarized photon. The tunability of the photon source is crucial to the success of the method. The core electron must be promoted into the empty part of the exchange-split d band within 1.3 to 2 eV of E_f. According to Strange *et al*, this will insure the core hole will be highly polarized, and the resulting emission spectrum will show a strong dichroic anisotropy. Experiments of this type will enable one to study how the spin polarization changes across the Fermi energy.

It will also be possible to study, in the same apparatus, the unoccupied spin states in the conduction band by measuring the magnetic circular dichorism (MCD) of the core absorption by observing the core-core transition as a function of the excitation energy. In fact the x-ray dichroic fluorescence and the MCD in absorption can be observed *simultaneously.*

The events of the past year have provided additional stimulation to a field that has a long and distinguished history. All of us involved in its development have been rewarded beyond our wildest dreams with new and interesting phenomena to study.

5. Acknowledgments:

This research was supported in part by National Science Foundation Grant No. DMR-8715430 and by a Science Alliance Center for Excellence Grant from the University of Tennessee, and by the U.S. Department of Energy (DOE) Contract No. DE-AC05-84OR21400 with Oak Ridge National Laboratory. The research was carried out at the National Synchrotron Radiation Laboratory at Brookhaven National Laboratory, with support from DOE Contract No. DE-AC02-76CH00016. One of the authors, DLE, was a visiting scientist at the Laboatoire de Spectroscopie Atomique et Ionique, Orsay, France, during a portion of this work. The authors are grateful to E. Morikawa of the Center for Advanced Microstructures and Devices for providing figure 3.

6. References:

* Present Address for W.L. O'Brien is the Synchrotron Radiation Center, Stoughton, WI.

1. W.C. Röntgen (Trans. by A. Stanton), *Nature*, **53** 274 (1896)
2. H.W.D. Skinner, *Phil. Trans of the Roy. Soc.(London)* **239**, 95 (1940).
3. T.A. Callcott, K.L. Tsang, C.H. Zhang, D.L. Ederer, and E.T. Arakawa, *Nucl., Instrum. and Methods* **A266**, 578 (1988).
4. T.A. Callcott, K.L. Tsang, C.H. Zhang, D.L. Ederer, and E.T. Arakawa, *Rev. Sci. Inst.* **57**, 2680 (1986).
5. T.A. Callcott, W.L. O'Brien, J.J. Jia, Q.-Y. Dong, D.L. Ederer, R.N. Watts, and D.R. Mueller, *Nucl., Instrum. and Methods* **A319**, 128 (1992).
6. T. Miyahara, T. Hanyu, H. Ishii, M. Yanasihara, T. Kamada, H. Kato, K. Naito, S. Suzuke, *J. Phys. Soc., Japan* **55**, 408 (1986).
7. T.B. Lucatorto, T.J. McIlrath, J. Sugar, S.M. Younger, *Phys. Rev. Letts.* **47** 1124 (1981).
8. P.Strange, P.J. Durham, and B.L. Gyorffy, *Phys. Rev. Lett.* **67**,3590 (1991).
9. J.-M. Mariot, and C.F. Hague, these proceedings
10. J.R. Chelikowskyand M.L. Cohen, *Phys Rev.* **B10**, 5095 (1974).

11. J.-E. Rubensson, D. Mueller, R. Shuker, D.L. Ederer, C.H. Zhang, J. Jia and T.A. Callcott, *Phys. Rev. Lett.* **64**, 1047 (1990).
12. Y. Ma, N. Wassdahl, P. Skytt, J. Guo, J. Nordgren, P.D. Johnson, J.-E. Rubensson, T. Boske, W. Eberhardt, S. D. Kevan, submitted to *Phys. Rev. Letts.*
13. K.E. Miyano, D.L. Ederer, T.A. Callcott, W.L. O'Brien, J.J. Jia, L. Zhou, Q.-Y. Dong, Y. Ma, J.C. Woicik, and D.R. Mueller, submitted to *Phys. Rev. Letts.*
14. W.L. O'Brien, J.J. Jia, T.A. Callcott, J.-E. Rubensson, D.R. Mueller, and D.L. Ederer, *Phys. Rev* **B44**, 1013, (1991).
15. W.L. O'Brien, J.J. Jia, Q.-Y. Dong, T.A. Callcott, D.R. Mueller, and D.L. Ederer, *Phys. Rev* **B45**, 3882, (1992).
16. W.L. O'Brien, J.J.Jia, Q.-Y. Dong, T.A. Callcott, K.E. Miyano, D.L. Ederer and D.R. Mueller, to be published in *Phys. Rev*
17. W.L. O'Brien, J.J.Jia, Q.-Y. Dong, T.A. Callcott, K.E. Miyano, D.L. Ederer and D.R. Mueller, submitted to *Phys. Rev Letts.*
18. C.-O. Almbladh, *Phys. Rev.* **B16**, 4343 (1977).
19. G.D. Mahan, *Phys. Rev.* **B15**, 4587 (1977).
20. T.A. Callcott, E.T. Arakawa, and D.L. Ederer, *Phys. Rev.* **B16**, 5185 (1977).
21. T.A. Callcott, E.T. Arakawa, and D.L. Ederer, *Phys. Rev.* **B 18**, 6622 (1978).
22. G.D. Mahan, *Phys. Rev.* **B21**, 4791 (1979).

EXCITATION DYNAMICS IN IONIC CRYSTALS

CHAOSHU SHI, YI DONG and XIAOPU LU
Department of Physics, University of Science and Technology of China,
Hefei Anhui 230026, P. R. China

ABSTRACT

In this paper the excitation dynamics of ionic crystals BaF_2 and BaF_2:RE(Ce, Dy, Eu), especially luminescence dynamical properties at the excitation with SR were summarily described. It includes the excitation characteristic of the electrons in valence- and core-band of the crystals, energy transfer and their temperature- and concentration-dependence. A possible new thermoluminescence mechanism was proposed.

1. Introduction

Ionic crystals have been studied for more than 20 years with synchrotron radiation. The crystals are mainly some alkali halides, alkaline-earth halides, transition metal halides, rare-earth fluorides and oxides etc.[1], which have been widely used for scintillator, X-ray storage, radiation dosimetry, laser, luminescence, and other optical materials etc. Their fundamental absorptions are distributed in the VUV range.

In the 1970's, most of the work concentrated on the absorption-, reflection-, modulation-, photoelectron-spectroscopy of the crystals for researching electronic structure. In the 1980's main investigations were on the luminescence and its excitation as well as decay time and temperature dependence etc. It is due to that much more sensitive luminescence technique and high resolved energy- and time-spectroscopy were employed. In recent years excitation dynamics of ionic crystals, especially luminescence dynamics is being studied in detail. It is taken into account not only in the energy region between the conduction- and valence-band, but also in that between valence- and core-band.

Among the investigations mentioned above the most interesting subjects are:

1) cross luminescence (CRL) from transition between valence- and core band[2-19].
2) Interactions between impurities and defects.

The BaF_2 crystal is the fastest scintillator known and a prepesentive with cross luminescence (CRL)[2-8]. Since discovering CRL of BaF_2 in 1982, more than 10 kinds of crystal with CRL have been studied[9-17].

BaFBr:Eu is an excellent material used for X-ray imaging storage and for researching interaction between the rare earth (Eu^{2+}) centers and defects such as F centers[18-21].

In order to study excitation process in the ionic crystals we combined both subjects mentioned above and studied mainly BaF_2 and BaF_2:RE(Ce,Dy,Eu). The rare earth dopants are used as a luminescence probe for studying excitation dynamics and for exploring new materials.

2. Experimental

The samples used were mainly single crystals including pure BaF_2 and RE (Ce, Dy, Eu, La) doped BaF_2 which were grown by Münster University in Germany, Shanghai Institute of Ceramics, Beijing Institute of Glass and Chungchun Institute of Fine Optical Mechanics in China. Microcrystal powder samples including BaFBr, BaFCl, BaF_2, $BaCl_2$, $BaBr_2$ doped with Eu^{2+} were also studied. They were prepared by using the solid phase reaction method in reduction gas.

The experiments using SR were performed at the SUPERLUMI station in HASYLAB at DESY in Germany. The SR is from DORIS II and DORIS III with an energy of 3.7 GeV and 5.25 GeV, respectively.

The spectral resolution of the monochromator dispersing SR can reach 0.007 nm. The monochromator used for luminescence analysis has an extremely large aperture with f number 1:2.8 and spectral resolution of 5 Å. Detectors used were XP2020Q and position sensitive detector. The decay time and time-resolved spectra were measured using single photon counting method.

X-ray with 35 keV, 20 mA and γ-ray from Cs^{137} were also used as excitation source. γ-ray from Co^{60} was used for irradiation effect. An additional UV source was Hg, Xe lamps. In this experimental set up there is a monochromator with a working range of 2000-8000 Å, a resolution of 2 Å and photomultiplier R456. The apparatus was used for measuring emission spectra at X-ray or UV excitation and thermoluminescence in the range 77 K - 460 K or 20 K - 300 K.

3. Main Results[5-8]

Under SR excitation the intrinsic emissions of BaF_2 can be classified into self trapped exciton luminescence (STE-L) and cross-luminescence (CR-L). They originate from different electronic excitation and can be quenched by impurities such as RE or defects including intrinsic or extrinsic defects.

3.1 The excitation of the electrons in valence band and its temperature dependence[5].

The STE-L efficiently origin from the electron excitation in valence band, especially valence exciton excitation (Fig. 1). The temperature effects is the most obvious characterisitic of the kind of excitation as following:

- The valence exciton peak at 9.8 eV under RT was much lower than that at low temperature (LHe) and shifted to higher energy (10.086 eV).
- The energy gap (Eg) was extended from 10.94 eV to 11.23 eV when the temperature was decreased from RT to LHe. Corresponding temperature coefficient is -9.5×10^{-4} eV/K [9].
- The temperature dependence of intensities of STE-L was very complex[5]. A maximum appeared at ≈ 175 K. Lower intensities on the side of lower T and that of higher T of the maximum are related to shallow traps in ionic crystals and thermoquenching, respectively.

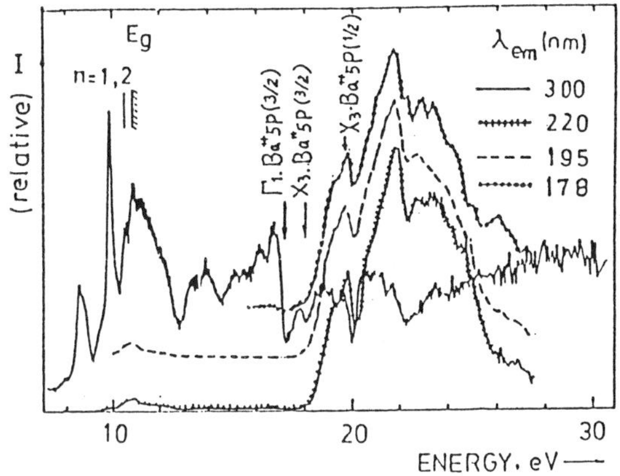

Fig. 1 The excitation spectra of the emissions at 300, 220, 195, 178 nm at 10 K.

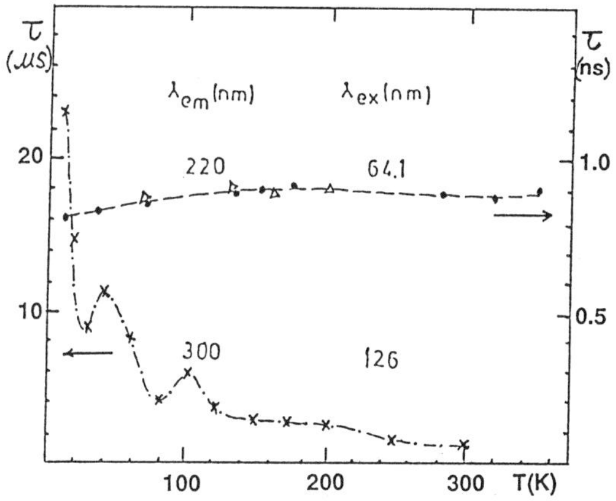

Fig. 2. The temperature dependence of the decay time of 300, 220 nm emission.

- In the temperature dependence of decay time ("τ-T" curve)[6] there are some structures including peaks at about 40 K, 100 K, and a shoulder in the range of 120 - 210 K (Fig. 2). The structures are associated with the traps in the crystal. The consideration has been elucidated by thermoluminescence curve[8].

3.2 The excitation of the electrons in core band and cross luminescence.
The excitation of core electrons is responsible for CRL in BaF_2. The CRL has the characteristic as following:

- The CRL in BaF_2 crystals contains three emission bands at 220, 195, and 178 nm. As shown in Fig. 1, they have the same excitation spectra with threshold energy 18 eV which corresponds to the excitation of the core electron ($Ba^{++}5p$) of most outshell[5]. The excitation threshold energy is the first criterion of the CRL.

- The decay time of 220 nm band which is a representative of the CRL in BaF_2 crystal, is 0.88 ns ± 0.05 ns. The decay time of 195 nm band was 1 ns order of magnitude. So the fast decay time about ns or sub-ns order of magnitude is second criterion of CRL.

- The decay time of 220 nm band was independent on temperature (Fig. 2). The intensities of the 220, 195 nm bands were increased in the range of 100 - 370 K with rising temperature[5]. It is shown that the luminescence without thermoquenching is the third criterion of CRL.

3.3 Energy transfer and its temperature-and concentration-dependence

- Energy transfer from FE to STE[7]
From the time resolved spectra and decay curve of STE-L at 300 nm band, one can see that after stopping excitations the STE-L occured in 3.4 - 7.4 ns, and was increased gradually to a maximum of 80 ns or so. The time to reach the maximum was shortened from 80 ns to 30 ns by increasing T from 300 K to 330 K. It indicates that a process of energy transfer exists. The most likely reason for the intensity increase is that the rate of energy transfer from free exciton (FE) to STE is faster than that of luminescence decay from STE-L. Namely in several tens ns FE transfer energy to STE. In several ns (3.4 - 7.4 ns) after the pulse excitation the STE was formed.

- Energy transfer from BaF_2 (host) to RE centers.
The luminescence of BaF_2:RE (Ce, Dy, Eu) includes intrinsic emissions mentioned above and the extrinsic emission from RE centers. The intensities of RE luminescence were enhanced while intrinsic luminscence either from STE-L at 300 nm or from CRL at 220 nm were decreased with rising RE concentration. It is due to partially overlapping between main intrinsic emission bands and excitation spectra of RE emissions. For example, in excitation spectra of Ce^{3+} emission at 332 and 302 nm bands in BaF_2:Ce (Fig. 3), one can see that besides valence exciton band at 1300 Å or so, intense excitation bands peaked at 2000, 2800 Å are distributed in the range of 1800 - 3100 Å, in which there are just main intrinsic luminescence bands at 220, 300 nm of BaF_2 host. In this case, strong resonance energy transfer can occur. The energy transfer was enhanced with increasing Ce^{3+} concentration as in Fig. 4. In the range of Ce^{3+} concentration 10^{-2} - 1

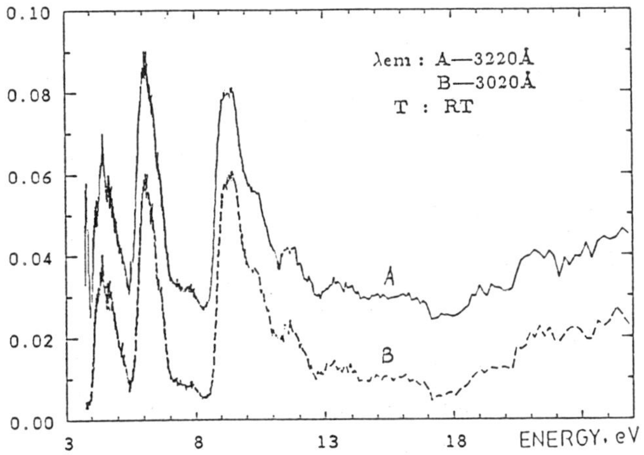

Fig. 3 The excitation spectra of 3020, 3220 Å emissions from $BaF_2:Ce^{3+}$.

mol.%, fast component (including 220, 195, 178 nm) I_1, and slow component (at 300 nm band) I_3 were quickly decreased to disappear meanwhile the emissions of Ce^{3+} were rapidly increased to maximum. These curves were measured at excitation with γ-ray (Cs^{137}) [27]. In the excitation spectra of Dy^{3+} emission (5800 Å) and Eu^{2+} emission (4500 Å) from BaF_2:Dy and BaF_2:Eu there are very intense excitation bands of valence exciton (≈ 1300 Å or so). It is a powerful evidence for energy transfer from host BaF_2 to RE. Other analogous studies[22-25] can also elucidate the energy transfer.

As shown in Fig. 5, which is the emission spectra of $BaF_2:Dy^{3+}$ with different Dy^{3+} concentration (1 ppm - 10^4 ppm) at X-ray excitation[28]. Intrinsic emission bands of STE at ≈ 300 nm and CRL at 220 nm were also decreased with rising Dy^{3+} concentration. (The emission band of STE shown was shifted to 320 nm. It is due to the dip of the sensitivity of the monochromator near the edge of shorter wavelength.) The intensities of Dy^{3+} emission lines (480, 580, 668 and 745 nm) were reached a maximum when Dy^{3+} concentration was 10^3 ppm. However, under UV 254 nm excitation they were always increased till to Dy^{3+} concentration 10^4 ppm. Because the series of the lines have the same excitation spectra and concentration dependence of the intensity, they were identified as having same initial state and belong to the transition of $^4F_{9/2} \rightarrow {}^6H_{15/2}$ $^6H_{13/2}$ $^6H_{11/2}$ and $^6H_{9/2}$ of Dy^{3+} respectively. Under UV excitation, in the BaF_2:Dy there are another series of emission lines (381, 436 and 542 nm) with the same excitation spectra. These lines disappeared at X-ray excitation or were very weak at SR excitation. It indicate that the emission spectra of BaF_2:RE can be changed as the excitation way changed. Under X-ray excitation the emission band of STE was much higher than that under VUV excitation.

4. Discussion

4.1 Temperature dependence of CRL at 220 nm.

According to origination of 220 nm, its intensities should be independent on temperature like "τ-T" curve of the band. However, with rising temperature intensities of 220 nm band were increased in the range of ≈ 100 - 370 K. The results were also observed at the excitation with X-ray, electron beam. The fact shows that the dip in 20 - 150 K may be related to existence of many core hole traps which distributed above core level. These core hole traps can be produced by intrinsic defects, which may origin from a excitation effect in CRL crystal as Toyozawa's idea[26]. When a core electron is excited by VUV or X-ray (radiation), in condensed matter the atoms as well as the electrons around the core holes are redistributed so as to decrease the total energy of the system concerned. It leads to localized core hole states above core band.

The core hole traps from intrinsic defects or trace impurity could change temperature dependence of CRL. It is similar to the results observed in CsCl:Br [29].

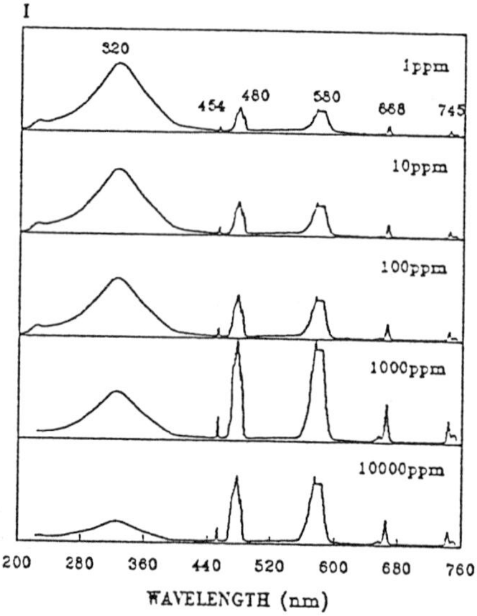

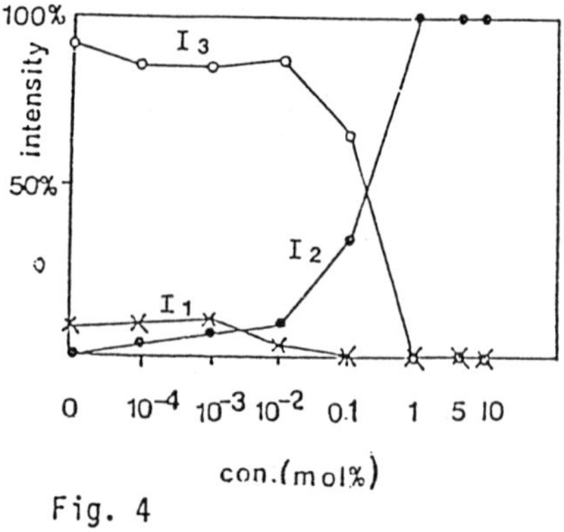

Fig. 4

The relative intensity of fast-, slow-component (I_1, I_3) and Ce^{3+} emission in BaF_2:Ce as a function of Ce^{3+} concentration.

Fig. 5 The emission spectra of $BaF_2:Dy^{3+}$ at X-ray excitation.

4.2 Interaction between RE centers and the defects.

From obtained experimental results, it seems to deduce a conclusion that the defects in ionic crystal occur easily near RE^{3+} center and combine with RE^{3+} to become a complex. In the complex there is energy transfer between RE^{3+} and defects by tunneling. Our evidence is as following:

- Thermoluminescence (TL) of $BaF_2:Ce^{3+}$ was observed at UV (254 nm) excitation[30] (Fig. 6). It is well known that the electrons (or holes) released from the traps must be through conduction band (or valence band) and induce thermoluminescence. However, the excitation energy with 254 nm photons was not enough to ionize Ce^{3+} in BaF_2, for the excited states of Ce^{3+} in BaF_2:Ce are far from conduction band (≈ 6 eV). As can be seen in excitation spectra of Ce^{3+} in BaF_2:Ce the ground states of Ce^{3+} are very near top of valence band (≈ 0.2 eV). In addition, photoconductivity was not found in BaF_2:Ce at UV excitation.

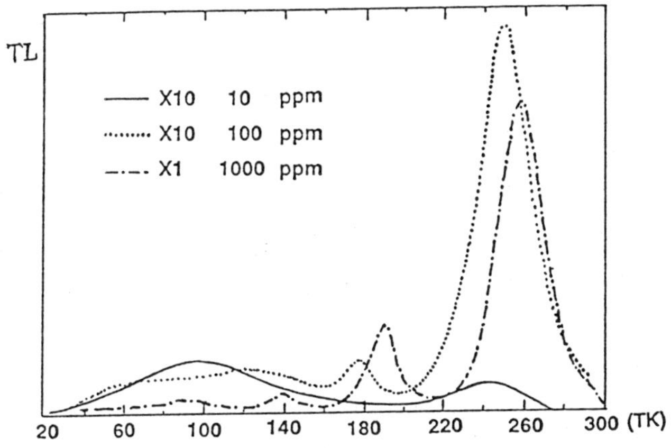

Fig. 6 The thermoluminescence of BaF_2:Ce at UV 254 nm exciton.

- Thermoluminescence of $BaF_2:Ce^{3+}$ only can be observed for BaF_2 with light doping (10, 100, 1000 ppm), no TL was measured in BaF_2 with heavy doping (10^4, 10^5 ppm). The TL data is in agreement with temperature dependence of Ce^{3+} emission in BaF_2. The intensities of Ce^{3+} emission were independent on temperature for heavy doped BaF_2. Only in a complex constructed by RE and defects it is possible that energy bands (4f, 5d) of Ce^{3+} were extended so as to disappear the traps from defects in BaF_2 at high Ce^{3+} concentrations.

- After irradiation with γ-, X-ray, BaF_2:RE (Ce, Dy, La, Eu) crystals were changed into red colour which was gradually obvious with increasing RE^{3+} concentration. The red colour is due to absorption band at ≈ 500 nm which seems to be from similar absorption like F centers in BaF_2. Therefore one can say that amount of the defects induced during excitation may be associated with RE^{3+} concentration.

- From thermoluminescence curve of $BaF_2:Ce^{3+}$ (Fig. 6), as Ce^{3+} concentration increased the peaks of TL curve shifted to higher temperature which means that the deepness of the the traps was increased. It is the reason why the decay time of Ce^{3+} emission as shown in table 1 was increased from 46 ns to 80 ns with rising Ce^{3+} concentration from 10^{-4} % to 1 % [29].

Table 1: The decay time of BaF_2:Ce emission[27]

Crystal	B0	B1	B2	B3	B4	B5	B6	B7
Ce (mol %)	0	10^{-4}	10^{-3}	10^{-2}	0.1	1	5	10
τ_1 (ps)	815	788	776	785	838	no	no	no
error	15	33	25	37	145	no	no	no
I_1 (%)	7.5	9.8	8.1	2.3	.6	no	no	no
τ_2 (ns)	no	46	50	42	58	80	56	44
error	no	5	8	4	10	15	1	4
I_2 (%)	no	4.5	7.2	10	34.1	100	100	100
τ_3 (ns)	604	680	640	530	640	no	no	no
error	100	40	40	40	40	no	no	no
I_3 (%)	925	85.7	84.7	87.7	65.3	no	no	no

5. Summary

(1) Intensities of intrinsic emission in BaF_2 depend on the excitation energy and excitation way. The emission band at 300 nm (STE) X-ray excitation at RT was much higher than under VUV excitation.

(2) Temperature dependence of emission intensities at 220 nm band may origin from core hole traps.

(3) Intrinsic emission (STE, CRL) of BaF_2 were quenched by the RE dopants. It is due to strong energy transfer from host (BaF_2) to RE. The transfer efficiency depends on the concentration of RE ion.

(4) The defects in ionic crystals may be produced near RE centers. Amount of the defects were increased with rising concentration of RE ions.

(5) RE ions and the defect centers such as F may form some complex. In the complex the energy transfer between RE-defects can occur by tunnel way. It is base of a new thermoluminescence mechanism.

6. Conclusion

The ionic crystal BaF_2, BaF_2:RE (Ce, Dy, Eu) are a very complicated system in which there are a lot of intrinsic effects. The characteristic of the intrinsics emission including both STE-L and CRL was strongly dependent on the excitation energy, sample temperature, the defects and the impurities such as RE and their concentration. Interaction between RE centers and intrinsic defects may lead to a complex. The complex is the base of a new thermoluminescence mechanism.

7. Acknowledgements

We would like to thank colleagues from the Universität Hamburg and HASYLAB in Germany. In particular, cooperation and help of Prof. G. Zimmerer, Mr. D. Varding, and Miss A. Schröder are kindly acknowledged.

References

1. G. Zimmerer, *in Defects in Insulating Crystals*, (ed. V. M. Tuchkevich and K. K. Shvarts) P. 503 (Springer, Berlin, 1981).
2. M. Laval, M. Moszynski, R. Allemand, E. Cormoreche, P. Guinet, R. Odru and J. Vacher, *Nucl. Industr. and Meth.* **206** (1982) 169.
3. Yu. M. Aleksandrov, V. N. Makhov, P. A. Rodnyi, Z. I. Syreishchikova and M. N. Yakimento, *Sol Phys State*, **26** (1984) 1734
4. S. Kubota, M. Itoh, J. Ruan (Gen), S. Sakuragi and S. Hashimoto; *Phys.Rev. Lett.* **60** (1988) 2319.
5. C. Shi, T. Kloiber and G. Zimmerer, J. Lumin **40/41** (1988) 189
6. C. Shi, T. Kloiber and G. ZImmerer, *proc. Int. Conf. of SR Appplication* (Hefei, China. 1988) p. 468
7. C. Shi, T. Kloiber and G. Zimmerer, *physica Scripta*, **41 No 6** (1990) 1002
8. C. Shi, T. KLoiber and G. Zimmerer, J. Lumin **48/49** (1991) 597
9. I. Kuusmann, T. Kloiber, W. Laasch and G. Zimmerer, *in Radiation effects and defects in solids*, **vol 119-121** (1991) pp. 21-26.
10. V. N. Makhov and N. M. Khaidukov, *Nucl. Instr. and Meth.* **A308** (1991) 205
11. P. A. Rodnyi and M. A. Terekhin, Phys. State Sol. (b) 166 (1991) 283
12. V. N: Makhov, *Nucl. Instr and Meth* **A 308** (1991) 187
13. E. N. Melchakov, V. A. Petrova, I. G. Podkolzina, P. A. Rodnyi and M. A. Terekhin, *Opt. Spectrosc*, (USSR) **69(4)** (1990) 481
14. O. T. Antonyak, A. S. Voloshinovskii, I. P. Pashuk, S. N. Pidzyrailo and P. A. Rodnyi, *Opt Spectrosc*, (USSR) **70(5)**, (1991) 607
15. Yu. M. Aleksandrov, I. L. Kuusmann, V. N: Makhov, S. B. Mirov, T. V. Uvarova and M. N. Yakimenko, *Nucl. Instr. and Meth*, **A 308** (1991) 208
16. I. A. Kamenskikh, M. A. Macdonald, V. V. Mikhailin, I. H. Munro and M. A. Terekhin, *Rev. Sci. Instrum*, **63(1)**, (1992) 1447
17. M. Itoh, S. Kubota, J. Ruan (Gen) and S. Hashimoto, *Rev Solid State Science*, **4 No 2/3** (1990) 467
18. J. Miyahara, K. Takahashi, Y. Amemiya and N. Kamiya and Y. Satow, *Nucl. Instr. and Meth*, A246 (1986) 572
19. H. von Seggern, T. Voigt, W. Knüpfer and G. Lange, *J. Appl. Phys*, **64(3)** (1988) 1405.
20. H. H. Rüter, H. von Seggern, R. Reininger and V. Saile, *HASYLAB Jahresbericht* (1988) p.141, idem (1989) p. 149

21. M. Casalboni and U. M. Grassano, *J. Phys. Chem. Solids*, **51(7)**, (1990) 805
22. A.V. Golovin, N.G. Zakharov and P.A. Rodnyi, *Opt. Spectrosc. (USSR)* **65(1)** (1988) 102
23. P. Schotanus, P. Dorenbos, C.W.E. Van Eijk and H.J. Lamfers, *Nucl. Instr. and Meth.* **A281** (1989) 162
24. E.N. Melchakov, P.A. Rodnyi, and M.A. Terekhin, *Opt. Spectrosc. (USSR)* **69(5)** (1990) 634
25. P.A. Rodnyi, M.A. Terekhin, and E.N. Melchakov, *J. Lumin.* **47** (1991) 281
26. Y. Toyozawa, *in proc. Taniguchi Symposium on Core Level Spectros. in Condensed Systems*, (Springer-Verlag, Berlin, 1987) pp. 231
27. Z. Xu, Z. Gong, J. Chang, C. Shi, D. Sheng et al. (to be published)
28. Y. Dong, J. Deng, Z. Xie, Y. Su, C. Shi, J. Zhao, and Z. Song (to be published)
29. A.N. Belskiy, V.V. Mikhailin, A.L. Rogalerev, Conference proc. vol. 25, "*2nd European Conference on Progress in X-ray SR Research*" (1990) 797
30. Y. Tang, C. Shi, Y. Dong, and Li Lou (to be published)

AUGER-FREE LUMINESCENCE FROM LARGE GAP INSULATORS

SHINZOU KUBOTA
Rikkyo University, Nishi-Ikebukuro 3
Tokyo, 178 Japan

ABSTRACT

Recent experimental results for the radiative decay of valence band electrons into outermost-core holes from large gap insulators such as cesium-halides, RbF and BaF_2 were reviewed comparing uv excitation, soft x-ray excitation and charged particle excitation of high-energy electrons and heavy ions.

1. Introduction

There is great current interest in high energy physics and nuclear medicine in the development of new scintillation materials having fast decay components, such as CsF and BaF_2. Central to further progress in this area is the need for an understanding of the fundamental processes following interaction of incident radiation or incident charged particles with the scintillation materials.

Recently, luminescence associated with interatomic radiative transitions of valence band electons to outermost-core holes has been observed from large gap insulators, such as BaF_2, CsF, CsCl, CsBr, and RbF.[1-15] In these crystals, the band gap energy E_g is larger than the energy difference E_{vc} between the tops of the valence band and the outermost-core hole state, so that the Auger electron emission is energetically impossible. Such an outermost-core hole is unaffected by Auger decay processes and recombines radiatively with a valence band electron. Such luminescence is called "Auger-free luminescence"[9] (sometimes also called "cross luminescence"[6] or "radiative core-valence luminescence"[14]). Hereafter, Auger-free luminescence is abbreviated as AF luminescence.

In this paper, relaxation of hot holes in the outermost-core band and competition process between AF luminescence and Auger electron emission are summarized. Experimental results are discussed on excitation spectra, decay curves for AF luminescence from large gap ionic crystals under excitation of photons from synchrotron radiation with energy up to 140 eV, as well as under excitation of ionizing charged particles of different specific ionization density.

2. Relaxation of hot holes in the outermost-core band

According to Kikas and Elango,[16] we introduce three relaxation times of hot holes in the outermost-core band. They are the luminescence decay time τ_d, the thermalization time τ_p of hot holes by scattering with phonons, and the time τ_a of the interatomic Auger process. The luminescence decay time is measured[12] to be about 10^{-9} sec that is much larger than τ_p and τ_a. The inequality between τ_p and τ_a is discussed in Section 4. The band width of the outermost-core band of alkali halides or alkali-earth fluorides is measured to be several eV by using the photoelectron spectroscopy.[17,18] The outermost-core holes are created throughout this band under incident photons or incident charged particles, except in the case that the holes are created in the upper part of this band by photons with energy just above the ionization threshold $E_{th} = E_g + E_{vc}$ of this band. We consider the relaxation of hot holes in this outermost-core band for four types LL, L, AL and A of crystals depending on the inequality between E_g and three energies $E_{vc} + \Delta E_c$, E_{vc} and $E_{vc} - \Delta E_v$. Here, ΔE_c and ΔE_v are the outermost-core band width and the valence-band width, respectively.

LL type of $E_{vc} + \Delta E_c < E_g$: Auger electron emission is energetically impossible even for the hottest hole at the bottom of the core band. A hot hole relaxes to the top of the core band and then recombines radiatively with a valence band electron.

L type of $E_{vc} < E_g < E_{vc} + \Delta E_c$: A hot hole with energy higher than $2E_g$ from bottom of the conduction band thermalizes, competing with Auger elctron emission. When energy of a hot hole becomes less than $2E_g$, a hole is free from Auger electron emission and relaxes to the top of the core band and recombines radiatively. Thus, the radiative intensity is determined by inequality between τ_p and τ_a. If $\tau_p >> \tau_a$, a hot hole with energy higher than $2E_g$ relaxes non-radiatively. If $\tau_p << \tau_a$, there is no difference in the features of Auger free luminescence between the crystals of LL type and L type.

AL type of $E_{vc} - \Delta E_v < E_g < E_{vc}$: Auger electron emission in which there is a hole in the valence band with energy range from E_g to E_{vc} in the final state is possible. Since $\tau_d >> \tau_p$ and τ_a, the Auger process is dominant and, therefore, the luminescence intensity is considerably reduced.

A type of $E_g < E_{vc} - \Delta E_v$. The Auger process is dominant and the luminescence intensity is considerably reduced.

Those four types are shown schematically in Fig.1. For example, CsF and BaF_2 are of LL type, CsCl, CsBr and RbF are of L type, and RbI, KI, and SrF_2 are of A type. CsI is of AL type and AF luminescence of intensity exceeding 10^{-3} of that of CsCl was not observed under outermost-core band excitation.[9,12] A correlation between the types and the photo-emission yield under outermost-core band excitation is reviewed in Ref. 12.

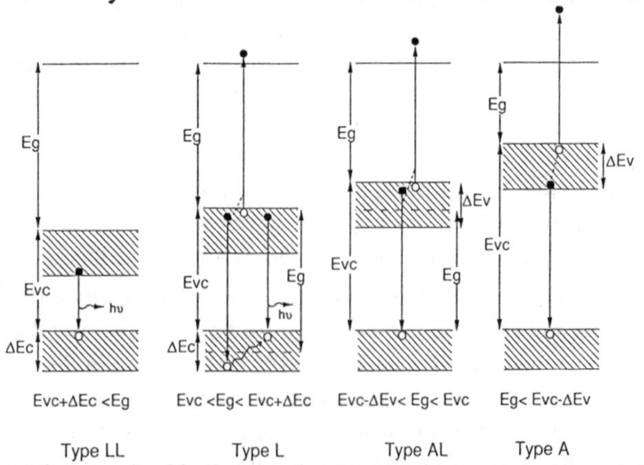

Fig. 1. Schematic energy level for the relaxation processes for the outermost-core hole in ionic crystals.

3. Excitation spectra

Figure 2 shows the excitation spectra for AF luminescence from CsF, CsCl, CsBr and RbF crystals,[9] plotted against incident photon energy, together with photo-emission yield spectra measured by Sugawara and Sasaki,[19] and absorption spectra (Ref.20) or reflectivity spectra (Ref.21). Synchrotron radiation from an electron storage ring (UVSOR) at Okazaki was used as a light source. All AF luminescences have threshold energies that correspond to the ionization thresholds of respective positive ions. They are at 14.0 and 17.1 eV for three cesium-halides and RbF, respectively.

The outstanding general features of the excitation spectra of AF luminescence are deep dips and their "anti-correlation" with respect to the photo-emission yield spectra in the photon energy regions of higher than about 23 eV in CsF, about 20 eV in CsCl, CsBr and RbF. In these energy regions, the transitions from F^- 2s, Cl^- 3s and Br^- 4s states and from

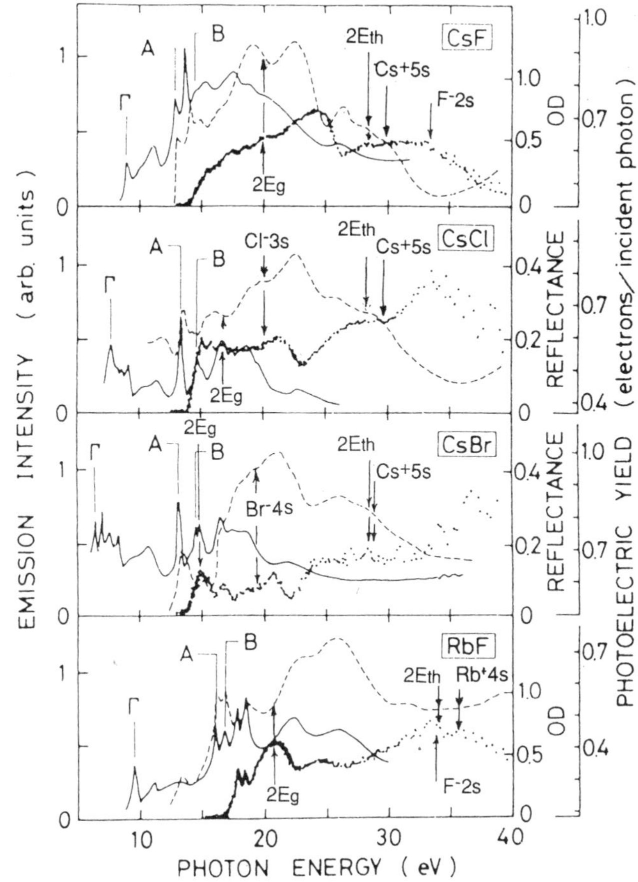

Fig.2. Excitation spectra (dotted points) for the 3.2 eV band (CsF), 4.5 eV band (CsCl), 4.9 eV band (CsBr) and 5.3 eV band (RbF). Photo-emission yield spectra (Ref. 19) are also shown by dashed curves. Absorption (Ref.20) or reflectivity spectra(Ref.21) are shown by solid curves. Arrows mark the positions of respective energies. Γ indicates the peak positions of the valence excitons, and A and B correspond to the core excitons of j = 3/2 and 1/2, respectively.

Cs^+ 5s and Rb^+4s states occur in three halides and RbF. The onsets of these levels, which are obtained by photoelectron spectroscopy data[22-24] and band gap energy,[18] are indicated by arrows. The energies corresponding to $2E_{th}$ and $2E_g$ are also shown by arrows.

Excitation spectrum for the AF luminescence from CsF starts to decrease at the binding energy of F^-2s state. Excitation spectra from CsCl and CsBr start to decrease at the photon energy corresponding to about 1 eV higher than the binding energies of Cl^-3s and Br^-4s states. A correlation between the onsets of the increase in photo-emission yield spectra and the F^-2s, Cl^-3s and Br^-4s levels is clearly observed for three cesium halides. Thus, deep gaps for excitation spectra at 40 eV in CsF, at 23 eV in CsCl and at 23 eV in CsBr can be interpreted in terms of switching in the H^-(ns) photoionization followed by H^-(ns)VV Auger transition, where H^-(ns) denotes initial ns-hole states of halogen ns state and V denotes a valence band hole in a final state. The H^-(ns)VV Auger does not create a hole in the Cs^+5p state, resulting in the decrease of AF luminescence and the increase in photo-emission yield spectra.

For three cesium-halides, the energy $2E_{th}$ nearly coincides with the binding energy of Cs^+5s state. If the photo-ionization of Cs^+5s state is dominant compared with the multiple process of outermost-core excitation, onsets of decrease in AF luminescence and of increase in photo-emission yield are expected at a photon energy corresponding to Cs^+5s level. This is because a Cs^+5s hole decays by 5sVV Auger transition without creating a 5p hole. If the multiple process of outermost-core excitation is dominant, an increase in excitation spectrum and a decrease in photo-emission yield spectrum are expected at photon energy higher than $2E_{th}$. As shown in Fig.2, increases in excitation spectra and decreases in photo-emission yield spectra are observed at photon energies corresponding to the binding energy of Cs^+5s state and $2E_{th}$ for three cesium-halides. The results indicate that the multiple process of outermost-core excitation is dominant.

In CsF, a dip in the excitation spectrum and a peak in the photo-emission yield spectrum are observed at the photon energy of 26 eV. This dip can be explained in terms of switching of an incident photon in excitation of a valence band electron to high conduction band states. This process results in an increase in the photo-emission yield. The origin of dips in the AF luminescence at photon energy of 16 eV for CsBr and of 22 eV for RbF is discussed in the next section.

4. Competition process of Auger electron emission and phonon relaxation

For L type crystals, such as CsCl, CsBr, and RbF, hot holes in the outermost-core band relax competing with Auger process and phonon relaxation. If $\tau_p >> \tau_a$, onsets of decrease(increase) in excitation spectrum for AF luminescence(photo-emission yield spectra) should occur at photon energy of $2E_g$. On the other hand, if $\tau_p << \tau_a$, there should be no difference in excitation spectra and photo-emission spectra around photon energy of $2E_g$. As shown in Fig.2, onsets of decreases in excitation spectra and of increases in photo-emission yield spectra are clearly observed in CsBr and RbF at $2E_g$. While an onset of decrease in the excitation spectrum in CsCl is not clear, an onset of increase in the photo-emission yield spectrum is clearly observed. From this result we can conclude that $\tau_p >> \tau_a$.

The inequality $\tau_p >> \tau_a$ means that quantum efficiency for AF luminescence for L type crystals increases with increasing energy difference $E_g - E_{vc}$. Since band-gap energy E_g in alkali halides at room temperature is smaller than that at low temperature up to approximately 0.2 eV,[18] $E_g - E_{vc}$ increases with decreasing temperature. As a result, an increase in luminescence intensity is expected for the outermost-core holes which are created throughout this band. Figure 3 shows emission spectra from CsCl crystal for various temperatures under high-energy electron excitation. CsCl has two luminescence components: fast and slow. Emission spectra for the fast component were obtained by measuring intensities of the decay curves for different energies of luminescence photons with constant excitation of 511 keV annihilation radiation from ^{22}Na source. The decay curves were measured by a single-photon counting method. The measured decay time obtained was 2.0 ns which was almost constant in the temperature range from 15 K to 300 K. This result agrees with the reported decay times for CsCl at 5K and 77K by Blair et al.[26] The fast luminescence component in the energy range from 3.9 eV to 6.0 eV was confirmed to be AF luminescence.[9] An increase in AF luminescence intensity of 50 % was measured at 15 K compared with that at 296 K.

The luminescence of energy range from 3.9 eV to 6.0 eV from CsCl crystal was reported by Pooley and Runciman[25] as a title of the paper, "Recombination luminescence in alkali-halides". It should be noted here that their observed luminescence was not recombination luminescence but AF luminescence.

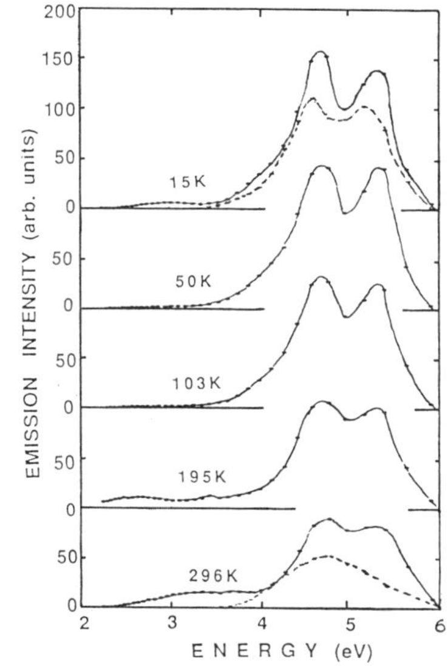

Fig. 3. Emission spectra from CsCl crystal for various temperatures under high-energy electron excitation(solid curves) and emission spectra for the fast component luminescence (broken curves).

5. 4d core excitation

Excitation spectra for AF luminescence from CsCl and CsBr were measured over photon energies from 65 to 140 eV with use of Synchrotron Radiation Source (SRS) at Daresbury Laboratory.[13] With this energy range, the 4d core of Cs^+ at 79 eV can be excited as well as the 3d core of Br^- at 69 eV. Figure 4 shows the excitation spectra for

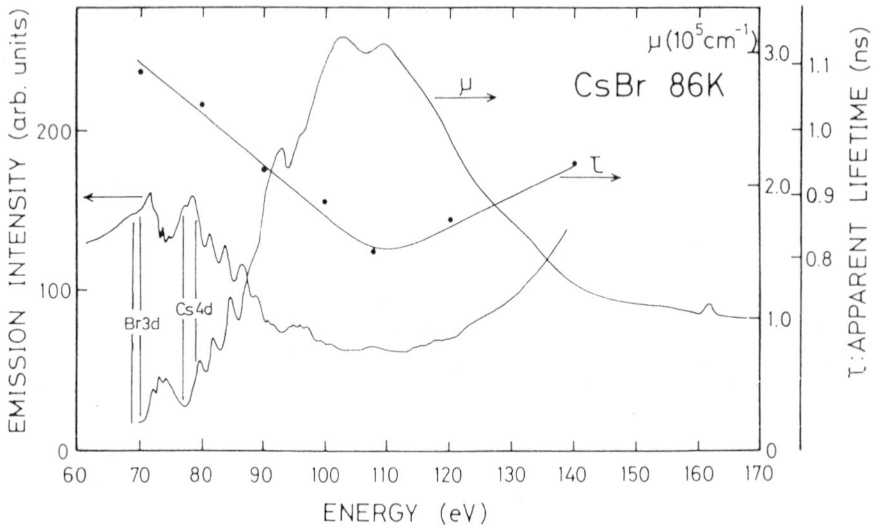

Fig.4. Excitation spectrum and the apparent lifetime for AF luminescence of the 4.9 eV band from CsBr at 86K. The absorption coefficient μ of CsBr is taken from Ref. 27.

AF luminescence of the 4.9 eV band from CsBr at 86 K, together with an absorption coefficient μ reported by Cardona et al.[27]

A slight dip is observed at the photon energy region from 70 to 77 eV that corresponds to the excitation of Br^-3d core. At this energy region, the process competing with AF luminescence is the photo-ionization of Br^-3d core. A Br^- 3d hole recombines with an electron in the Br^-4s core state or with an electron in the valence band ejecting an Auger electron without producing a Cs^+5p hole. Thus the slight dip is observed.

The onset of remarkable decrease in the excitation spectrum is observed at photon energy corresponding to Cs^+4d core excitation. At the incident photon energy of the maximum μ the luminescence intensity decreases to about 50 % of its value compared with the intensity at the energy just below the Br^-4s core excitation. The general feature of the excitation spectrum is its "anti-correlation" with respect to the absorption spectrum. Similar decreases of up to 50 % in the AF luminescence were observed from CsCl and BaF_2 at room temperature as well as at 86 K.[28]

A variation of decay curves for differing incident photon energies was observed in CsCl and CsBr crystals.[13] Differing incident photon energy relates to the differing absorption length μ^{-1}. Figure 5 shows typical decay curves for the 4.9 eV band from CsBr at 86 K for incident photon energies of 70 eV ($\mu^{-1} = 5 \times 10^{-5}$ cm), 108 eV ($\mu^{-1} = 3 \times 10^{-6}$ cm) and 140 eV ($\mu^{-1} = 8 \times 10^{-6}$ cm), respectively. The decay curves are fitted to a single exponential decay down to intensities of 0.01, 0.1 and 0.02 of the peak intensity for incident photon energies of 70, 108 and 140 eV, respectively. The fitted apparent lifetimes

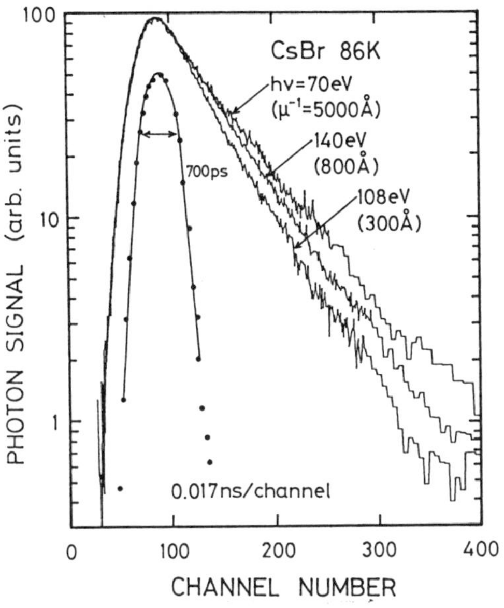

Fig.5. Decay curves for the 4.9 eV AF luminescence from CsBr at 86K for differing incident photon energies. The system instrumental response is also shown.

are shown in Fig.4. It is seen that the apparent lifetime has a minimum at 108 eV, where μ^{-1} is minimum, and is correlated with respect to the luminescence intensity. It should be noted that measured apparent lifetimes through Cs^{+}4d core excitation are apparently smaller than the decay time of 1.3 ns under the Cs^{+} 5p outermost-core excitation.[11]

Since the photo-ionization cross section of the Cs^{+}5p state is more than two orders of magnitude smaller than that of the 4d state around 100 eV, incident photons are almost absorbed at the 4d state. Kamada et al. studied the non-radiative decay of excited 4d states in CsCl, CsBr and BaF_2 by photoelectron spectroscopy.[22,23] It is shown that excited 4d states in those crystals predominantly decay through the N_{45}-$O_{23}O_{23}$ Auger process, i.e. a 4d hole and a 5p electron recombine with the energy transferred to another 5p electron that is excited into the continuum. The final state after the decay through the N_{45}-$O_{23}O_{23}$ Auger process has two holes at the Cs^{+} 5p level. Therefore, these two holes in the 5p state may be responsible for the decrease in the luminescence intensity and the faster decay in decay curves with increasing μ.

Kayanuma,[29] and Shin[30] independently proposed the non-radiative decay of one of the correlated holes in 5p state with the energy transfer to a valence band electron by taking into account of the correlation energy U_c about 5 eV.[32] After the above process another hole in 5p level is subject to the attractive potential that was induced when two holes were in 5p level. This potential should be much stronger than that induced by the one hole in the 5p level. It is not known whether this other 5p hole recombines with a valence band electron with the emission of the photon whose energy is the same as that due to the radiative decay of a valence band electron into the outermost-core hole produced by direct photo-ionization at the 5p level.[29,31]

As shown in Figs. 4 and 5, the luminescence intensity decays more rapidly when incident photons are absorbed at 4d level about 3 x 10^{-6} cm from the surface than when they are absorbed in bulk at > 5 x 10^{-5} cm. Furthermore, it should be noted that the absorption length of uv photons is smaller than that of soft x-rays. This observation together with the anticorrelation relationship described above strongly suggest that 5p holes produced through 4d excitation near to the surface relax nonradiatively more rapidly than do those produced by direct photo-ionization at the 5p level.

6. Charged particle excitation

6.1 High-energy electron excitation

Variation of decay curves under uv photon excitation and under high-energy electron excitation was observedfor the AF luminescence from CsCl and BaF_2. High-energy electrons were obtained under 511 keV annihilation radiation from ^{22}Na source. Typical results were shown in Fig. 6. Under high-energy electron excitation, decay for the 4.5 eV band from CsCl shows a single exponential decay down to intensities of 10^{-2} of the peak intensity. The fitted decay time is measured to be 2.0 ± 0.1 ns. Under excitation of 24 eV photons, the decay curve shows no single exponential decay. The measured decay

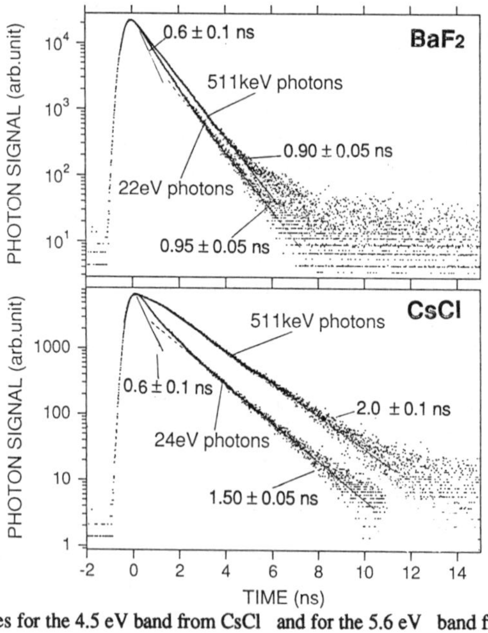

Fig.6. Decay curves for the 4.5 eV band from CsCl and for the 5.6 eV band from BaF2 under high-energy electron excitation by using 511 keV annihilation photons and uv photons.

curve is tentatively fitted as a sum of two exponential decay components down to an intensity of 10^{-3} of the peak intensity. The two fitted decay times are 0.6 ± 0.1 ns and 1.50 ± 0.05 ns which are smaller than that of 2.0 ns for high-energy electron excitation.

Under high-energy electron excitation, the decay for the 5.6 eV band from BaF_2 shows a single exponential decay down to the intensity of 10^{-2} of the peak intensity with decay time of 0.90 ± 0.05 ns. Under excitation of 22 eV photons, the decay curve can be fitted with two exponential decays, giving decay times of 0.6 ± 0.1 ns and 0.95 ± 0.05 ns. The latter value agrees with that under high-energy electron excitation.

The absorption length for 511 keV annihilation radiation is about 10 cm, and that for uv photons is less than 10^{-6} cm. As a result under uv photon excitation, outermost-core holes are produced within 10^{-6} cm from the crystal surface. On the other hand, high-energy electrons produce outermost-core holes in crystal bulk. A possible explanation for the faster decay component in the decay curve for uv excitation is that outermost-core holes near crystal surface relax with non-radiative process. Measurement of quantum efficiency for AF luminescence is necessary to confirm this explanation.

7.2 Heavy ion excitation

It was reported in our earlier paper that no photons were detected for the 5.6 eV band from BaF_2 crystal under 5.6 MeV alpha particle excitation.[8,12] An alpha particle is characterized by its high ionization density of about 10^3 times compared with that of high-energy electrons. Electrons, holes and excitons are densely created in the alpha-particle track together with outermost-core holes. Self-trapped excitons are produced either through recombination process of electrons and holes or exctions. Very weak AF luminescence under 5.6 MeV alpha particle excitation can be explained in terms of interaction between outermost-core holes and high-density electrons, holes and self-trapped excitons, in which outermost-core holes relax without emitting 5.6 eV photons.

Very recently, Kimura and Kumagai[33] observed decrease in decay time for the 5.6 eV AF luminescence from BaF_2 crystal with increasing energy deposit of heavy ions. For example, under Ar^{4+} ion excitation, the decay curve shows two exponential decay components: very fast and fast. Those decay times are 240 ps and 800 ps, respectively. An explanation by Kimura and Kumagai for their result is that under the heavy ion excitation, outermost-core holes recombine non-radiatively with high density electrons in conduction band (perhaps high Rydberg states). This results in the decrease in decay time with increasing energy deposit at an early stage of excitation.

7. Conclusion

Experimental results for the radiative decay of valence band electrons into outermost-core holes from large gap insulators were reviewed comparing uv photon excitation, soft x-ray excitation and charged particle excitation of high-energy electrons and heavy ions. From those experimental results it was found that AF luminescence is related to the following processes for outermost-core holes: thermal relaxation, lattice relaxation, two-hole states produced through 4d core photo-ionization, non-radiative decay near crystal surface and non-radiative decay induced by interaction of valence band electrons and self-trapped excitons that are produced in high density. Further work on AF luminescence is of value for understanding the above processes as well as practical application of development of fast scintillation materials.

8. Acknowledgements

I would like to thank J. Gen(Run), N. Nunoya, S. Sakuragi, S. Hashimoto, M. Itoh, M. MacDonald and I. H. Munro for working with me on the topics for this article, to K. Fukui for his close cooperation during UVSOR experiment, and to M. Watanabe, S. Shin, M. Kamada, Y. Kayanuma Y.Toyozawa for their stimulating discussions.

9. References

1. Yu. M. Aleksandrov, V. N. Makhov, P. A. Rodnyi, T. I. Syreishcikova and M. N. Yakimenko, Sov. Phys. Solid State **26** (1984) 1734.
2. Yu. M. Aleksandrov, V. N. Makhov, T. I. Syreishcikova and M. N. Yakimenko, Nucl. Instr. Methods **A261** (1987) 153.
3. Yu. M. Aleksandrov, I. L. Kuusmann, P. Kh. Liblik, Ch. B. Luschchik, V. N. Makhov, T. I. Syreishcikova and M. N. Yakimenko, Sov. Phys. Solid State **29** (1987) 587
4. Yu. M. Aleksandrov, I. L. Kuusmann, V. N. Makhov, S. B. Mirov, T. V. Uvarova and M. N. Yakimenko, Nucl. Instr. Methods **A308** (1991) 208.
5. Ya. A. Valbis, Z. A. Rachko, Ya. L. Yansons, Opt. Spectrosc. (USSR) **60** (1986) 679.
6. J. L. J ansons, V. J. Krumins, Z. A. Rachko and J. A. Valbis, Phys. Status Solidi (b) **144** (1987) 835.
7. J. L. Jansons, V. J. Krumins, Z. A. Rachko and J. A. Valbis, Solid State Comm. **67**(1988) 183.
8. S. Kubota, N. Kanai and J. Raun(Gen), Phys. Status Solidi (b) **139** (1987) 635.
9. S. Kubota, M. Itoh, J. Ruan(Gen), S. Sakuragi and S. Hashimoto, Phys. Rev. Letters **60** (1988) 2319.
10. M. Itoh, S. Hashimoto, S. Sakuragi and S. Kubota, Solid State Commun. **65** (1988) 523.
11. M. Itoh, S. Kubota, J. Ruan(Gen), and S. Hashimoto, Rev. Sol. State Science **4** (1990)467.
12. S. Kubota, J. Ruan(Gen), M. Itoh, S. Sakuragi and S. Hashimoto, Nucl. Instr. Methods **A289** (1990) 253.
13. S. Kubota, M. MacDonald and I. H. Munro, J. Lum. **48** & **49** (1991) 589.
14. E. N. Mel'chakov, P. A. Rodnyi, B. V. Rybakov, A. Ya. Smakov and M. A. Terekhin, Sov. Phys. Solid State **31** (1989) 878.
15. P. A. Rodnyi, M. A. Terekhin and S. V. Petrov, Sov. Phys. Solid State **32** (1990) 1844.
16. A. Kikas and M. Elango, Nucl. Instr. Methods **A308** (1991) 191.
17. C. S. Inouye and W. Pong, Phys. Rev. **B15** (1977) 2265. J. A. Smith and W. Pong, Phys. Rev. **B12** (1975) 5931.
18. R. T. Poole, J. G. Jenkin, J. Liesegang and R. C. G. Leckey, Phys. Rev. **B11** (1975) 5179. R. T. Poole et al. Phys. Rev. **B12** (1975) 5872.
19. H. Sugawara and T. Sasaki, J. Phys. Soc. Jap. **46** (1979) 132.
20. M. Watanabee, H. Nishida and A. Ejiri, in *Proc. of the Fourth Inter. Conf. on VU radiation physics*, ed. E. E. Koch, R. Haensel and C. Kunz (Pergamon-Vieweg, 1974), p. 370.
21. G. W. Rubloff, Phys. Rev. **B 5** (1972) 662.
22. M. Kamada, O. Aita, K. Ichikawa, M. Okusawa and K. Tsutsumi, Phys. Rev. **B45** (1992) 12725.
23. K. Ichikawa, O. Aita, K. Aoki, M. Kamada and K. Tsutsumi, Phys. Rev. **B45** (1992) 3221.
24. M. Okusawa, K. Soda, T. Mori and T. Ishii, J. Phys. Soc. Jap. **60** (1991) 768.
25. D. Pooley and W. A. Runciman, J. Phys. **C 3** (1970)1815.
26. I. M. Blair, D. Pooley and D. Smith, J. Phys. **C 5** (1972) 1537.
27. M. Cardona, R. Haensel, D. W. Lynch and B. Sonntag, Phys. Rev. **B2** (1970) 1117.
28. S. Kubota, M. MacDonald and I. H. Munro, to be published.
29. Y. Kayanuma, private communication.
30. S. Shin, private communication.
31. M. Kamada, K. Ichikawa and K. Tsutsumi, Phys. Rev. **B28** (1983) 7225.
32. M. Kamada, private communication.
33. K. Kimura and H. Kumagai, Rad. Eff. Def. Solids (1992) in press.

X. MICROSCOPY

RECENT DEVELOPMENTS IN SCANNING SOFT X-RAY MICROSCOPY

H. Ade, C.Jacobsen, J.Kirz, C.-H. Ko, S. Lindaas, S. Williams, X. Zhang
Dept. of Physics, SUNY@Stony Brook, Stony Brook, NY11794

C. Buckley
King's College, Strand, London, WC2R 2LS

E. Anderson, K. Goncz, and S. Rothman
Lawrence Berkeley Laboratory, Berkeley, CA 94720

We review work performed with the Scanning Transmission X-ray Microscope at the National Synchrotron Light Source. Significant progress has been made in recent years, and higher spatial resolution and increased count rates have been achieved. The microscope is in routine operation, and has been used by a number of investigators to examine a variety of samples, including chromosomes, zymogen granules, fibroblasts, cartilage, and thin sections of polymers. Studies also address radiation damage in fixed and hydrated specimens. We continue to explore new operating modes of the microscope and recently added the capabilities to excite visible light luminescence from fluorescent dyes, X-ray Absorption Near Edge Structure (XANES) imaging to achieve chemical state specific mapping, and XANES spectroscopy from areas as small as 0.01 μm^2.

1. Introduction

Since their discovery, X-rays have always been recognized for their potential in microscopy due to their short wavelengths and their distinct interaction with matter. Progress was hampered for the most part of this century due to the low brightness of available x-ray sources and the inability to focus x-rays. The development of microfabrication techniques in the 1970s has made it possible to produce high resolution zone plates and the advent of high brightness synchrotron radiation sources and lasers has provided sufficiently bright sources. Owing to these developments, progress in x-ray microscopy has been enormous over the last decade. This growth should continue with the impending completion of third generation synchrotron sources, and ever improving optical elements.

In this manuscript we outline the progress of the field since the VUV9 conference[1,2] and describe our own work at the National Synchrotron Light Source (NSLS). Section 1.1 provides a short synopsis of the motivation to develop x-ray microscopy, which lead to a variety of instruments aimed at biological as well as material science applications. Section 2 focuses on the hardware and performance of the Scanning Transmission X-ray Microscope (X1-STXM) at the NSLS, while Section 3 details applications of the X1-STXM. Some of the other imaging efforts at beamline X1A of the NSLS are described in section 4.

1.1 Motivation for x-ray microscopy

The dominant interaction of x-rays in the 50 to 1500 eV energy range with matter is photoelectric absorption. The cross section is a strong function of the photon energy, particularly at or near the so called absorption edges when the photon energy becomes larger than the binding energy of a particular electronic shell of an atom: the cross section increases dramatically, by often more than an order of magnitude. This provides the contrast mechanism to map low to medium Z elements via differential absorption imaging, in which two images, above and below an edge are acquired and appropriately processed (see section 3.3). Furthermore, the X-ray Absorption Near Edge Structure (XANES) is sensitive to the chemical environment[3], and can be used to provide chemical state imaging (see section 3.4). The photoelectrons themselves can be energy analyzed with X-ray Photoelectron Spectroscopy (XPS) techniques, thereby providing a wealth of information on both the elemental composition and chemical state of a thin layer of a few nm near the surface of a sample.

The so-called "water window" (the energy region between the O K-edge at 540 eV and C K-edge at 290 eV) where water is relatively transparent while organic material is not, has been identified to be particularly promising to image hydrated, unsectioned, and unstained biological samples in their natural environment. While X-rays are generally produced and transported in a vacuum enclosure, the cross sections are small enough to allow with tolerable losses the penetration of thin membranes, several microns of water, or even the transport through an atmospheric pressure environment. This underscores the relative flexibility of the specimen environment in x-ray microscopy, in that the sample can either be in ultra-high-vacuum, in air, or in a wet cell/environmental chamber at atmospheric pressure.

Imaging of biological samples has been the initial driving force behind the developments in x-ray microscopy. Full field imaging, as well as scanning instruments, have been developed for transmissive samples[4]. Examples include the microscopes operated by the Göttingen group at BESSY[5], the King's College, London STXM at Daresbury Laboratory[6], X-ray microscopes at the photon factory[7], and our own work at the NSLS. Most of these microscopes use high resolution zone plates as the crucial x-ray optical element. Recent efforts of the Göttingen group are described by Rudolf et al. in these proceedings. Additional techniques such a contact microscopy, Gabor holography, Fourier transform holography, and large angle diffraction have been developed for biological imaging[4].

Applications of X-ray Microscopy are not restricted to biological samples and problems in the life sciences. There are in fact several efforts underway to utilize x-ray microscopy for material and surface science problems of bulk samples with the techniques of XPS and XANES imaging. Due to the generally increased demand of these instruments on the source brigthness when compared to transmission x-ray microscopes, XPS instruments have been lagging in development. A recent review emphasizing micro-probe based XPS and XANES instruments can be found by Ade[8], while the reviews of Mundschau[9] and Tonner[10] are emphasizing parallel imaging techniques.

2. Scanning Transmission X-ray Microscope (X1-STXM) at the NSLS

While the X1-STXM undergoes continuous development, its conceptual design has not changed much over the years. The x-ray beam exits the vacuum through a thin Si_3N_4 membrane into an atmospheric pressure environment, and illuminates a zone plate to form a micro probe. In order to suppress unwanted diffraction orders from the zone plate, we employ an opaque stop in the center of the ZP and a strategically placed slightly smaller pinhole which acts as an Order Selecting Aperture (OSA) (see Fig. 1a). The sample is mechanically scanned with piezo electric transducers in raster fashion under computer control. The transmitted flux is detected in a gas flow counter in pulse counting mode and recorded digitally under CAMAC control. Recent improvements include a new scanning stage, better zone plates, a modified beamline, and the ability to acquire XANES spectra from small areas. We will describe these below.

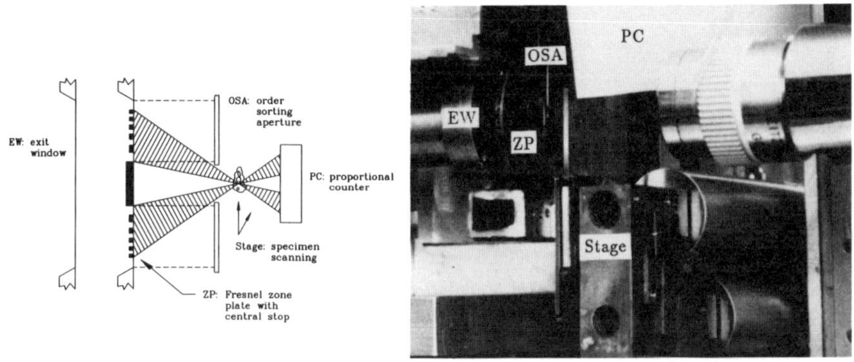

Figure 1: The X1 Scanning Transmission Microscope: (a) schematic of the major components. (b) A photograph of the microscope in which the components indicated in (a) have been separated to distances much larger than during operations to facilitate identification of the various parts. The proportional counter has been raised above the optical axis, which is typically done to pre-align the specimen with the visible light microscope at the right.

Our new fine scanning stage is a flexure stage driven by capacitance micrometer controlled piezo electric transducers (see Figure 1b). The range of the transducers is 75 μm, and the capacitance micrometer provides positioning with a minimum step size of 4.8 nm via closed loop feedback. This fine stage is mounted from a stepping motor driven X-Y stage that provides coarse positioning with 500 nm step size over a range of 25 mm.

A new generation of Ni phase zone plates, together with the new scanning stage, provides for a vastly improved performance over the one reported at the VUV9 meeting. The ZPs are 90 μm in diameter, have a central stop of 40 μm, an outermost zone width of 45 nm, and 100 nm thick zones. The diffraction efficiency into the first order is above 10% over the entire water window range, and can be as high as 14%. This efficiency results in an intense micro probe and allows fast data acquisition times. To determine the imaging characteristics and the resolution limit with the new zone plates, we imaged a star-like test pattern. Fig. 2a shows the theoretical Modulation Transfer Function (MTF), and the MTF as derived from a comparison of the X1-STXM micrograph with an STEM micrograph of the test pattern. The good agreement between experiment and theory allows the deconvolution of X1-STXM images with the theoretical point spread function and the deconvolved micrograph of the test pattern is shown in Fig. 2b. While the Rayleigh resolution of the X1-STXM is 55 nm, features < 35 nm in size can be observed. More details about the performance of the imaging system can be found by by Zhang et al.[11], and by Jacobsen et al.[12]

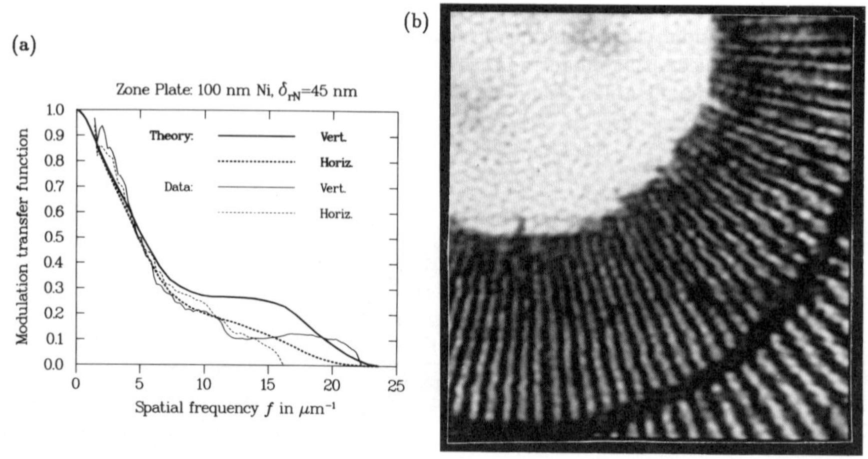

Figure 2: (a) theoretical and measured Modulation Transfer Functions of the X1-STXM (for details see ref. 12). (b) Deconvolved micrograph of the lower right quadrant of a Au test pattern with 128 spokes and a radial marker 2 μm from the center. This yields a nominal feature width of 49 nm at the location of the radial marker and a nominal width of about 25 nm for the innermost features. (At the time of data acquisition there were glitches in the stage motion, which have since been fixed.)

A recent upgrade of the beamline provides for vertical refocusing which results in an increase in X1-STXM flux by about a factor of 6 over the old configuration. This implies estimated countrates in excess of 10 Mcps. Our gas flow counter can not count at rates above about 3-4 Mcps in pulse counting mode and we are contemplating various analog detector schemes. Currently, we stop down the beam to be within the counting range of our detection system. Even at the reduced count rates, it takes no more than a few minutes to acquire a 400x400 pixel image, and 30-50 images are recorded on a typical day. More information about the technical aspects of the instrumentation, particularly about the beamline, can be found in Rarback et al[13], and Kirz et al.[14].

3. X1-STXM applications

3.1 Protein release from granules

One major advantage of x-ray microscopy is that biological specimens can be imaged whole, i.e. not sliced or sectioned, in a hydrated environment. The pH, ionic strength, or other characteristics of the suspending solution can be changed in a controlled manner using a special sample chamber, or wet-cell. Goncz and Rothman fully exploited this technique in time-resolved studies of protein release of single secretion vesicles (zymogen granules) isolated from the acinar cells of rat pancreas[15]. The natural contrast in the "water window" allows quantitative measurements of the protein density in these objects with femtogram sensitivity. Goncz and Rothman observed the release of protein from isolated zymogen granules resuspended in a variety of solutions over a period of hours. The granules were also observed to shrink and the size distributions of 50 individual granules, before and after changes in size were induced by perifusion of water, are shown in Fig. 3b. When examined on a individual basis, most if not all of the granule have shrunk. An increase in size was observed with the addition of exogenous chymotrypsinogen A (a protein similar to the ones inside the granule) to the wet-cell at a concentration of 250 mg/ml. Lysis, the rupture of the membrane, was not observed in these studies. Protein transport across membranes is still a much debated issue as discussed by Goncz and Rothman[15]. The X1-STXM results provide a direct visual and quantitative measure of the protein density and distribution in the granules and their dependence on the physiological environment, which in turn might answer still open questions about protein release and transport mechanisms. In addition when imaged in an aqueous environment, these granules do not always appear as uniformly dense spherical objects (see Fig. 3a) as they do in electron micrographs. The existence of heterogeneous granule morphology in x-ray micrographs suggests that thin sectioning and staining needed for electron micrographs produce an artificially high density.

(a)

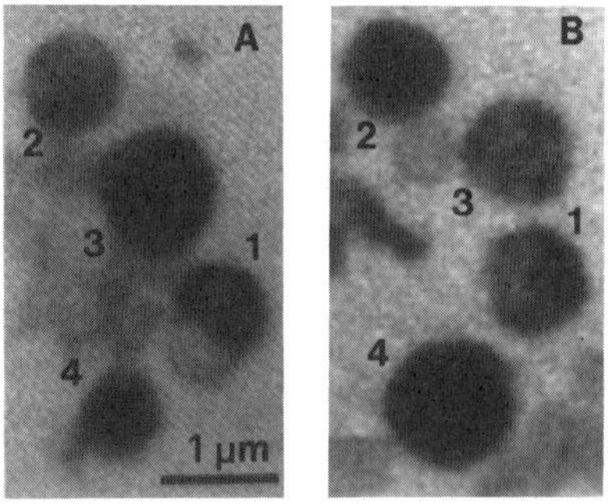

(b)

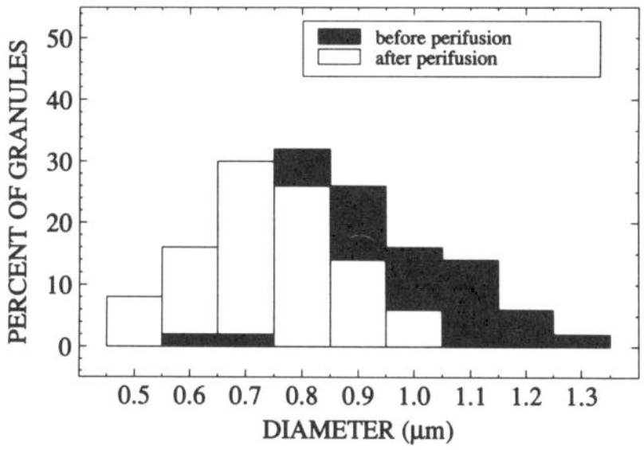

Figure 3. (a) zymogen granules showing inhomogeneous distribution of their protein content. (b) size distribution of 50 individual granules before and after a change in size (and protein density) was induced by flowing a 0.3 M sucrose solution in a 5mM $NaHPO_4$ buffer (pH 6.0) continually past the granules in the wet-cell.

3.2 Chromosome structure and radiation damage studies

In order to take full advantage of X-ray microscopy and image biological samples in a hydrated state, it is necessary to understand the limitations imposed by radiation damage, understand the damage mechanisms, and its dependance on the sample and sample preparation. Williams et al characterized radiation damage, measured as mass loss, for *V. faba* chromosomes in the X1-STXM for a variety of imaging protocols and sample preparation conditions. Freeze-dried chromosomes show no apparent damage after doses up to 150 Mrad. The total mass loss seen after several low dose images is greater than that seen in a single image with the same total dose, suggesting that the rate of data collection in a single image exceeds the rate of radiation damage (see Fig. 4a). Radiation damage depends sensitively on the fixation protocol employed, with short fixation times at higher concentrations of glutaraldehyde producing the most radiation resistant samples. Radiation damage to chromosomes mounted in wet-cells was independent of ionic strength of the buffer (Na_2HPO_4, NaH_2PO_4, pH 7.0) at 65 mM ionic strength and above, but increased for ionic strengths below 65 mM. Because this ionic strength is close to that of the phase transition for chromatin fibers, it is assumed that the configuration of these fibers plays a role in the radiation sensitivity of the chromosome. The free radical scavengers dithiothreitol and ethanol were shown to reduce the amount of radiation damage incurred by 3 to 10 fold, as did the reduction of air in the wet cell[16]. More than a twenty fold reduction in damage was observed using an improved wet cell, one which sandwiches the sample in a thin buffer layer between two silicon nitride windows. This reduction is much greater than that predicted by simply eliminating the available oxygen in the previous wet-cell. A greater understanding of the mechanism of radiation damage and mass loss in the X1-STXM is required to explain the damage reduction observed with the new wet cell, and to formulate strategies for further decreasing the radiation sensitivity of wet specimens.

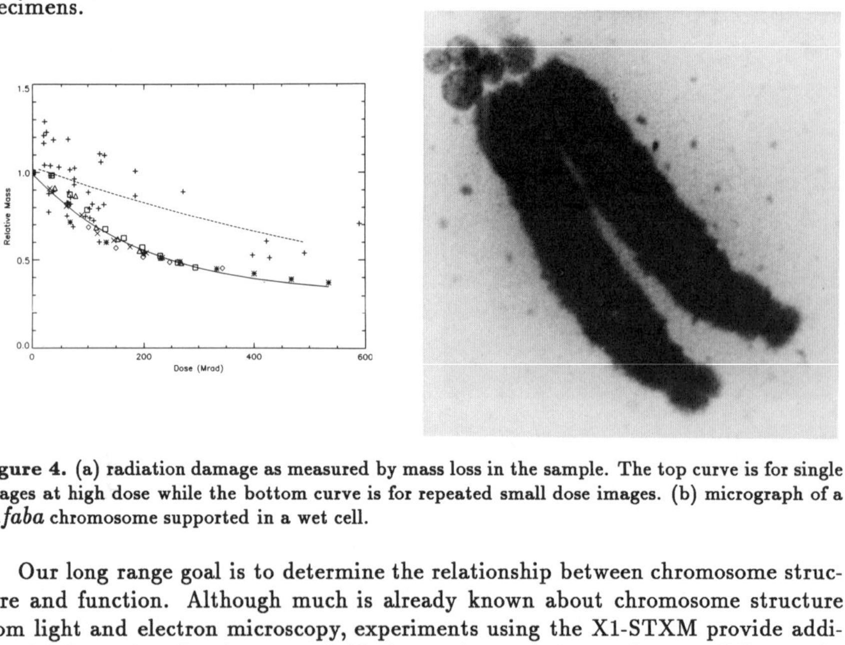

Figure 4. (a) radiation damage as measured by mass loss in the sample. The top curve is for single images at high dose while the bottom curve is for repeated small dose images. (b) micrograph of a *V.faba* chromosome supported in a wet cell.

Our long range goal is to determine the relationship between chromosome structure and function. Although much is already known about chromosome structure from light and electron microscopy, experiments using the X1-STXM provide additional information that is not available from other techniques. One possibility is the determination of chromosome composition with 50 nm resolution through XANES imaging (section 3.5). Using this procedure we intend to determine the spatial distribution of protein and DNA within metaphase chromosomes. Another advantage of imaging with x-rays is the ability to examine chromosome structure in situ, thereby avoiding artifacts caused by harsh isolation procedures. Because chromosomes need not be isolated or stained in the X1-STXM, it will be possible to investigate the chromosome condensation process during prophase. The folding of the 30 nm chromatin fiber into the metaphase chromosome is not known, and it is this level of structural organization that can be best addressed by the X1-STXM. The ability to examine intact nuclei provides the opportunity to correlate the activity of specific genes with their position in the nucleus (see section 4.1), by examining cells at different times in the cell cycle and in different tissues.

3.3 Elemental mapping of calcium with x-ray differential absorption imaging

Due to the large change in cross section across absorption edges it is possible to map the distribution of an element by acquiring an image above and below the edge. A signal proportional to the optical density of an element can then be computed by subtracting the log signals of the individual images. The large signal in x-ray differential absorption imaging allows for rapid data acquisition and 2D mapping, rather than just point analysis. Buckley et al. have used this technique to map the calcium distribution in sections of human articular cartilage[17] (see Fig. 5). The cartilage can contain deposits of calcium microcrystals, which are concentrated in a band about 5-50 μm from the articular surface and in clusters around chondrocyte cells. These studies are aimed at understanding the origin and function of these deposits and how they are related to the chondrocyte cells. Many elemental maps similar to the one presented in Fig.5 were evaluated and indicate that the crystals are not present in cells, but clustered around their periphery. This suggests that the microcrystals do not seed inside the cell, but that the cells may export the biochemistry required for seeding into the matrix surrounding it[17].

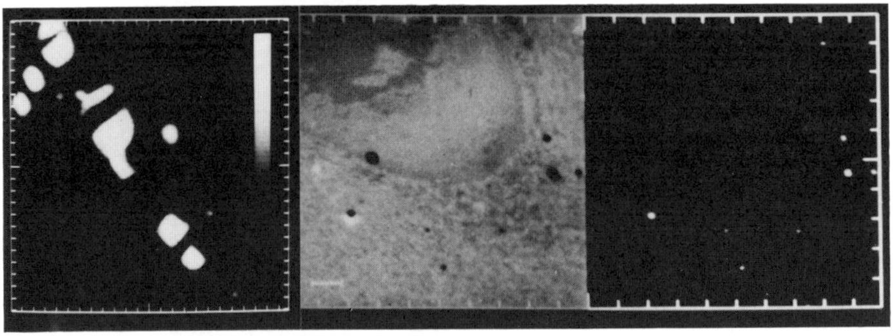

Figure 5. (a) Calcium mass thickness of calcium phosphate crystals as found in a thin section of cartilage (bar width = 0.5 μm). (b) Transmission image of a section of cartilage showing a chondrocyte cell at 350.6 eV, and (c) Ca mass thickness of the same region as in (b) produced by using image (a) and a corresponding image at 351.6 eV. Images (b) and (c) and similar images suggest that calcium deposits do not form inside the chondrocyte cell.

3.4. C-XANES imaging and micro-spectroscopy with X1-STXM

XANES spectroscopy is routinely used with great success to determine unoccupied densities of states in molecules and solids. The electronic levels probed provide much information about the chemistry, atom coordination, and bonding, and with polarized sources can furthermore be used to determine bond orientation. We have used this chemical sensitivity at the carbon K-edge[18], and have modified X1-STXM to acquire XANES spectra from areas as small as 0.01 μm^2. As an example of the chemical sensitivity of C-XANES relevant to biology, we show C-XANES spectra of DNA and albumin, a common protein, in Fig. 6a. The primary spectral features are associated with unsaturated bonding of carbon with oxygen, nitrogen and carbon. These spectral differences allow us to map specific compounds and bonds in a sample, which could be used to image the DNA/protein ratio in biological structures. We have used C-XANES imaging to map distinct phases in polymer blends, such as the thin section of a 50-50 weight% polystyrene/polypropylene blend shown in Fig. 6b. The micrograph was acquired at a photon energy of 285.5 eV, where polystyrene exhibits strong absorption due to a π^* resonance, where as polypropylene does not. Analogous information could in principle be achieved with core level Parallel Electron Energy Loss Spectroscopy (PEELS) imaging for very thin sections. PEELS is, however, so far only used for point analysis or elemental imaging, and the sample damage for organic materials is orders of magnitudes larger due to the high valence band cross section in EELS, which is negligible in XANES. XANES imaging can furthermore be easily adopted to imaging of bulk materials. Since C-XANES imaging of thin sections as presented here relies on absorption contrast the present resolution achievable with this technique is the same as exemplified by Figs.2a and 2b.

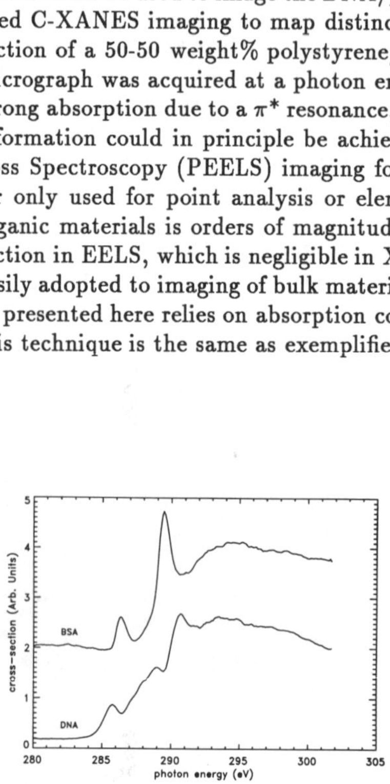

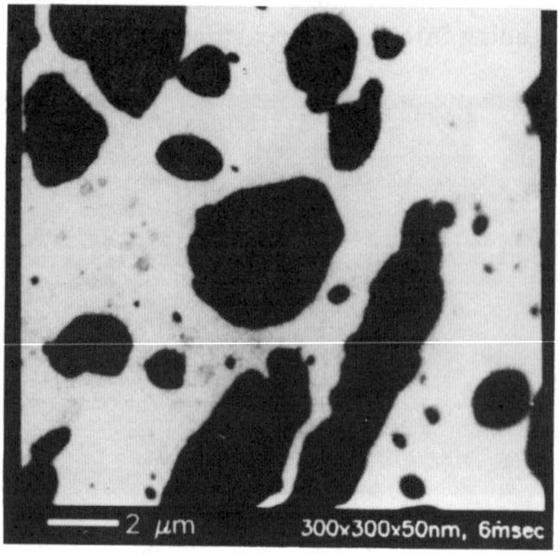

Figure 6. (a) C-XANES spectra of DNA and protein. (b) C-XANES imaging of a polystyrene/polypropylene (PS/PP) blend at a photon energy of 285.5 eV. PS is highly absorbing due to a π^* resonance at this energy, while PP is not.

3.5 Stereo microscopy and other X1-STXM applications

The well known technique of stereo microscopy can also be adopted for the X1-STXM to allow for 3-D imaging. A special stage allows the sample to be rotated by $\pm$ 30°, and the internal structure of human spermatozoa has been successfully observed[19] at a longitudinal resolution of 214 nm. The ability to rotate the sample will also be helpful to investigate polarization dependance in XANES imaging. We furthermore used our microscope to inspect the uniformity of thin, free standing High T_C films with a thickness sensitivity of a single unit cell[20]. Gilbert and Pine have used the X1-STXM to assess radiation damage in cell cultures of chick embryo fibroblasts[21].

4. Additional microscopes at X1A

4.1 Scanning Luminescence X-ray Microscope (SLXM)

Scanning luminescence x-ray microscopy[22] is based on the use of the focussed probe of the X1-STXM to stimulate visible light emission from dyes. The signal is collected by an optical microscope and detected with a photomultiplier. Jacobsen et al. have obtained 50–75 nm resolution images of P31 phosphor grains in first tests, and have imaged luminescence from polystyrene spheres loaded with 50–100 μmol/g of a fluorescent dye as shown in Fig. 7. If specific labels are available that fluoresce in the X-ray microscope this technique would make it possible to image dye-tagged sites of biochemical activity at the resolution of the x-ray microscope in wet, unsectioned, and perhaps even initially living cells.

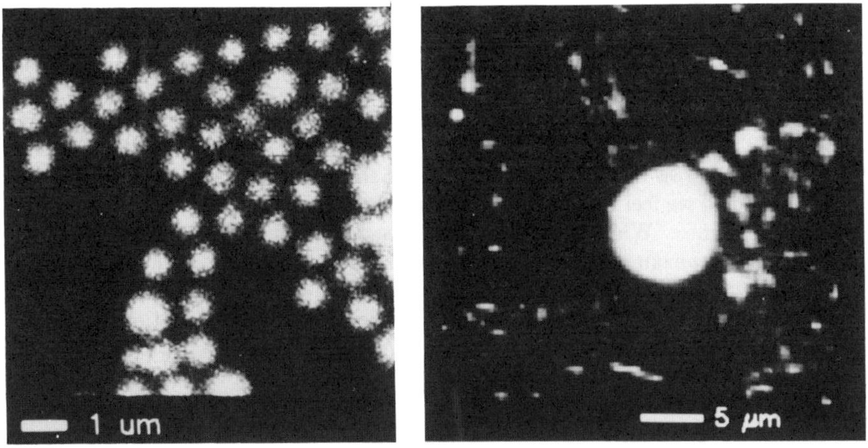

Figure 7. (a) Scanning Luminescence X-ray Microscope (SLXM) micrograph of dye-loaded polystyrene spheres. (b) Photon beam mediated oxidation of aluminum surface as observed with the X1-SPEM by monitoring the oxygen KVV Auger intensity. Only the central region of the sample that was exposed to the photon beam for 14 hours showed enhanced oxidation.

4.2 Scanning Photoemission Microscope

Ade et al. have furthermore developed a Scanning Photoemission Microscope (X1-SPEM) in order to be able to characterize and investigate surfaces of bulk material. The X1-SPEM is conceptually rather similar to the X1-STXM, in that it also uses a ZP/OSA arrangement to produce a small micro-probe and a mechanical stage to provide 2-D data acquisition. However, the sample resides in a UHV chamber and the primary signal is derived from energy analyzed photoelectrons. It can be used to determine and measure the density of occupied electronic levels via XPS. Since these energy levels are characteristic for a given element and exhibit so called chemical shifts depending the chemical state of the atom probed, the instrument can be utilized to map the chemical state of a given element on the surface. This has been demonstrated for the first time with photoemission microscopy at sub micron resolution utilizing a $Si/Al/SiO_2$ test pattern, in which contrast could be reversed between Si and SiO_2 regions even though the same atomic energy level (Si 2p) has been utilized[23]. More recently we have imaged microelectronic components and undertaken initial studies of the chemistry of the photon beam mediated oxidation of aluminum surfaces[24]. The resolution is presently approaching 0.1 μm [25], and expected to reach 60 nm in the near future[26].

4.3. Complementary efforts at X1

The scanning micro-probe instruments at beamline X1 described above, are complemented by imaging efforts utilizing large angle diffraction[27], Gabor holography[28] and Fourier transform holography[29].

5. Concluding remarks

We have presented the diverse imaging program and some of its applications at beamline X1A at the NSLS. It is carried out with a variety of instruments and distinct techniques, whose number might grow in the near future with additions such as differential phase contrast microscopy[30]. It is possible to add further capabilities with the installation of an x-ray fluorescence detector to either the X1-STXM or the X1-SPEM. Progress in x-ray microscopy in general since the VUV9 meeting has been substantial and we feel that the new techniques of SLXM and XANES imaging show particular promise. While we are in the process to substantially upgrade some of our instruments (for example the X1-SPEM[26]), and our own efforts at the NSLS will see further growth and broader applications, we are looking forward to the growth of the field as a whole with the commissioning of several microscopes at third generation storage rings presently being built or planed (ALS, Elettra, MAX-II, Bessy-II, DAPS, etc.) and our involvement in some of the projects there.

6. Acknowledgements

We are grateful to the many people who contributed to our efforts at X1A. H. Rarback, M. Rivers, N. Arnot have been instrumental in bringing the X1-STXM to life. We acknowledge the contributions of I. McNulty, M. Oversluizen, S. Rock, S. Wirick, M. Howells to our facilities at X1. We thank D. Kern and D. Attwood for their support in zone plate fabrication, and J. Van't Hoff and S. Lamm for their help with chromosome preparations. The NSLS management is continuing to support our efforts vigorously and we are grateful for it. The X1-SPEM has been built in collaboration with S. Hulbert and E. Johnson (NSLS). The calcium differential absorption work has been performed in collaboration with Y. Ali (Institute of Orthopedics, London), the polymer C-XANES imaging with S. Cameron, and C. Costello (EXXON, Annandale), and stereo microscopy in collaboration with B. Loo (LBL, Berkeley). We thank the National Science Foundation for support under Grant No. DIR-9005893, and DMR-8815043, and the Department for Energy for support under Grant No DE-FG02-89ER60858. Much of this work was carried out at the NSLS which is supported by the Department of Energy, Office of Basic Energy Sciences.

7. References

[1] for the review of X-ray Microscopy at the VUV9 conference, see J. Kirz et al., *Physica Scripta* **T31** (1990) 12-17.

[2] H Ade et al., *Physica Scripta* **41:6** (1990) 737-737, Rothman et al., *Physica Scripta* **T31** (1990), G.R. Harp, Z.L. Han, and B.P. Tonner *Physica Scripta* **T31** (1990) 23-27, P.L. King et al., *Physica Scripta* **41:4** (1990), and W.Ng et al., *Physica Scripta* **41** (1990) 758-760.

[3] J.Stöhr, *NEXAFS spectroscopy and the structure of Adsorbed Molecules*, Springer Series in Surface Science, Vol. 25 (Springer, Berlin, 1992)

[4] for a general overview of the established methods and techniques of the field, see A. Michette, G. Morrison, C. Buckley, eds., *X-Ray Microscopy III* (Springer, Berlin, 1992).

[5] D. Rudolph et al., and B. Niemann in ref.4, pp 392–396, and 143–146, respectively.

[6] G.R. Morrison et al. in ref.4, pp 139–142.

[7] N. Watanabe et al., Y. Kagoshima at al., and S. Ohsuka et al., in ref. 4.

[8] H.Ade, *Nucl. Instr. Meth. in Phys. Res.* **A**319 (1992) 311-319.

[9] M. Mundschau, Synchr. Rad. News Vol. 4(3), (1991).

[10] B.Tonner, Synchr. Rad. News Vol. 4(2), (1991) 27-31.

[11] X.Zhang et al., *SPIE* 1741.

[12] C.Jacobsen et al., *Opt. Commun.* **86** (1991) 351-364.

[13] H. Rarback et al., *J. X-ray Sci. Technol.* **2** (1990) 274.

[14] J. Kirz et al., *Rev. Sci. Instrum.* **63** (1990) 557-563.

[15] K.K. Goncz, and S.S. Rothman *Biochim. Biophys. Acta* (in press).

[16] S.Williams et al. *SPIE* 1741, and S. Williams et al, (submitted to *J. Microscopy.*

[17] C. Buckley et al. in Ref.4, pp. 423-426.

[18] H. Ade et al. (submitted to *Science*).

[19] B. Loo et al., *J.Micros.* **166(2)** (1992) RP5-RP6.

[20] C. Jacobsen, J. Kirz, and S. Williams (to be published Special Issue of *Ultramicroscopy*, EMSA Annual Meeting, Aug. 1991, San Jose).

[21] J.Gilbert, and J. Pine, in Ref.4, and J.Gilbert, Ph.D. thesis (Caltech 1992).

[22] C. Jacobsen et al. (submitted to *J. Microscopy*), and C. Jabobsen et al., *SPIE* 1741.

[23] H. Ade et al. *App. Phys. Lett.* **56** (1990) 1841–1843.

[24] H. Ade et al., *Surf. Interf. Anal.* **19** (1992) 17-22.

[25] H. Ade, C.H. Ko, and E. Anderson, *Appl. Phys. Lett.* **60** (1992) 1040–1042.

[26] C.-H. Ko et al., *SPIE* 1741.

[27] D. Sayre (to be published).

[28] M. Howells et al. *Science* **238** (1987) 514, and C. Jacobsen, M. Howells, J. Kirz, and S. Rothman, *J. Opt. Soc. Am.* **A 7** (1990) 1847.

[29] I. McNulty et al. *Science* **256** (1992) 1009-1012.

[30] J.R. Palmer and G.R. Morrison, in Ref.4.

PHOTOEMISSION SPECTROMICROSCOPIES OF NEURONS

Gelsomina De Stasio and P. Perfetti
Istituto di Struttura della Materia del CNR, Via Enrico Fermi 38, 00044 Frascati, Roma, Italy

Delio Mercanti and M. Teresa Ciotti
Istituto di Neurobiologia del Consiglio Nazionale delle Ricerche, Viale Marx 15, 00100 Roma, Italy

S. Hardcastle, S. F. Koranda and B. P. Tonner
Department of Physics, University of Wisconsin-Milwaukee, Milwaukee, WI 5321, USA

C. Capasso, W. Ng, A. K. Ray-Chaudhuri, S. Liang, S. Singh and F. Cerrina
Synchrotron Radiation Center, University of Wisconsin, Madison, Wisconsin 53706, USA

and G. Margaritondo
Institut de Physique Appliquée, Ecole Polytechnique Fédérale, PH-Ecublens, CH-1015 Lausanne, Switzerland

ABSTRACT

We present the first photoemission spectromicroscopy results on biological specimens. The data, obtained with imaging photoelectron spectromicroscopy, and by scanning photoelectron spectromicroscopy, study individual elements in micron and sub-micron size portions of neuron networks. We have found, in particular, evidence of inhomogeneous distribution of elements physiologically present in neural cells and of toxic trace elements uptaken in neuron networks, confirming the need for microscopic analysis of this kind in the life sciences.

1. Introduction

Photoemission spectroscopy has been for many years a leading probe of the chemical and electronic structure in materials science[1-4]. Unfortunately, its applications to biological systems have been drastically limited by insufficient spatial resolution. Recently, this obstacle began to be removed by dramatic improvements in the instrumentation.[1-8] High-quality photoelectron microimages of neuron networks have been obtained with resolution

a b

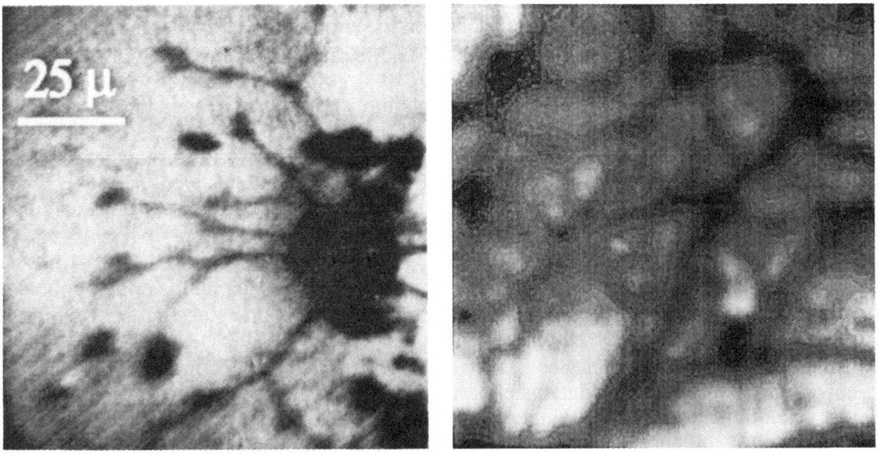

Fig. 1 -- (a) XSEM photoelectron micrograph of a portion of neuron network, taken with unmonochromatized synchrotron light with a lateral resolution of the order of 0.5 micron. In the right-hand part of the image there is an aggregate of neurons, with several individual cells connected by axons and dendrites. In the central upper part, we see a growth cone. (b) Scanning photoelectron micrograph obtained by MAXIMUM of a portion of neuron network, taken with monochromatic synchrotron light at 95 eV photon energy. A partial yield image (1.3 Ev kinetic energy) with a lateral resolution of the order of 0.5 micron. Cell bodies, axons and dendrites neural structures are distinguishable in the chemical/intensity contrast.

in the half-micron range, both with the electron-imaging instrument XSEM (X-ray Secondary electron Emission Microscope)[3] and with the scanning instrument MAXIMUM at Wisconsin[6]. Figure 1a shows an example of total yield electron-imaging photoelectron micrograph. Fig. 1b shows an example of partial yield scanning phoetoelectron micrograph, obtained by MAXIMUM, of a similar neuron network. Both these images reveal fine details of the cultured neuron networks.

2. X-ray Secondary electron Emission Microscopy (XSEM) Experiments

We present the first results of a novel type of experiments in the life science: photoemission chemical analysis on a microscopic scale.[1] A newly developed secondary-emission photoelectron microscope[2,3] was used to obtain microimages of several neuron networks with a lateral resolution of 0.5 micron. Then, the same XSEM instrument analyzed chemical elements and their status on the same microscopic scale. The test was performed on oxygen, and on traces of two uptaken toxic elements, cobalt and manganese. The results show for the first time evidence of a inhomogeneous distribution of such elements in a biological system. In general, they demonstrate the feasibility of using electron spectroscopy for microscopic chemical analysis in the life sciences.

The present work goes beyond the mere imaging, and performs chemical analysis. The experiments were conducted on fixed and dehydrated neuron networks. Cerebellar granule cells from 7-day-old rats were allowed to grow on a gold substrate. The cells were obtained [9] by enzymatical and mechanical dissociation of the cerebellar tissue and plated at a density of 10^5 cells/cm^2 Basal Medium (Eagle's salt), containing 10% fetal calf serum. Dissociated cells were seeded in Petri dishes on a gold coated stainless steel substrate, treated with a 5 microg/ml of poly-L-lysine solution, and cultured in an incubator at 37 C in a 5% CO_2 humidified atmosphere.

Afterwards, we followed either one of two final procedures in the specimen preparation process. In the first procedure, after 6 days the cultures were fixed for 10 min in a 4% solution of glutaraldehyde in phosphate buffered saline solution (PBSS), carefully washed with DD water, and dehydrated at a pressure of 10^{-3} mbar, at room temperature for 24 hours.[3,6]

The second procedure led to an uptake of toxic elements, either cobalt or manganese. The 8-days old specimens were washed twice in an uptake buffer[10] to remove the culture medium, incubated for 20 minutes with $CoCl_2$ or $MnCl_2$ (5 mM in 1 ml of the same buffer) and in the presence of a neurotransmitter: 100 mM kainate or glutamate for cobalt uptake and 100 mM glutamate for manganese uptake. These neurotransmitters excite and swell the neurons and facilitate the penetration of toxic elements through the sodium and calcium membrane channels. Based on many tests performed with kainate and glutamate as well as with NMDA (N-Methyl-D-Aspartate) and with no neurotransmitter, we confirmed the results of Ref. 11, that kainate is the most effective in enhancing the cobalt uptake.

Afterwards, the manganese-doped specimens were twice washed in the uptake buffer and placed for 5 minutes in a solution of Na_2S plus NH_4Cl to induce precipitation of manganese sulphate (whose solubility is quite low), fixed for 10 minutes in para-formaldehyde (4% in PBSS), washed once in the uptake buffer and twice in DD water and dehydrated as in the first procedure. For cobalt-doped specimens, the double washing in the uptake buffer was followed instead by fixing for 10 minutes with para-formaldehyde in the presence of a-nitrose-b-napthol to induce precipitation of the toxic element, and then by washing and dehydratation.

The specimens doped with toxic elements were characterized with induced coupled-plasma (ICP) emission spectroscopy[12], after detaching the neurons from the substrate and suspending them in NH_3. ICP is the most sensitive type of quantitative chemical analysis, capable of detecting elements in concentrations as low as 1 ppb. The ICP tests enabled us to prove the presence of the toxic elements in the neuron network, demonstrating in particular the effectiveness of kainate in stimulating the Co uptake[11] The limit of ICP with respect to spectromicroscopy is, of course, the inpossibility to study space distributions.

The XSEM used in the present study produces spatially-resolved chemical information by measuring the secondary (photoelectron) emission intensity as a function of the photon energy[1-3]. This requires using a tunable synchrotron-light source, in the present case the storage ring Aladdin at the Wisconsin Synchrotron Radiation Center, filtered by either a 10 or a 3-meter toroidal grating monochromator. The XSEM, described in detail in Refs. 2 and 3, achieves high lateral resolution by means of an electron-optics system, including an objective lens to accelerate the electrons and form a first focussed image of the emitting surface, a contrast aperture in the back-focal plane (diffraction plane) of the objective lens to define the accepted solid-angle of electron rays leaving the sample surface, a projective lens to provide additional, variable magnification, and a two-stage chevron micro-channelplate electron multiplier array (CEMA) and phosphor screen assembly.Images formed on the phosphor screen are recorded using a TV-camera with an

18 mm diameter silicon-target array low light-level vidicon tube --- either at standard 30 Hz TV rates on magnetic tape, or digitized into frames of 512 horizontal by 480 vertical pixels using a Data Translation frame-grabber and real-time image processing hardware (DT2861 and DT2858) in an 80386-based micro computer.

A scheme of the XSEM instrument is shown in Fig. 2.

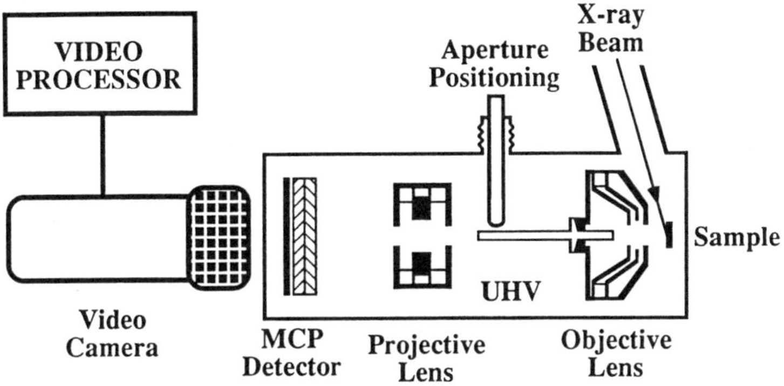

Fig. 2 -- Schematic diagram of the XSEM instrument used in the experiments.

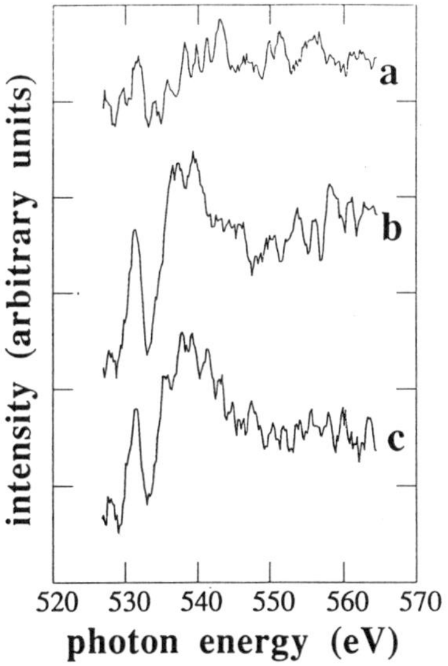

Fig. 3 -- Oxygen XSEM spectra[2] of a specimen without toxic elements, from three different areas of approximately 2 by 2 square micron: (a) from a substrate area; (b) and (c) from two different dendrite areas. The two typical optical absorption features of oxygen[13] can be observed. The data were normalized to the monochromator's output spectrum.

Images like that of Fig. 1a are quite typical as far as overall quality and resolution are concerned. As to the chemical analysis, the XSEM can take emission intensity vs photon energy spectra from selected microscopic areas of the specimen[2]. A first example is shown in Fig. 3 for specimens produced by the first procedure.

We see the well-known optical absorption spectral features of oxygen[13] from two different axon areas (curves b and c); the oxygen signal is much weaker in the substrate curve a. In the latter case, the oxygen is most probably contained in the poly-L-lysine that covers the gold substrate.

The two oxygen spectral features are due to O1s->p transitions (≈531 eV) and to O1s->Rydberg states transitions.[13] Note that the relative intensity of the two peaks is very different in curve a and in the other two spectra, with the higher-energy peak almost indistinguishable from the noise in the first. This indicates [13] that the molecular state of oxygen is different in the substrate poly-L-lysine and in the dendrites, as it should be expected.

A second typical example of microchemical analysis is shown in Fig. 4. Curves b, c and d were all taken from cobalt-doped neuron systems, whereas curve a is a reference spectrum from a Co wire. Curves b and d were taken on a specimen produced with

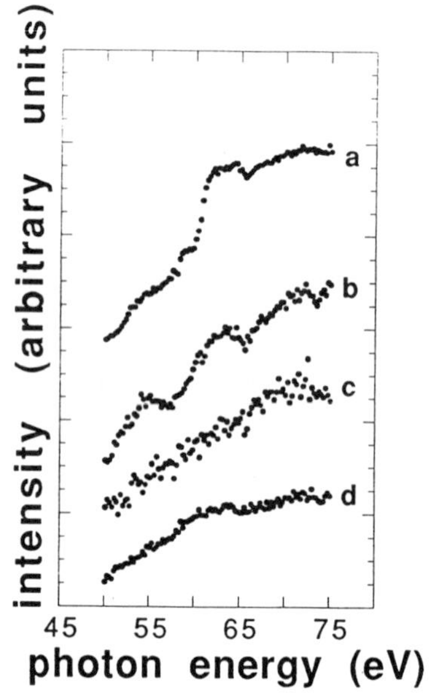

Fig. 4 -- Cobalt XSEM spectra of (a) a reference Co wire, and from three different 1 by 1 square micron areas of Co-doped specimens: (b) from a dendrite area and (d) from a cell body in a specimen treated with kainate; (c) from a dendrite area of a specimen treated with glutamate. Note the inhomogeneity in the Co distribution, revealed by the comparison of curves (b) and (d).

exposure to kainate, and curve c from a specimen produced with exposure to glutamate. Both curves b and c were taken on axon areas, and curve d was taken on a cell body.

From the results of Fig. 4 we can extract the following information. First, there is clear evidence of Co signal in curve b, weaker evidence in curve d and no evidence in curve c. This once again confirms the results of Ref. 11 as far as the superior effectiveness of kainate is concerned.

Second, the difference in the Co signal strength between curves b (axon) and d (cell body) reveals inhomogeneity in the distribution of the toxic element in the neuron network. To the best of our knowledge, this is the first time that evidence of this kind is obtained for a biological system, since previous chemical analysis was based on approaches that could not produce lateral resolution. The fact that inhomogeneities were observed is quite important, since it reveals a basic limitation of space-averaging chemical analysis in the study of toxic elements in biological systems.

Similar tests were performed on a very large number of specimen areas, with consistent results. We also found evidence for manganese signal in the Mn-doped samples. The experiments in this case were complicated by the volatility of the element, that made it difficult to correlate the XSEM signal intensity to the original distribution of the element.

In summary, we have seen that XSEM microchemical analysis can detect elements in neuron networks, and reveal inhomogeneities in their distribution. On one hand, this makes the use of chemical analysis on a microscopic scale all the more important for biological systems. On the other hand, it opens the way to this analysis by demonstrating its feasibility.

3. Scanning Photoelectron Spectromicroscopy with MAXIMUM

The advent of bright and ultrabright synchrotron radiation sources opens exciting new opportunities in X-ray microscopy. In particular, it becomes feasible to achieve lateral resolution in traditional spectroscopic techniques such as photoemission.[1] Scanning photoemission spectromicroscopy can, in fact, provide information on the chemical status of elements in the specimen, with a lateral resolution that has recently reached the 0.1 micron. Thus, photoemission microscopy is emerging as an important surface-sensitive tool, for the investigation of the *local* chemistry in biological specimens.

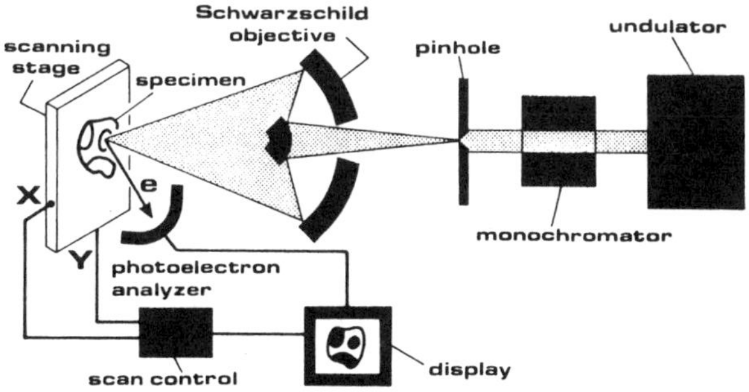

Fig. 5 -- An artist's view of the scanning photoelectron microscope MAXIMUM.

The undulator scanning spectromicroscope MAXIMUM (Multiple Application X-ray IMaging Undulator Microscope) has the overall structure shown in Fig. 5. The main components of the instrument are: (i) an undulator source of synchrotron radiation; (ii) a monochromator; (iii) an X-ray focussing system; (iv) a specimen scanning system; (v) a photoelctron analyzer (Cylindrical Mirror Analyzer); (vi) data-processing hardware and software. The key element is the focussing device: this consists of a Schwarzschild objective[14] working at normal incidence; the efficiency of the objective's mirrors in reflecting X-rays is increased by the use of Mo-Si multilayer coatings, to about 40% at the peak wavelength. The multilayer coating of the Schwarzschild objective used in this experiments has a photon energy bandpass centered at 95 eV.

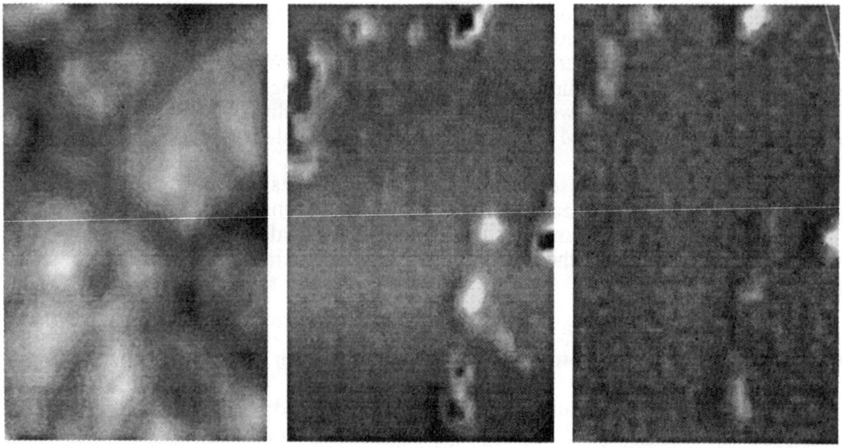

Fig. 6 -- (a) A partial yield image of a neuron network; (b) image of the same region acquired at 70 eV kinetic energy, sitting on the K3p core level emission. This image is a potassium distribution two-dimensional map; (c) the distribution map of Ca and Na (overlapped because not resolved) acquired at 58.5 eV kinetic energy.

The undulator i s a permanent-magnet array with variable gap, inserted in one of the straight sections of the Aladdin electron storage ring at the Wisconsin Synchrotron Radiation Center.[15] This device, developed by the Stanford synchrotron radiation laboratory and by the Lawrence Berkeley Laboratory, emits high-brightness radiation over a narrow spectral band, whose photon energy position can be controlled by changing the magnet gap. The radiation is further monochromatized by a 6m spherical grating monochromator, and a pinhole is used to create a small area source. Then the radiation is focussed by the objective onto the sample , extracting photoelectrons from a small area. The sample is mounted on a two-dimensional scanning mechanism with a coarse stage and a fine stage, based on piezo-actuators.

Maximum can perform experiments on uncoated and unstained neuron networks with a lateral resolution as low as 0.5 micron, producing images acquired at an emission photoelectron kinetic energy conveniently chosen and fixed, thus rendering two-dimensional distribution maps of specific elements in biological specimens.

Fig. 6a is a typical example of several micrographs of a neuron network obtained with the first procedure mentioned above. This image has been acquired at 95 eV photon energy, and at a fixed CMA kinetic energy at 1.3 eV, thus giving a partial yield micrograph of the secondary electrons photoemitted by the sample surface. Minute details of the network, such as cell bodies, axons and dendrites, are clearly visible in the photoemission intensity inhomogeneity rising from the chemical contrast between the cells and the gold substrate.

In Fig. 6b the same area of a neuron network has been scanned with 70 eV electron detection kinetic energy. This image represents the potassium 3p core level two-dimensional distribution on the specimen surface. The presence of K on the cells is evident from the micrograph, as it was expected. In fact, potassium is, together with calcium, accumulated in the cell membrane ion channels.

We have been able to detect evidence for sodium and calcium (see Fig. 6c) on the same detection area at 58.5 eV kinetic energy, where the Na2p and Ca3p core levels were not distinguishable, and appeared overlapped in the EDC's, given the relatively low energy resolution. In Fig. 7, an Energy Distribution Curve acquired on a 1 by 1 square micron area on a cell body of the same region imaged in Fig. 6 is reported, where the K and Ca/Na core levels emission peaks are visible.

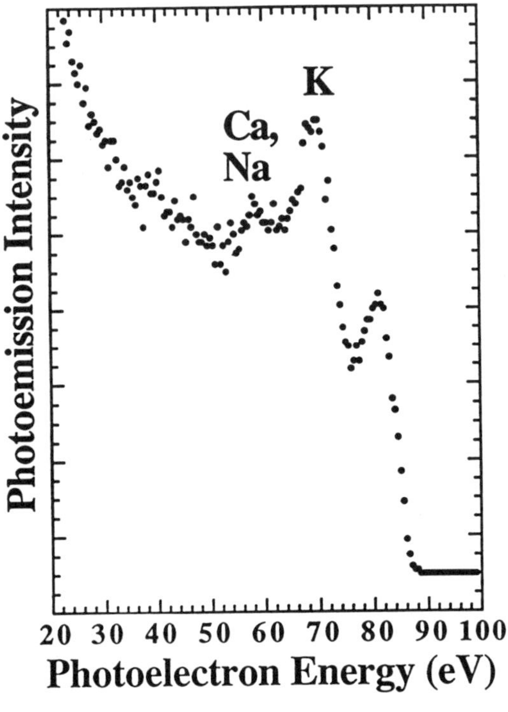

Fig. 7 -- The energy distribution curve acquired at 95 eV photon energy, with 50 pass energy, from a 1 micron by 1 micron area on a cell body of the region imaged in figure 6. The EDC shows the K3p peak at 70 eV, and the Ca3p/Na2p peaks at 58.5 eV.

These membrane elements in neuron networks are extremely important in the comprehension of the trans-membrane bias, that the cell exploits to regulate the input and output of the compounds necessary for its survival. This bias has also been demonstrated to play a fundamental role in nerve pulse transmission, that is still one of the most investigated, and less understood phenomena in the life sciences.

Experiments performed on similar specimens by ESCA spectroscopy (De Stasio et al., unpublished) could not detect the presence of K, Ca and Na. In the ESCA technique, in fact, the photon source was an Al Kα rotating anode whose photon energy is 1.4 KeV, much more penetrating than the soft X-ray radiation used by MAXIMUM. This demonstrates that MAXIMUM's surface sensitivity opens the way to the understanding of a number of cell membrane problems in biology.

With the advent of the ultra-bright sources the lateral resolution and the energy resolution of scanning photoelectron spectromicroscopy will be improved, rendering MAXIMUM a unique tool for the investigation of the most interesting open problems in the life sciences.

4. Acknowledgments

We are grateful to the ISA Italia and in particular to Drs. Oddo, Antonangeli and Galli for giving us the possibility to perform the ICP tests. The expert help of Mario Capozi, Marino Marsi and Tiziana Dell'Orto was essential to the success of the spectromicroscopy experiments. The work was supported by the Fonds National Suisse de la Recherche Scientifique, by the Consiglio Nazionale delle Ricerche, by the National Science Foundation and by the Ecole Polytechnique Fédérale de Lausanne. The spectromicroscopy experiments were performed at the Wisconsin Synchrotron Radiation Center, that is supported by the National Science Foundation.

5. References

1. For a recent review of photoelectron spectromicroscopy, see for example: G. Margaritondo and F. Cerrina, Nucl. Instr. Meth. **A291**(1990).p. 26
2. B. P. Tonner, Nucl. Instrum. Methods **291**(1990).p.60
3. Gelsomina De Stasio, S. Koranda, G. Harp, B. Tonner, D. Mercanti, M. T. Ciotti and G. Margaritondo, Europhys. Lett., in press.
4. G. Margaritondo: *Introduction to Synchrotron Radiation* (Oxford, New York, 1988).
5. F. Cerrina, B. Lai, C. Gong, A. Ray-Chaudhuri, G. Margaritondo, M. A. Green, H. Höchst, R. Cole, D. Crossley, S. Collier, J. Underwood, L. J. Brillson and A. Franciosi, Rev. Sci. Instrum. **60**(1989) p.2249; F. Cerrina, S. Crossley, D. Crossley, C. Gong, J. Guo, R. Hansen, W. Ng, A. Ray-Chaudhuri, G. Margaritondo, J. H. Underwood, R. Perera and J. Kortright, J. Vac. Sci. Technol. **A8**(1990)p.2563; W. Ng, A. K. Ray-Chadhuri, R. K. Cole, S. Crossley, D. Crossley, C. Gong, M. Green, J. Guo, R. W. C. Hansen, F. Cerrina, G. Margaritondo, J. H. Underwood, J. Korthright and R. C. C. Perera, Phys. Scripta **41** (1990)p.758; C. Capasso, A. K. Ray-Chaudhuri, W. Ng, S. Liang, R. K. Cole, J. Wallace, F. Cerrina, G. Margaritondo, J. H. Underwood, J. B. Kortright and R. C. C. Perera, J. Vac. Sci. Technol. **A9**(1991).p.1248.
6. Gelsomina De Stasio, W. Ng, A. K. Ray-Chaudhuri, R. K. Cole, Z. Y. Guo, J. Wallace, G. Margaritondo, F. Cerrina, J. Underwood, R. Perera, J. Kortright, Delio

Mercanti and M. Teresa Ciotti, Nucl. Instrum. Methods **A294** (1990)p.351; Delio Mercanti, Gelsomina De Stasio, M. Teresa Ciotti, C. Capasso, W. Ng, A. K. Ray-Chaudhuri, S. H. Liang, R. K. Cole, Z. Y. Guo, J. Wallace, G. Margaritondo, F. Cerrina, J. Underwood, R. Perera and J. Kortright, J. Vac. Sci. Technol. **A9** (1991).p.1320; Gelsomina De Stasio, C. Capasso, W. Ng, A. K. Ray-Chaudhuri, S. H. Liang, R. K. Cole, Z. Y. Guo, J. Wallace, F. Cerrina, G. Margaritondo, Underwood, R. Perera, J. Kortright, Delio Mercanti, M. Teresa Ciotti and Alessandro Stecchi, Europhysics Lett. **16** (1991).p.411.

7. H. Ade, J. Kirz, S. Hulbert, E. Johnson, E. Anderson and D. Kern, Nucl. Instrum. Methods **291** (1990).p.126.
8. P. L. King, A. Borg, C. Kim, P. Pianetta, I. Lindau, G. S. Knapp, M. Keenlyside and R. Browning, Nucl. Instrum. Methods **291** (1990).p.19.
9. G. Levi, F. Aloisi, M. T. Ciotti and V. Gallo, Brain Res. **290** (1984) p.77; M. T. Ciotti, D. Mercanti and G. Levi, in *A Dissection and Tissue Culture Manual of the Nervous System*, A. Shahar, J. De Vellis, A. Vernadakis and B. Haber, eds. (Alan R. Liss, 1989), p. 291.
10. The uptake buffer composition was: 139 mM sucrose, 57.5 mM NaCl, 5 mM KCl, 2 mM $MgCl_2$, 1 mM $CaCl_2$, 12 mM glucose, 10 mM HEPES with a pH of 7.2. The final pH of the buffer was 7.5.
11. R. M. Pruss, R. L. Akeson, M. M. Racke and J. L. Wilburn, Neuron **7** (1991) p.509.
12. R. M. Barnes, Chemia Analityczna, **28** (1983) p.179; S. Caroli, E. Coni, A. Alimonti, E. Beccaloni, E. Sabbioni and R. Pietra, Analusis, **16** (1988) p.75; L. Bourrier-Guerin, Y. Mauras, J. L. Truelle and P. Allain, Trace Element in Medicine, **2** (1985) p.88; C. De Martino, S. Caroli, A. Alimonti, F. Petrucci, G. Citro and A. Nista, J. Exp. Clin. Cancer Res. **10** (1991) p.1.
13. Y. Ma, C. T. Chen, G. Meigs, K. Randall and F. Sette, Phys. Rev. **A44** (1991) p.1848.
14 K. Scwarzschild, Goettingen Mitt. **10** (1905) p.2; E. Spiller, in: Scanned Image Microscopy, ed. E.A. Ash (Plemum, New York, 1980) p. 365; R.P.Haelbich, ibid., p. 413; I. Lovas, W. Santy, E. Spiller, R. Tibbets and J Wilczynsky, Proc. SPIE **316** (1980) p.90; H. P. Bruegmann, Conic Mirrors (Focal Press, London, 1986); J. A. Trail and R. L. Byer, Proc. SPIE **563** (1985) p.80; F. Cerrina, Imaging Science 30 (1986) p.80; B. Lai, F. Cerrina and J. H. Underwood, Proc. SPIE **563** (1985).p.173.
15 K. Halbach, J. Chin, E. Hoyer, H. Winick, R. Cronin, J. Yang and Y. Zambre, IEEE Trans. Nucl. Sci. **NS-28** (1981) p.31; H. Winick, R. Boyce, G. Brown, N. However, Z. Hussain, T. Pate and E. Umbach, Nucl. Instr. and Meth. 208 (1983) p.127.

X-RAY MICROSCOPY: CURRRENT STATUS

D. Rudolph, G. Schmahl, B. Niemann, P. Guttmann, J. Thieme, G. Schneider
C. David, M. Diehl, T. Wilhein

Forschungseinrichtung Röntgenphysik, Universität Gottingen,
Geiststrasse 11, 3400 Gottingen, Germany

Abstract

X-ray microscopy provides higher resolution than optical microscopy and higher penetration ability than electron microscopy ancl allows the investigation of hydrated samples. X-ray microscopy studies of hydrated biological and colloidal chemical systems were performed with an X-ray microscope, which is - together with a scanning X-ray microscope - installed at the BESSY electron storage ring in Berlin. Structures as small as 30 nm have been imaged using germanium phase zone plates generated by electron beam lithography. An X-ray microscope with a pulsed plasma X-ray source is under development. The aim is to take high resolution X-ray images with one pulse of 5 nanoseconds.

1. Introduction

X-ray photons in the wavelength range of about 0.3 nm - 5 nm are best suited for X-ray microscopy studies. X-ray microscopy provides higher resolution than optical microscopy and higher penetration ability than electron microscopy. Most importantly, X-ray microscopy has the potential for imaging hydrated specimens with high resolution.

The two dominating processes causing the contrast are photoelectric absorption and phase shift. In the wavelength range $\lambda \geq 2.4$ nm, the absorption cross-sections are about one order of magnitude smaller than the cross-sections for electrons in the energy range in which electron microscopy is performed. These absorption cross-sections for soft X-rays are appropriate for high-resolution investigation of specimens in a water layer of up to 10 μm thickness.The wavelength range between the K absorption edges of oxygen ($\lambda =$ 2.34 nm) and carbon ($\lambda = 4.38$ nm) is especially interesting because in this wavelength range the radiation is weakly absorbed by water but strongly absorbed by organic matter resulting in a good amplitude contrast of wet specimens [1].

In addition to the amplitude contrast, caused by photoelectric absorption, phase contrast can also be used for X-ray microscopy experiments [2,3], especially in the shorter wavelength at about $\lambda = 0.3$ nm.

2. X-Ray Microscopes

Up to now, different types of microscopes have been developed, namely transmission X-ray microscopes at synchrotron sources, transmission X-ray microscopes at laboratory sources, scanning transmission X-ray microscopes and photoelectric X-ray microscopes.

X-ray microscopes with zone plates as micro-objectives at synchrotron sources exist at BESSY in Berlin, Germany, at UVSOR in Okazaki, Japan, and at the Photon Factory in Tsukuba, Japan. Since August 1992, an X-ray microscope at ISA in Aarhus, Denmark started operation [37]. Furthermore, an X-ray microscope is planned at the ALS in Berkeley, USA [38].

X-ray microscopy with scanning X-ray microscopes is performed at several laboratories. Such instruments, equipped with a micro-zone plate as the high resolution optical element, are installed at the Brookhaven National Laboratory [28] in Stony Brook, USA, at the Daresbury Synchrotron Source, England [29], and at the BESSY storage ring in Berlin [30]. The scanning X-ray microscopes at Stony Brook and Daresbury are installed at high brilliance undulator beam lines. The scanning X-ray microscope at BESSY is installed at a bending magnet beamline.

Detailed information on the state of the art can be found in proceedings of symposia on X-ray microscopy held in New York in 1979, in Göttingen in 1983, in Taipei in 1986, in Brookhaven National Laboratory, NY, in 1987, in Toshigi in 1988 and in London in 1990 [4-9].

The Forschungseinrichtung Röntgenphysik, Universität Gottingen, has developed two X-ray microscopes, an X-ray microscope, and a scanning X-ray microscope with zone plate optics. The two microscopes are installed at the electron storage ring BESSY in Berlin. In addition, an X-ray microscope with a pulsed plasma X-ray source is under development. This system has an ellipsoidal mirror condenser and a micro zone plate as a high resolution X-ray objective. In the following only the x-ray microscope at BESSY and the x-ray microscope with a pulsed plasma source, and results obtained with these microscopes will be discussed.

2.1. Zone Plates

Diffraction gratings of different dispersive and imaging properties can be termed as diffractive optics. A special type of diffraction gratings, namely circular gratings with radially increasing line density are zone plates. Zone plates for use in transmission consist of alternating transparent and absorbing, respectively, phase-shifting zones. Their focusing properties have long been well-known [10]. Imaging with zone plates with zone numbers $n \geq 100$ obeys the same laws as imaging with thin refractive lenses.

With r_1 the radius of the innermost zone, r_n the radius of the nth zone, and n the zone number, the focal length f of a zone plate used in the first diffractive order is approximately given by:

$$f = r^2_1/\lambda \quad (1)$$

As f is proportional to λ^{-1}, a zone plate has to be used with radiation with a bandwidth $\Delta\lambda$ given by $\lambda/\Delta\lambda = cn$, where c is a constant. Theoretical investigations of imaging properties of zone plates as a function of the reciprocal bandwidth $\lambda/\Delta\lambda$ have shown that with c = 0.5–1 practically no contrast in the image is lost compared to imaging with quasimonochromatic radiation [11] .

The width of the outermost zone of a zone plate is approximately given by

$$dr_n = r_n/2n \quad (2)$$

According to the Rayleigh criterion, the smallest distance of two point sources (Δ) which can be resolved with a zone plate is given by $\Delta = 1.22 dr_n$.

An important parameter characterizing a zone plate is the efficiency with which it collects incident photons into its first order focus. For a zone plate of thickness t and laminar zone profiles the groove efficiency is given by:

$$E = 1/\pi^2\, (1 + e^{-2\eta\Phi} - 2e^{-\eta\Phi} \cos\Phi) \quad (3)$$

with $\Phi = 2\pi\delta\tau/\lambda$ and $\eta = \beta/\delta$. δ and β are given by the refractive index $n = 1 - \delta - i\beta$ [12].

If one only considers the absorption in the material of the absorbing zones ($\delta = 0$), the two exponential terms vanish and one gets the efficiency of an amplitude zone plate which is E = 10.1%. The efficiency can be considerably improved by choosing an appropriate material with a thickness t such that $\delta t = 1/2\lambda$. If absorption is ignored, i.e., $\beta = 0$, the groove efficiency is given by $E = 4\pi^{-2}$, resulting in E = 40.4%. In practice, absorption cannot be ignored, but materials can be found with $\eta \cong 0.2$ resulting in theoretical groove efficiencies $\geq$ 20%.

The absolute efficiency E_a of the zone plate is given by $E_a = ET$ with T the transmission of the support foil of the zone plate.

For X-ray microscopy two types of zone plates are needed. First, the object has to be illuminated by a condenser or a collector-condenser system which has to collect as much radiation as possible from the X-ray source. For X-ray microscopy at electron storage rings, condensers with diameters of 2 to 9 mm and rather large zone numbers of 10^3 to about 4×10^4 are required. Secondly, to image the object, high-resolution zone plates are necessary, i.e., zone plates with outermost zone widths dr_n as small as possible. These zone plates have only one hundred to a few hundred zones in order not to restrict the usable X-ray bandwidth too much. As a consequence, high-resolution zone plates have very small diameters in the region of 10 to 100 μm and are, therefore, called micro-zone plates.

Up to now, micro-zone plates with up to a few hundred zones are the only X-ray optical elements for diffraction limited imaging using soft X-rays and this will probably remain the case in the forseeable future.

Two methods of zone plate fabrication have up to now provided zone plates for X-ray microscopy experiments. Zone plate patterns are generated either by holographic methods with visible and UV laser light [13,14] or by electron beam writing [15–18]. These patterns are then transferred into absorbing and/or phase-shifting structures by, e.g., reactive ion etching [19]. Additional information on zone plates made by electron beam lithography can be found in [5, 7, 9, 20].

The optimum thickness t of a zone plate made of, e.g., germanium is 380 nm for λ = 2.4 nm. For a zone plate with an outermost zone width of dr_n = 30 nm, this leads to an aspect ratio t/dr_n = 12.7. As the pattem transfer of structures with this aspect ratio is extremely demanding, zone plates with dr_n = 30 nm have been made up to now with germanium layers of up to 280 nm [18].

For zone plates with outermost zone widths of 30 nm to 50 nm groove efficiencies of about 10% have been achieved, resulting in absolute efficiencies of about 5% [17-19]. Even higher efficiencies have been achieved for zone plates with coarser outermost zones [21, 22]. Figure 1 shows a SEM photograph of the central part of a germanium phase zone plate.

Fig.1 SEM photograph of the central part of a germanium phase zone plate.

For optically thick zone plates, i.e., zone plates with structures of even higher aspect ratios than discussed, the efficiency can be evaluated by application of coupled wave theory [23]. Such zone plates - which will have very high resolution and high efficiencies - are under construction [24, 25].

Table I lists the parameters of condenser zone plates and of micro-zone plates used for X-ray microscopy studies with the Göttingen X-ray microscopes. The condenser zone plates KZP5 and KZP7 and the micro-zone plate MZP3 are made by holographic lithography, the micro zone plates M1, M2, and M4 by electron-beam lithography.

Table 1. Parameters of the zone plates used in the Göttingen X-ray microscopes

	Condenser zone plates		Micro-zone plates			
	KZP5	KZP 7	MZP 3	M I	M 2	M 4
Radius of first zone r_1/µmm	36	22	1.76	1.06	1.02	1.25
Zone plate diameter D/µm	2500	9000	55.6	37.2	34.5	42
Number of zones N	1200	42000	251	305	285	282
Width of outermost zone dr_N/µm	0.46	0.054	0.055	0.030	0.030	0.037
Focal length $f_{\lambda=2.4nm,\ 1.order}$/mm	544	201.4	1.3	0.470	0.435	0.653

2.2. The X-Ray Microscope at the electron Storage Ring BESSY

Electron storage rings provide very intense synchrotron radiation in a broad spectral range from far infrared to X-radiation. Therefore, a storage ring is a well suited source of radiation for X-ray microscopy. The electron storage ring BESSY operates at an electron energy of 800 MeV. For a bending magnet source, the brilliance is $2x10^{13}$ photons sec^{-1} mm^{-2} $mrad^{-2}$ at 0.1% bandwidth at 0.5 keV and 500 mA beam current [26].

The optical setup of the X-ray microscope - which is located at a distance of 15 m from a bending magnet - is shown in Fig.2a. The condenser images the source into the object plane. The object field is limited by a pinhole with a diameter of about 20 µm. Due to the wavelength dependence of the focal length of a zone plate the condenser - in combination with the pinhole - acts as a linear monochromator. It reduces the bandwidth of the polychromatic synchrotron radiation. The monochromaticity is given by $\lambda/\Delta\lambda \cong D/2d$ with D = diameter of the condenser zone plate and d = diameter of the pinhole [27]. The micro zone plate as a high resolution X-ray objective generates a magnified image of the object in the image field. The central stop of the condenser zone plate prevents that the object is illuminated by broad-band zero-order radiation of the condenser. In addition, due to the stop the image field is free of zero order radiation of the micro-zone plate.

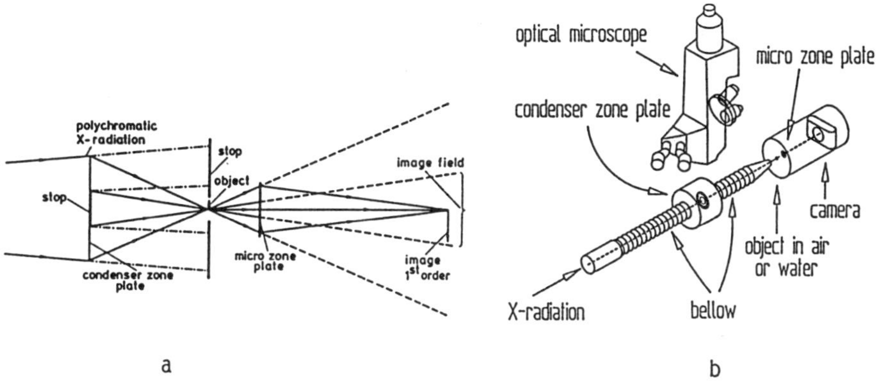

Fig. 2 The x-ray optical arrangement (a) and the scheme (b) of the x-ray microscope at BESSY

Figure 2b shows the scheme of the X-ray microscope. The condenser zone plate is mounted in a bellow which ends in an Al_2O_3-foil transparent to X-radiation. This foil separates the vacuum in the bellow from air. The object is normally mounted in an environmental object chamber - not shown in Fig. 2b - where it can be prepared in its natural state which may be air, water or some other liquid. This chamber is inserted in the air gap indicated by the pointer as "object in air or water". The next compartment contains

Fig. 3 The X-ray microscope at BESSY

the micro zone plate and the camera under prevacuum conditions. The enlarged X-ray images can be recorded either with photographic plates or with a CCD-detector. In addition, a micro-channel plate is installed so that the image may be viewed directly. To adjust and to prefocus the object an optical microscope is installed with a swivel-mount. The bellow containing the condenser zone plate can be bend downwards to swivel the optical microscope into position. A photograph of the X-ray microscope is shown in Fig.4.

2.3. The X-Ray Microscope with a Pulsed Plasma X-Ray Source

To perform X-ray microscopy experiments independent of electron storage rings as X-ray sources, an X-ray microscope with a pulsed plasma source is under development. This is a joint effort of the institute in Göttingen, the Fraunhofer Institut für Lasertechnik, Aachen, and the Carl Zeiss company [28,29,30].

Figure 4 shows the optical arrangement of the X-ray microscope. The plasma X-ray source operates as a gas discharge system with nitrogen as working gas. The 1s $\rightarrow$ 2p transition of N VII at $\lambda = 2.48$ nm and the closely neighboured 1s2 $\rightarrow$ 1s3p of N VI at $1 = 2.49$ nm are chosen. Due to their small separation $\Delta\lambda = 0.1\ 17$ nm ($\lambda/\Delta\lambda > 200$) these lines can be used simultaneously. Since the plasma X-ray source emits mainly line radiation, a reduction of the bandwidth of the radiation is not necessary and it can be used directly without a monochromator. Oxygen as beamline gas is used to filter undesired radiation below $\lambda = 2.33$ nm. For about 5 nanoseconds electron densities $n_e \geq 10^{20}$ cm^{-3} and temperatures of $kT_e > 200$ eV are achieved in the gas discharge within a cylindrical volume with 0.5 mm diameter and 10 mm length leading to a strong emission of soft X-rays. The X-ray condensor is an ellipsoidal mirror that images the X-ray source into the object plane. All rays which emerge from one focal point of an ellipsoid of revolution and which are reflected by the inner surface of this ellipsoid are focused in the other focal point. The condenser is a 115 mm long part of an ellipsoid of revolution. Said fragments is chosen in such a way that the opening angle of the focused radiation fits to the aperture of the micro zone plate M2 which is used to create the magnified image of the object. Images were

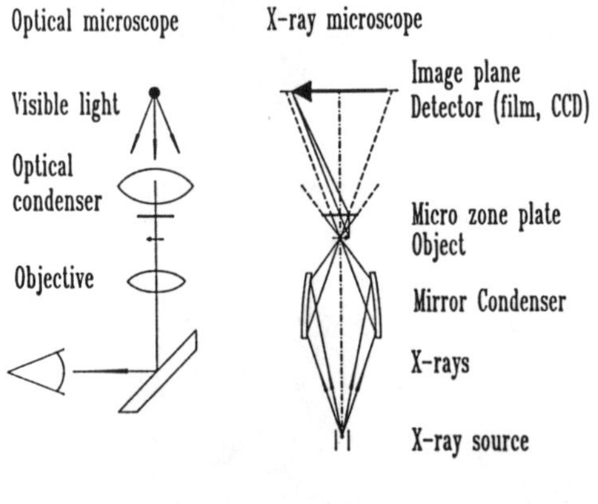

Fig.4. The X-ray optical arrangement of the X-ray microscope with a pulsed plasma source

Fig. 5 X-ray microscope with a pulsed plasma X-ray source.

recorded with nuclear emulsion photo-plates and with a CCD-camera. Object and micro-zone plate together can be swiveled into an inverse optical microscope. This allows adjustment and prefocusing. Figure 7 shows a photograph of the microscope.

Images of gratings, diatoms, test structures and dried cells were taken. Figure 6 shows, as an example, the image of a diatom, taken with a CCD-camera with a single X-ray pulse. The X-ray magnification was 380 x. The CCD was covered with an X-ray to visible converter Gd_2O_2S:Tb with 8 µm grains.

In future experiments the X-ray source, the X-ray optical elements and the detector can be improved considerably. This will enable imaging of wet specimens in their natural state with one pulse. The main advantage of exposure times as short as 5 nanoseconds is that blurring of the image by any kind of object motion will be avoided.

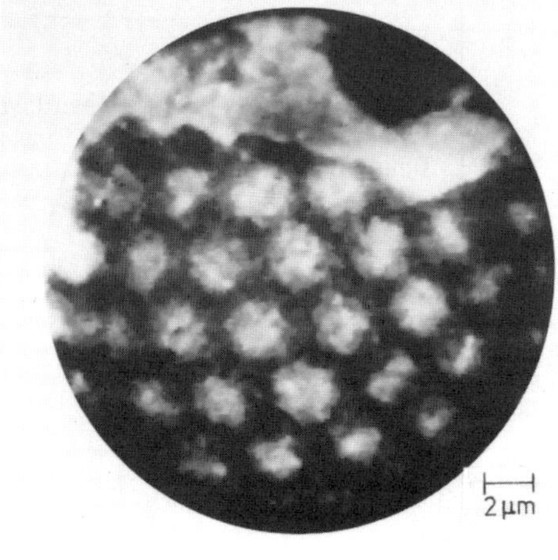

Fig.6. X-ray image of a diatome taken with one pulse with a CCD-camera.

3. X-Ray Microscopy Studies with the Microscope at BESSY

A great advantage of X-ray microscopy is the possibility to investigate specimens in a wet state. In the wavelength range between the K-absorption edges of oxygen and carbon, namely 2.34 nm $\leq \lambda \leq$ 4.38 nm, the transmission of water is considerably higher than the transmission of a multitude of different materials, e.g., organic material and silicates. This provides a natural contrast between the material and the surrounding water. However, taking into account the transmission of water - see Fig.7 - it is evident that specimens of reasonable thickness can best be investigated close to λ = 2.4 nm. As an example for wet cell imaging Fig. 8 shows the x-ray image of a basophilic leukemia cell (RBL-3H3-cell fixed with glutaraldehyde 1.5%) from a joined program that is just runnig together with D. Jovin and T. M. Jovin, Max-Planck-Institut für biophysikalische Chemie, Göttingen.

To describe the capabilities of X-ray microscopy, examples of some different areas of interest are presented. The structure of chromosomes in a range between the 10 nm scale and the µm scale is of great functional importance. However, there is little knowledge about the relation between form and function in this range. X-ray microscopy as a well suited tool to investigate biological specimens in their natural wet state may contribute to close this gap. X-ray microscopy studies of polytene chromosomes isolated from salivary glands of *Chironomus thummi* larvae have been performed in collaboration with M. Robert-Nicoud and T. M. Jovin. Details and x-ray images of those chromosomes are given in [31].

Resurrection plants or parts of them are an example for objects that can be investigated with X-rays in their inactive dry state and in their different states between dry and becoming active by soaking up water. X-ray microscopy studies of spores of the australian moss *Dawsonia superba* have been made in collaboration with V. Sarafis, University of Western Sydney, Australia. Figure 9 shows an X-ray image of such a spore in a state where it just develops a germ. This image and the following image (Fig. 10) were taken with the micro-zone plate M4.

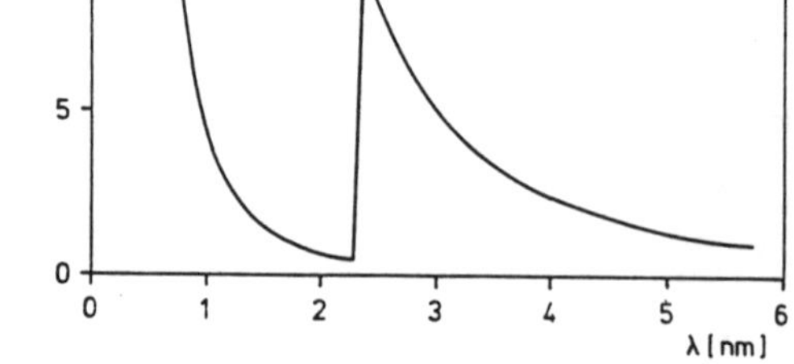

Fig. 7 Thickness $t_{1/e}$ of a water layer with a transmission T = 36.8 % as a function of the wavelength.

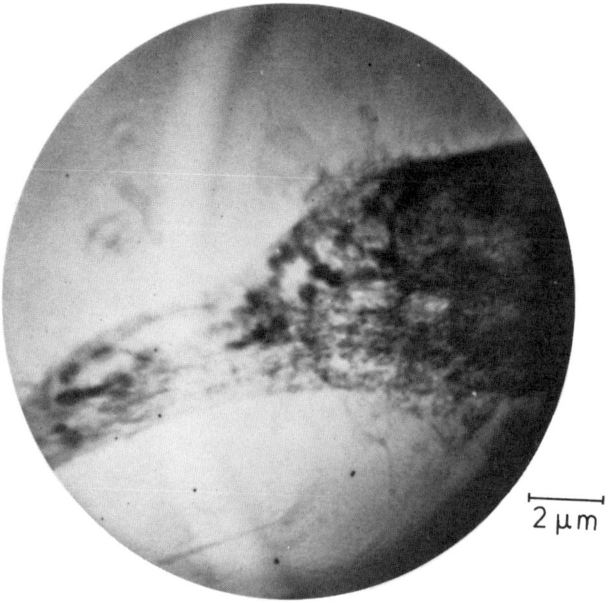

Fig. 8 Part of a basophilic leukemia cell (RBL-3H31, fixed with glutaraldehyde 1.5%)

The third example shows the examination of colloidal chemical systems. Three-layer minerals as the montmorillonit are characterized by their large specific surface, their high adsorption power, and their ability to form a gel. This property is used to change the fluidity of aqueous or other non-aqueous solutions. From measurements of the rheologic properties it has been deduced that the clay platelets build up so called "card-house"- or "book-house"- structures. They are characterized by an edge-to-face linkage of some platelets ("card-house") or of stacks of those platelets ("book-house"). In these configurations cations are the bridges between the negatively charged clay-platelets [32]. Figure 10 shows an X-ray image of a 1 % suspension of montmorillonit coated with Ca^{2+} cations. The "book-house"-structure is seen clearly between the large dense particles. In addition, the edge-to-face linkage can be seen. Direct imaging of this configuration of clay platelets in aqueous solution could not be performed till now by other means [33].

The images of the biological specimens shown in Fig. 8–10 were recorded on Ilford L4 plates with exposure times between 10 seconds and 35 seconds.

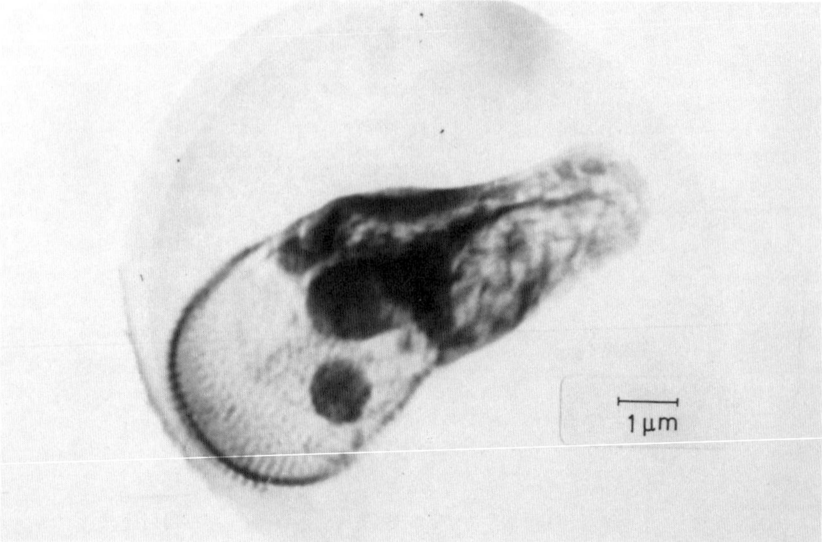

Fig.9. Germinating spore of the australian moss *Dawsonia superba*, $\lambda = 2.4$ nm.

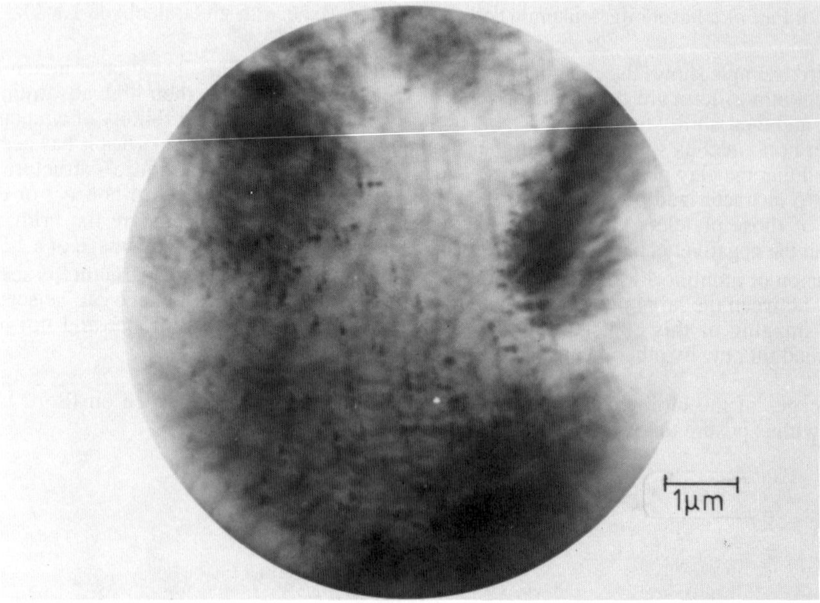

Fig.10. X-ray image of the "book-house"-structure of a montmorillonit gel, $\lambda = 2.4$ nm.

4. Improvements

For future studies the following improvements are envisaged. The absolute efficiencies of the micro-zone plates will be improved from 5% to better than 10% by two approaches. The transmission of the support foils will be increased by using thiner silicon foils with low absorption. In addition, the lithography and the pattern transfer of electron beam written zone plates will be improved to obtain a closer approach to the theoretical figure of the groove efficiency. The exposure time will be further reduced by about one order of magnitude by use of a camera with a back thinned CCD. Both improvements will lead to a reduced radiation dose given to the specimens. This is of great importance, especially for the investigation of initially live specimens as the radiation dose applied to the specimens under investigation at the moment is in the region of approximately 10^7 Gy which is at the limit where structural damage of wet biological specimens occurs at a noticeable range [34, 35, 36]. Furthermore, higher radiation doses may be tolerable when cryofixation is used. The application of this method to x-ray microscopy is under development in the Göttingen Institute.

The field of X-ray microscopy will derive benefit from the construction of third generation synchrotron sources, such as the Advanced Light Source in Berkeley, ELETTRA in Trieste, and BESSY II in Berlin. This is especially valid for a scanning X-ray microscope and also for an X-ray microscope. For example, at BESSY II, the exposure time will be reduced by more than one order of magnitude compared to the microscope installed at BESSY. These developments will lead to X-ray images with an X-ray microscope at BESSY II with exposure times of less than 0.1 second.

5. Acknowledgements

The current research is being funded by the German Federal Minister of Research and Technology (BMFT) under contract numbers 05 5MGDXB 6 and 13 N 5328 A0.

6. References

[1] H. Wolter, Ann. Phys. **6**. Folge, Bd. 10 (1952) p.94.
[2] G.Schmahl, D.Rudolph, in ref. [6], p. 231.
[3] G.Schmahl, D.Rudolph, P.Guttman, in ref. [7], p. 228.
[4] D.F.Parsons, ed., *Annals of the New York Academy of Sciences,* Vol.**342** (1980).
[5] G.Schmahl, D.Rudolph, eds., *X-Ray Microscopy,* Springer Series in Optical Sciences, Vol.**43**, Springer Verlag, Berlin (1984).
[6] P.C.Cheng, G.J.Jan, eds., *X-ray Microscopy-Instrumentation and Biological Applications,* Springer Verlag, Berlin (1987).
[7] D. Sayre, M. Howells, J. Kirz, H. Rarback, eds., *X-Ray Microscopy II,* Springer Series in Optical Siences, Vol. **56**, Springer Verlag, Berlin (1988).
[8] K. Shinohara, K.Yada, H.Kihara, T.Saito, eds., *X-Ray Microscopy in Biology and Medicine,* Japan Scientific Societies Press, Tokyo, Springer Verlag, Berlin (1990).

[9] A. G. Michette, G. R. Morrison, C. J. Buckley, eds., *X-Ray Microscopy III,* Springer Series in Optical Sciences, Vol. **67**, Springer Verlag, Berlin (1992).
[10] J. L. Soret, Annal. Phys. Chem. **156** (1875) 99.
[11] J. Thieme, in ref.[7], p. 70.
[12] J. Kirz, *J. Opt. Soc. Am.,* Vol. **64** (1974) 301.
[13] G. Schmahl, D. Rudolph, P. Guttmann, and O. Christ, in ref.[5], p. 63.
[14] D. Rudolph, G. Schneider, P. Guttmann, G. Schmahl, B. Niemann, and J. Thieme, in ref.[9], p. 392.
[15]V. Bogli, B. Unger, and H. Beneking, in ref. [20], p. 449.
[16] Y. Vladimirsky, D. Kern, T.H.P. Chang, D. Attwood, H. Ade, J. Kirz, I. McNulty, H. Rarback, and D. Shu, *J. Vac. Sci. Technol. B* **6** (1981) 311.
[17] E. Anderson and D. Kern, in ref. [9], p. 75.
[18] C. David, J. Thieme, P. Guttmann, G. Schneider, D. Rudolph, and G. Schmahl, *Optik,* in press.
[19] J. Thieme, D. Rudolph, G. Schmahl, P. Guttmann, B. Greinke, and M. Diehl, in ref. [9], p. 83.
[20] E. E. Koch, G. Schmahl, eds., *Soft X-Ray Optics and Technology,* SPIE Proc., Vol. **733** (1986).
[21] R. Hilkenbach, and J. Thieme, in ref.[20], p. 464.
[22] C. David, in ref.[9], p. 87.
[23] J. Maser, in ref.[9], p. 104.
[24] D. Rudolph, B. Nieman , and G. Schmahl, *SPIE Proc.,* Vol. **316** (1981), 103 .
[25] R. Hilkenbach, in ref.[9], p. 94.
[26] *BESSY II , Eine optimierte Undulator/Wiggler-Speicherring-Lichtquelle für den VUV- und XUV-Spektralbereich,* 2.Teil: Technische Studie, Hrsg. BESSY-GmbH, Berlin (1989).
[27] B. Niemann, D. Rudolph, and G. Schmahl, *Optics Communications,* **12**, (1974), 160.
[28] D. Rudolph, B. Niemann, G. Schmahl, and J.Thieme, in ref.[7], p. 216.
[29] B. Niemann, D. Rudolph, G. Schmahl, M. Diehl, J. Thieme, W. Meyer-Ilse, W. Neff, R. Holz, R. Lebert, F. Richter, and G. Herziger, *Optik* **84** (1990) 35.
[30] G. Schmahl, B. Niemann, D. Rudolph, M. Diehl, J. Thieme, W. Neff, R. Holz, R. Lebert, F. Richter, and G.Herziger, in ref.[9], p. 66.
[31] P. Guttmann, G. Schneider, M. Robert-Nicoud, B. Niemann, D. Rudolph, J. Thieme, T. M.J ovin, and G. Schmahl, in ref. [9], p. 404.
[32] H. Van Damme, P. Levitz, J. Fripiat, J. Alcover, L. Gatineau, and F. Bergaya, in *Physics of Finely Divided Matter,* **5** (1985) 24.
[33] J. Thieme, P. Guttmann, J. Niemeyer, G. Schneider, C. David, B. Niemann, D. Rudolph, and G. Schmahl, *Nachr.Chem.Tech.Lab.,* **40** (1992) 562.
[34] D. Sayre, J. Kirz, R. Feder, D. Kim, and E. Spiller, *Ultramicroscopy* **2** (1977) 337.
[35] P.Golz, in ref.[9], p. 313.
[36] G. Schneider, *Röntgenmikroskopie mit Synchrotronsstrahlung an wässrigen biologischen Systemen - experimentelle und theoretische Untersuchungen,* Dissertation, Universität Göttingen (1992).
[37] R. Medenwald, private communication.
[38] D. Attwood, private communication.